Vision Group
P9-BHX-687

Vision Group
11/2002

Ten Lectures on Statistical and Structural Pattern Recognition

Computational Imaging and Vision

Managing Editor

MAX A. VIERGEVER
Utrecht University, Utrecht, The Netherlands

Editorial Board

RUZENA BAJCSY, *University of Pennsylvania, Philadelphia, USA*
MIKE BRADY, *Oxford University, Oxford, UK*
OLIVIER D. FAUGERAS, *INRIA, Sophia-Antipolis, France*
JAN J. KOENDERINK, *Utrecht University, Utrecht, The Netherlands*
STEPHEN M. PIZER, *University of North Carolina, Chapel Hill, USA*
SABURO TSUJI, *Wakayama University, Wakayama, Japan*
STEVEN W. ZUCKER, *McGill University, Montreal, Canada*

Volume 24

Vision Group
11/2002

Ten Lectures on Statistical and Structural Pattern Recognition

by

Michail I. Schlesinger

Ukranian Academy of Sciences,
Kiev, Ukraine

and

Václav Hlaváč

Czech Technical University,
Prague, Czech Republic

KLUWER ACADEMIC PUBLISHERS
DORDRECHT / BOSTON / LONDON

A C.I.P. Catalogue record for this book is available from the Library of Congress.

ISBN 1-4020-0642-X

Published by Kluwer Academic Publishers,
P.O. Box 17, 3300 AA Dordrecht, The Netherlands.

Sold and distributed in North, Central and South America
by Kluwer Academic Publishers,
101 Philip Drive, Norwell, MA 02061, U.S.A.

In all other countries, sold and distributed
by Kluwer Academic Publishers,
P.O. Box 322, 3300 AH Dordrecht, The Netherlands.

This is a completely revised and updated translation of *Deset prednasek z teorie statistickeho a strukturniho rozpoznavani*, by M.I. Schlesinger and V. Hlaváč. Published by Vydavatelstvi CVUT, Prague 1999. Translated by the authors. This book was typeset by Vit Zyka in Latex using Computer Modern 10/12 pt.

Printed on acid-free paper

All Rights Reserved
© 2002 Kluwer Academic Publishers
No part of this work may be reproduced, stored in a retrieval system, or transmitted
in any form or by any means, electronic, mechanical, photocopying, microfilming, recording
or otherwise, without written permission from the Publisher, with the exception
of any material supplied specifically for the purpose of being entered
and executed on a computer system, for exclusive use by the purchaser of the work.

Printed in the Netherlands.

Contents

Preface

Preface to the English edition

This monograph *Ten Lectures on Statistical and Structural Pattern Recognition* uncovers the close relationship between various well known pattern recognition problems that have so far been considered independent. These relationships became apparent when formal procedures addressing not only known problems but also their generalisations were discovered. The generalised problem formulations were analysed mathematically and unified algorithms were found.

The book unifies of two main streams in pattern recognition—the statistical and structural ones. In addition to this bridging on the uppermost level, the book mentions several other unexpected relations within statistical and structural methods.

The monograph is intended for experts, for students, as well as for those who want to enter the field of pattern recognition. The theory is built up from scratch with almost no assumptions about any prior knowledge of the reader. Even when rigorous mathematical language is used we make an effort to keep the text easy to comprehend. This approach makes the book suitable for students at the beginning of their scientific career. Basic building blocks are explained in a style of an accessible intellectual exercise, thus promoting good practice in reading mathematical text. The paradoxes, beauty, and pitfalls of scientific research are shown on examples from pattern recognition. Each lecture is amended by a discussion with an inquisitive student that elucidates and deepens the explanation, providing additional pointers to computational procedures and deep rooted errors.

We have tried to formulate clearly and cleanly individual pattern recognition problems, to find solutions, and to prove their properties. We hope that this approach will attract mathematically inclined people to pattern recognition, which is often not the case if they open a more practically oriented literature. The precisely defined domain and behaviour of the method can be very substantial for the user who creates a complicated machine or algorithm from simpler modules.

The computational complexity of some of the proposed algorithms was reduced, with important practical consequences. For the practitioners we provide WWW addresses of MATLAB toolboxes written by our students, which implement many algorithms from the book.

Both authors come from Eastern Europe and still live there. The book builds on the Eastern European tradition and gives references to several works which have appeared there, and many of them have remained unknown in other parts of the globe. This view might be of interest for the wider scientific community.

We are interested in readers' feedback. Do not hesitate to send us an email (`schles@image.kiev.ua`, `hlavac@fel.cvut.cz`). We wish the reader enjoyable and profitable reading.

M. I. Schlesinger, V. Hlaváč, May 2001

A letter from the doctoral student Jiří Pecha prior to publication of the lectures

Dear Professor Hlaváč, *Český Krumlov, November 25, 1996*

I am a doctoral student of the Electrical Engineering Faculty, Czech Technical University, in my first year of PhD studies. I learned that in the summer term of 1996 a course of lectures on the mathematical theory of pattern recognition had been delivered at our faculty by a professor from Ukraine. Unfortunately, I found it out too late to be able to attend his lectures. Other doctoral students refer to those lectures very often and I am only too sorry to have missed them. I was told that the lectures were not published. It occurred to me whether the lecturer had left any texts related to the lectures with you. If it is so, please, allow me to make a copy of them.

I myself would like to deal with image processing and pattern recognition in my PhD research. I admit I have already gathered some not very positive experience, as well. I have written a relatively complicated program recognising characters from their digital images, without much reading and studying. My work was, rather, based on my own considerations, which seemed natural to me. I am quite good at programming and was surprised by bad results of my program.

I can see now that pattern recognition is an immense field which is hard to enter, and after entering it one can easily lose one's way. I have tried to learn more about pattern recognition, and so I looked at some textbooks and journal articles. Some publications I have come across refer to pattern recognition in a popular way. And on the other hand, I am not at home with more profound books and articles. They seem to me as if they were a part of a novel that I have not begun reading from the very beginning but from somewhere in the middle. Perhaps in reading I have missed the important fundamental knowledge which is no longer quoted in those publications. That is natural because those publications have not been intended for novices such as myself.

My professors tell me that the mathematical foundations of pattern recognition are more extensive today than they were thirty years ago, when pat-

tern recognition was based almost entirely on mathematical statistics in multi-dimensional linear spaces. At present the pattern recognition theory is developing further and its mathematical foundations are much richer. Modern pattern recognition makes use of the results of graph theory, formal languages and grammars, automata theory, Markovian chains and fields, mathematical programming, computational and discrete geometry, and, moreover, pattern recognition applies new algebraic constructs. But in these domains I am far less at home than is needed for an active orientation in the present day field.

In studying, I was also looking up the missing knowledge in scientific publications which had been referred to. They often were either inaccessible publications of one and the same author, or a formal mathematical monograph, in which I rarely found a small and simple piece explaining what the subject was. I have acquired the impression that in the present day extensive bibliography on the theory and practice of pattern recognition a monograph, textbook, or other publication are missing which would be an introduction to pattern recognition. In such a book I would like to find the most needed concepts as well as the solution of fundamental tasks which would make possible my further exploration through pattern recognition on my own. The book should introduce the subject matter with minimum assumptions about the reader's knowledge. In brief, I lack a book that would be written just for me.

I am told by my supervisor that my search for such a book may be in vain. I have been advised that the book I am looking for can be, to some extent, compensated for by the lectures delivered by the professor from Kiev. Could I, perhaps, ask you to kindly sending me a copy of notes from the lecture.

Sincerely,
Jiří Pecha

A letter from the authors to the doctoral student Jiří Pecha

Dear colleague Pecha, Kiev, January 5, 1997

We agree to a considerable extent with your evaluation of the situation in contemporary pattern recognition. Pattern recognition was born in the fifties, and in the nineties it has enjoyed a new enhancement in popularity. It is applied in so many diverse fields of human activities that there is hardly a domain where it is not used. The domain of applying pattern recognition methods extends from the micro-world to the universe.

Such a popularity naturally stimulates everybody who is engaged in pattern recognition. The popularity is, at the same time, a source of justified fears that pattern recognition will never be able to fulfil the expectations set upon it. A reason for discomfort could be that pattern recognition has acquired the reputation of being a magic wand, which solves practical tasks without their detailed and painstaking research. Naturally, this reputation calls for vigilance with sensible users. But for the more credulous users, this reputation will become a source of disappointment when it will be found out that a mere application of the words pattern recognition, neural nets, artificial intelligence, and the like, does not guarantee the solution of an application task.

Therefore it is the highest time for the users and authors of pattern recognition methods to be capable of separating from reality the beautiful dreams about the miraculous powers of pattern recognition. Present day pattern recognition offers enough knowledge so that its authority could be based on real, and not on imaginary values.

The lectures you are hunting for contain just what you miss, to an extent. But the professor's notes, which he prepared before each lecture, will hardly be of any use for you. They contain formulations of major concepts and theorems. And this is only a small part of what was being explained at the lecture. Furthermore, the lectures contained a motivating part, a critical (and at times sharp) analysis of methods currently applied in pattern recognition and warning against different pitfalls, such as seemingly common sense solutions, but in reality erroneous ones. Today, we know no longer whether these parts were prepared beforehand or whether they originated immediately in the lecturing hall. And it were just these commentaries after which we began to comprehend how little we know about what we are assumed to know a great deal. The lectures had also an opposite, but still a positive aspect. We saw that a certain group of pattern recognition methods, which we had regarded as isolated islets dispersed in the ocean of our ignorance, formed at least an archipelago, which could be overlooked at a glance. We kept asking ourselves if we should consider publishing this interesting but not very academic considerations. Doing so seemed to us impossible.

We increasingly regret it because others appear to be interested in the lectures as you are. That is why we decided to put them in order and publish them. It will not be the sort of book as you have imagined, but there will be a little more to it. If we understood you properly you would prefer to have a textbook like a reference book, where chapters on different mathematical disciplines used in pattern recognition would be described. We have thought of publishing a reference book like this. But we have changed our opinion and it was mostly your letter that has suggested this to us. Even if you evaluate your skill in the mathematical apparatus of modern pattern recognition rather modestly, you have found that it is usually only a matter of applying some fundamental and simple mathematical concepts. Your troubles do not lie in that these concepts are too complicated, but that they are too many and come from different mathematical disciplines. A novice in pattern recognition may also feel embarrassed by the fact that some mathematical concepts have acquired a rather different meaning within the pale of pattern recognition than the original one in mathematics. Present day pattern recognition has not only taken over concepts from different domains of mathematics but it has brought them into a mutual relation that resulted in coming into existence of new concepts, new task formulations and new issues that already belong to pattern recognition and not to the mother disciplines from which they originated.

The lectures we are revising have been focused on pattern recognition tasks and not on the tasks of linear programming, mathematical statistics, or the graph theory. The required mathematical means has not been presented mutually isolated, but in a context necessary for solving a particular pattern recognition task. Mathematical results have taken their part in one mechanism, which could be referred to as the mathematical apparatus for pattern recognition. Our opinion is that it is an all round view which you do not possess.

Unfortunately, we are only at the beginning of setting up the lectures into a form which is publishable. About two years of work lie before us. We understand that you are not willing to wait for such a long time. We have already started writing the lectures down. It will naturally take a while to supplement bare comments with all that is necessary for the text to be readable and understandable. Not to leave you idle during all this, we would like to ask for your collaboration. We could send you the text of separate lectures, as we assume that you need not have all the lectures in once. It would be perfect if you, after having gone through a part of the subject matter, could write down your critical findings, your ideas on the subject matter and questions, if any. We would consider such a feedback a necessary condition of our collaboration. You will obtain the next lecture only when we get a thorough analysis of the previous lecture from you.

We are now sending the first lecture to you and are looking forward to our collaboration.

Michail I. Schlesinger, Václav Hlaváč

Basic concepts and notations

Let us introduce the notation of sets. If a *set* has only a few members it will be presented as a list of elements in curly brackets, e.g., $\{A, B, C\}$. Sets with many elements will be presented in the form $\{x \mid \varphi(x)\}$, where x denotes a general element and $\varphi(x)$ specifies properties of all elements of a set. For instance, $\{x \mid 0 \leq x \leq 100\}$ is a set of numbers satisfying the statement $0 \leq x \leq 100$.

A *conjunction* (a logical multiplication) is denoted by $\wedge$ and a *disjunction* (a logical addition) is expressed as $\vee$.

The usual *set operations* will be used, union $X \cup Y$, intersection $X \cap Y$, Cartesian product $X \times Y = \{(x, y) \mid (x \in X) \wedge (y \in Y)\}$ and set difference $X \setminus Y = \{x \mid (x \in X) \wedge (x \notin Y)\}$.

Let X and Y be two sets. The denotation $f\colon X \to Y$ represents a *function* which assigns to each member $x \in X$ a member $f(x) \in Y$. The set X is a (definition) domain of the function f, Y gives range of function values and $f(x)$ is a value which the function f assumes in the point $x \in X$.

Let the denotation $I(n)$ means a subset of whole numbers (integers, denoted as $\mathbb{Z}$) $\{i \mid 0 \leq x \leq n\}$ corresponding to an integer n, X is a set and x is a function $x\colon I(n) \to X$. Such a function x represents a *sequence*. We will denote the i-th element of a sequence by $x(i)$ and sometimes as x_i. The sequence will be given also as a list of its elements, e.g., $x_1, x_2, \ldots, x_n$.

Both the function and the sequence can be understood as an *ensemble* of values which will be denoted by $(x_y, y \in Y)$.

Furthermore the set, the concept of the *multi-set* (sometimes called bag) is useful. A set with a finite number of members can be expressed as a list of its members in which members do not repeat. A finite multi-set can be expressed by the list of its members in which the same member can occur more than once. Even though members in the list can permute they still represent the same set or multi-set. The list of multi-set members is given in rounded brackets. Let us give an example. Let X be the set $\{1,2,3,4,5,6\}$ and $\varphi\colon X \rightarrow X$ is a function defined is such a way that $\varphi(1) = 1$, $\varphi(2) = 1$, $\varphi(3) = 2$, $\varphi(4) = 2$, $\varphi(5) = 6$, $\varphi(6) = 1$. Then $\{1,6,2\}$ is the set $\{\varphi(x) \mid x \in X\}$ and $(1,2,6,1,2,1)$ is a multi-set $(\varphi(x) \mid x \in X)$.

Let X be a finite set and $\mathbb{R}$ be a set of real numbers. The function $p\colon X \rightarrow \mathbb{R}$ will be called a *probability distribution* on the set X if $\sum_{x \in X} p(x) = 1$ holds and if $p(x) \geq 0$ holds for any $x \in X$. The value of the function $p(x)$ is called the probability of the element x.

We will deal with probability distributions $p\colon X \rightarrow \mathbb{R}$ for finite sets X almost everywhere in this book. Although this assumption narrows the range where the results can be applied, the unnecessary complications that would occur if infinite sets were taken into account are avoided. In such a way the important results are highlighted. Because the set considered are finite we let the denotation $|X|$ express the number of members of the set X.

We will study various probabilities with the domain represented by diverse sets, but all probabilities will be denoted by p. If the need to distinguish probabilities occurs, the notation p will be enriched by a lower index representing the appropriate domain. For instance, p_X is a function $X \rightarrow \mathbb{R}$ and p_Y is a function $Y \rightarrow \mathbb{R}$.

Let X, Y be two sets and p_{XY} a probability distribution $X \times Y \rightarrow \mathbb{R}$. The number $p_{XY}(x,y)$ is called a *joint probability*, of the event, that the corresponding random variables will assume the values x and y.

The $p_{X|Y}$ denotes the function $X \times Y \rightarrow \mathbb{R}$, whose value in the element in the point $x \in X$ and $y \in Y$ is given by the following expression

$$p_{X|Y}(x|y) = \frac{p_{XY}(x,y)}{\sum\limits_{y \in Y} p_{XY}(x,y)} .$$

The function $p_{X|Y}$ of two variables introduced in the preceding equation is called a *conditional* (also *a posteriori*) *probability distribution*. The value of the function $p_{X|Y}(x|y)$ is the conditional probability of the value $x \in X$ on the condition $y \in Y$.

It is more appropriate in some cases to express the function of two variables $p_{X|Y}$ as the ensemble of functions $(p_{X|y}, y \in Y)$ of a single variable x. The function $p_{X|y}$ itself depends on the parameter y. The conditional probability of the random value $x \in X$ under condition $y \in Y$ will be denoted as $p_{X|y}(x)$ in this case. In other cases, the value x will be fixed instead of y and the function $p_{X|Y}$ will be understood as the ensemble of functions $(p_{x|Y}, x \in X)$ of a single variable y. The function $p_{x|Y}\colon Y \rightarrow \mathbb{R}$ is parameterised by the value x. Even when $p_{x|Y}(y)$ is a conditional probability of the variable x under condition y then the function $p_{x|Y}\colon Y \rightarrow \mathbb{R}$ does not determine the probability distribution on the set Y because the sum $\sum_{y \in Y} p_{x|Y}(y)$ is not necessarily equal to one. Therefore the number $p_{x|Y}(y)$ is called the *likelihood* of the value y.

The function $p_X : X \to \mathbb{R}$ is called an *a priori probability distribution* on the set X for a given joint probability distribution $p_{XY} : X \times Y \to \mathbb{R}$ and it is defined as

$$p_X(x) = \sum_{y \in Y} p_{XY}(x, y) .$$

Let X be a set, $p_X : X \to \mathbb{R}$ be a probability distribution, and $f : X \to \mathbb{R}$ be any function. The number

$$\sum_{x \in X} f(x)\, p_X(x)$$

is called the *mathematical expectation* of a random variable $f(x)$ and it is denoted $E(f)$.

Let X be a set and $f : X \to \mathbb{R}$ be a function. The denotation $\operatorname{argmax}_{x \in X} f(x)$ will be used for any $x^* \in X$, for which $f(x^*) = \max_{x \in X} f(x)$ holds. The denotation $\operatorname{argmin}_{x \in X} f(x)$ has an analogous meaning.

The *scalar product* (also called inner or dot product) of a, b will be denoted as $\langle a, b \rangle$. Sometimes the assignment statement will be needed and it will be denoted as ::=. To ease the reader's orientation in the text, the formulation of theorems, lemmata, examples, and remarks is finished by the symbol ▲ at the right margin of the text. The corresponding symbol for finishing proofs is ■.

The book consists of ten lectures, which are numbered decimally. The lectures are divided into sections (numbered, e.g., 2.3). The section consists of subsections (e.g., 2.3.1). The last section of each lecture is the discussion with the student Jiří Pecha.

Acknowledgements

This book should express our deep respect and appreciation to our teachers Prof. V. A. Kovalevski and Prof. Z. Kotek. We would consider it the highest honour if our teachers were to place this monograph among their own achievements, because without their contribution it would not occur.

The work on the Czech version of a manuscript lasted much longer than we initially expected. T. N. Barabanuk (Ms.) helped us in typing the manuscript. We thank several of our co-workers and students for stimulating discussions over two years, in which the Czech manuscript was in preparation. These discussions continued for another almost two years while the edition in English was in preparation. Assoc. Prof. M. Tlalková (Ms.) helped us with the translation of the Czech version into English.

The manuscript in preparation was read and commented on by Dr. J. Grim (the reviewer of the Czech edition), colleagues Assoc. Prof. M. Navara, Dr. J. Matas, Prof. B. Melichar, Assoc. Prof. C. Matyska, Prof. H. Bischof, doctoral students P. Bečvář, A. Fitch, L. Janků (Ms.), B. Kurkoski, D. Průša, D. Beresford. Students V. Franc and J. Dupač wrote as diploma theses the Matlab toolboxes implementing many algorithms given in our monograph. The doctoral student V. Zýka helped us a lot with typography of the monograph in LaTeX.

We thank the director of the Czech Technical University Publishing House Dr. I. Smolíková (Ms.), who helped the Czech edition to be published as well as with the issue of transferring the copyright for the English and intended Russian editions of the monograph. We acknowledge Prof. M. Viergever, the Managing Editor of Kluwer Academic Publishers series Computational Imaging and Vision, who very positively accepted the idea of publishing the monograph in English in Spring 1999 and kept stimulating us to do so. The support of the series editors Dr. P. Roos, Mr. J. Finlay, and their assistant Inge Hardon (Ms.) was of great help too.

Co-authors are grateful to each other for cooperation that allowed the work to be finished in spite of age, geographical, language, and other differences.

Lecture 1

Bayesian statistical decision making

1.1 Introduction to the analysis of the Bayesian task

The Bayesian theory belongs to the building blocks which constitute the basis of statistical pattern recognition. We shall introduce fundamental concepts of the theory, formulate the Bayesian task of statistical decision making, and prove the most important properties of the task. These properties are regarded as being generally known. We would be glad if this was really so, but proposals can quite often be seen which contradict the results of Bayesian theory, although they look natural at first glance. This testifies that the knowledge of Bayesian theory is only illusory. Such partial knowledge can be worse than entire ignorance. Someone was certainly right when he said that he preferred to communicate with a person who did not read books at all to communicating with someone who had read one single book only.

Incomprehension of Bayesian decision making tasks and partial knowledge of them is caused by diverse results in probability theory and statistical decision making being associated with the name of T. Bayes. The formula according to which conditional probabilities are calculated is known as Bayes formula. A recommendation how to select, under certain conditions, the most probable value of a random variable is also called Bayesian. Also, statistical decision making tasks based on risk minimisation and penalties are called Bayesian.

We shall introduce the basic concepts and their notation which will be used in the later lectures too.

1.2 Formulation of the Bayesian task

Let an object be described by two parameters x and k. The first parameter is observable and the second is hidden, i.e., inaccessible to direct observation. The parameter x will be called a *feature of an object* or *observation* or observable parameter. The feature x assumes a value from a set X. The second parameter will be named a *state of an object* or a *hidden parameter*. Let us denote by the symbol K a set of values of a hidden parameter k. The symbol D will denote a *set of possible decisions*.

Let us note that the state of the object k is sometimes understood as a class from a finite set of classes K. We will not constrain ourselves to such an understanding of the state k because the concept of the class of objects does not have a natural meaning in many applications. For instance, the hidden parameter k can be a position of an object in the robot working space, the observation x being an image captured by a camera. In this case it is somewhat unnatural to identify the hidden state k with any class of objects.

In the set $X \times K$ of all possible pairs of observations $x \in X$ and states $k \in K$, the probability distribution $p_{XK}: X \times K \to \mathbb{R}$ is given so that for each state $k \in K$ and observation $x \in X$ the number $p_{XK}(x,k)$ represents a joint probability that the object is in the state k and the feature corresponding to it assumes the value x.

Let $W: K \times D \to \mathbb{R}$ be a *penalty function*, where $W(k,d)$, $k \in K$, $d \in D$, denotes a penalty which is paid when an object is in a state k and the decision d is taken. The function W is also called a loss function. Let $q: X \to D$ denote a function which assigns to any $x \in X$ the decision $q(x) \in D$. The function q is called a decision strategy and also a decision function. The mathematical expectation of the penalty which has to be paid if the strategy q is used is considered as a quality of the strategy and is called a *risk*. The risk of the strategy q is denoted by $R(q)$.

The Bayesian statistical decision making task consists in that for given sets X, K, D, given functions $p_{XK}: X \times K \to \mathbb{R}$, $W: K \times D \to \mathbb{R}$, a strategy $q: X \to D$ has to be found which minimises the Bayesian risk

$$R(q) = \sum_{x \in X} \sum_{k \in K} p_{XK}(x,k)\, W\big(k, q(x)\big)\,.$$

The solution of the Bayesian task is the strategy q that minimises the risk $R(q)$ and it is called the Bayesian strategy.

Let us stress the immense generality of the Bayesian task formulation outlined. Nothing has been said so far about how the set of observations X, states K, and decisions D ought to be understood. In other words, there has not been any constraint on what mathematical shape the elements of those sets are to have. The observation $x \in X$, depending on the application, can be a number or a non-numerical mathematical object. The example can be a symbol in an abstract alphabet, it can be a vector or an ensemble of characters, it can be a function of a single variable (a process) or of two variables (an image). We can further think of a function with a more complicated domain than a set of values of one or two numerical values has, i.e., of a graph or another algebraic structure. The sets of states K and decisions D can be similarly diverse.

Various concretizations of Bayesian tasks which have proved to be useful in pattern recognition will be carefully analysed in this course. Before we get down to it, we shall list several properties which are valid, in entire generality, for the whole class of Bayesian tasks. There are not many such properties whose validity does not depend on a specific application. These properties can be quite easily formulated and, moreover, they are important, as they allow us to avoid severe errors.

1.3 Two properties of Bayesian strategies

The Bayesian task was formulated as a search for the deterministic strategy $q\colon X \to D$. This means that it is required from the very beginning that the same decision $d = q(x)$ must be made every time a given x is observed even though the unobservable state k might be different. This fact can stimulate the desire to extend somehow the set of strategies among which the best one is sought. The set of deterministic strategies could be extended in such a way to cover not only all possible functions of the form $q\colon X \to D$ but also various probability distributions $q_r(d \,|\, x)$. These distributions could be understood as randomised strategies where for each observation x the appropriate decision d is selected randomly according to the probability distribution $q_r(d \,|\, x)$.

The following theorem will show that it is hopeless to look forward to miraculous possibilities of random strategies.

Theorem 1.1 Deterministic character of Bayesian strategies. *Let X, K, D be finite sets, $p_{XK}\colon X \times K \to \mathbb{R}$ be a probability distribution, $W\colon K \times D \to \mathbb{R}$ be a penalty function. Let $q_r\colon D \times X \to \mathbb{R}$ be a stochastic strategy. Its risk is*

$$R_{\text{rand}} = \sum_{x \in X} \sum_{k \in K} p_{XK}(x,k) \sum_{d \in D} q_r(d \,|\, x)\, W(k,d)\,. \tag{1.1}$$

In such a case there exists the deterministic strategy $q\colon X \to D$ with the risk

$$R_{\text{det}} = \sum_{x \in X} \sum_{k \in K} p_{XK}(x,k)\, W\big(k, q(x)\big)$$

which is not greater than R_{rand}. ▲

Proof. Let us rewrite equation (1.1) in another form

$$R_{\text{rand}} = \sum_{x \in X} \sum_{d \in D} q_r(d \,|\, x) \sum_{k \in K} p_{XK}(x,k)\, W(k,d)\,.$$

The equality $\sum_{d \in D} q_r(d \,|\, x) = 1$ holds for any $x \in X$ and $q_r(d \,|\, x) \geq 0$ holds for any $d \in D$ and $x \in X$. Thanks to it the inequality

$$R_{\text{rand}} \geq \sum_{x \in X} \min_{d \in D} \sum_{k \in K} p_{XK}(x,k)\, W(k,d) \tag{1.2}$$

holds for all $x \in X, d \in D$.

Let us denote by $q(x)$ any value d that satisfies the equality

$$\sum_{k \in K} p_{XK}(x,k)\, W\big(k, q(x)\big) = \min_{d \in D} \sum_{k \in K} p_{XK}(x,k)\, W(k,d)\,. \tag{1.3}$$

The function $q\colon X \to D$ defined in such a way is a deterministic strategy which is not worse than the stochastic strategy q_r. In fact, when we substitute

Equation (1.3) into the inequality (1.2) then we obtain the inequality

$$R_{\text{rand}} \geq \sum_{x \in X} \sum_{k \in K} p_{XK}(x,k)\, W\big(k, q(x)\big) .$$

The risk of the deterministic strategy q can be found on the right-hand side of the preceding inequality. It can be seen that $R_{\text{det}} \leq R_{\text{rand}}$ holds. ■

We have seen that the introduction of the stochastic strategy (it is also called a randomisation) cannot improve Bayesian strategy from the point of view of the mathematical expectation of a penalty.

Let first us explain the second important property of Bayesian strategy in an example. Let the hidden parameter assume two values only, $K = \{1, 2\}$. Let us assume that from all data needed to create the Bayesian strategy $q\colon X \to D$, only conditional probabilities $p_{X|1}(x)$ and $p_{X|2}(x)$ are known. The *a priori* probabilities $p_K(1)$ and $p_K(2)$ and penalties $W(k,d)$, $k \in \{1,2\}$, $d \in D$, are not known. In this situation the Bayesian strategy cannot be created. However, it can be shown that the strategy cannot be an arbitrary one any more. The strategy should belong to a certain class of strategies, i.e., it should have certain properties.

If the *a priori* probabilities $p_K(k)$ and the penalty $W(k,d)$ were known then the decision $q(x)$ about the observation x ought to be

$$\begin{aligned} q(x) &= \operatorname*{argmin}_d \big(p_{XK}(x,1)\, W(1,d) + p_{XK}(x,2)\, W(2,d)\big) \\ &= \operatorname*{argmin}_d \big(p_{X|1}(x)\, p_K(1)\, W(1,d) + p_{X|2}(x)\, p_K(2)\, W(2,d)\big) \\ &= \operatorname*{argmin}_d \left(\frac{p_{X|1}(x)}{p_{X|2}(x)}\, p_K(1)\, W(1,d) + p_K(2)\, W(2,d)\right) \\ &= \operatorname*{argmin}_d \big(\gamma(x)\, c_1(d) + c_2(d)\big) . \end{aligned} \tag{1.4}$$

The notation $c_1(d) = p_K(1)\, W(1,d)$, $c_2(d) = p_K(2)\, W(2,d)$ was used in the last line of Equation (1.4) and the *likelihood ratio* $\gamma(x) = p_{X|1}(x)/p_{X|2}(x)$ was introduced which is a well known and an important concept. It can be seen from equation (1.4) that the subset of observations $X(d^*)$, for which the decision d^* should be made, is the solution of the system of inequalities

$$\gamma(x)\, c_1(d^*) + c_2(d^*) \leq \gamma(x)\, c_1(d) + c_2(d) , \quad d \in D \setminus \{d^*\} .$$

Each inequality in the system is linear with respect to the likelihood ratio $\gamma(x)$ and therefore the subset $X(d^*)$ corresponds to a convex subset of the values of the likelihood ratio $\gamma(x)$. As $\gamma(x)$ are real numbers, their convex subsets correspond to the numerical intervals. We have arrived at the important property of Bayesian strategy, valid in the particular case in which the hidden parameter can assume only two values. (However, there can be more than two decisions.)

Any Bayesian strategy divides the real axis from 0 to ∞ into $|D|$ intervals $I(d)$, $d \in D$. The decision d is made for observation $x \in X$ when the likelihood ratio $\gamma = p_{X|1}(x)/p_{X|2}(x)$ belongs to the interval $I(d)$.

In a more particular case, when only two decisions $D = \{1, 2\}$ are possible, the generally known result is obtained. In this case the *Bayesian strategy is characterised by a single threshold value θ. For an observation x the decision depends only on whether the likelihood ratio is larger or smaller than θ.*

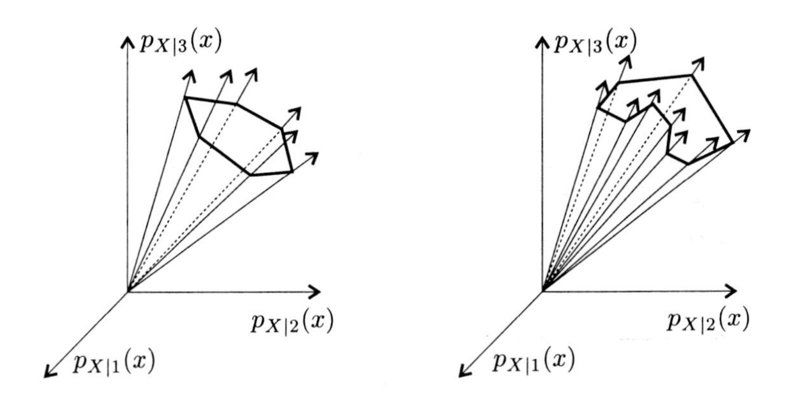

Figure 1.1 Convex cone (left) and non-convex one (right).

Let us express this property of the Bayesian strategy in a more general case in which the hidden parameter k can assume more than two values. The likelihood ratio does not make any sense in such a case. Let us recall the previous case for this purpose and give an equivalent formulation of Bayesian strategies for the case in which $|K| = 2$ and $|D| \geq 2$. Each observation $x \in X$ will be represented by a point on a plane with coordinates $p_{X|1}(x)$ and $p_{X|2}(x)$. In this a way the set X is mapped into the upper right quadrant of the plane. Each set $X(d)$, $d \in D$, is mapped into a sector bound by two lines passing through the origin of the coordinate system. The sectors are convex, of course.

Let us proceed now to a more general case in which $|K| > 2$. Let Π be a $|K|$-dimensional linear space. The subset Π' of the space Π is called a *cone* if $\alpha\,\pi \in \Pi'$ holds for an arbitrary $\pi \in \Pi'$ and an arbitrary real number $\alpha > 0$. If the subset is a cone and, in addition, it is convex then it is called a convex cone, see Fig. 1.1.

Let us map the set of observations X into the positive hyperquadrant of the space Π, i.e., into the set of points with non-negative coordinates. The point $\pi(x)$ with coordinates $p_{X|k}(x)$, $k \in K$, corresponds to the observation $x \in X$.

Any Bayesian strategy can be formed by decomposition of the positive hyperquadrant of the space Π into $|D|$ convex cones $\Pi(d)$, $d \in D$, in such a way that the decision d is taken for observation x when $\pi(x) \in \Pi(d)$. Some of the cones can be empty.

Let us express this general property of Bayesian strategies in the following theorem.

Theorem 1.2 Convex shape of classes in the space of probabilities. *Let X, K, D be three finite sets and let $p_{XK}: X \times K \to \mathbb{R}$, $W: K \times D \to \mathbb{R}$ be two functions. Let $\pi: X \to \Pi$ be a mapping of the set X into a $|K|$-dimensional linear space Π; $\pi(x) \in \Pi$ is a point with coordinates $p_{X|k}(x)$, $k \in K$.*

Let any decomposition of the positive hyperquadrant of the space Π into $|D|$ convex cones $\Pi(d)$, $d \in D$, define the strategy q for which $q(x) = d$ if and only if $\pi(x) \in \Pi(d)$. Then a decomposition $\Pi^*(d)$, $d \in D$, exists such that corresponding strategy q^* minimises a Bayesian risk

$$\sum_{x \in X} \sum_{k \in K} p_{XK}(x, k)\, W\big(k, Q(x)\big)\,.$$

▲

Proof. Let us create cones which are referred to in the theorem being proved. Let us enumerate every decision from the set D in such a way that $n(d)$ is the number of the particular decision d. Let us state one of the possible strategies that minimises the risk. It is going to be the strategy that makes a decision d^* when such x is observed that

$$\sum_{k \in K} p_{X|K}(x)\, p_K(k)\, W(k, d^*) \le \sum_{k \in K} p_{X|K}(x)\, p_K(k)\, W(k, d)\,, \quad n(d) < n(d^*)\,,$$

$$\sum_{k \in K} p_{X|K}(x)\, p_K(k)\, W(k, d^*) < \sum_{k \in K} p_{X|K}(x)\, p_K(k)\, W(k, d)\,, \quad n(d) > n(d^*)\,.$$

The system of equations given above can be expressed by means of coordinates of the point $\pi(x) \in \Pi$, i.e., numbers $\pi_k = p_{X|k}(x)$. The point π with coordinates π_k, $k \in K$, has to be mapped into the set $\Pi(d^*)$, if

$$\sum_{k \in K} \pi_k\, p_K(k)\, W(k, d^*) \le \sum_{k \in K} \pi_k\, p_K(k)\, W(k, d)\,, \qquad n(d) < n(d^*)\,,$$

$$\sum_{k \in K} \pi_k\, p_K(k)\, W(k, d^*) < \sum_{k \in K} \pi_k\, p_K(k)\, W(k, d)\,, \qquad n(d) > n(d^*)\,. \tag{1.5}$$

From the given system of inequalities it is obvious that the set expressed in such a way is a cone, because if the point with coordinates π_k, $k \in K$, satisfies the inequalities then any point with coordinates $\alpha\, \pi_k$, $\alpha > 0$, satisfies the system too.

The system of inequalities (1.5) is linear with respect to variables π_k, $k \in K$, and thus the set of its solutions $\Pi(d)$ is convex. ■

An important consequence results from the proved property of Bayesian strategies. It is known that two disjoint convex sets can be separated by a hyperplane. This means that there is a vector α and number θ such that $\langle \alpha, \pi \rangle < \theta$ holds for all elements π from the first set and $\langle \alpha, \pi \rangle \ge \theta$ holds for all points from the second set. Theorem 1.2 about the convex shape in the space of probabilities does not only state that classes in the space of probabilities are convex but, in addition, they are cones, too. It follows from this that there is a linear function

$\langle \alpha, \pi \rangle$ such that $\langle \alpha, \pi \rangle < 0$ for one cone and $\langle \alpha, \pi \rangle \geq 0$ for the second one. Such sets are called *linearly separable*, or that there exists a linear discriminant function separating these two sets. This property has been popular in pattern recognition so far. One of the later lectures will be devoted to it. The theorem provides a certain basis and explanation of this popularity as it states that the Bayesian strategy surely decomposes the space of probabilities into classes that are linearly separable.

1.4 Two particular cases of the Bayesian task

1.4.1 Probability of the wrong estimate of the state

In most cases the pattern recognition task is to estimate the state of an object. This means that a set of decisions D and a set of states K are the same. The decision $q(x) = k$ means that an object is in the state k. Of course, the estimate $q(x)$ not always is equal to the actual state k^*. Thus the probability of the wrong decision $q(x) \neq k^*$ is required to be as small as possible. We will demonstrate that such a requirement can be expressed as a special case of a Bayesian task.

Indeed, let us imagine that the classifier has to pay a unit penalty when the situation $q(x) \neq k^*$ occurs and it does not pay any penalty otherwise. This means that $W\big(k^*, q(x)\big) = 1$ if $q(x) \neq k^*$ and $W\big(k^*, q(x)\big) = 0$ if $q(x) = k^*$. The mathematical expectation

$$R = \sum_{x \in X} \sum_{k^* \in K} p_{XK}(x, k^*)\, W\big(k^*, q(x)\big) \tag{1.6}$$

is thus the probability of the situation $q(x) \neq k^*$. The Bayesian task consists, just as in the general case, in determining the strategy $q\colon X \to K$ which minimises the mathematical expectation given by equation (1.6), i.e.,

$$q(x) = \operatorname*{argmin}_{k \in K} \sum_{k^* \in K} p_{XK}(x, k^*)\, W(k^*, k)\,. \tag{1.7}$$

Let us modify equation (1.7) using following equivalent transformations

$$\begin{aligned}
q(x) &= \operatorname*{argmin}_{k \in K} \sum_{k^* \in K} p_{XK}(x, k^*)\, W(k^*, k) \\
&= \operatorname*{argmin}_{k \in K} p_X(x) \sum_{k^* \in K} p_{K|X}(k^* \mid x)\, W(k^*, k) \\
&= \operatorname*{argmin}_{k \in K} \sum_{k^* \in K} p_{K|X}(k^* \mid x)\, W(k^*, k) \\
&= \operatorname*{argmin}_{k \in K} \sum_{k^* \in K \setminus \{k\}} p_{K|X}(k^* \mid x) \\
&= \operatorname*{argmin}_{k \in K} \Big(\sum_{k^* \in K} p_{K|X}(k^* \mid x) - p_{K|X}(k \mid x) \Big) \\
&= \operatorname*{argmin}_{k \in K} \big(1 - p_{K|X}(k \mid x)\big) = \operatorname*{argmax}_{k \in K} p_{K|X}(k \mid x)\,.
\end{aligned}$$

The result is that the *a posteriori* probability of each state k is to be calculated for the observation x and it is to be decided in favour of the most probable state.

Even if the abovementioned rule is entirely clear and natural, there are sometimes suggestions to 'improve' it. For example, it is suggested that one selects the class k randomly in accordance with probability distribution $p_{K|X}(k \mid x)$. Suggestions of this kind are not often based on logical reasoning but on misty poetic analogies, as in the tale in which Jill seeks a straying Jack and she has to look for him not only in the pub, where he would often sit, but also in places where he goes only from time to time. These kinds of suggestions are erroneous because they contradict Theorem 1.1 about the deterministic nature of Bayesian strategies. Let us examine more thoroughly why such pseudo-solutions are wrong.

Let us assume that it was found, for a certain observation, that a probability of the first state, say $k = 1$, is equal to 0.9 and the probability of the second state is 0.1. If the most probable state is chosen, the error occurs in 10% of cases. If the decision is made randomly, it means that in 90% of cases it will be decided that the first state occurred and in the second state it will be indicated in 10% of cases. In this case the probability of error will be 0.1 in 90% of cases and it will be 0.9 in the rest 10% of cases. The total probability of error will be 0.18 which is almost twice as much as the minimal possible value 0.1.

Let us mention one more pseudo-solution which, regrettably, occurs even more often than the previous case. Assume that a device or a computer program is created that implements a strategy $q\colon X \to D$ which for a given observation x decides about a state k, which can assume one of four values: 1, 2, 3 and 4. Assume that this strategy is optimal from the standpoint of the probability of the wrong decision. Let us now imagine that it appears that it is not needed to provide such a detailed information about the state. It is sufficient to decide whether the state is smaller than 3 (or not). It is obvious that the task is modified. In the first mentioned case, the set D consists of four decisions $k = 1$, $k = 2$, $k = 3$ and $k = 4$. In the second, new, task, the set D' contains two decision $k \in \{1, 2\}$ and $k \in \{3, 4\}$ only. It is thus needed to replace the previous strategy $q\colon X \to D$ by the new strategy $q'\colon X \to D'$. It could appear that the new task is simpler than the previous one and that (watch out, the error follows!) the existing strategy q can be used when designing a new strategy q'. Then it has be to decided that the state is smaller than 3, if $q(x) = 1$ or $q(x) = 2$ and the state is not smaller than 3, if $q(x) = 3$ or $q(x) = 4$. Theorem 1.2 about the convex shape of classes in the space of probabilities Π provides good reasons to doubt about the proposed solution of the described task.

When the first task is solved, the space of probabilities is separated into four classes $\Pi(1)$, $\Pi(2)$, $\Pi(3)$ and $\Pi(4)$. The strategy q' is constructed for the new task in such a way that the space Π is divided into two classes $\Pi(1) \cup \Pi(2)$ and $\Pi(3) \cup \Pi(4)$. But Theorem 1.2 states that each of these six sets has to be a convex cone. When the strategy q' is created in such a simple way it may easily happen that the classes corresponding to the strategy are not convex because a union of convex sets can be a non-convex set, see Fig 1.2. When this happens it will mean that the new strategy q' does not only reach the minimal probability of a wrong decision but moreover it does not solve any Bayesian task.

Let us show as an example how it can happen that the strategy q' created in the abovementioned way is not the best one. Assume that for some observation x the *a posteriori* probabilities of the states 1, 2, 3 and 4 correspond to 0.3; 0.3; 0.4 and 0.0, respectively. The strategy q decides in this case that the state $k = 3$ has occurred. It is the best decision from the point of view of the minimal probability of a wrong state. The strategy q' explores the previous decision and determines that the state is not smaller than 3. Indeed, it is not the best decision. The probability of error in this case is equal to 0.6. If the opposite answer was given then the probability of error would be 0.4, i.e., the smaller one.

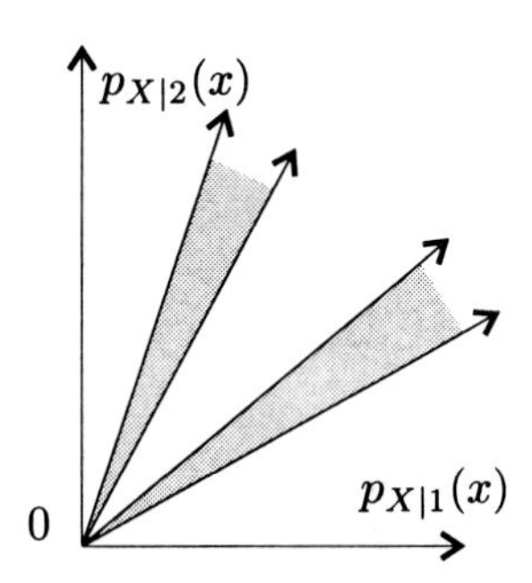

Figure 1.2 The union of two convex cones does not necessarily have to be a convex cone.

1.4.2 Bayesian strategy with possible rejection

Let us denote the conditional mathematical expectation of penalty by $R(x, d)$. It is obtained under the condition of an observation x and a decision d

$$R(x,d) = \sum_{k \in K} p_{K|X}(k \mid x)\, W(k,d)\,. \tag{1.8}$$

The value $R(x, d)$ will be called the ***partial risk***. Naturally, a decision d has to be set for each observation x in such a way that the partial risk is minimal. However, it can happen for some observations that this minimum will be quite large. It would be appropriate if the set of all decisions contained also a particular decision corresponding to the answer `not known`. The decision `not known` is given in the case in which the observation x does not contain enough information to decide with a small risk. Let us formulate this task within the Bayesian approach.

Let X and K be sets of observations and states, $p_{XK}\colon X \times K \to \mathbb{R}$ be a probability distribution and $D = K \cup \{\texttt{not known}\}$ be a set of decisions. Let us determine penalties $W(k, d)$, $k \in K$, $d \in D$, according to the following rule:

$$W(k,d) = \begin{cases} 0, & \text{if } d = k\,, \\ 1, & \text{if } d \neq k \text{ and } d \neq \texttt{not known}\,, \\ \varepsilon, & \text{if } d = \texttt{not known}\,. \end{cases} \tag{1.9}$$

Let us find the Bayesian strategy $q\colon X \to D$ for this case. The decision $q(x)$ corresponding to the observation x has to minimise the partial risk. This means that

$$q(x) = \operatorname*{argmin}_{d \in D} \sum_{k^* \in K} p_{K|X}(k^* \mid x)\, W(k^*, d)\,. \tag{1.10}$$

Definition (1.10) is equivalent to the definition

$$q(x) = \begin{cases} \operatorname*{argmin}_{d \in K} R(x,d)\,, & \text{if } \min_{d \in K} R(x,d) < R(x, \texttt{not known})\,, \\ \texttt{not known}\,, & \text{if } \min_{d \in K} R(x,d) \geq R(x, \texttt{not known})\,. \end{cases} \tag{1.11}$$

There holds for $\min_{d \in K} R(x,d)$

$$\begin{aligned} \min_{d \in K} R(x,d) &= \min_{d \in K} \sum_{k^* \in K} p_{K|X}(k^* \mid x)\, W(k^*, d) \\ &= \min_{k \in K} \sum_{k^* \in K \setminus \{k\}} p_{K|X}(k^* \mid x) \\ &= \min_{k \in K} \left(\sum_{k^* \in K} p_{K|X}(k^* \mid x) - p_{K|X}(k \mid x) \right) \\ &= \min_{k \in K} \left(1 - p_{K|X}(k \mid x)\right) = 1 - \max_{k \in K} p_{K|X}(k \mid x)\,. \end{aligned}$$

There holds for $R(x, \texttt{not known})$

$$\begin{aligned} R(x, \texttt{not known}) &= \sum_{k^* \in K} p_{K|X}(k^* \mid x)\, W(k^*, \texttt{not known}) \\ &= \sum_{k^* \in K} p_{K|X}(k^* \mid x)\, \varepsilon = \varepsilon\,. \end{aligned} \tag{1.12}$$

The rule (1.11) becomes

$$q(x) = \begin{cases} \operatorname*{argmax}_{k \in K} p_{K|X}(k \mid x)\,, & \text{if } 1 - \max_{k \in K} p_{K|X}(k \mid x) < \varepsilon\,, \\ \texttt{not known}\,, & \text{if } 1 - \max_{k \in K} p_{K|X}(k \mid x) \geq \varepsilon\,. \end{cases} \tag{1.13}$$

The description of the strategy $q(x)$ may be put in words as: The state k first has to be found which has the largest *a posteriori* probability. If this probability is larger than $1 - \varepsilon$ then it is decided in favour of the state k. If its probability is not larger than $1 - \varepsilon$ then the decision `not known` is provided. Such a strategy can also be understood informally. If $\varepsilon = 0$ then the best strategy is not to decide. It is understandable since the penalty corresponding to the answer `not known` is zero. If it is decided in favour of any state then the error is not excluded and thus the non-zero penalty is not excluded either. Conversely, for $\varepsilon > 1$ the partial risk of decision `not known` will be greater than the partial risk of any other decision. In this case the answer `not known` will be never used.

This correct strategy is quite natural, but pseudo-solutions are suggested even in this case. For example, the following one: it is suggested to use the answer `not known` for an observation that has a small probability for any state k, i.e., if the probability $p_{X|K}(x \mid k) < \theta$ for any $k \in K$. The number θ indicates a threshold that says what is understood as a small probability observation. It follows from Theorem 1.2 on the convex shape of classes in the space of probabilities that this strategy is not a solution of our task but, moreover, it is not a solution of any Bayesian task. The set of $|K|$-dimensional vectors, i.e., the set of $|K|$-tuplets $\big(p_{X|K}(x \mid k),\, k \in K\big)$, which constitutes the class `not known` in the space of probabilities, is convex but it is not a cone. The case is illustrated in Fig. 1.3. The grey shaded square illustrates observations with small probabilities. Though the square is convex but it is not a cone due to its spatial limitations.

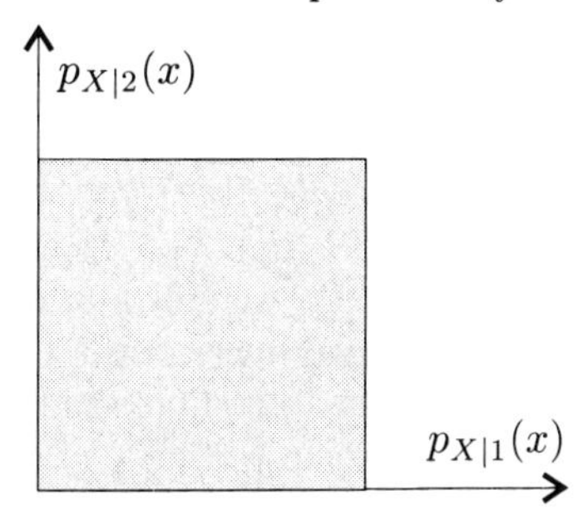

Figure 1.3 Decision `not known` for observations with small probability is not a cone and thus it does not correspond to any Bayesian strategy.

1.5 Discussion

I have come across Bayesian theory several times and each time it seemed to me that the theory was explained too generally. It appears to me that only the particular case is useful, i.e., the minimisation of the wrong decision. I do not see the applicability of another approach yet; perhaps my horizon is not wide enough. Is the general theory not a mere mental exercise? Does the lecture provide the lesson that only the last part of the lecture dealing with the particular case has a practical significance?

We do agree with you that the probability of a wrong decision has to be minimised in many practical tasks. On the other hand, let us not forget that the set of practical tasks has an inexhaustible variability. It is hopeless to attempt to squeeze the possible tasks into any theoretical construction based on the optimisation of a single criterion even if it were as respected as the probability of a wrong decision is. We shall see in Lecture 2 that many practical tasks cannot be compressed even into such a rich theoretical construct as the Bayesian approach is.

If the Bayesian approach, though if it is so general, does not suffice for some practical tasks then one cannot think that it is too rich. Imagine that you come across an applied problem which cannot be squeezed into a part of the Bayesian approach that selects the most probable state. In the better case it will raise your doubts and you will try to find a better solution. In the worse case you will distort your task so that no one could recognise it any more, and will try to modify it to match that part of the theory you are familiar with. Finally, you will solve a totally different task than was needed.

The recognition task is often understood in a constrained manner as a maximisation of the *a posteriori* probability $p_{K|x}(k)$. Indeed, the Bayesian recognition was understood in your question in this way too. Such an approach follows from a simplified assumption that all errors have the same significance even if it is natural to penalise them differently.

The Bayesian approach covers a more realistic case which takes into account that errors are of different significance. This is expressed by the dependence of a penalty on the deviation from the correct answer. This is the reason why it is often needed that not only the *a posteriori* probability $p_{K|x}(k)$ has to be a part of the maximised criterion but the penalty function as well which takes into account the significance of the error.

As an exercise we would like you to solve an example which can serve as a model of many practical situations. Assume that a set X is a set of images of digits ranging from 0 to 9 which were written on a sheet of paper. The set of possible states is consequently $K = \{0, 1, 2, \ldots, 9\}$. Let us assume that the *a priori* probabilities on the set K are known. For example, $p_K(k) = 0.1$ for all $k \in K$. In addition, let us presume that conditional probabilities $p_{X|K}(x \mid k)$ are known as well, even when they are enormously complicated. Even so, let us imagine that we already have a program in hand that yields 10 probabilities $p_{X|K}(x \mid k)$ for each image $x \in X$. This program maps the analysed image into a space of probabilities. Everything that is needed to solve any Bayesian task is at one's disposal at this moment. The popular task estimating the digit k, which secures the smallest probability of error, belongs to these tasks. A digit k with the largest *a posteriori* probability is searched for. The *a posteriori* probabilities are calculated according to Bayes' formula

$$p_{K|X}(k \mid x) = \frac{p_K(k)\, p_{X|K}(x \mid k)}{\sum\limits_{k=0}^{9} p_K(k)\, p_{X|K}(x \mid k)}\,.$$

Simply speaking, it is decided in favour of k, for which the probability $p_{X|K}(x \mid k)$ is the largest because *a priori* probabilities $p_K(k)$ are equal for all digits k.

Assume now that not only a single image is submitted to the recognition procedure but 20 images independent of each other. Your task is to estimate the sum of digits $k_1, k_2, \ldots, k_{20}$, which are depicted on images $x_1, x_2, \ldots, x_{20}$.

A person who thinks that only the formula $k^* = \operatorname{argmax}_k p_{K|X}(k \mid x)$ makes sense in the entire Bayesian theory starts solving the task resolutely and at once. For each $i = 1, 2, \ldots, 20$ she or he calculates

$$k_i^* = \operatorname*{argmax}_k \; p_{K|X}(k \mid x_i)$$

and estimates the sum $S = \sum_{i=1}^{20} k_i^*$. If the question is asked why is it done exactly in this way then there is an answer prepared for all cases: 'it follows from the Bayesian theory that such an algorithm yields the best results'. Nothing of this kind follows from the Bayesian theory, of course. We are looking forward to you proposing a correct solution of a specified task which actually follows from the Bayesian theory.

I have developed an opinion of my own of how this task has to be solved. At the same time I can imagine what you want me to say. First, let me write what follows from the lecture.

It is obvious that the set of possible decisions D is $\{0, 1, \ldots, 180\}$. The sum of twenty digits can be one of the listed values. The set of observations $\widetilde{X}$ is

$$\underbrace{X \times X \times \ldots \times X}_{20 \text{ times}}$$

which is the set of sequences which consists of 20 images each. The set of states $\widetilde{K}$ is

$$\underbrace{K \times K \times \ldots \times K}_{20 \text{ times}},$$

i.e., the set of sequences consisting of 20 digits. Each digit can be 0, 1, ..., 9. Let us denote by $\tilde{x}$ the sequence $x_1, x_2, \ldots, x_{20}$ that is submitted to recognition. In addition, let us denote by $\tilde{k}$ the sequence $k_1, k_2, \ldots, k_{20}$ of digits that are really shown on the recognised images. The probabilities $p_{\widetilde{X}\widetilde{K}} \colon \widetilde{X} \times \widetilde{K} \to \mathbb{R}$ are clearly

$$p_{\widetilde{X}\widetilde{K}}(\tilde{x}, \tilde{k}) = \prod_{i=1}^{20} p_K(k_i)\, p_{X|K}(x_i \mid k_i)$$

because the images in the analysed sequence are mutually independent. The penalty function $W \colon \widetilde{K} \times D \to \mathbb{R}$ has the value either zero or one. If $\sum_{i=1}^{20} k_i = d$ then $W(\tilde{k}, d) = 0$. If $\sum_{i=1}^{20} k_i \neq d$ then $W(\tilde{k}, d) = 1$. The risk R of the strategy $q \colon \widetilde{X} \to D$ will be

$$R(q) = \sum_{\tilde{x} \in \widetilde{X}} \sum_{\tilde{k} \in \widetilde{K}} p_{\widetilde{X}\widetilde{K}}(\tilde{x}, \tilde{k})\, W(\tilde{k}, d)\,. \tag{1.14}$$

From the above, the following Bayesian strategy results

$$\begin{aligned}
q(\tilde{x}) &= \operatorname*{argmin}_{d \in D} \sum_{\tilde{k} \in \widetilde{K}} p_{\widetilde{X}\widetilde{K}}(\tilde{x}, \tilde{k})\, W(\tilde{k}, d) \\
&= \operatorname*{argmin}_{d \in D} p_{\widetilde{X}}(\tilde{x}) \sum_{\tilde{k} \in \widetilde{K}} p_{\widetilde{K}|\widetilde{X}}(\tilde{k} \mid \tilde{x})\, W(\tilde{k}, d) \\
&= \operatorname*{argmin}_{d \in D} \sum_{\tilde{k} \notin \widetilde{K}(d)} p_{\widetilde{K}|\widetilde{X}}(\tilde{k} \mid \tilde{x}) \\
&= \operatorname*{argmin}_{d \in D} \left(1 - \sum_{\tilde{k} \in \widetilde{K}(d)} p_{\widetilde{K}|\widetilde{X}}(\tilde{k} \mid \tilde{x})\right) \\
&= \operatorname*{argmax}_{d \in D} \sum_{\tilde{k} \in \widetilde{K}(d)} p_{\widetilde{K}|\widetilde{X}}(\tilde{k} \mid \tilde{x})\,.
\end{aligned}$$

In the last three steps of the derivation, $\widetilde{K}(d)$ denotes the set of such sequences $k_1, k_2, \ldots, k_{20}$, whose sum $\sum_{i=1}^{20} k_i = d$. Furthermore, the expression for a *derived Bayesian strategy can be made more specific because images in the observed sequence are independent. Thus*

$$q(\tilde{x}) = \operatorname*{argmax}_{d \in D} \underbrace{\sum_{k_1 \in K} \sum_{k_2 \in K} \dots \sum_{k_{19} \in K} \sum_{k_{20} \in K}}_{k_1+k_2+\dots+k_{19}+k_{20}=d} \prod_{i=1}^{20} p_{K|X}(k_i \,|\, x_i), \tag{1.15}$$

where the expression

$$\underbrace{\sum_{k_1 \in K} \sum_{k_2 \in K} \dots \sum_{k_{19} \in K} \sum_{k_{20} \in K}}_{k_1+k_2+\dots+k_{19}+k_{20}=d}$$

means the summation along all sequences $(k_1, k_2, \dots, k_{20})$, whose sum $\sum_{i=1}^{20} k_i = d$.

I believe that I honestly did all that was recommended in the lecture and obtained the Bayesian strategy suitable to our task. Nevertheless, I have doubts about the value of the result. The expression obtained is not likely to be used in practice. The maximisation is not a problem any more because it is necessary to find the largest number out of only 181 numbers (the value 181 is the number of possible values of the sum $\sum_i k_i$). What matters is the fantastically difficult calculation of those 181 numbers from which the maximal one is to be selected. Indeed, it is required to calculate a sum of so many summands; roughly speaking such that is equal to the number of all possible sequences $\tilde{k}$, i.e., 10^{20} summands.

It can be clearly concluded from the example mentioned why the theoretical recommendation typically is not followed. It seems to me that the reason does not lie in not understanding the theory, but in the difficulties when implementing the theoretical recommendations in practice. I prefer to solve the task using the method you laughed at.

The summation is not so dreadful if a certain computational trick is used. We shall explore similar tricks later as well. Let us have a more careful look at them. Let us denote that dreadful function the values of which are to be calculated as $F(d)$. Then

$$F(d) = \underbrace{\sum_{k_1 \in K} \sum_{k_2 \in K} \dots \sum_{k_{19} \in K} \sum_{k_{20} \in K}}_{k_1+k_2+\dots+k_{19}+k_{20}=d} \prod_{i=1}^{20} p_{K|X}(k_i \,|\, x_i). \tag{1.16}$$

Let us express 20 auxiliary functions of a similar form:

$$F_1(d) = \underbrace{\sum_{k_1 \in K}}_{k_1=d} \prod_{i=1}^{1} p_{K|X}(k_i \,|\, x_i)\,,$$

$$F_2(d) = \underbrace{\sum_{k_1 \in K} \sum_{k_2 \in K}}_{k_1+k_2=d} \prod_{i=1}^{2} p_{K|X}(k_i \mid x_i) \,,$$

$$\vdots$$

$$F_j(d) = \underbrace{\sum_{k_1 \in K} \sum_{k_2 \in K} \dots \sum_{k_j \in K}}_{k_1+k_2+\dots+k_j=d} \prod_{i=1}^{j} p_{K|X}(k_i \mid x_i) \,,$$

$$F_{j+1}(d) = \underbrace{\sum_{k_1 \in K} \sum_{k_2 \in K} \dots \sum_{k_j \in K} \sum_{k_{j+1} \in K}}_{k_1+k_2+\dots+k_j+k_{j+1}=d} \prod_{i=1}^{j+1} p_{K|X}(k_i \mid x_i) \,,$$

$$\vdots$$

$$F_{20}(d) = \underbrace{\sum_{k_1 \in K} \sum_{k_2 \in K} \dots \sum_{k_{19} \in K} \sum_{k_{20} \in K}}_{k_1+k_2+\dots+k_{19}+k_{20}=d} \prod_{i=1}^{20} p_{K|X}(k_i \mid x_i) \,.$$

It is obvious that the function F_{20} is the same as the function F. It is clear too that values $F_1(d)$ are easy to calculate. Actually, they even need not be computed. Indeed, the product $\prod_{i=1}^{1} p_{K|X}(k_i \mid x_i)$ consists of a single multiplicative term $p_{K|X}(k_1 \mid x_1)$. The sum

$$F_1(d) = \underbrace{\sum_{k_1 \in K}}_{k_1=d} p_{K|X}(k_1 \mid x_1)$$

consists of one single summand and thus $F_1(d) = p_{K|X}(d \mid x_1)$. We will show how to calculate the values $F_{j+1}(d)$ provided that $F_j(d)$ are already available:

$$\begin{aligned}
F_{j+1}(d) &= \underbrace{\sum_{k_1 \in K} \sum_{k_2 \in K} \dots \sum_{k_j \in K} \sum_{k_{j+1} \in K}}_{k_1+k_2+\dots+k_j+k_{j+1}=d} \prod_{i=1}^{j+1} p_{K|X}(k_i \mid x_i) \\
&= \underbrace{\sum_{k_{j+1} \in K} p_{K|X}(k_{j+1} \mid x_{j+1}) \sum_{k_1 \in K} \sum_{k_2 \in K} \dots \sum_{k_j \in K}}_{k_1+k_2+\dots+k_j+k_{j+1}=d} \prod_{i=1}^{j} p_{K|X}(k_i \mid x_i) \\
&= \sum_{k_{j+1} \in K} p_{K|X}(k_{j+1} \mid x_{j+1}) \underbrace{\sum_{k_1 \in K} \sum_{k_2 \in K} \dots \sum_{k_j \in K}}_{k_1+k_2+\dots+k_j=d-k_{j+1}} \prod_{i=1}^{j} p_{K|X}(k_i \mid x_i) \\
&= \sum_{k_{j+1} \in K} p_{K|X}(k_{j+1} \mid x_{j+1}) \, F_j(d - k_{j+1}) \,.
\end{aligned} \tag{1.17}$$

If the derivation is drawn into a single expression we obtain

$$F_{j+1}(d) = \sum_{k_{j+1} \in K} p_{K|X}(k_{j+1} \,|\, x_{j+1}) \; F_j(d - k_{j+1}) \,. \tag{1.18}$$

When calculating the value of the function $F_j(d)$ for one d, we have to perform 10 multiplications and 9 additions. There are not more than 181 such values. The transformation of the functions F_j to F_{j+1} should be done 19 times before we obtain the function F_{20}. Consequently, we do not need 10^{20} calculations but only $10 \times 181 \times 19$ multiplications at most and nearly the same number of additions. It is not worth mentioning from the computational complexity point of view.

Expression (1.17) surprised me. I believe that it is not a mere trick. There is probably some depth in it when you said that similar modifications would be needed several times.

Let me come back to my persuasion that nothing but maximisation of a posteriori probability is needed. The previous example did not disprove my persuasion, perhaps just the converse is true. If I look at calculated functions carefully then it becomes clear that the function $F(d)$ is nothing else but an a posteriori probability that the sum of random numbers $k_1, k_2, \ldots, k_{20}$ is equal to d. By the way, the recursive expression (1.18) is nothing but the known formula which we learned at college to calculate the probability distribution of the sum of two independent variables. One is k_{j+1} and the other has the probability distribution F_j. It seems that the best strategy again means the search for the most probable value which is calculated in a non-trivial way, I have to admit. Maximisation of some probabilities was avoided but we reached the probability maximisation via a detour through more general Bayesian considerations. What differed were the events.

It is good you realised that you have been familiar with formula (1.18) for long time. If you had recalled it earlier, you would have overcome by yourself the troubles that blocked your way. We did not get to the bottom of maximisation of *a posteriori* probability together but you did it yourself when you formulated the penalty function $W(\tilde{k}, d)$ in a way that seemed to you the only possible one. We do not object to this function, neither do we like to consider it as the only possible one or the most natural one. It is quite unnatural to pay the same penalty if instead of the actual value 90 the sum is estimated to be 89 or 25. It would be more natural if the penalty was larger in the second case than that in the first one. What about analysing various penalty functions that may suit the formulated task?

Let $d^*(k^*)$ denote the true result $\sum_{i=1}^{20} k_i^*$. The first penalty function could be: $W(\tilde{k}, d)$ is zero if $|d^*(k^*) - d|$ is not greater than an acceptable error (tolerance) Δ. The unit penalty is used when the difference $|d^*(k^*) - d|$ is greater than the tolerance Δ. Minimisation of the risk in this case means minimisation of a probability that the difference of estimated d from the correct value $d^*(k^*)$ will be greater than Δ.

Let the second penalty function be the difference $|d^*(k^*) - d|$ and the third one be proportional to the squared difference $\big(d^*(k^*) - d\big)^2$.

Although the second penalty function worried me before, I hope that I have mastered it now. The algorithms for the three formulations of Bayesian tasks are rather similar. All three comprise a calculation of a posteriori *probabilities $F(d)$ for each value of the sum. When the function $F(d)$ is available then the decision is made for each formulation of the penalty function differently. The simplest situation occurs in the case of quadratic penalty for which the Bayesian strategy is approved in the following manner. The decision d has to minimise the partial risk*

$$R(d) = \sum_{k^* \in \widetilde{K}} p_{\widetilde{K}|\widetilde{X}}(k^* \mid \tilde{x}) \left(\sum_{i=1}^{20} k_i^* - d \right)^2 = \sum_{k^* \in \widetilde{K}} p_{\widetilde{K}|\widetilde{X}}(k^* \mid \tilde{x}) \big(d^*(k^*) - d\big)^2 .$$

This means that the decision d is a solution of the equation that requires the derivative of function $R(d)$ be equal to zero, i.e.,

$$\begin{aligned}
0 &= -2 \sum_{k^* \in \widetilde{K}} p_{\widetilde{K}|\widetilde{X}}(k^* \mid \tilde{x}) \big(d^*(k^*) - d\big) \\
&= 2\,d - 2 \sum_{k^* \in \widetilde{K}} p_{\widetilde{K}|\widetilde{X}}(k^* \mid \tilde{x})\, d^*(k^*) \\
&= 2\,d - 2 \sum_{d^* \in D} \sum_{k^* \in \widetilde{K}(d^*)} p_{\widetilde{K}|\widetilde{X}}(k^* \mid \tilde{x})\, d^* \\
&= 2\,d - 2 \sum_{d^* \in D} d^* \sum_{k^* \in \widetilde{K}(d^*)} p_{\widetilde{K}|\widetilde{X}}(k^* \mid \tilde{x}) \\
&= 2\,d - 2 \sum_{d^* \in D} d^*\, F(d^*) \,.
\end{aligned}$$

It follows that $d = \sum_{d^* \in D} d^*\, F(d^*)$, as one could expect. The decision d will be in favour of the a posteriori *mathematical expectation of the correct sum d^*.*

Let us return to the first penalty function with a tolerance Δ. The penalty $W(k^*, d)$ is now either one or zero. The first option occurs when the error $|d^*(k^*) - d|$ is greater than Δ. The second option applies otherwise. Let us denote by symbol $g(d^*, d)$ a function of two variables. Its value is equal to 0 if $|d^* - d| \le \Delta$, or it is equal to 1 if $|d^* - d| > \Delta$. In such a case the decision has to be

$$\begin{aligned}
&\operatorname*{argmin}_{d \in D} \sum_{k^* \in \widetilde{K}} p_{\widetilde{K}\,|\,\widetilde{X}}(k^* | \tilde{x})\, W(k^*, d) \\
&= \operatorname*{argmin}_{d \in D} \sum_{k^* \in \widetilde{K}} p_{\widetilde{K}|\widetilde{X}}(k^* \mid \tilde{x})\, g\big(d^*(k^*), d\big) \\
&= \operatorname*{argmin}_{d \in D} \sum_{d^* \in D} \sum_{k^* \in \widetilde{K}(d^*)} p_{\widetilde{K}|\widetilde{X}}(k^* \mid \tilde{x})\, g(d^*, d)
\end{aligned}$$

$$= \operatorname*{argmin}_{d \in D} \sum_{d^* \in D} g(d^*, d) \sum_{k^* \in \widetilde{K}(d^*)} p_{\widetilde{K}|\widetilde{X}}(k^* \mid \tilde{x}) = \operatorname*{argmin}_{d \in D} \sum_{d^* \in D} g(d^*, d)\, F(d^*)$$

$$= \operatorname*{argmin}_{d \in D} \sum_{|d^* - d| > \Delta} F(d^*) = \operatorname*{argmin}_{d \in D} \left(\sum_{d^* \in D} F(d^*) - \sum_{|d^* - d| \le \Delta} F(d^*) \right)$$

$$= \operatorname*{argmin}_{d \in D} \left(1 - \sum_{|d^* - d| \le \Delta} F(d^*) \right) = \operatorname*{argmax}_{d \in D} \sum_{d^* = d - \Delta}^{d + \Delta} F(d^*) \,.$$

So minimization of the risk has been reduced to the maximization of the sum

$$R'(d) = \sum_{d^* = d - \Delta}^{d + \Delta} F(d^*) \,, \tag{1.19}$$

which depends on the decision d. This summation has to be calculated for each value $d \in D$, i.e., 181 times in total. The largest sum has to be selected out of 181 sums. Certainly, none of the sums $R'(d)$ should be calculated using formula (1.19) but using the recurrent formula

$$R'(d) = R'(d-1) + F(d+\Delta) - F(d - \Delta - 1) \,.$$

It is more favourable if the acceptable error tolerance Δ is great enough.

The most difficult task was to find the solution for the second penalty function $W(\tilde{k}, d) = |d^*(\tilde{k}) - d|$. The task resembles (at first, erroneous glance) the previous task. A wrong reasoning could look like this. For the given 181 numbers $F(d^*)$, $d^* = 0, \ldots, 180$, the Bayesian decision d has to be found that minimises the sum

$$R(d) = \sum_{d^* \in D} F(d^*) |d^* - d| \,. \tag{1.20}$$

I will decompose the set $D = \{0, \ldots, 180\}$ into three subsets:

$$\begin{aligned} D^+ &= \{d^* \mid (d^* \in D) \wedge (d^* > d)\} \,, \\ D^- &= \{d^* \mid (d^* \in D) \wedge (d^* < d)\} \,, \\ D^= &= \{d^* \mid (d^* \in D) \wedge (d^* = d)\} \,. \end{aligned}$$

I can write the expression for $R(d)$

$$R(d) = \sum_{d^* \in D^+} F(d^*)(d^* - d) - \sum_{d^* \in D^-} F(d^*)(d^* - d) \,, \tag{1.21}$$

which (watch out, an error follows!) depends linearly on d. The derivative of the expression is

$$- \sum_{d^* \in D^+} F(d^*) + \sum_{d^* \in D^-} F(d^*) \,.$$

If the derivative is assumed equal to zero then I get the constraint

$$\sum_{d^* \in D^+} F(d^*) = \sum_{d^* \in D^-} F(d^*) \,, \tag{1.22}$$

which says that the minimal risk is achieved when the total probability of the event that the random variable d^* is smaller than d is equal to the total probability that a random variable d^* is larger than d. The value d is the median of the random variable d^*.

Although this result is correct (which I will show later), the procedure that led me to it is incorrect. First, the function $R(d)$ in expression (1.21) only seems to be linear. The value $R(d)$ depends on d not only explicitly but also implicitly by means of the sets D^+ and D^- because the sets themselves depend on d. Second, it may happen that the condition (1.22) will not be satisfied for any d. That would mean that the function (1.21) does not have the minimum, which is not true. It is obvious that some more complicated considerations are needed for the minimisation of $R(d)$. I have to confess that I have not yet solved the task. Nevertheless, I am convinced that the solution corresponds to the median of random variable d^* and I back up it by the following mechanical model.

Let us imagine that a thin wooden board is at hand, a straight line is drawn on it, and 181 holes regularly spaced, 1 cm apart, are drilled in it. The holes will match values $d^* = 0, 1, \ldots, 180$. Let us raise the board to a horizontal position 180 cm above the ground. Let us get 181 strings, each 180 cm long. The free ends of strings on one side will be tied together into a single knot. The other free ends of the strings remain free. Each string will be led through a hole so that the knot remains above the wooden board and the free ends hang below. A weight will be tight on to the free end of each string. From the string passing through the d^*-th hole a weight of the mass $F(d^*)$ will be hanging.

The strings will stretch out owing their weights and will reach a steady position. We are now interested in what position the knot remains at above the wooden board. The knot cannot get under the board. It could do so only by passing through a hole in the board. Thus one of the weights would have to lie on the ground, since the length of the string exactly corresponds to the distance between the board and the ground. The other weights would pull the knot on the top side of the board. Now we know that the knot must remain above the wooden board. The knot has to lie on the straight line connecting the holes on the board. If the knot got off the straight line then the resultant force of all the weights would pull it back. The knot is steady. Consequently the sum of weights pulling leftward cannot be greater than 0.5, i.e., greater than the sum of weights pulling rightwards.

This means that the knot positions d, for which $\sum_{d^*<d} F(d^*) > 0.5$, cannot be stable positions. By the same reasons, the positions for which $\sum_{d^*>d} F(d^*) > 0.5$ have to be also excluded. So only one position d remains, i.e., that for which $\sum_{d^*<d} F(d^*) \leq 0.5$ and $\sum_{d^*>d} F(d^*) \leq 0.5$ simultaneously hold. This corresponds to the median of a random variable with probability distribution $F(d^*)$.

I have to make sure that the sum $\sum_{d^*\in D} F(d^*)|d^* - d|$ achieves the minimal value at the steady position of the knot. The mechanical system of ours cannot be in any other state than in that where the potential energy $\sum_{d^*\in D} F(d^*)h(d^*)$ is minimal. The value $h(d^*)$ corresponds to height of the d^*-th weight from the *ground. The total length 180 cm of each string can be divided into two parts: the length $l_1(d^*)$ lying on the board, and the length $l_2(d^*)$ below the board. The distance 180 cm from the ground to the d^*-th hole consists of two parts too: the length $l_2(d^*)$ and the height $h(d^*)$ of d^*-th weight above the ground. This means that 180 cm $= l_1(d^*) + l_2(d^*) = l_2(d^*) + h(d^*)$. From this it follows that $l_1(d^*) = h(d^*)$. But the length $l_1(d^*)$ is equal to the distance of d^*-th hole from the knot which is $l_1(d^*) = |d^* - d|$. This means that the potential energy in our mechanical system is $\sum_{d^* \in D} F(d^*)|d^*(\tilde{k}) - d|$. And it is this value that is minimised.*

The idea with a mechanical interpretation of the task is neat. What pleased us most was when you analysed the quadratic penalty function and unwittingly uttered 'one can expect that the decision will be in favour of the *a posteriori* mathematical expectation of the correct sum'. You did not recall that it was not that long ago when you had not wanted to approve anything but the decision in favour of the most probable sum. You were not puzzled when even the mathematical expectation was not an integer. However, only integer sums have non-zero *a posteriori* probabilities. This means that as a rule the decision is made in favour of the states with zero *a posteriori* probability, not with maximal.

I think that all three tasks became solvable only because the sum $\sum_{i=1}^{20} k_i$ assumed a finite and, in particular, a small number of values on the set $\tilde{K}$. Thus the probability distribution can be calculated for every value of the sum. Then it is easy to determine the best value from 181 probability values.

Assume that we want to solve a little more complicated task. The aim of the previous simpler task was to determine a sum of twenty individual digits under uncertainty in recognising digits from images. Let us imagine a more complicated case in which twenty digits express a single decimal number. The rightmost digit corresponds to units, the digit one position to the right matches tens, etc.. The digits are $k_1 k_2 k_3 \ldots k_{20}$. We are supposed to determine the value of the decimal number, i.e., the sum

$$\sum_{i=1}^{20} a_i k_i = \sum_{i=1}^{20} 10^{i-1} k_i \,.$$

The aim of the new more general task is to estimate the value of a decimal number with the smallest possible quadratic error. The complication of the task is caused by coefficients a_i. Owing to them the sum $\sum_{i=1}^{20} a_i k_i$ can take not 181 but tremendously many values on the set $\tilde{K}$

Let us analyse the newly formulated task from the very beginning, i.e., starting already from the expression for the risk (1.14). If the function F cannot be computed, let us try to avoid its computation. The decision d^* about the sum $\sum_{j=1}^{20} a_j\, k_j$ has to minimise the expected quadratic error $(d - \sum_{j=1}^{20} a_j\, k_j)^2$.

Thus

$$
\begin{aligned}
d^* &= \operatorname*{argmin}_{d \in D} \sum_{\tilde{k} \in \tilde{K}} \prod_{i=1}^{20} p_{K|X}(k_i \mid x_i) \left(d - \sum_{j=1}^{20} a_j \, k_j \right)^2 \\
&= \sum_{\tilde{k} \in \tilde{K}} \prod_{i=1}^{20} p_{K|X}(k_i \mid x_i) \sum_{j=1}^{20} a_j \, k_j = \sum_{j=1}^{20} a_j \sum_{\tilde{k} \in \tilde{K}} k_j \prod_{i=1}^{20} p_{K|X}(k_i \mid x_i) \\
&= \sum_{j=1}^{20} a_j \sum_{k_1 \in K} \sum_{k_2 \in K} \cdots \sum_{k_j \in K} \cdots \sum_{k_{20} \in K} k_j \prod_{i=1}^{20} p_{K|X}(k_i \mid x_i) \\
&= \sum_{j=1}^{20} a_j \sum_{k_j \in K} p_{K|X}(k_j \mid x_j) \, k_j \sum_{k_1 \in K} \cdots \sum_{k_{j-1} \in K} \sum_{k_{j+1} \in K} \cdots \sum_{k_{20} \in K} \prod_{i=1, i \neq j}^{20} p_{K|X}(k_i \mid x_i) \\
&= \sum_{j=1}^{20} a_j \sum_{k_j \in K} k_j \, p_{K|X}(k_j \mid x_j) = \sum_{j=1}^{20} a_j \hat{k}_j \,.
\end{aligned}
$$

This proves the generally known result that the expected value of the sum of random variables is equal to the sum of expected values of individual summands. We can see that algorithms estimating the linear function $\sum_{i=1}^{20} a_i \, k_i$, which are the best in a quadratic sense, do not depend to a great extent on the coefficients a_i. The 20 *a posteriori* mathematical expectations $\hat{k}_i$, $i = 1, \ldots, 20$, have to be calculated at first. This is the most difficult part of the task because the function $p_{K|X}(k \mid x)$ can be quite complicated. This most difficult calculation does not depend on the coefficients a_i. Only after that the coefficients a_i are used to compute the best estimate according to the extremely simple expression $d = \sum_{i=1}^{20} a_i \, \hat{k}_i$.

This result is so significant that it deserves to be looked at it from a different perspective. Assume that the aim of recognition is not to estimate the sum $\sum_{i=1}^{20} a_i \, k_i$ but to estimate the whole sequence k_i, $i = 1, 2, \ldots, 20$. The penalty function is given by the expression

$$
W(\tilde{k}^*, \tilde{k}) = \left(\sum_{i=1}^{20} a_i \, k_i^* - \sum_{i=1}^{20} a_i \, k_i \right)^2 . \tag{1.23}
$$

We have already learned that the optimal estimate for the sequence $\tilde{k} = (k_1, k_2, \ldots, k_{20})$ is the sequence of the mathematical expectations of values k_i,

$$
\sum_{k_i \in K} p_{K|X}(k_i \mid x_i) \, k_i \,,
$$

which, as can be seen, does not depend on the coefficients $a_i, i = 1, 2, \ldots, 20$, at all. In this case the risk can be even minimised under the condition that the penalty function is not known. It has to be certain only that the penalty function is quadratic, i.e., it has the form of (1.23). If this result is used another step forward can be made. The strategy minimising the mathematical expectation of penalty of the form (1.23) independently of coefficients $a_i, i = 1, 2, \ldots, 20$, is suitable for any penalty function that is defined as a sum of quadratic penalty functions of the form (1.23), i.e., also for a penalty function of the form

$$W(\tilde{k}^*, \tilde{k}) = \sum_{j=1}^{20} \left(\sum_{i=1}^{20} a_{ij} \, (k_i^* - k_i) \right)^2 . \tag{1.24}$$

Under the condition that the matrix containing coefficients a_{ij} is positive semi-definite, the same strategy is suitable for any function of the form

$$W(\tilde{k}^*, \tilde{k}) = \sum_{j=1}^{20} \sum_{i=1}^{20} a_{ij} \, (k_i^* - k_i) \, (k_j^* - k_j) . \tag{1.25}$$

It is because any positive semi-definite function of the form (1.25) can be expressed as (1.24).

We want to emphasise an important result. You need not to care about specific coefficients a_{ij} for a task with a positive semi-definite quadratic penalty function (1.25). Not knowing them you can create a strategy that minimises the mathematical expectation of such a penalty that is not fully known.

Let us return to the task aiming to estimate the sum $\sum_{i=1}^{20} k_i^*$ that we started with. We can see now that for the certain penalty function the task can be solved easier than you did it earlier. Namely, for the optimal estimate of the sum $d = \sum_{i=1}^{20} k_i^*$ it is not needed for all values of the sum to calculate the probability $F(d)$ at all.

Indeed, quadratic penalty functions are beautiful!

You are not the only one of this opinion. But we would like to warn you against being taken by the beauty and using the quadratic penalty function where it is not appropriate. There are many such cases as well.

We can finish the analysis of the Bayesian task and proceed to the next lecture which will be devoted to non-Bayesian statistical decision making.

December 1996.

1.6 Bibliographical notes

Bayesian statistics is named in memory of a clergyman T. Bayes [Bayes, 1763], who suggested informally early in 18th century how to deal with conditional frequencies (probabilities). Statistics was and is the great inspiration for pattern recognition, taken it all round. The pattern recognition tasks are mostly not explicitly seen in the statistical literature. The reader who likes to learn about the bases of the Bayesian decision tasks in the original mathematical literature is mainly recommended [Wald, 1950], where the tasks are formulated in a very general form. The decision tasks are given in book [Anderson, 1958] for Gaussian random quantities.

As soon as statistical pattern recognition was formulated, many researchers connected it closely with learning. It has been like that till now and the majority of publications about statistical pattern recognition are concerned with learning. We will almost not touch two areas of pattern recognition in this book, namely the feature selection and the nearest neighbourhood decision rule which is analysed in detail in [Devroye et al., 1996].

The study of pattern recognition methods without connection to learning is also very valuable, less attention has been devoted to it in the literature. Let us mention Kovalevsky [Kovalevski, 1965] among works which formulated the pattern recognition tasks in the Bayesian framework. Chow [Chow, 1965] studied the concrete decision making strategy under uncertainty which we also analysed.

The outstanding book [Duda and Hart, 1973] includes the statistical pattern recognition in the whole wideness and it is recommended to the reader who meets pattern recognition with serious interest, and perhaps not only for the first time. The considerably revised second edition of the book was published recently [Duda et al., 2001].

Some general books about pattern recognition are for instance [Devijver and Kittler, 1982], [Fukunaga, 1990], [Chen et al., 1993], [Nadler and Smith, 1993], [Pavel, 1993], [Young, 1994], [Bishop, 1996], [Theodoridis and Koutroumbas, 1999].

Lecture 2

Non-Bayesian statistical decision making

2.1 Severe restrictions of the Bayesian approach

The enormous generality of the Bayesian approach has been emphasised in the first lecture several times. It follows from the fact that the problem formulation and some of its properties are valid for the diverse set structure of the observations X, states K, and decisions D. It is surely desirable to master the whole richness of Bayesian tasks, and not to identify it with a special case. We already know that the class of Bayesian tasks is more than minimisation of the probability of a wrong decision.

Despite the generality of the Bayesian approach there exist many tasks which cannot be expressed within its framework. The more general class of non-Bayesian decision methods is needed and this lecture is devoted to it. One needs to know non-Bayesian decision methods as it is necessary to choose the most suitable formalisation of the task for each specific application and not the other way round. The primary application task should not be twisted in an attempt to squeeze it into a certain formal framework even if the formalism itself is highly respected.

Bayesian tasks have already been restricted by key concepts on which the approach is based. The first is the *penalty function* $W: K \times D \rightarrow \mathbb{R}$. The second is the *probability distribution* $p_{XK}(x, k)$, $x \in X$, $k \in K$, which comprises additional independent concepts: an *a priori probability* $p_K(k)$ of the situation $k \in K$ which should be recognised, and *conditional probabilities* $p_{X|K}(x \mid k)$ of observation $x \in X$ under the condition of situation $k \in K$. Let us have a more careful look at how the acceptance of the mentioned concepts narrows the set of application tasks that can be formalised by the Bayesian approach. The look will be informal in the coming Subsections 2.1.1–2.1.3 for the time being.

2.1.1 Penalty function

As soon as the 'minimisation of the mathematical expectation of the penalty' is stated it is implicitly acknowledged that the penalty assumes the value in the totally ordered set in which, besides the relation $<$ or $\geq$, multiplication by a real number and addition are defined. It is essential for the Bayesian task that the penalty is defined as a real number. However, there are many applications in which the assignment of the penalty values to the real numbers deforms the original task. They do not belong to the Bayesian construction.

Speaking not in a very serious tone, you may know of the hero from Russian fairy tales. When he turns to the left, he loses his horse, when he turns to the right, he loses his sword, and if he turns back, he loses his beloved girl. Even if the hero knew Bayesian methods it would be of no use for him because it is impossible to find out in any reasonable way if the sum of p_1 horses and p_2 swords is less or more than p_3 beloved girls. In addition, it is not easy to grasp what the total sum of p_1 horses and p_2 swords means, just as adding up 30 metres and 80 seconds has no sense.

The situation is not rare in which *various losses cannot be measured by the same unit* even in one application. Let us recall, more seriously now, an extended application field such as a diagnosis of a complex device. It is to be found out if the device is in a regular state or in a state that starts to be dangerous. Each wrong decision causes certain damage which can be assessed by a penalty. But a damage caused by the situation where the regular state is assessed as a dangerous one (false positive, also error of the first type) is of quite a different sort than any damage due to an overlooked danger (false negative, also error of the second type). The unnecessary preventive check of the device is performed in the first case, while, in the second case, something can be destroyed which cannot be restored by any number of preventive checks. Something which nature took millions years to create can be ruined in such a case. The incomparability of different penalties mentioned is even more distinct in tasks where human society or one particular person is evaluated whatever diagnostics, medical or legal, is used.

2.1.2 *A priori* probability of situations

If the task is to be formulated in the Bayesian framework it is necessary that the *a priori* probabilities $p_K(k)$ for each state $k \in K$ are assigned. It is noticeable at first glimpse that it can be difficult to find these probabilities. More detailed study reveals that the difficulties are of much deeper nature. Let us sort out the possible situations to three groups and order them from the easiest to the more difficult ones.

1. The object state k is random and *a priori* probabilities $p_K(k)$, $k \in K$, are known. These situations are well mastered by the Bayesian approach.
2. The object state is random but *a priori* probabilities $p_K(k)$, $k \in K$, are not known. It is the case in which an object has not been analysed sufficiently. The user has two possibilities: (a) She or he can try to formulate the task not in the Bayesian framework but in another one that does not require statistical properties of the object which are unknown. (b) She or he will start analysing the object thoroughly and gets *a priori* probabilities which are inevitable for the Bayesian solution.

3. The object state is not random and that is why the *a priori* probabilities $p_K(k)$, $k \in K$, do not exist and thus it is impossible to discover them by an arbitrary detailed exploration of the object. Non-Bayesian methods must be used in this situation. They are treated in this lecture. Let us illustrate such a situation on an example.

Example 2.1 Task not belonging to the Bayesian class. *Let us assume that x is a signal originating from the observed airplane. On the basis of the signal x it is to be discovered if the airplane is an allied one ($k = 1$) or an enemy one ($k = 2$). The conditional probability $p_{X|K}(x|k)$ can depend on the observation x in a complicated manner. Nevertheless, it is natural to assume at least that there exists a function $p_{X|K}(x \mid k)$ which describes dependence of the observation x on the situation k correctly. What concerns a priori probabilities $p_K(k)$, these are not known and even cannot be known in principle because it is impossible to say about any number α, $0 \leq \alpha \leq 1$, that α is the probability of occurrence of an enemy plane. In such a case probabilities $p_K(k)$ do not exist since the frequency of experiment result does not converge to any number which we are allowed to call probability. In other words, k is not a random event.* ▲

One cannot speak about the probability of an event which is not random just as one cannot speak either about the temperature of a sound or about the sourness or bitterness of light. A property such as probability is simply not defined on the set of non-random events. Application tasks in which it is needed to estimate the value of a non-random variable do not belong to the Bayesian tasks. Their formalisation needs a theoretical construction in which the concept of the *a priori* probability does not arise at all.

Let us show a spread pseudo-solution of an applied task that is similar to that mentioned in Example 2.1. If *a priori* probabilities are unknown the situation is avoided by supposing that *a priori* probabilities are the same for all possible situations. In our case, it should mean that an occurrence of an enemy plane has the same probability as the occurrence of an allied one. It is clear that it does not correspond to the reality even if we assume that an occurrence of a plane is a random event. Logical reasons, which should prove such an assumption, are difficult to find. As a rule, logical arguments are quickly substituted by a pseudo-argument by making a reference to some renowned person. In the given case, this would be, e.g., to C. Shannon thanks to the generally known property that an equally distributed probability has the highest entropy. It happens even if this result does not concern the studied problem in any way.

2.1.3 Conditional probabilities of observations

Let us have a look at the following application task. Let X be a set of pictures. A letter is written in each picture. The letter name (label) will be marked by symbol k, the set of all letter names will be denoted by K. Let us assume that letters were written by three people. The set identifying all writers will be denoted by Z. The variable $z \in \{1, 2, 3\}$ determines which of the persons wrote the letter. It is not typically important in character recognition who wrote the letter and it is often impossible to find out. On the other hand, the writer influences what the letter looks like, and it affects its recognition too. As can be seen, the third parameter $z \in \{1, 2, 3\}$, a so called unobservable *intervention*, was added to observable parameters $x \in X$ and hidden parameters $k \in K$.

The goal of the task is to answer the following question for each picture x. Which letter is written in the picture? It is possible to speak about the penalty function $W(k, d)$ and about the *a priori* probabilities $p_K(k)$ of the individual letters but it is not possible to talk about conditional probabilities $p_{X|K}(x \mid k)$ in this application. The reason is that the appearance of the specific letter x depends not only on the letter label but also on a non-random intervention, i.e., on the fact who wrote the letter. We can speak only about conditional probabilities $p_{X|K,Z}(x \mid k, z)$, i.e., about how a specific character looks like if it was written by a certain person. If the intervention z would be random and the probability $p_Z(z)$ would be known for each z then it would be possible to speak also about probabilities $p_{X|K}(x \mid k)$, because they could be calculated using the formula

$$p_{X|K}(x \mid k) = \sum_{z=1}^{3} p_Z(z) p_{X|K,Z}(x \mid k, z) .$$

But preconditions for applying an algorithm do not provide any evidence to assume how often it will be necessary to recognise pictures written by this or that person. Rather, it is not excluded during the whole period of the algorithm application that only pictures written by only one single writer are used but it will be unknown by whom. Under such uncertain statistical conditions an algorithm ought to be created that will secure the required recognition quality of pictures independently on the fact who wrote the letter. This means that the task should be formulated in the way that the concept of *a priori* probabilities $p_Z(z)$ of the variable z will not be used because this variable is not random and such a feature as probability is not defined for it.

Let us introduce the most famous formulations of non-Bayesian tasks and their solutions here. In addition, we introduce new modifications of these known tasks. We shall see that the whole class of non-Bayesian tasks has common features in spite of the variety of non-Bayesian tasks. These allow us to analyse and solve them by the same procedure. Later on, we shall see that there is not any crucial gap between the class of Bayesian tasks and all non-Bayesian ones. We shall show that the strategy solving any non-Bayesian task can be realised similarly as, that for the Bayesian tasks in the space of probabilities. The strategy divides the space of probabilities into convex cones in the same manner as in the Bayesian tasks. This means that their solution, in spite of all basic difference between Bayesian and non-Bayesian tasks, is found within the same set of strategies.

2.2 Formulation of the known and new non-Bayesian tasks

2.2.1 Neyman–Pearson task

Let some object be characterised by the feature x which assumes a value from the set X. The probability distribution of the feature x depends on the state

k, to which the object belongs. There are two possible states—the normal one $k = 1$ and the dangerous one $k = 2$. The set of states K is thus $\{1, 2\}$. The probability distributions are known and defined by a set of conditional probabilities $p_{X|K}(x \mid k)$, $x \in X$, $k \in K$.

The goal of recognition is to decide according to the observed feature x if the object is in the normal or dangerous state. The set X is to be divided into two such subsets X_1 and X_2 that for an observation $x \in X_1$ is being decided the normal state and for an observation $x \in X_2$ the dangerous state.

In view of the fact that some values of the feature x can occur both in the normal and in the dangerous state of the object, there is no faultless strategy and it is characterised by two numbers. The first number is a probability of an event that the normal state will be recognised as a dangerous one. Such an event is called a *false alarm* or a false positive. The second number is the probability of the event that the dangerous state will be recognised as a normal one and it is called an *overlooked danger* or a false negative. These two faults are sometimes called the error of the first and second type, respectively. The conditional probability of the false alarm is given by the sum $\sum_{x \in X_2} p_{X|K}(x \mid 1)$ and the conditional probability of the overlooked danger is $\sum_{x \in X_1} p_{X|K}(x \mid 2)$.

Such a strategy is sought in the Neyman–Pearson task [Neyman and Pearson, 1928; Neyman and Pearson, 1933] (we shall call it simply Neyman task hereafter), i.e., a decomposition of the set X into two subsets $X_1 \subset X$ and $X_2 \subset X$, $X_1 \cap X_2 = \emptyset$, that, firstly, the conditional probability of the overlooked danger is not larger than a predefined value ε,

$$\sum_{x \in X_1} p_{X|K}(x \mid 2) \le \varepsilon \, . \tag{2.1}$$

Secondly, a strategy is to be chosen from all strategies satisfying the above condition for which the conditional probability of the false alarm is the smallest. This means that classes X_1 and X_2 are to minimise the sum

$$\sum_{x \in X_2} p_{X|K}(x \mid 1) \tag{2.2}$$

under the conditions

$$\sum_{x \in X_1} p_{X|K}(x \mid 2) \le \varepsilon \, , \tag{2.3}$$

$$X_1 \cap X_2 = \emptyset, \quad X_1 \cup X_2 = X \, . \tag{2.4}$$

The fundamental result of Neyman–Pearson states that for sets X_1 and X_2, which solve a given optimisation task, there exists such a threshold value θ that each observation $x \in X$, for which the *likelihood ratio*

$$\frac{p_{X|K}(x \mid 1)}{p_{X|K}(x \mid 2)}$$

is smaller than θ, belongs to the set X_2. And also, vice versa, the assignment $x \in X_1$ is made for each $x \in X$ with likelihood ratio larger than θ. Let us put the case of equality aside for pragmatic reasons. This case occurs so rarely in practical situations that we do not need to deal with it. The theoretical analysis might be interesting, but it is complicated and not necessary for our purpose.

The known solution of Neyman task is not proved easily, it means that it is not easy to show that it follows from the formulation given by relations (2.2)–(2.4). Therefore, presumably, it is not the knowledge but more a belief based on Neyman's and Pearson's authority. This belief suffices when exactly Neyman task is to be solved. As soon as a task is met, which differs from the Neyman task in some trifle, a mere belief is not sufficient.

For instance, let us have a look at the following tiny modification of Neyman task. Let the number of states of the recognised object be not two but three. For each state $k \in \{1,2,3\}$ and for each observation x the conditional probability $p_{X|K}(x\,|\,k)$ is determined. Only one state $k = 1$ is normal, whereas the other two states are dangerous. In the same way as in the original Neyman task, the aim is to find out a reasonable strategy which has to determine for each observation if the state is normal or one of these two dangerous ones.

Several suggestions how to solve this task occur at a folklore level. For instance, two likelihood ratios are being computed

$$\gamma_{12} = \frac{p_{X|K}(x\,|\,1)}{p_{X|K}(x\,|\,2)} \quad \text{and} \quad \gamma_{13} = \frac{p_{X|K}(x\,|\,1)}{p_{X|K}(x\,|\,3)}$$

and two threshold values θ_{12} and θ_{13} are set. The situation is considered as normal, if $\gamma_{12} > \theta_{12}$ and $\gamma_{13} > \theta_{13}$. Other suggestions are based on the effort to invent such a generalisation of the likelihood ratio concept which should suit even in the case in which it concerns the 'ratio' not of two but of three quantities, for instance

$$\frac{p_{X|K}(x\,|\,1)}{\max\big(p_{X|K}(x\,|\,2),\, p_{X|K}(x\,|\,3)\big)} \quad \text{or} \quad \frac{p_{X|K}(x\,|\,1)}{\sum\limits_{k\in K} p_{X|K}(x\,|\,k)}$$

or other similar figments. Then it is decided for the normal or dangerous state by comparing the mythical 'ratio' with a certain threshold value.

Suggestions of a similar sort demonstrate the effort to find out the algorithm of the solution at once without formulating the task which that algorithm is to solve. That is why such proposals are not convincing enough. Of course, such suggestions are not supported by Neyman's authority as he was not interested in this modification at all.

In the effort to manage a task, even if it is merely a slight generalisation of Neyman task, it is not possible to start with a direct endeavour to alter Neyman's strategy to a slightly modified task. It is correct to begin from the formulation of a corresponding generalised task and then pass the whole way through from the task formulation to the algorithm similarly as Neyman did with his task.

2.2.2 Generalised Neyman task with two dangerous states

The following formulation of generalised Neyman task can be helpful in our case. Observations will be classified into two sets, which correspond to two events:

$k = 1$ corresponds to the set denoted X_1;
$k = 2$ or $k = 3$ corresponds to the set denoted X_{23}.

It is necessary to find such a strategy for which the conditional probability of the overlooked dangerous state both $k = 2$ and $k = 3$ is not larger than the beforehand given value. Simultaneously, the strategy minimising the false alarm is to be selected from all the strategies that satisfy such a condition. Formally, two searched sets X_1 and X_{23} have to minimise the total sum

$$\sum_{x \in X_{23}} p_{X|K}(x \mid 1) \tag{2.5}$$

under conditions

$$\sum_{x \in X_1} p_{X|K}(x \mid 2) \le \varepsilon \,, \tag{2.6}$$

$$\sum_{x \in X_1} p_{X|K}(x \mid 3) \le \varepsilon \,, \tag{2.7}$$

$$X_1 \cap X_{23} = \emptyset \,, \tag{2.8}$$

$$X_1 \cup X_{23} = X \,. \tag{2.9}$$

The formulated optimisation task will be thoroughly analysed later when the other non-Bayesian tasks will be formulated too. In addition, it will be seen that the whole series of non-Bayesian tasks can be solved in a single constructive framework.

2.2.3 Minimax task

Let X be a set of observations and let K be a set of object states as before. The probability distribution $p_{X|K}(x \mid k)$ on the set X corresponds to each state k. Let the strategy be determined by the decomposition $X(k)$, $k \in K$, that decides for each observation $x \in X$ that the object is in the state k when $x \in X(k)$. Each strategy is characterised by $|K|$ numbers $\omega(k)$, $k \in K$, which stand for the conditional probability of a wrong decision under the condition that the actual state of the object was k,

$$\omega(k) = \sum_{x \notin X(k)} p_{X|K}(x \mid k) \,.$$

The minimax task requires a decomposition of the set of observations X into subsets $X(k)$, $k \in K$, such that they minimise the number $\max_{k \in K} \omega(k)$.

The nature of this task can be imagined more practically when we consider the following situation. Let us assume that the creation of the recognition algorithm was ordered by a customer who demanded in advance that the algorithm would be evaluated by two tests: the preliminary test and the final one. The customer himself would perform the preliminary test and would check what the probability of a wrong decision $\omega(k)$ was for all states k. The customer selects the worst state $k^* = \text{argmax}_{k \in K} \omega(k)$. In the final test, only those objects are checked that are in the worst state.

The result of the final test will be written in the protocol and the final evaluation depends on the protocol content. The algorithm designer aims to achieve the best result of the final test.

It is known for the task with two states as well as for Neyman task that the strategy solving a minimax problem is based on the comparison of the likelihood ratio with some threshold value. Similarly as in Neyman task, it is more a belief than knowledge, because hardly anybody is able to derive a solution of this minimax problem. That is why the solution of the problem has not been widely known for the more general case, i.e., for the arbitrary number of object states.

2.2.4 Wald task

The task, which is going to be formulated now, presents only a tiny part of an extensive scientific area known as Wald sequential analysis [Wald, 1947; Wald and Wolfowitz, 1948].

When the formulation of Neyman task is recalled, its lack of symmetry with respect to states of the recognised object is apparent. First of all, the conditional probability of the overlooked danger must be small, which is the principal requirement. The conditional probability of the false alarm is a subsidiary, inferior criterion in this respect. It can be only demanded to be as small as possible even if this minimum can be even big.

It would certainly be excellent if such a strategy were found for which both probabilities would not exceed a predefined value ε. These demands can be antagonistic and that is why the task could not be accomplished by using such a formulation. To exclude this discrepancy, the task is not formulated as a classification of the set X in two subsets X_1 and X_2 corresponding to a decision for the benefit of the first or the second state, but as a classification in three subsets X_0, X_1 and X_2 with the following meaning:

if $x \in X_1$, then $k = 1$ is chosen;

if $x \in X_2$, then $k = 2$ is chosen; and finaly

if $x \in X_0$ it is decided that the observation x does not provide enough information for a safe decision about the state k.

A strategy of this kind will be characterised by four numbers:

$\omega(1)$ is a conditional probability of a wrong decision about the state $k = 1$,

$$\omega(1) = \sum_{x \in X_2} p_{X|K}(x \mid 1) ;$$

$\omega(2)$ is a conditional probability of a wrong decision about the state $k = 2$,

$$\omega(2) = \sum_{x \in X_1} p_{X|K}(x \mid 2) ;$$

$\chi(1)$ is a conditional probability of a indecisive situation under the condition that the object is in the state $k = 1$,

$$\chi(1) = \sum_{x \in X_0} p_{X|K}(x \mid 1) ;$$

$\chi(2)$ is a conditional probability of the indecisive situation under the condition that the object is in the state $k = 2$,

$$\chi(2) = \sum_{x \in X_0} p_{X|K}(x \mid 2) .$$

For such strategies the requirements $\omega(1) \leq \varepsilon$ and $\omega(2) \leq \varepsilon$ are not contradictory for an arbitrary non-negative value ε because the strategy $X_0 = X$, $X_1 = \emptyset$, $X_2 = \emptyset$ belongs to the class of allowed strategies too. Each strategy meeting the requirements $\omega(1) \leq \varepsilon$ and $\omega(2) \leq \varepsilon$ is, moreover, characterised by how often the strategy is reluctant to decide, i.e., by the number $\max\big(\chi(1), \chi(2)\big)$.

Wald task seeks among the strategies satisfying the requirements $\omega(1) \leq \varepsilon$, $\omega(2) \leq \varepsilon$ for a strategy which minimises the value $\max\big(\chi(1), \chi(2)\big)$. It is known that the solution of this task is based on the calculation of the likelihood ratio

$$\gamma(x) = \frac{p_{X|K}(x \mid 1)}{p_{X|K}(x \mid 2)} .$$

For certain threshold values θ_1 and θ_2, $\theta_1 \leq \theta_2$, it is decided for the benefit of one or the other state, or the state is left undecided on the basis of comparing likelihood ratio $\gamma(x)$ with the previous two threshold values. Later on, a solution will be proved that will allow us to make a decision in a case of Wald task of a more general character where the number of states can be arbitrary, and not merely two.

2.2.5 Statistical decision tasks with non-random interventions

The previous non-Bayesian tasks took into account the property that in a certain application the penalty function or *a priori* probabilities of the state did not make sense. Statistical decision tasks with non-random intervention make one step further and they are concerned with the situations in which, in addition, the conditional probabilities $p_{X|K}(x \mid k)$ do not exist.

The statistical decision task with non-random intervention was formulated by Linnik [Linnik, 1966]. The task relates to the situation in which the feature x is a random quantity that depends not only on the object state but also on some additional parameter z of the object. This additional parameter z is not directly observable either. Moreover, the end user is not interested in the value of the parameter z and thus need not be estimated. In spite of it, the parameter z must be taken into account because conditional probabilities $p_{X|K}(x \mid k)$ are not defined. There exist only conditional probabilities $p_{X|K,Z}(x \mid k, z)$. The problem of statistical testing of objects, when the observations are influenced by non-random and unknown interventions are known as *statistical decisions with non-random interventions* or *evaluations of complex hypotheses* or *Linnik tasks*.

We speak in plural about statistical decision tasks in the presence of non-random interventions because the specific form of the task depends on whether or not the state k is or is not random, whether or not the penalty function is determined, etc.. Let us mention two examples from a large set of possible tasks.

Testing of complex hypotheses with random state and with non-random intervention

Let X, K, Z be finite sets of of possible values of the observation x, state k and intervention z, $p_K(k)$ be the *a priori* probability of the state k and $p_{X|K,Z}(x \mid k, z)$ be the conditional probability of the observation x under the condition of the state k and intervention z. Let $X(k)$, $k \in K$, be the decomposition of the set X which determines the strategy how to estimate the states k. The probability of the incorrect decision ω depends not only on the strategy itself but also on the intervention z,

$$\omega(z) = \sum_{k \in K} p_K(k) \sum_{x \notin X(k)} p_{X|K,Z}(x \mid k, z) .$$

The *quality* ω^* *of a strategy* $(X(k),\ k \in K)$ is defined as the probability of the incorrect decision obtained in the case of the worst intervention z for this strategy, that is

$$\omega^* = \max_{z \in Z} \omega(z) .$$

The task consists in the decomposition of the observation set X into classes $X(k)$, $k \in K$, in such a way that ω^* is minimised, i.e.,

$$\big(X^*(k), k \in K\big) = \operatorname*{argmin}_{(X(k), k \in K)} \max_{z \in Z} \sum_{k \in K} p_K(k) \sum_{x \notin X(k)} p_{X|K,Z}(x \mid k, z) .$$

Testing of complex hypotheses with non-random state and with non-random interventions

The second task conforms to the case in which neither the state k nor intervention z can be considered as a random variable and consequently *a priori* probabilities $p_K(k)$ are not defined. In this situation the probability of the wrong decision ω obtained by a strategy, depends not only on the intervention z but also on the state k

$$\omega(k, z) = \sum_{x \notin X(k)} p_{X|K,Z}(x \mid k, z) .$$

The quality ω^* of the strategy $\big(X(k),\ k \in K\big)$ will be defined as

$$\omega^* = \max_{k \in K} \max_{z \in Z} \omega(k, z) ,$$

and the task will be formulated as a search for the best strategy in this sense, i.e., as a search for decomposition

$$(X^*(k), k \in K) = \operatorname*{argmin}_{(X(k), k\in K)} \max_{k\in K} \max_{z\in Z} \sum_{x\notin X(k)} p_{X|K,Z}(x \mid k, z) .$$

2.3 The pair of dual linear programming tasks, properties and solution

The non-Bayesian tasks mentioned can be analysed in a single formal framework because they are particular cases of the linear programming optimisation task. The reader has probably come across linear programming tasks in the canonical formulation. This formulation is suitable for solving a specific task, e.g., by a simplex method (which is a simple linear optimisation technique used commonly in economics and elsewhere). There is a good reason for starting the theoretical analysis with a more general formulation. The canonical formulation results from it.

The basis of the analysis of a linear programming task is constituted by two theorems about dual tasks which are known in several equivalent formulations. A necessary piece of knowledge about dual tasks is mentioned here in the form which is the most suitable for our case. It is possible to comprehend this section as an information guide which can be omitted by the reader experienced in the transition from primal linear programming tasks to dual ones and vice versa. He or she can proceed to Section 2.4.

Let $\mathbb{R}^m$ be an m-dimensional linear space, each point $x \in \mathbb{R}^m$ of which has coordinates $x_1, x_2, \ldots, x_m$. Let I denote the set of indices $\{1, 2, \ldots, m\}$. Let the vector x satisfy the constraint that some of its coordinates cannot be negative and let us denote by $I^+ \subset I$ the set of indices of those coordinates that comply with this constraint. The index set $I^0 = I \setminus I^+$ corresponds to all other coordinates which can be both positive or negative. Let us introduce the set X of points in $\mathbb{R}^m$ which comply with the introduced constraint, i.e., $X = \{x \in \mathbb{R}^m \mid x_i \geq 0, i \in I^+\}$.

Let $\mathbb{R}^n$ be an n-dimensional linear space, J be the set of indices of its coordinates, J^+ be the set of indices of those coordinates that cannot be negative and $J^0 = J \setminus J^+$. Let us introduce a set Y as a set of points in $\mathbb{R}^n$ in which for all $j \in J^+$ the coordinate y_j is not negative, i.e., $Y = \{y \in \mathbb{R}^n \mid y_j \geq 0, j \in J^+\}$.

Let the function $f\colon X \times Y \to \mathbb{R}$ be defined on the set $X \times Y$. The function $f(x, y)$ itself need not be linear, but the function $f(x^*, y)$ is a linear function of one variable y for any fixed value $x^* \in X$. The opposite holds too, i.e., the $f(x, y^*)$ is a linear function of one variable x for any value $y^* \in Y$. Functions with these properties are called *bilinear functions*. Each bilinear function can be expressed in the following general form

$$f(x, y) = \sum_{i\in I} a_i x_i - \sum_{i\in I} \sum_{j\in J} b_{ij} x_i y_j + \sum_{j\in J} c_j y_j . \tag{2.10}$$

The point $x' \in X$ will be called an acceptable one in respect to the function (2.10) if the function $f(x', y)$ of one variable y is bound maximally on the set Y. Let us denote by $\widetilde{X} \subset X$ the set of all acceptable points x. It follows from this definition directly that the function $\varphi \colon \widetilde{X} \to \mathbb{R}$ with the values

$$\varphi(x) = \max_{y \in Y} f(x, y)$$

can be defined on the set $\widetilde{X}$.

The *primal task of linear programming* aims to detect if the set of all acceptable points $\widetilde{X}$ is non empty. If $\widetilde{X} \neq \emptyset$ then it should be verified if the function $\varphi \colon \widetilde{X} \to \mathbb{R}$ is bound minimally. If the prior conditions are satisfied then the $x^* \in \widetilde{X}$ should be found which minimises φ. The *primal linear programming task* is called *solvable* in this case.

The *dual linear programming task* is defined symmetrically. The set $\widetilde{Y}$ is a set of such $y' \in Y$ for which the function $f(x, y')$ is bound minimally on the set X. Let us denote for each point $y \in \widetilde{Y}$

$$\psi(y) = \min_{x \in X} f(x, y) .$$

The dual task checks if $\widetilde{Y} \neq \emptyset$ is satisfied. If the function ψ is bound maximally on $\widetilde{Y}$ then such the point $y^* \in \widetilde{Y}$ is looked for that maximises ψ. In this case the *dual linear programming task* is called solvable.

The relation between the primal and dual linear programming tasks is given in the following theorem.

Theorem 2.1 The first duality theorem, also the Kuhn–Tucker theorem. *If the primal linear programming task is solvable then the dual linear programming task is solvable also. Moreover, the following equation holds*

$$\min_{x \in \widetilde{X}} \varphi(x) = \max_{y \in \widetilde{Y}} \psi(y) . \tag{2.11}$$

▲

Proof. The proof is not short and can be found in books about mathematical programming in different formulations. The proof in the form matching our explanation is in [Zuchovickij and Avdejeva, 1967]. ■

The set $\widetilde{X}$ and the function $\varphi \colon \widetilde{X} \to \mathbb{R}$ can be expressed explicitly as it is shown in the lemma.

Lemma 2.1 Canonical form of the linear programming task. *The set $\widetilde{X}$ is a set of solution of the system of equation and inequalities*

$$\sum_{i \in I} b_{ij}\, x_i \geq c_j, \quad j \in J^+ , \tag{2.12}$$

$$\sum_{i \in I} b_{ij}\, x_i = c_j, \quad j \in J^0 , \tag{2.13}$$

$$x_i \geq 0, \quad i \in I^+ ,$$

and for each $x \in \widetilde{X}$ holds

$$\varphi(x) = \sum_{i \in I} a_i\, x_i . \tag{2.14}$$

▲

Proof. Let us rewrite the function f in the following form

$$f(x,y) = \sum_{i\in I} a_i x_i + \sum_{j\in J^0} y_j \left(c_j - \sum_{i\in I} b_{ij}\, x_i\right) + \sum_{j\in J^+} y_j \left(c_j - \sum_{i\in I} b_{ij}\, x_i\right) \quad (2.15)$$

Let us prove first that if some x' does not satisfy the system of inequalities (2.12) and equations (2.13) then $x' \notin \widetilde{X}$. In other words, the function $f(x',y)$ of one variable y is not bound maximally.

Let us assume that for such an x' one of the inequalities from the system (2.12), say the j'-th one, does not hold. The function f can thus have an arbitrarily large value. If this is to happen it suffices that the coordinate $y_{j'}$ is large enough. There is nothing that can prevent the growth of the coordinate $y_{j'}$ since the matching coordinate is limited only by the property that it cannot be negative.

Let us permit now that $x' \notin \widetilde{X}$, because for some $j'' \in J^0$ the equation from the system (2.13) is not satisfied. In such a case the function f can achieve an arbitrary large value again. It suffices if the absolute value of $y_{j''}$ is large enough. The coordinate $y_{j''}$ itself can be either positive or negative depending on the difference $c_{j''} - \sum_{i\in I} b_{ij''}\, x_i$ being positive or negative. From the contradiction we can see that any $x \in \widetilde{X}$ complies with the conditions (2.12) and (2.13).

Let us demonstrate furthermore that for each x' satisfying the relations (2.12) and (2.13), the function $f(x',y)$ is bound maximally on the set Y, i.e., $x' \in \widetilde{X}$.

The function value $f(x',y)$ comprises three additive terms, cf. (2.15). The first term is independent of y. The second term is independent of y too, since the difference $c_j - \sum_{i\in I} b_{ij}\, x'_i$ is zero for all $j \in J^0$ as it follows from the condition (2.13). The third term is not positive as none of the additive terms constituting it is positive. Indeed, there holds $y_j \geq 0$ and $c_j - \sum_{i\in I} b_{ij}\, x'_i \leq 0$ for any $j \in J^+$, which implies from the condition (2.12). It follows that the third term in equation (2.15) is bound maximally and this upper limit is zero. It implies too, that the whole expression (2.15) has an upper limit as well and it is $\sum_{i\in I} a_i\, x'_i$. In such a way it is proved that the set $\widetilde{X}$ is identical with the set of solutions of the system of inequalities (2.12) and equalities (2.13).

As the proof of the equation (2.14) is concerned, it is obvious that the upper limit $\sum_{i\in I} a_i x'_i$ of the function $f(x',y)$ is achievable on the set Y, because $f(x',0) = \sum_{i\in I} a_i\, x'_i$ and $0 \in Y$. Then $\max_{y\in Y} f(x',y) = \sum_{i\in I} a_i\, x'_i$ and the equation (2.14) is satisfied too. ∎

Having the proved lemma in mind, the primal linear programming task can be expressed in the following canonical form. The

$$\min_{x\in X} \sum_{i\in I} a_i\, x_i \quad (2.16)$$

is to be found under the conditions

$$\sum_{i\in I} b_{ij}\, x_i \geq c_j\,, \quad j \in J^+\,, \tag{2.17}$$

$$\sum_{i\in I} b_{ij}\, x_i = c_j\,, \quad j \in J^0\,, \tag{2.18}$$

$$x_i \geq 0\,, \quad i \in I^+\,. \tag{2.19}$$

Properties of the dual task can be proved in a similar manner. Let us repeat the thoughts which were used when proving Lemma 2.1. It can be shown that in the dual linear programming task,

$$\max_{y\in Y} \sum_{j\in J} c_j\, y_j \tag{2.20}$$

is sought under the conditions

$$\sum_{j\in J} b_{ij}\, y_j \leq a_i\,, \quad i \in I^+\,, \tag{2.21}$$

$$\sum_{j\in J} b_{ij}\, y_j = a_i\,, \quad i \in I^0\,, \tag{2.22}$$

$$y_j \geq 0\,, \quad j \in J^+\,. \tag{2.23}$$

The function f should be expressed in the form

$$f(x,y) = \sum_{i\in I^0} x_i \left(a_i - \sum_{j\in J} b_{ij}\, y_j\right) + \sum_{i\in I^+} x_i \left(a_i - \sum_{j\in J} b_{ij}\, y_j\right) + \sum_{j\in J} c_j\, y_j\,. \tag{2.24}$$

If one from the inequalities from the system (2.21) does not hold for some y' and for some $i' \in I^+$ then the second additive term in relation (2.24) can become arbitrarily small by selecting the big enough value of $x_{i'}$ and the same is valid for the value $f(x, y')$. If one of the equations from the system (2.22) does not hold for some $i'' \in I^0$ then the first sum in the right-hand side of (2.24) can reach arbitrarily small value if the coordinate $x_{i''}$ is selected appropriately. It is obvious that for any y, which satisfies conditions (2.21) and (2.22), the value $f(x,y)$ on the set X has lower bound given by the number $\sum_{j\in J} c_j\, y_j$. This lower bound is achieved in point $x = 0$ which belongs to the set X.

Although expressions (2.16)–(2.23) unambiguosly define the pair of dual tasks *it might be worthwhile to remember the following formulation in terms of words.* It is not so concise and uniquely declared as expressions (2.16)–(2.23), but it is more comprehensible.

1. The primal linear programming task requires the minimisation of the linear function dependent on a certain group of variables under linear constraints. The dual linear programming task demands maximisation of the linear function dependent on other linearly constrained variables.

2. Each constraint in the primal task corresponds to the variable in the dual task and each variable in the primal task matches the constraint in the dual task.
3. Constraints in the primal task are either linear equations or linear inequalities of the form $\geq$. Constraints in the dual task are either linear equations or linear inequalities of the form $\leq$.
4. Values of some variables in both the primal and the dual task can be positive and negative. These variables are called *free variables.* There are other variables in the primal and dual tasks which are not allowed to be negative. Such variables are called *non-negative variables.*
5. To each equality among the primal task constraints the free variable in the dual task corresponds. To each inequality among the primal task constraints the non-negative variable in the dual task matches. To each free variable in the primal task the constraint in the dual task in the equality form corresponds.
6. Coefficients a_i, which express the minimised function in the primal task, are present as threshold values on the right-hand side of equalities or inequalities in the dual task. Thresholds c_j appearing on the right-hand side of equalities or inequalities of the primal task appear as coefficients of the linear function being minimised in the dual task.
7. The coefficient matrix in the system of equalities or inequalities of the primal task corresponds to the transposed matrix of coefficients of the system of equalities and inequalities in the dual task.

In our exposition we can proceed to the next theorem which is particularly important when analysing the pair of dual linear programming tasks. Namely, this theorem will help us when analysing non-Bayesian decision tasks that are formulated in the form of linear programming tasks.

Theorem 2.2 Second duality theorem, also called theorem on mutual non-movability. *Let the solution of primal linear programming task be $x^* = (x_i^*,\ i \in I)$, let the solution of the dual task be $y^* = (y_j^*,\ j \in J)$.*

Unless some coordinate x_i^* of the point x^* is equal to zero, the corresponding constraint of the dual task for $i \in I^+$ is then satisfied by the equation $\sum_{j \in J} b_{ij}\, y_j^* = a_i$ (although it was an inequality in the task formulation).

If the j-th constraint in the primal task is satisfied in the point x^* as a strict inequality, i.e., if $\sum_{i \in I} b_{ij}\, x_i^* > c_j$ holds then the corresponding value y_j^* in the dual task is equal to zero. ▲

Proof. The theorem actually says that for all $i \in I$ there holds

$$x_i^* \left(a_i - \sum_{j \in J} b_{ij}\, y_j^* \right) = 0\,, \tag{2.25}$$

and for all $j \in J$ holds

$$y_j^* \left(c_j - \sum_{i \in I} b_{ij}\, x_i^* \right) = 0\,. \tag{2.26}$$

The equation

$$\sum_{i\in I} a_i\, x_i^* + \sum_{j\in J} y_j^* \left(c_j - \sum_{i\in I} b_{ij}\, x_i^* \right) = \sum_{i\in I} x_i^* \left(a_i - \sum_{j\in J} b_{ij}\, y_j^* \right) + \sum_{j\in J} c_j\, y_j^* \quad (2.27)$$

is apparently valid.

The first duality theorem states that $\sum_{i\in I} a_i\, x_i^* = \sum_{j\in J} c_j\, y_j^*$ which implies

$$\sum_{j\in J} y_j^* \left(c_j - \sum_{i\in I} b_{ij}\, x_i^* \right) = \sum_{i\in I} x_i^* \left(a_i - \sum_{j\in J} b_{ij}\, y_j^* \right) . \quad (2.28)$$

As x^*, respectively y^*, are solutions of the primal, respectively dual task, the constraints (2.17)–(2.19) and (2.21)–(2.23) are satisfied for them and it implies that for any $j \in J$ there holds

$$y_j^* \left(c_j - \sum_{i\in I} b_{ij}\, x_i^* \right) \le 0, \quad (2.29)$$

and for any $i \in I$ there holds

$$x_i^* \left(a_i - \sum_{j\in J} b_{ij}\, y_i^* \right) \ge 0 . \quad (2.30)$$

Equation (2.28) states that the sum of non-positive additive terms is the same as the sum of non-negative additive terms. This is possible only in that case in which all additive terms equal zero. The validity of equations (2.25) and (2.26) follows from that. ■

2.4 The solution of non-Bayesian tasks using duality theorems

We will show how the formalism of dual linear programming tasks can be used to solve non-Bayesian tasks (and Bayesian ones too) in pattern recognition.

So far, the recognition strategies have corresponded to the decomposition of the set of observations X into $X(k)$, $k \in K$. For each pair of sets $X(k')$ and $X(k'')$, the relation $X(k') \cap X(k'') = \emptyset$ was satisfied. In addition, for all sets it held that $\bigcup_{k\in K} X(k) = X$. This means that each observation $x \in X$ belonged just to one set $X(k)$. The strategy was based on such a decomposition, which decided for the observation $x \in X(k)$ whether the observed objects are in the state k.

The same strategy can be expressed equivalently by a function $\alpha \colon X \times K \to \mathbb{R}$, that $\alpha(x, k)$ is a non-negative number and, moreover, $\sum_{k\in K} \alpha(x, k) = 1$ for all $x \in X$. Every decomposition of the set X, i.e., any *deterministic strategy*, can be implemented as an integer function from the mentioned class so that if $x \in X(k)$ then $\alpha(x, k) = 1$.

When non-integer functions are allowed then the set of all possible functions α will be more extensive. Any function α can then be understood as a *randomised strategy*. The value $\alpha(x, k)$ is a probability of the event that having observed x, it is decided in favour of the state k. *After this generalisation all non-Bayesian tasks formulated earlier can be expressed as a particular case of linear programming and can be analysed in a single formal framework.*

We will analyse common properties of non-Bayesian tasks keeping in mind the properties of the linear optimisation tasks. The most important property claims that the solution of any non-Bayesian task differs only fractionally from the Bayesian strategy. Saying it more precisely, the strategy solving the arbitrary non-Bayesian and Bayesian tasks is implementable in the space of probabilities. Each decision corresponds to a convex cone in the space of probabilities. Deterministic decisions match the inner points of the cone. The random decisions need not appear always, and if they occur then they occur in points which lie at the boundary of the convex cone.

2.4.1 Solution of the Neyman–Pearson task

The Neyman–Pearson task was formulated as an optimisation task, see relations (2.2)–(2.4). The task, which for brevity we call Neyman task, can be formulated in another form now. The decision strategy will be expressed by means of a function $\alpha\colon X \times K \to \mathbb{R}$ instead of the sets X_1 and X_2 used earlier. The task is not changed at all, because it is necessary to minimise the conditional probability of the false alarm as before provided that the conditional probability of overlooked danger is not greater than ε. The probability of the false alarm is

$$\sum_{x \in X} \alpha(x, 2)\, p_{X|K}(x \mid 1) \tag{2.31}$$

and the probability of the overlooked danger is

$$\sum_{x \in X} \alpha(x, 1)\, p_{X|K}(x \mid 2)\,. \tag{2.32}$$

Variables α have to satisfy conditions $\alpha(x, 1) + \alpha(x, 2) = 1$, $\alpha(x, 1) \geq 0$, $\alpha(x, 2) \geq 0$ for all $x \in X$. These demands can be expressed in the form of a linear programming task

We will need linear programming tasks many times in this lecture and therefore let us introduce the concise notation. The task will be expressed as several linear expressions. The expression in the first line is the criterion which will be maximised or minimised. The optimised variables will not be presented explicitly in the criterion, as it is obvious that it is optimised with respect to all the variables in the given task. Next equalities and non-equalities below the criterion express constraints. For every constraint the corresponding dual variable will be defined. Dual variables are written on the left side of the corresponding primal constraint and are separated from constraints by a vertical line. In a similar way, if the dual task is expressed then the variables of the primal task will be given on the left-hand side of the dual task constraints.

Quantities $\alpha(x,k)$, $x \in X$, $k \in K$, which determine the strategy, will be treated as variables in further tasks. Namely, these quantities are to be determined when solving the problem, i.e., through this choice the expression written in the first line is optimised. Additional auxiliary variables will be introduced in the minimax task in particular. Their meaning will be explained when its turns come. Here we consider necessary to forewarn, that in the coming expressions the quantities x and k are not any more variables according to which the optimisation is performed. They are variables, according to which we sum up in linear expressions, which are to be optimised, or which express constraints of the task. In this sense the quantities x and k have the same meaning as indices i and j in the previous Section 2.3, where linear programming tasks were analyzed in a general form. Various probabilities $p_{X|K}(x,k)$, $p_K(k)$ etc.. are considered as known constant numbers. These probabilities will play the role of multiplicative coefficients in linear expressions which are to be optimised or which express constraints in the appropriate linear programming task.

Minimisation of (2.31) under the condition (2.32) and additional self-explaining conditions will be expressed in the following form of linear programming task

$$\left.\begin{array}{rr}
\min \sum\limits_{x\in X} \alpha(x,2)\, p_{X|K}(x\,|\,1)\,, & \text{(a)}\\
\sum\limits_{x\in X} \alpha(x,1)\, p_{X|K}(x\,|\,2) \le \varepsilon\,, & \text{(b)}\\
\alpha(x,1)+\alpha(x,2)=1,\;\; x\in X\,, & \text{(c)}\\
\alpha(x,1)\ge 0\,,\;\; x\in X\,, & \text{(d)}\\
\alpha(x,2)\ge 0\,,\;\; x\in X\,. & \text{(e)}
\end{array}\right\} \qquad (2.33)$$

In the optimisation task (2.33) the variables are the values $\alpha(x,1)$, $\alpha(x,2)$ for all $x \in X$. The constants are values ε and $p_{X|K}(x\,|\,k)$ for all $x \in X$, $k = 1,2$. Let us rewrite the task to convert the inequality (2.33b) into a standard form with the relation $\ge$ as is required in the primal linear programming task.

Let us take into account that the line (2.33c) represents not just one but $|X|$ constraints. There is a dual variable $t(x)$ corresponding to each of these $|X|$ constraints. We obtain the primal task

$$\left.\begin{array}{r|rr}
 & \min \sum\limits_{x\in X} \alpha(x,2)\, p_{X|K}(x\,|\,1)\,, & \text{(a)}\\
\tau & -\sum\limits_{x\in X} \alpha(x,1)\, p_{X|K}(x\,|\,2) \ge -\varepsilon\,, & \text{(b)}\\
t(x) & \alpha(x,1)+\alpha(x,2)=1\,,\; x\in X\,, & \text{(c)}\\
 & \alpha(x,1)\ge 0\,,\; x\in X\,, & \text{(d)}\\
 & \alpha(x,2)\ge 0\,,\; x\in X\,. & \text{(e)}
\end{array}\right\} \qquad (2.34)$$

The following dual task corresponds to the previous primal task

$$
\left.
\begin{array}{r|rr}
 & \max \left(\sum\limits_{x \in X} t(x) - \varepsilon\, \tau \right) , & \text{(a)} \\
\alpha(x,1) & t(x) - \tau\, p_{X|K}(x \,|\, 2) \leq 0 \,,\; x \in X \,, & \text{(b)} \\
\alpha(x,2) & t(x) \leq p_{X|K}(x \,|\, 1) \,,\; x \in X \,, & \text{(c)} \\
 & \tau \geq 0 \,. & \text{(d)}
\end{array}
\right\} \qquad (2.35)
$$

We will explain the mechanism of deriving the dual task more thoroughly in the first non-Bayesian task analysed. The following tasks will be described less cautiously. The line (2.35a) exhibits a linear function depending on dual variables τ and $t(x)$. Each variable $t(x)$, $x \in X$, is multiplied by a unit coefficient as the number 1 appears on the right-hand side of the inequality (2.34c) of the primal task to which the dual variable $t(x)$ corresponds. The variable τ is multiplied by $-\varepsilon$ because the threshold $-\varepsilon$ occurs on the right-hand side of the inequality (2.34b).

The line (2.35b) specifies $|X|$ constraints. Each of them corresponds to the variable $\alpha(x,1)$ in the primal set. The constraints are expressed as an inequality due to the property that $\alpha(x,1)$ is non-negative, cf., constraint (2.34d). The value 0 is on the right-hand side of the constraint as the variable $\alpha(x,1)$ is not present in the function (2.34a) which is to be minimised in the primal task. It is the same as if the variable was present and multiplied by the coefficient 0. The left-hand side of the constraints is composed of two additive terms since the variable $\alpha(x,1)$ occurs in two constraints of the primal task only, i.e., in one constraint from the group (2.34c) and in the constraint (2.34b). The variable $t(x)$ in the constraint (2.35b) is multiplied by 1, because also $\alpha(x,1)$ in the constraint (2.34c) is multiplied by 1. The variable τ in (2.35b) is multiplied by $-p_{X|K}(x \,|\, 2)$. The variable $\alpha(x,1)$ is multiplied by $-p_{X|K}(x \,|\, 2)$ in the constraint (2.34b).

The line (2.35c) specifies $|X|$ constraints corresponding to group of variables $\alpha(x,2)$. There is the probability $p_{X|K}(x \,|\, 1)$ on the right-hand side of the constraints since this coefficient multiplies the variable $\alpha(x,2)$ in the linear function (2.34a) which is minimised in the primal task. There is a single variable $t(x)$ on the left-hand side of the constraint (2.35c) since the variable $\alpha(x,2)$ occurs only in a single constraint (2.34c) of the primal task. The variable $t(x)$ in (2.35c) is multiplied by 1 since the variable $\alpha(x,2)$ is multiplied by 1 in the constraint (2.34c) too. The constraint (2.35c) is the inequality as the variable $\alpha(x,2)$ in the primal task is defined as an non-negative variable, cf. the constraint (2.34e).

The constraint (2.35d) requires the variable τ to be non-negative as it corresponds in the primal task to the constraint (2.34b) which is expressed by an inequality. Dual variables $t(x)$, $x \in X$, can be both positive and negative, because the matching constraints to it in the primal task are expressed as equalities.

Only the reader not confident in transforming primal tasks of linear programming to the dual ones is likely to need this explanation. These transformations are more or less automatic. Here, the primal task (2.34) was transformed into the dual task (2.35). These explanations are superfluous with respect to the proof of Neyman task. In fact, the pair of tasks (2.34) and (2.35) is a pair of dual tasks of linear programming.

Thanks to that, it is possible to find out Neyman strategy based on the Second Duality Theorem (Theorem 2.2) after the following simple consideration. The task (2.35) cannot be solved for such values of τ and $t(x)$, $x \in X$, for which the both constraints (2.35b) and (2.35c) were strictly satisfied. Having in mind the Second Duality Theorem $\alpha(x, 1) = \alpha(x, 2) = 0$ should have held, this would be in contradiction with the constraint (2.34c). Thus the equality should hold for each $x \in X$ in one of inequalities (2.35b) and (2.35c) at least. This means that

$$t(x) = \min \left(p_{X|K}(x \mid 1),\; \tau\, p_{X|K}(x \mid 2)\right) .$$

It implies that if

$$p_{X|K}(x \mid 1) < \tau\, p_{X|K}(x \mid 2) \tag{2.36}$$

then $t(x) < \tau\, p_{X|K}(x \mid 2)$ and the strict inequality (2.35b) is satisfied, and therefore $\alpha(x, 1) = 0$. Due to the constraint (2.34c) $\alpha(x, 2) = 1$ and the state k has to be labelled as dangerous. If

$$p_{X|K}(x \mid 1) > \tau\, p_{X|K}(x \mid 2) \tag{2.37}$$

then $t(x) < p_{X|K}(x \mid 1)$. As the inequality (2.35c) is satisfied strictly, $\alpha(x, 2) = 0$, $\alpha(x, 1) = 1$ and the state k is assessed as a normal one. The conditions (2.36) and (2.37) can be expressed in the known form in which the likelihood ratio

$$\gamma(x) = \frac{p_{X|K}(x \mid 1)}{p_{X|K}(x \mid 2)} \tag{2.38}$$

is calculated and this ratio is compared to the non-negative threshold value τ.

We showed that Neyman tasks in the form of the dual task pair (2.34) and (2.35) can be quite briefly expressed and solved in a transparent way. This briefness is based on the theory of dual linear programming tasks. In the given case it is based on the Second Duality Theorem which helps to solve not only Neyman task in an easier way, but also other non-Bayesian tasks.

2.4.2 Solution of generalised Neyman–Pearson task with two dangerous states

Let us show more briefly now how to deduce strategies for solving modified Neymans task which we formulated in Subsection 2.2.2 by the conditions (2.5)–(2.9).

Let the object be in one of three possible states, $k = 1$, 2 or 3. The state $k = 1$ is considered as normal and other two states as dangerous ones. The aim is to find two functions $\alpha_1(x)$ and $\alpha_{23}(x)$ with the following meaning. If it is to be decided according to observation x for the state $k = 1$ then $\alpha_1(x) = 1$. If it is to be decided for the state $k = 2$ or $k = 3$ then $\alpha_{23}(x) = 1$. The functions $\alpha_1(x)$ and $\alpha_{23}(x)$ sought have to minimise the sum

$$\sum_{x \in X} \alpha_{23}(x)\, p_{X|K}(x \,|\, 1) \tag{2.39}$$

under the conditions

$$\sum_{x \in X} \alpha_1(x)\, p_{X|K}(x \,|\, 2) \le \varepsilon \,, \tag{2.40}$$

$$\sum_{x \in X} \alpha_1(x)\, p_{X|K}(x \,|\, 3) \le \varepsilon \,, \tag{2.41}$$

$$\alpha_1(x) + \alpha_{23}(x) = 1 \,, \quad x \in X \,, \tag{2.42}$$

$$\begin{aligned} \alpha_1(x) &\ge 0 \,, \quad x \in X \,, \\ \alpha_{23}(x) &\ge 0 \,, \quad x \in X \,. \end{aligned} \tag{2.43}$$

The expression (2.39) represents probability of the false alarm. The condition (2.40) means that the probability of the overlooked dangerous state $k = 2$ has to be small and the condition (2.41) requires the same for the state $k = 3$. Constraints (2.42)–(2.43) are the standard conditions the strategy has to meet. We rewrite the conditions for the pair of dual tasks now.

Primal task:

$$\left.\begin{array}{r|l} & \min \sum_{x \in X} \alpha_{23}(x)\, p_{X|K}(x \,|\, 1) \,, \\ \tau_2 & -\sum_{x \in X} \alpha_1(x)\, p_{X|K}(x \,|\, 2) \ge -\varepsilon \,, \\ \tau_3 & -\sum_{x \in X} \alpha_1(x)\, p_{X|K}(x \,|\, 3) \ge -\varepsilon \,, \\ t(x) & \alpha_1(x) + \alpha_{23}(x) = 1 \,, \quad x \in X \,, \\ & \alpha_1(x) \ge 0 \,, \quad x \in X \,, \\ & \alpha_{23}(x) \ge 0 \,, \quad x \in X \,. \end{array}\right\} \tag{2.44}$$

Dual task:

$$\left.\begin{array}{r|ll} & \max \left(\sum_{x \in X} t(x) - \varepsilon(\tau_2 + \tau_3) \right) , & \\ \alpha_1(x) & t(x) - \tau_2\, p_{X|K}(x \,|\, 2) - \tau_3\, p_{X|K}(x \,|\, 3) \le 0 \,, \quad x \in X \,, & \text{(a)} \\ \alpha_{23}(x) & t(x) \le p_{X|K}(x \,|\, 1) \,, \quad x \in X \,, & \text{(b)} \\ & \tau_2 \ge 0 \,, & \\ & \tau_3 \ge 0 \,. & \end{array}\right\} \tag{2.45}$$

From conditions (2.45a), (2.45b) and the fact that $\alpha_1(x)$ and $\alpha_{23}(x)$ cannot be equal to zero simultaneously, it implies that for $x \in X$ and the variable $t(x)$ there must hold

$$t(x) = \min\left(p_{X|K}(x \mid 1),\ \tau_2\, p_{X|K}(x \mid 2) + \tau_3\, p_{X|K}(x \mid 3)\right) . \tag{2.46}$$

This means that if

$$p_{X|K}(x \mid 1) < \tau_2\, p_{X|K}(x \mid 2) + \tau_3\, p_{X|K}(x \mid 3) \tag{2.47}$$

then $\alpha_1(x)$ must equal 0, $\alpha_{23}(x)$ must equal 1, and x signifies a dangerous state. If

$$p_{X|K}(x \mid 1) > \tau_2\, p_{X|K}(x \mid 2) + \tau_3\, p_{X|K}(x \mid 3) \tag{2.48}$$

then x is a sign of the normal state. The strategy solving this task has the following form: for certain non-negative numbers τ_2 and τ_3 the likelihood ratio is computed

$$\gamma(x) = \frac{p_{X|K}(x \mid 1)}{\tau_2\, p_{X|K}(x \mid 2) + \tau_3\, p_{X|K}(x \mid 3)}\,, \tag{2.49}$$

and then it has to be decided for either a normal or a dangerous state, if the likelihood ratio is higher or smaller than 1, respectively.

2.4.3 Solution of the minimax task

We will show a standard procedure which allows the reformulation of the minimax task as a minimisation task with no maximisation. Let $f_j(x)$, $j \in J$, be a set of functions of one variable x. The value x^* that minimises $\varphi(x) = \max_{j \in J} f_j(x)$ has to be found. This minimisation problem can be written as a mathematical programming task not only of one variable x, but of two variables, i.e., the original x and a new auxiliary variable c. In this new formulation, the pair (x, c) is sought which minimises c under the constraint $c \geq f_j(x)$, $j \in J$. The task

$$\min_x \max_{j \in J} f_j(x) \tag{2.50}$$

is thus equivalent to the task

$$\left.\begin{array}{l} \min c\,, \\ c - f_j(x) \geq 0\,, \quad j \in J\,. \end{array}\right\} \tag{2.51}$$

This procedure is demonstrated in Fig. 2.1 for the case $J = \{1, 2, 3\}$. The shaded area in the figure shows the set of pairs (x, c) which satisfy conditions $c - f_1(x) \geq 0$, $c - f_2(x) \geq 0$, $c - f_3(x) \geq 0$. The task (2.51) requires that a point with the minimal coordinate c must be found in the area.

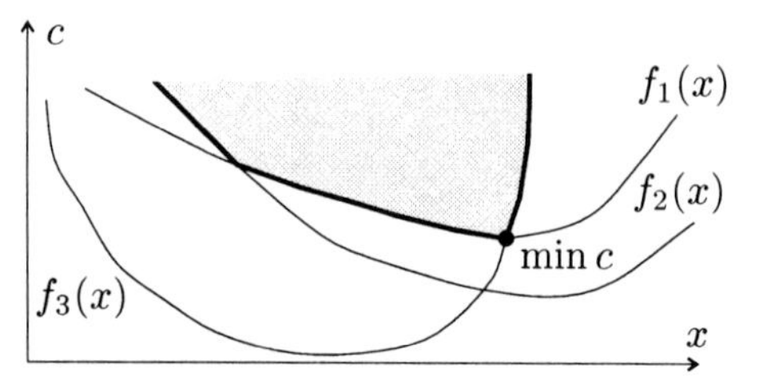

Figure 2.1 Minimax task for three functions f.

Apparently, it is the point denoted by a filled circle in Fig. 2.1. The function $\max\big(f_1(x), f_2(x), f_3(x)\big)$ is shown as a bolded curve. The task (2.50) requires that the point with the smallest coordinate c has to be found on the bold curve. It is the same point denoted by the filled circle.

Because of the equivalence of the tasks (2.50) and (2.51), the minimax task formulated in Subsection 2.2.3 can be expressed as the following linear programming task. The variables are $\alpha(x,k)$ and the auxiliary variable c,

$$\left.\begin{array}{ll} \min c\,, & \\ c - \sum\limits_{x\in X}\Big(\sum\limits_{k^*\neq k}\alpha(x,k^*)\,p_{X|K}(x\,|\,k)\Big) \geq 0, \quad k\in K\,, & \text{(a)} \\ \sum\limits_{k\in K}\alpha(x,k) = 1, \quad x\in X\,, & \text{(b)} \\ \alpha(x,k)\geq 0, \quad x\in X, \quad k\in K\,. & \end{array}\right\} \tag{2.52}$$

Thanks to (2.52b), the sum $\sum_{k^*\neq k}\alpha(x,k^*)$ is equal to $1-\alpha(x,k)$, and therefore the inequality (2.52a) will be transcribed as $c+\sum_{x\in X}\alpha(x,k)\,p_{X|K}(x\,|\,k)\geq 1$. Our task will be expressed as a pair of dual tasks.

Primal task:

$$\left.\begin{array}{r|l} & \min c\,, \\ \tau(k) & c + \sum\limits_{x\in X}\alpha(x,k)\,p_{X|K}(x\,|\,k)\geq 1\,, \quad k\in K\,, \\ t(x) & \sum\limits_{k\in K}\alpha(x,k) = 1\,, \quad x\in X\,, \\ & \alpha(x,k)\geq 0\,, \quad x\in X\,, \quad k\in K\,. \end{array}\right\} \tag{2.53}$$

Dual task:

$$\left.\begin{array}{r|ll} & \max\Big(\sum\limits_{x\in X}t(x) + \sum\limits_{k\in K}\tau(k)\Big)\,, & \\ \alpha(x,k) & t(x) + \tau(k)\,p_{X|K}(x\,|\,k)\leq 0\,, \quad x\in X\,,\ k\in K\,, & \text{(a)} \\ c & \sum\limits_{k\in K}\tau(k) = 1\,, & \\ & \tau(k)\geq 0, \quad k\in K\,. & \end{array}\right\} \tag{2.54}$$

It follows from the condition (2.54a) and the requirement to obtain the largest $t(x)$ that for any x the variable $t(x)$ is equal to

$$\min_{k\in K}\big(-\tau(k)\,p_{X|K}(x\,|\,k)\big)\,.$$

For certain $x\in X$ and $k^*\in K$ satisfying the condition

$$k^*\neq \operatorname*{argmax}_{k\in K}\big(\tau(k)\,p_{X|K}(x\,|\,k)\big)\,,$$

$\alpha(x, k^*)$ must equal to zero. Furthermore it follows that $\alpha(x, k^*)$ can be non-zero only when

$$k^* = \operatorname*{argmax}_{k \in K} \left(\tau(k)\, p_{X|K}(x \mid k) \right) .$$

2.4.4 Solution of Wald task for the two states case

The Wald task was formulated in Subsection 2.2.4. We will express the strategy solving Wald task with the help of three functions $\alpha_0(x)$, $\alpha_1(x)$ and $\alpha_2(x)$ which are defined on the set X. If $\alpha_1(x)$ or $\alpha_2(x)$ have value one then it is decided for the first or second state. The value $\alpha_0(x) = 1$ indicates that it is not decided for any of two states. The Wald task can be expressed as a pair of dual task and solved using the unified approach we already know.

Primal task:

$$\left.\begin{array}{r|l}
 & \min c \\
q_1 & c - \sum\limits_{x \in X} \alpha_0(x)\, p_{X|K}(x \mid 1) \geq 0\,, \\
q_2 & c - \sum\limits_{x \in X} \alpha_0(x)\, p_{X|K}(x \mid 2) \geq 0\,, \\
\tau_2 & - \sum\limits_{x \in X} \alpha_1(x)\, p_{X|K}(x \mid 2) \geq -\varepsilon\,, \\
\tau_1 & - \sum\limits_{x \in X} \alpha_2(x)\, p_{X|K}(x \mid 1) \geq -\varepsilon\,, \\
t(x) & \alpha_0(x) + \alpha_1(x) + \alpha_2(x) = 1\,, \quad x \in X\,, \\
 & \alpha_0(x) \geq 0\,, \quad \alpha_1(x) \geq 0\,, \quad \alpha_2(x) \geq 1\,, \quad x \in X\,.
\end{array}\right\} \qquad (2.55)$$

Dual task:

$$\left.\begin{array}{r|ll}
 & \max \left(\sum\limits_{x \in X} t(x) - \varepsilon(\tau_1 + \tau_2) \right), & \\
\alpha_1(x) & t(x) - \tau_2\, p_{X|K}(x \mid 2) \leq 0\,, \quad x \in X\,, & \text{(a)} \\
\alpha_2(x) & t(x) - \tau_1\, p_{X|K}(x \mid 1) \leq 0\,, \quad x \in X\,, & \text{(b)} \\
\alpha_0(x) & t(x) - q_1\, p_{X|K}(x \mid 1) - q_2\, p_{X|K}(x \mid 2) \leq 0\,, \quad x \in X\,, & \text{(c)} \\
c & q_1 + q_2 = 1\,, & \\
 & q_1 \geq 0\,, \quad q_2 \geq 0\,, \quad \tau_1 \geq 0\,, \quad \tau_2 \geq 0\,. &
\end{array}\right\} \qquad (2.56)$$

From conditions (2.56a), (2.56b), (2.56c) and a form of the function, which is minimised in the dual task, it is implied that $t(x)$ has to be the smallest from three values $\tau_1\, p_{X|K}(x \mid 1)$, $\tau_2\, p_{X|K}(x \mid 2)$ and $q_1\, p_{X|K}(x \mid 1) + q_2\, p_{X|K}(x \mid 2)$. Let us study the *three corresponding cases* according to which of these three values is the smallest one.

Case 1. If

$$\left. \begin{array}{l} \tau_1\, p_{X|K}(x \,|\, 1) < \tau_2\, p_{X|K}(x \,|\, 2)\,, \\ \tau_1\, p_{X|K}(x \,|\, 1) < q_1\, p_{X|K}(x \,|\, 1) + q_2\, p_{X|K}(x \,|\, 2) \end{array} \right\} \tag{2.57}$$

then the strict inequalities (2.56a) and (2.56c) hold. It implies that $\alpha_1(x) = 0$, $\alpha_0(x) = 0$ and consequently $\alpha_2(x) = 1$.

Case 2. If

$$\left. \begin{array}{l} \tau_2\, p_{X|K}(x \,|\, 2) < \tau_1\, p_{X|K}(x \,|\, 1)\,, \\ \tau_2\, p_{X|K}(x \,|\, 2) < q_1\, p_{X|K}(x \,|\, 1) + q_2\, p_{X|K}(x \,|\, 2) \end{array} \right\} \tag{2.58}$$

then $\alpha_1(x) = 1$, because the strict inequalities (2.56b) and (2.56c) have to hold, and therefore $\alpha_2(x) = 0$ and $\alpha_0(x) = 0$ has to hold.

Case 3. Finally if

$$\left. \begin{array}{l} q_1\, p_{X|K}(x \,|\, 1) + q_2\, p_{X|K}(x \,|\, 2) < \tau_1\, p_{X|K}(x \,|\, 1)\,, \\ q_1\, p_{X|K}(x \,|\, 1) + q_2\, p_{X|K}(x \,|\, 2) < \tau_2\, p_{X|K}(x \,|\, 2) \end{array} \right\} \tag{2.59}$$

then $\alpha_0(x) = 1$ holds, as the strict inequalities (2.56a) and (2.56b) are satisfied, and thus $\alpha_1(x) = 0$ and $\alpha_2(x) = 0$ must hold.

Conditions (2.57)–(2.59) express the strategy solving Wald task explicitly.

We will show that the strategy found is in accordance with the known and commonly used for of the strategy by Wald [Wald, 1947]. It can be seen that the decision depends on the fact which one of the three quantities

$$\tau_1\, p_{X|K}(x \,|\, 1)\,, \quad \tau_2\, p_{X|K}(x \,|\, 2)\,, \quad q_1\, p_{X|K}(x \,|\, 1) + q_2\, p_{X|K}(x \,|\, 2)$$

is the smallest. It is the same as to ask, which of the three variables

$$\frac{\tau_1}{\tau_2}\, \gamma(x)\,, \quad 1\,, \quad \frac{q_1}{\tau_2}\, \gamma(x) + \frac{q_2}{\tau_2} \tag{2.60}$$

is the smallest, where $\gamma(x)$ is the likelihood ratio $p_{X|K}(x \,|\, 1)/p_{X|K}(x \,|\, 2)$. Let us draw in Fig. 2.2 how these three functions (2.60) depend on the likelihood ratio $\gamma(x)$. Two thresholds θ_1, θ_2 are represented on the horizonal axis. It can be seen that the condition $\gamma(x) < \theta_1$ is equivalent to the condition (2.57). The condition $\gamma(x) > \theta_2$ is the same as the condition (2.58), and finally, the condition $\theta_1 < \gamma(x) < \theta_2$ corresponds to the condition (2.59). This means that it is decided for the second state in the first case, for the first state in the second case, and no decision is made in the third case. This is, namely, the solution of Wald task. Fig. 2.2 demonstrates an additional interesting property. It can be seen that the subset X_0 can be empty in some tasks. This means that at some tasks the optimal strategy never says `not known` though such response is allowed.

The solution of Wald generalised task in the case, in which the number of states is greater than two, is not so illustrative. That might be the reason why so many pseudo-solutions occur here, similar to thoses which spin around Neyman strategy. We will show a reasonable formulation and a solution of this generalised task in the following paragraph. We will see that even when it is not possible to use the likelihood ratio, it is possible to master the task easily.

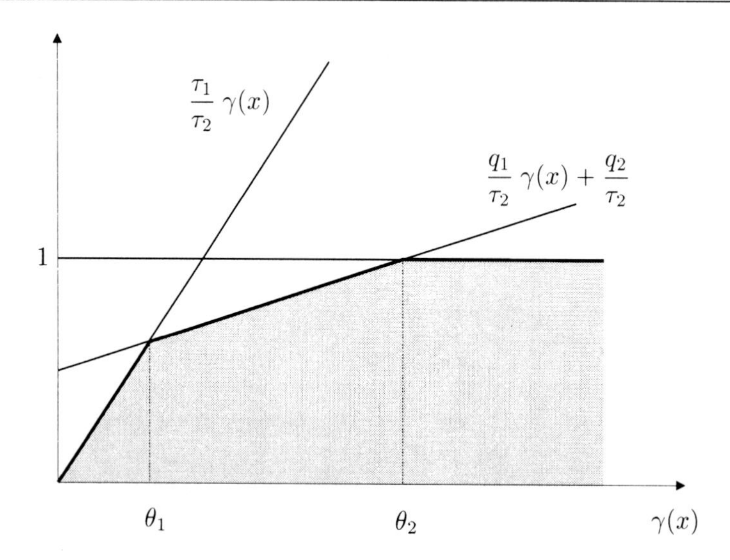

Figure 2.2 On the solution of Wald task.

2.4.5 Solution of Wald task in the case of more states

If $K = \{1, 2, \ldots, n\}$ then the decision strategy can be expressed using the function $\alpha(x, k)$, where $k \in K$ or $k = 0$ holds. A certain strategy can be characterised by $2|K|$ probabilities $\omega(k)$ and $\chi(k)$, $k \in K$. The variable $\omega(k)$ is the probability of a wrong decision under the condition that the actual state is k. The variable $\chi(k)$ is the probability of the answer `not known` under the condition that the actual state is k.

The strategy sought should satisfy the conditions $\omega(k) \leq \varepsilon$, $k \in K$, with an inaccuracy ε set in advance. The strategy that minimises $\max_{k \in K} \chi(k)$ has to be selected out of the strategies that comply with the abovementioned conditions.

There holds for the probability $\omega(k)$, $k \in K$,

$$\begin{aligned}
\omega(k) &= \sum_{x \in X} \left(\sum_{k^* \in K \setminus \{k,0\}} \alpha(x, k^*) \right) p_{X|K}(x \mid k) \\
&= \sum_{x \in X} \Big(1 - \alpha(x, k) - \alpha(x, 0) \Big)\, p_{X|K}(x \mid k) \\
&= 1 - \sum_{x \in X} \Big(\alpha(x, k) + \alpha(x, 0) \Big)\, p_{X|K}(x \mid k)\,.
\end{aligned}$$

There holds for the probability $\chi(k)$, $k \in K$,

$$\chi(k) = \sum_{x \in X} \alpha(x, 0)\, p_{X|K}(x \mid k)\,.$$

Generalised Wald task can be expressed as the pair of dual tasks

Primal task:

$$\left.\begin{array}{r|l}
 & \min c\,, \\
q(k) & c - \sum\limits_{x\in X} \alpha(x,0)\, p_{X|K}(x\,|\,k) \geq 0\,, \quad k \in K\,, \\
\tau(k) & \sum\limits_{x\in X} \big(\alpha(x,k) + \alpha(x,0)\big)\, p_{X|K}(x\,|\,k) \geq 1-\varepsilon\,, \quad k \in K\,, \\
t(x) & \alpha(x,0) + \sum\limits_{k\in K} \alpha(x,k) = 1\,, \quad x \in X\,, \\
 & \alpha(x,0) \geq 0\,, \quad \alpha(x,k) \geq 0\,, \quad k \in K\,, \quad x \in X\,.
\end{array}\right\} \qquad (2.61)$$

Dual task:

$$\left.\begin{array}{r|l}
 & \max \left(\sum\limits_{x\in X} t(x) + (1-\varepsilon) \sum\limits_{k\in K} \tau(k) \right), \\
\alpha(x,0) & t(x) + \sum\limits_{k\in K} \tau(k)\, p_{X|K}(x\,|\,k) - \sum\limits_{k\in K} q(k)\, p_{X|K}(x\,|\,k) \leq 0,\; x \in X, \\
\alpha(x,k) & t(x) + \tau(k)\, p_{X|K}(x\,|\,k) \leq 0\,, \quad x \in X\,, \quad k \in K\,, \\
c & \sum\limits_{k\in K} q(k) = 1\,, \\
 & q(k) \geq 0\,, \quad \tau(k) \geq 0\,, \quad k \in K.
\end{array}\right\} \qquad (2.62)$$

It is obvious from the dual task that the quantity $t(x)$ has to be equal to the smallest value from the following $|K|+1$ values: the first $|K|$ values represent values $-\tau(k)\, p_{X|K}(x\,|\,k)$ and the $(|K|+1)$-th value is $\sum_{k\in K} \big(q(k) - \tau(k)\big)$ $p_{X|K}(x\,|\,k)$. The smallest value determines which decision is chosen: $\alpha(x,k) = 1$ for some $k \in K$ or $\alpha(x,0) = 1$. More precisely, this rule is as follows. The following quantity has to be calculated,

$$\max_{k\in K} \big(\tau(k)\, p_{X|K}(x\,|\,k)\big)\,.$$

If

$$\max_{k\in K} \big(\tau(k)\, p_{X|K}(x\,|\,k)\big) < \sum_{k\in K} \big(\tau(k) - q(k)\big)\, p_{X|K}(x\,|\,k)$$

then the decision is $k^* = 0$. In the opposite case it is decided for

$$k^* = \operatorname*{argmax}_{k\in K} \big(\tau(k)\, p_{X|K}(x\,|\,k)\big)\,.$$

The strategy solving the generalised Wald task is not as simple as we have seen in the previous tasks. It would hardly be possible to guess the strategy only on the basis of mere intuition without thorough formulation of the task, and without its expression in the form of the pair of dual linear programming tasks (2.61), (2.62), followed by a formal deduction.

2.4.6 Testing of complex random hypotheses

The task of testing of complex random hypotheses was formulated in Subsection 2.2.5. The aim of the task is to find a strategy that minimises a probability of the wrong decision $\omega(z)$ under the worst possible intervention $z \in Z$. This means that the quantity $\max_{z \in Z} \omega(z)$ has to be minimised. There holds for the probability $\omega(z)$ that is achieved by the strategy $\alpha(x,k)$, $k \in K$, $x \in X$,

$$\begin{aligned}
\omega(z) &= \sum_{k \in K} p_K(k) \sum_{x \in X} \left(\sum_{k^* \neq k} \alpha(x, k^*) \right) p_{X|K,Z}(x \mid k, z) \\
&= \sum_{k \in K} p_K(k) \sum_{x \in X} \Big(1 - \alpha(x,k) \Big) p_{X|K,Z}(x \mid k, z) \\
&= \sum_{k \in K} p_K(k) \left(1 - \sum_{x \in X} \alpha(x,k)\, p_{X|K,Z}(x \mid k, z) \right) \\
&= 1 - \sum_{k \in K} p_K(k) \sum_{x \in X} \alpha(x,k)\, p_{X|K,Z}(x \mid k, z) \,.
\end{aligned} \tag{2.63}$$

The requirement to minimise $\max_{z \in Z} \omega(z)$ can be expressed as a pair of dual linear programming tasks.

Primal task:

$$\left.\begin{array}{rl}
 & \min c\,, \\
\tau(z) & c + \sum\limits_{k \in K} p_K(k) \sum\limits_{x \in X} \alpha(x,k)\, p_{X|K,Z}(x \mid k, z) \geq 1\,, \quad z \in Z\,, \\
t(x) & \sum\limits_{k \in K} \alpha(x,k) = 1\,, \quad x \in X\,, \\
 & \alpha(x,k) \geq 0\,, \quad k \in K\,, \quad x \in X\,.
\end{array}\right\} \tag{2.64}$$

Dual task:

$$\left.\begin{array}{rl}
 & \max \left(\sum\limits_{x \in X} t(x) + \sum\limits_{z \in Z} \tau(z) \right), \\
\alpha(x,k) & t(x) + \sum\limits_{z \in Z} p_K(k)\, \tau(z)\, p_{X|K,Z}(x \mid k, z) \leq 0\,, \; x \in X,\; k \in K\,, \\
c & \sum\limits_{z \in Z} \tau(z) = 1\,, \\
 & \tau(z) \geq 0\,, \quad z \in Z\,.
\end{array}\right\} \tag{2.65}$$

It results from the form of the maximised function in the dual task and from the upper limit of the quantity $t(x)$ that the optimum occurs when $t(x)$ is the minimal value out of the following $|K|$ possible numbers,

$$- \sum_{z \in Z} p_K(k)\, \tau(z)\, p_{X|K,Z}(x \mid k, z)\,, \quad k \in K\,.$$

The decision $k \in K$, for which $\alpha(x,k) = 1$, depends on which of these $|K|$ numbers is the minimal one, or which of the quantities

$$\sum_{z \in Z} p_K(k)\, \tau(z)\, p_{X|K,Z}(x \mid k, z), \quad k \in K\,,$$

is the maximal one. The decision k^* has to be equal to

$$\operatorname*{argmax}_{k \in K} \sum_{z \in Z} p_K(k)\, \tau(z)\, p_{X|K,Z}(x \mid k, z)\,.$$

2.4.7 Testing of complex non-random hypotheses

The task evaluating complex non-random hypotheses was formulated in Subsection 2.2.5. It is possible to solve the task with the help of the pair of dual linear programming tasks. These tasks are composed as follows.

Primal task:

$$\left.\begin{array}{r|l}
 & \min c\,, \\
\tau(z,k) & c + \sum\limits_{x \in X} \alpha(x,k)\, p_{X|K,Z}(x \mid k,z) \geq 1\,,\ z \in Z\,,\ k \in K, \\
t(x) & \sum\limits_{k \in K} \alpha(x,k) = 1\,, \quad x \in X\,, \\
 & \alpha(x,k) \geq 0\,, \quad k \in K\,, \quad x \in X\,.
\end{array}\right\} \tag{2.66}$$

Dual task:

$$\left.\begin{array}{r|l}
 & \max \left(\sum\limits_{x \in X} t(x) + \sum\limits_{z \in Z} \sum\limits_{k \in K} \tau(z,k) \right)\,, \\
\alpha(x,k) & t(x) + \sum\limits_{z \in Z} \tau(z,k)\, p_{X|K,Z}(x \mid k,z) \leq 0\,,\ x \in X\,,\ k \in K\,, \\
c & \sum\limits_{z \in Z} \sum\limits_{k \in K} \tau(z,k) = 1\,, \\
 & \tau(z,k) \geq 0\,, \quad z \in Z\,, \quad k \in K\,.
\end{array}\right\} \tag{2.67}$$

It follows from the abovementioned pair of tasks that, if for some $x \in X$ and some $k^* \in K$ and all $k \in K \setminus \{k^*\}$ the following inequality is satisfied,

$$\sum_{z \in Z} \tau(z,k^*)\, p_{X|K,Z}(x \mid k^*, z) > \sum_{z \in Z} \tau(z,k)\, p_{X|K,Z}(x \mid k, z)$$

then $\alpha(x,k^*) = 1$.

2.5 Comments on non-Bayesian tasks

We have seen a variety of non-Bayesian tasks. We have introduced several examples of these tasks and it would be possible to continue. It can be seen, in spite of all this variety, that all tasks do not appear any more as isolated islands in the ocean of the unknown. They constitute an archipelago at least, to which the Bayesian class of tasks also belongs.

We could have noticed earlier that these tasks are related. In case of two states only, the strategy always has the same single form for any Bayesian task, as well as for Wald, minimax, or Neyman–Pearson tasks. It is necessary to calculate the likelihood ratio and then the decision corresponds to a certain contiguous interval of the likelihood ratio.

The analysis presented generalises the likelihood ratio to the case in which there are more than two states k. In such a case the characteristic, such as the likelihood ratio, is not defined and consequently it cannot be applied. The core of generalisation is the knowledge that the solution of any Bayesian, as well as non-Bayesian, tasks, as we already know, corresponds to the decomposition of the space of probabilities into convex cones. Namely, the space of probabilities constitutes this dividing line. On one side of this dividing line all tasks have their own application specificity. On the other side, provided we express the tasks in the space of probabilities, all the tasks become similar. The solution of each task is given by the decomposition of the space of probabilities into classes using linear discriminant functions. Lecture 5 is devoted to this topic.

2.6 Discussion

Even when I studied all the non-Bayesian tasks formulated and their formal solution thoroughly, I cannot say that I mastered the explained material well and understood the presented results in the same way as if I had discovered them myself. I doubt that I would be able to solve a new practical task by myself.

Could you elucidate your difficulties a little to allow us to help you?

Nearly all tasks formulated in the lecture are new to me. I have known only about the Neyman–Pearson task so far. I did not think about the proof of the strategy and its solution too much. An extensive and unfamiliar landscape of non-Bayesian tasks unfolded in front of me. It was presented in the lecture that all these tasks could be analysed and solved by using the linear programming gadget. But I regret, I do not know this gadget.

It is unbelievable. Did you not learn linear programming at university?

Yes, of course I did. But it was in a quite different spirit from your lecture. The main emphasis was placed to computational procedures optimising linear functions with linear constraints. The focus was mainly on the simplex method which I understood quite well. The linear programming gadget now plays a quite different role. It is not just a tool for writing the program for optimisation tasks, but it is a tool for theoretical analysis of the task with pencil and paper. Such a use of linear programming is new to me.

If I understood non-Bayesian tasks well and informally, without using the dual tasks gadget, then the theoretical matter from the lecture could support my current comprehension now and simultaneously understand the dual task tool, too. On the other hand, if I understood quite clearly the dual tasks then I could also understand better non-Bayesian tasks. Now I see two areas, I am not well orientated in either of them, and I know, in addition, that they are closely connected. It is of little help to me if I wanted to penetrate into each of them. There is no other way out but to orientate myself independently in non-Bayesian tasks or to dig deeply into the scientific literature concerning mathematical programming. When all this is done I can return to non-Bayesian tasks. What would you recommend me?

Both options, but the second one a little more. You should be familiar with the theoretical basis of mathematical programming (namely the theory, and not merely a calculational procedure) whether you are engaged in pattern recognition or not. These pieces of knowledge are useful in any area of applied informatics. We do not want to leave you standing in front of problems alone. We will continue in the direction of your remark. You have said that you would understand the dual tasks principle much better if there was something that followed from formal theorems, something you also knew without duality theorems. We can go through one example of that kind with you but it will require a bit of patience on your side. O.K.?

Do not doubt my patience. I consider it important to be familiar with dual tasks. Please do not use examples from Economics. In nearly every linear programming textbook, the tasks are demonstrated with the help of concepts from Economics which is of less help for my understanding. I can understand if it concerns the maximisation of profit. The explanation of the dual task substance by using the examples of mutual relation between supply and demand, production effectiveness, pricing, etc. says little to me, because I understand economic concepts even less than non-Bayesian tasks.

Do not be afraid. Our example demonstrates the substance of dual tasks in electrical circuits. We hope that almost everyone is familiar with it. All three of us have graduated from electrical engineering, some of us earlier and some not so long ago, and so we remember something from electrical circuit theory. These circuits are likely to be understandable to everyone, as high school knowledge should suffice. If we were speaking about something you do not understand, do interrupt us without any hesitation.

In our example numerical quantities will be represented by alternating currents in a circuits branches or alternating voltages between the points of the same circuit. Thus, for instance in Fig. 2.3a, the quantity x labelling the branch of the circuit represents the current $x \cdot \sin t$. The arrow on the branch shows the direction of the current which is considered positive. There are two possible cases: (a) The quantity $x = 3$, for example, corresponds to the alternating current $3 \sin t$, i.e., the current flowing from the point 1 to the point 2 in the

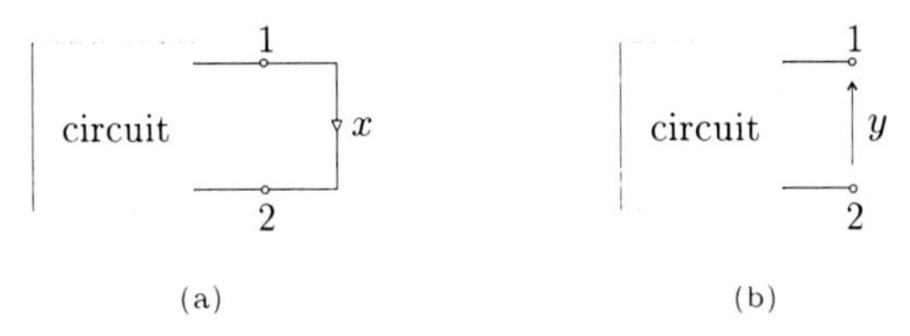

Figure 2.3 Electrical analogy. (a) Current x, (b) voltage y.

case of $\sin t \geq 0$ as well as the current flowing from the point 2 to the point 1 for $\sin t \leq 0$. (b) The quantity $x = -3$ will correspond to the current $-3 \sin t$ which means the current from the point 2 to the point 1 for $\sin t \geq 0$, and conversely from the point 1 to the point 2 for $\sin t \leq 0$.

Similarly, we will show some quantities with the help of voltage which is the difference of electrical potential between two points. E.g., the quantity y between the points 1 and 2 of the circuit (cf., Fig. 2.3b) representing potentials $\varphi_1 - \varphi_2$ between points 1 and 2 is $y \sin t$. The arrow between points 1 and 2 shows the voltage direction which is given as positive, i.e., $y \sin t$ is the difference $\varphi_1 - \varphi_2$ and not the other way round.

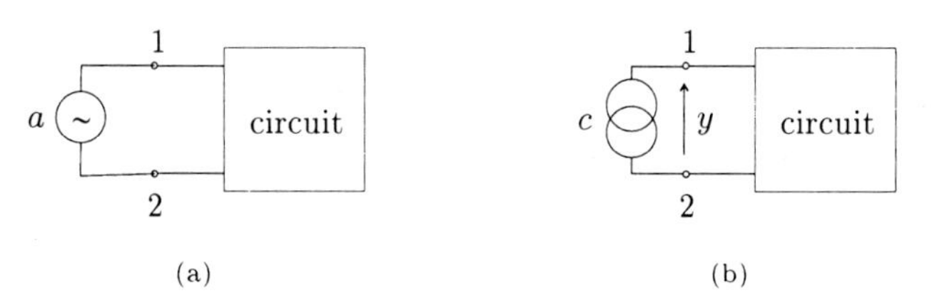

Figure 2.4 (a) Voltage sources. (b) Current sources.

Currents and voltages arise in the circuit thanks to sources of electrical energy. There are two types of sources: ideal voltage sources and ideal current sources. The ideal voltage source a (see Fig. 2.4a) provides the voltage $\varphi_1 - \varphi_2 = a \sin t$ independently of load which is connected to the source terminals. Only the current x, which is taken from the source, depends on the load. The ideal current source c (see Fig. 2.4b) provides the current, which is always $c \sin t$ regardless of the load connected to its terminals. Only the voltage y between the source terminals depends on the load.

Another component which will be needed is a special device called a phase rectifier. When it is connected to the circuit branch (cf. Fig. 2.5a) then it prevents the current $x \sin t$ in the branch with negative x and the voltage $\varphi_1 - \varphi_2 = y \sin t$ between points 1 and 2 with positive y. The phase rectifier has one more important property, i.e., that the product xy can be equal to zero only. This means that if there is a negative voltage across the phase rectifier then the corresponding current must be equal to zero. The converse is valid too, if a positive current flows through the phase rectifier then there must be zero a voltage across it. The situation when both current and voltage equal to zero is not excluded, of course.

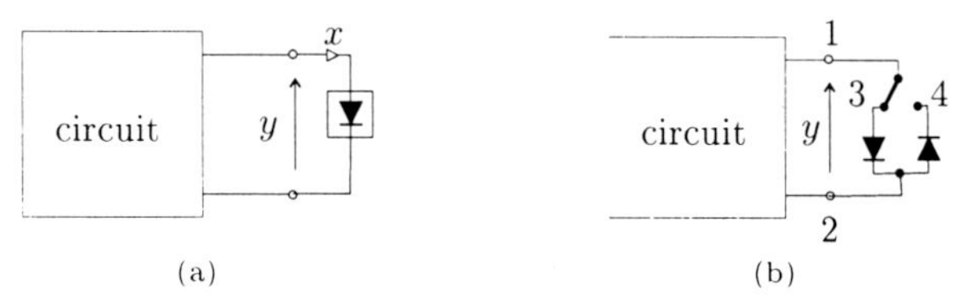

Figure 2.5 (a) Phase rectifier. (b) Possible implementation of a phase rectifier.

I must interrupt you here. Do I properly understand that the quantities x and y are not represented with the help of alternating voltages and currents, but in some other way? When I connect the rectifier (I assume it is an ordinary diode) to the circuit in Fig. 2.5a then the current in this branch cannot alternate. This must be so for the voltage at the diode too.

You perhaps did not understand the explanation properly. It is good that you interrupted us. You cannot realise the phase rectifier in Fig. 2.5a as an ordinary diode. That is why we called it the phase rectifier. It is an idealised device such that the current passing through it can be only $x \sin t$, where $x \geq 0$. The phase rectifier can be implemented, e.g., by the circuit in Fig. 2.5b. The controlled switch is connected to the point 1 which alternatively connects the point 1 to the point 3 when $\sin t \geq 0$ holds, and to the point 4 when $\sin t < 0$ holds.

Is the phase rectifier in Fig. 2.5b implemented by means of ordinary diodes and some switch?

Yes, it is.

I understand it now. Only the current $x \sin t$, $x \geq 0$ or the voltage $y \sin t$, $y \leq 0$, $xy = 0$ can occur in the circuit branch in Fig. 2.5b.

The unexplained last component, which is needed in our circuit, is a common transformer, i.e., several windings on a shared core of ferromagnetic material. Each winding is characterised by the number of turns and the winding direction. E.g., the transformer in Fig. 2.6 consists of five windings with the number of turns given from the left 3, 3, 0, −4 and 1. The zero number of turns of the third winding means that there is no winding on the core. The turn number

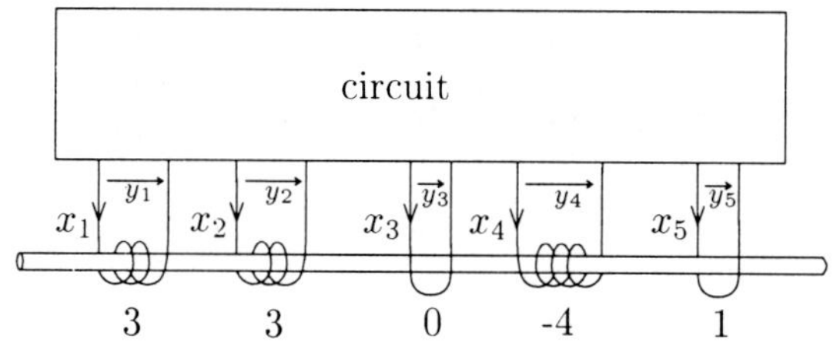

Figure 2.6 A transformer diagram for varying number of turns in the coil and directions of winding.

-4 of the fourth winding says that the winding direction is the opposite of the first, second and fifth winding, whose direction is considered as positive.

Thanks to the interaction of transformer winding with the shared magnetic field in the ferromagnetic core, the currents $x_1, x_2, \ldots, x_m$ in m coils must satisfy the equation

$$\sum_{i=1}^{m} x_i b_i = 0\,,$$

where b_i is the number of turns of the i-th winding and the used signs $+$ or $-$ agree to the rule given above. The currents $x_1, x_2, \ldots, x_5$ in the transformer windings in Fig. 2.6 can thus reach only the values satisfying the constraint

$$3x_1 + 3x_2 - 4x_4 + x_5 = 0\,.$$

The second property of the transformer is that voltages of the windings are proportional to the number of turns. Do not forget that the number of turns is considered both positive and negative according to the winding direction. In Fig. 2.6, the voltages y_1, y_2, y_3, y_4 are uniquely constrained by the voltage y_5, i.e., $y_1 = 3y_5$, $y_2 = 3y_5$, $y_3 = 0$, $y_4 = -4y_5$.

We will need to know what is represented by the current source, voltage source, phase rectifier and transformer when considering electrical analogy of the dual linear programming tasks.

I think that I understand the components of the electrical circuit quite well now.

Create a circuit from the given components with the currents $x_1 \sin t$, $x_2 \sin t$, $x_3 \sin t$, $x_4 \sin t$, $x_5 \sin t$, in its branches which comply with the following system of equalities and inequalities,

$$x_1 \ge 0\,, \quad x_2 \ge 0\,, \quad x_3 \ge 0\,, \tag{2.68}$$

$$\left.\begin{array}{rl}
2x_1 - 3x_2 + 4x_3 + 5x_4 - 2x_5 & \ge c_1\,, \\
x_1 + 2x_2 + x_3 + x_5 & \ge c_2\,, \\
-x_1 + 3x_3 & \ge c_3\,, \\
x_1 + x_2 + x_3 + x_4 + x_5 & = c_4\,, \\
-x_1 + 2x_2 + 3x_3 + x_4 - x_5 & = c_5\,, \\
3x_1 + 2x_2 - 4x_3 - 2x_4 + 3x_5 & = c_6\,.
\end{array}\right\} \tag{2.69}$$

The previous system can be understood as constraints of the primal linear programming task. The linear function that is to be minimised in the task will not be considered for the moment.

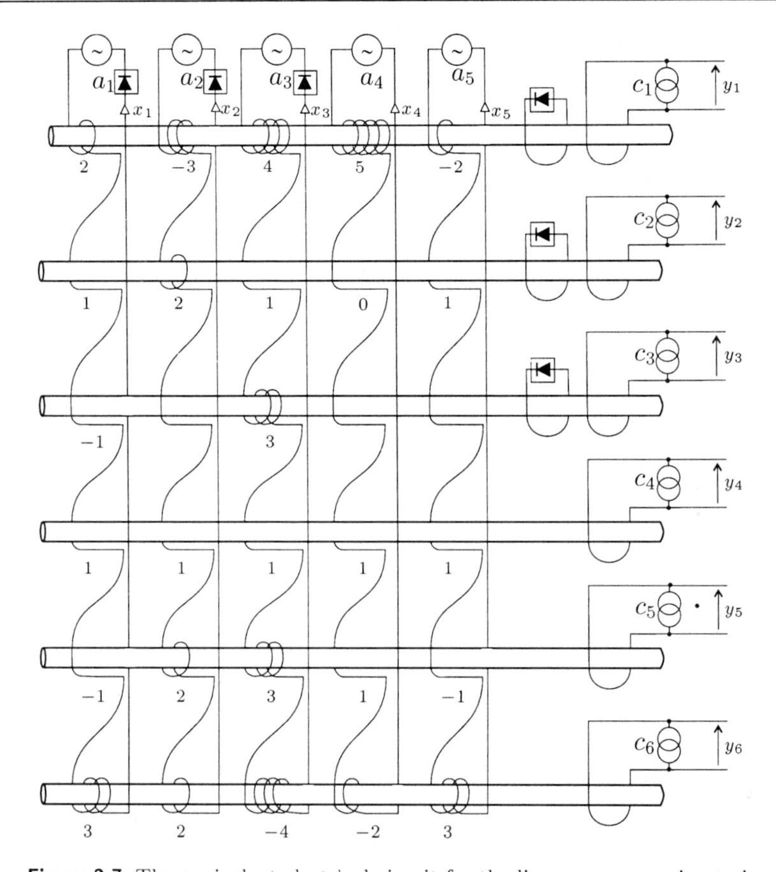

Figure 2.7 The equivalent electrical circuit for the linear programming task.

It was not easy for me to draw the required electrical circuit even when I am familiar with electrical concepts. The most difficult was to satisfy the three conditions in the system (2.69) which have the form of inequalities. I succeeded eventually and the resulting circuit is drawn in Fig. 2.7. It is ensured in the circuit that the currents $x_1, \ldots, x_5$ *satisfy the conditions (2.68) and (2.69). I hope that you expected a similar result, because the circuit is quite straightforward. The circuit shape resembles the shape of a matrix with the constraints (2.69). My circuit consists of six transformers which matches six constraints from the system (2.69), and five branches whose currents correspond to the variables* $x_1, \ldots, x_5$. *Each* i*-th branch,* $i = 1, \ldots, 5$, *turns around the core of the* j*-th transformer,* $j = 1, 2, \ldots, 6$, *and the number of turns can be positive, negative or zero. The number of turns corresponds to the coefficient* b_{ij} *by which the variable* x_i *is multiplied in the* j*-th constraint of the system (2.69).*

I ensured the conditions (2.68) easily by connecting the phase rectifiers to the first, second and third branch. Thanks to it, negative currents cannot occur in these branches. I satisfied the three last conditions in the form of equations from the system (2.69) by inserting the coil on each j*-th,* $j = 4, 5, 6$, *transformer which makes* -1 *turn around the particular core and to which the current* c_j

flows from an ideal current source. The following equation is satisfied thanks to it and thanks to the transformer properties for $j = 4,5,6$

$$\sum_{i=1}^{5} b_{ij} x_i - c_j = 0$$

which corresponds to the three last constraints in the system (2.69).

The first three constraints in the form of inequalities from the system (2.69) were satisfied as follows. I put an additional coil with an arbitrary but negative number of turns on the j-th, $j = 1,2,3$, transformer and connected the phase rectifier to it. Thanks to that, the current x_{0j} cannot be negative in these additional coils. Having the properties of the transformer, I can write for $j = 1,2,3$

$$\sum_{i=1}^{5} x_i b_{ij} - x_{0j} - c_j = 0\,, \quad x_{0j} \geq 0$$

which corresponds to the first three constraints in the system (2.69). That is, perhaps, all.

You did everything right and you even got ahead of it a little bit. In Fig. 2.7 we can see the voltage sources $a_1, \ldots, a_5$, which are in fact not needed to satisfy constraints (2.68) and (2.69). Why did you introduce them?

It was a mere intuition which I cannot prove, but I can support it with some general thoughts. It seems to me, that when I introduced the voltage sources my diagram became not only a tool to satisfy constraints (2.68) and (2.69), but also a tool to minimise the sum $\sum_{i=1}^{5} a_i x_i$ on the set given by systems (2.68) and (2.69). Indeed, the sum $\sum_{i=1}^{5} a_i x_i$ corresponds to the power that the voltage sources $a_1, \ldots, a_5$ obtain from surroundings. But the voltage sources $a_1, \ldots, a_5$ are sources of energy, not consumers of it. That is the reason why these sources interact with the surroundings in such a way that they get rid of energy as soon as possible, i.e., they pass it to the surroundings with the highest possible power. This means, till now only intuitively, that such currents get stabilised in the diagram in Fig. 2.7 that maximise the total power $-\sum_{i=1}^{5} a_i x_i$ in which the voltage sources are deprived of energy. It is clear enough that the suggested diagram constitutes a physical model of the linear programming task. Actually, an arbitrary linear programming task can be modelled using the analogy. I do not claim that I can prove these thoughts.

We will find that out later.

I would like to proceed to the main issue now for which I needed the electrical interpretation of linear programming tasks. First of all, I wanted to clarify the relation between the pair of dual tasks. I have not observed anything like this in the proposed diagram yet, it even seems to me that I understand the diagram quite well.

You will see this in a while. You analysed the properties of currents $x_1, \ldots, x_5$ which the diagram can generate, and have proved (and it was not difficult) that the currents satisfy the constraints (2.68) and (2.69). But beside the current, the voltages are generated by the diagram too, and these are not only voltages $a_1, \ldots, a_5$ at voltage sources but also voltages at the current sources $c_1, \ldots, c_6$.

I have got it! I am surprised that I did not notice it earlier. Indeed, these voltages correspond to the variables in the task which is dual with respect to the task minimising sum $\sum_{i=1}^{5} a_i x_i$ *under constraints (2.68) and (2.69). They are, namely, dual variables!*

Are you able to prove it?

Yes, of course, I am! By transformer properties the voltage $b_{ij} y_j$ *is induced on the coil which is in the* i*-th branch,* $i = 1, \ldots, 5$ *and is put on the core of the* j*-th transformer,* $j = 1, 2, \ldots, 6$*. Here* b_{ij} *is the number of winding turns and* y_i *is the voltage on the current source* c_j*. Moreover, it is well known that the sum of voltages in any closed circuit is equal to zero. It follows immediately that the voltage on the coils has a certain relationship with the voltages* $a_1, \ldots, a_5$ *on the sources of voltages. These relations are extremely simple for the fourth and fifth branch to which the phase rectifier is not connected. The relation for the fifth branch is written as*

$$-2y_1 + y_2 + y_4 - y_5 + 3y_6 = a_5 \ .$$

Similarly, for the fourth branch

$$5y_1 + y_4 + y_5 - 2y_6 = a_4 \ .$$

Because the phase rectifiers are connected to the first, second and the third branch, the sum $\sum_{j=1}^{6} b_{ij} y_j$ *of voltages on coils in the* i*-th branch,* $i = 1, 2, 3$*, is smaller than voltage* a_i*. For the first branch there holds*

$$2y_1 + y_2 - y_3 + y_4 - y_5 + 3y_6 \leq a_1 \ .$$

For the second branch there holds

$$-3y_1 + 2y_2 + y_4 + 2y_5 + 2y_6 \leq a_2$$

and for the third one I write

$$4y_1 + y_2 + 3y_3 + y_4 + 3y_5 - 4y_6 \leq a_3 \ .$$

Earlier, I applied additional coils to the first, second and third cores and connected the phase rectifier to it. I did it to satisfy the first three conditions (inequalities) in the system (2.69). I did not think of it at that time, but now I see that additional coils on the first, second and third transformer secure that voltages y_1, y_2, y_3 *on the corresponding voltage sources cannot be negative.*

In fact, the following happened. When designing the diagram, my original intention was to generate currents x_1, x_2, x_3, x_4, x_5 in agreement with conditions

$$\left.\begin{array}{l} x_1 \ge 0, \quad x_2 \ge 0, \quad x_3 \ge 0\,, \\ 2x_1 - 3x_2 + 4x_3 + 5x_4 - 2x_5 \ge c_1\,, \\ x_1 + 2x_2 + x_3 \qquad\qquad x_5 \ge c_2\,, \\ -x_1 \qquad + 3x_3 \qquad\qquad \ge c_3\,, \\ x_1 + x_2 + x_3 + x_4 + x_5 \ge c_4\,, \\ -x_1 + 2x_2 + 3x_3 + x_4 - x_5 \ge c_5\,, \\ 3x_1 + 2x_2 - 4x_3 - 2x_4 + 3x_5 \ge c_6\,, \end{array}\right\} \tag{2.70}$$

which I understood as constraints in a certain linear programming task. I can see now, and did not anticipate it before, that the same diagram also generates the voltages $y_1, y_2, y_3, y_4, y_5, y_6$ in agreement with constraints of the dual task

$$\left.\begin{array}{l} y_1 \ge 0, \quad y_2 \ge 0, \quad y_3 \ge 0, \\ 2y_1 + y_2 - y_3 + y_4 - y_5 + 3y_6 \le a_1\,, \\ -3y_1 + 2y_2 \qquad + y_4 + 2y_5 + 2y_6 \le a_2\,, \\ 4y_1 + y_2 + 3y_3 + y_4 + 3y_5 - 4y_6 \le a_3\,, \\ 5y_1 \qquad\qquad + y_4 + y_5 - 2y_6 = a_4\,, \\ -2y_1 + y_2 \qquad + y_4 - y_5 + 3y_6 = a_5\,. \end{array}\right\} \tag{2.71}$$

In this way I came to the idea that any physical system which is a model of a linear programming task is also inevitably a model of a dual task. It seems that I am starting to understand dual tasks slowly, but in spite of that there are still more unclear than clear facts. I expressed my hypothesis earlier that in the diagram in Fig. 2.7. I cannot implement arbitrary solutions of the system (2.70), but only those that maximise the total power $-\sum_{i=1}^{5} a_i x_i$, by which the voltage sources $a_1, \ldots, a_5$ dissipate energy. I presume also that the voltages $y_1, \ldots, y_6$ cannot correspond to arbitrary solutions of the system (2.71), but only to those that maximise the total power $\sum_{j=1}^{6} c_j y_j$ of current sources $c_1, \ldots, c_6$. I cannot prove my hypothesis properly. Probably you can help me with it?

Yes, with pleasure, of course. First of all we will prove an auxiliary statement. We will do it in an abstract manner not referring to electrical analogies and after that we will have a look at what this statement means for our electrical diagram.

Let $x = (x_1, x_2, \ldots, x_m)$ be a vector confirming conditions

$$\sum_{i=1}^{m} b_{ij} x_i \ge c_j\,, \quad j = 1, \ldots, n^*\,, \tag{2.72}$$

$$\sum_{i=1}^{m} b_{ij}x_i = c_j\,, \quad j = n^* + 1, n^* + 2, \ldots, n\,, \tag{2.73}$$

$$x_i \geq 0\,, \quad i = 1, \ldots, m^*\,, \tag{2.74}$$

where m, n, m^*, n^* are integers, $m^* \leq m$, $n^* \leq n$. Let $y = (y_1, y_2, \ldots, y_n)$ be a vector that satisfies

$$\sum_{j=1}^{n} b_{ij}y_j \leq a_i\,, \quad i = 1, 2, \ldots, m^*\,, \tag{2.75}$$

$$\sum_{j=1}^{n} b_{ij}y_j = a_j\,, \quad i = m^* + 1, m^* + 2, \ldots, m\,, \tag{2.76}$$

$$y_j \geq 0\,, \quad j = 1, \ldots, n^*\,. \tag{2.77}$$

In such a case there holds

$$\sum_{i=1}^{m} a_i x_i \geq \sum_{j=1}^{n} c_j y_j\,. \tag{2.78}$$

The proof of the previous statement is rather simple. It follows from inequalities (2.72) and (2.77)

$$\sum_{j=1}^{n^*} \left(\sum_{i=1}^{m} b_{ij}x_i - c_i \right) y_j \geq 0\,.$$

The equation (2.73) implies

$$\sum_{j=n^*+1}^{n} \left(\sum_{i=1}^{m} b_{ij}x_i - c_i \right) y_j = 0\,.$$

If the two latter expressions are added we get

$$\sum_{j=1}^{n} \left(\sum_{i=1}^{m} b_{ij}x_i - c_i \right) y_j \geq 0\,. \tag{2.79}$$

It follows from inequalities (2.75) and (2.74)

$$\sum_{i=1}^{m^*} \left(a_i - \sum_{j=1}^{n} b_{ij}y_j \right) x_i \geq 0\,.$$

It follows from equation (2.76) that

$$\sum_{i=m^*+1}^{m} \left(a_i - \sum_{j=1}^{n} b_{ij}y_j \right) x_i = 0\,.$$

Summing two latter expressions we obtain

$$\sum_{i=1}^{m} \left(a_i - \sum_{j=1}^{n} b_{ij} y_j \right) x_i \geq 0 \,. \tag{2.80}$$

Summing inequalities (2.79) and (2.80) we can write

$$\sum_{i=1}^{m} a_i x_i - \sum_{j=1}^{n} c_j y_j \geq 0$$

which is only another form of inequality (2.78) that was being proved. The proof is finished.

As all the constraints (2.72)–(2.77), which you created in Fig. 2.7, are satisfied, the inequality (2.78) is satisfied, too.

Inequality (2.78) can be easily understood without proof when one has in mind our electrical diagram. The total energy $\sum_{j=1}^{n} c_j y_j$*, which current sources* c_j*,* $j = 1, \ldots, n$*, dissipate at any instant, cannot be larger than the total energy* $\sum_{i=1}^{m} a_i x_i$*, which is received from the surrounding by the voltage sources* a_i*,* $i = 1, \ldots, m$*. Otherwise, the energy conservation law would be violated.*

I can see now that an even stricter relation than (2.78) can be proved on the basis of purely physical considerations, which is the equation

$$\sum_{i=1}^{m} a_i x_i = \sum_{j=1}^{n} c_j y_j \,, \tag{2.81}$$

which is equal to the first duality theorem (cf. Theorem 2.1). Its analytical proof is rather difficult. Equation (2.81) holds simply because that the total energy $\sum_{i=1}^{m} a_i x_i$*, which occurs on voltage sources, cannot be larger than the total energy* $\sum_{j=1}^{n} c_j y_j$ *produced by current sources. Otherwise, it would not be clear from where energy could originate. Equation (2.81) confirms my earlier hypothesis that only such currents* x_i *appear in the diagram as are solutions of the primal linear programming task. Similarly, only voltages* y_i *are generated that are solutions of the dual task.*

Only a slight effort is needed to give the electrical interpretation to the second duality theorem too. When that is done we will understand it entirely and informally. Are you able to do that on your own?

Yes, indeed. The electrical interpretation of the second duality theorem is so obvious that the formal proof can make it only unclear. The interpretation can be formulated as follows.

If the current passing through a phase rectifier in our diagram is not zero then the voltage on the same rectifier is zero. And conversely, too, if the voltage on a phase rectifier is not zero the current through this rectifier is equal to zero.

That might be all we can help you with today to clarify the dual tasks of linear programming and allow you to manipulate it freely when analysing practical tasks.

The electrical analogy of the abstract duality theorems is quite obvious. There are some questions left which concern non-Bayesian tasks.

Go ahead, ask!

I have understood from the lecture that in many practical tasks a penalty function, as occurs in Bayesian tasks, does not make any sense in reality. I noticed that the concept of a penalty function has not been used at all in the listed non-Bayesian tasks.

Assume that the statistical model of an object is entirely known in an application, but the penalty function is not defined. This means that both a priori *probabilities $p_K(k)$, $k \in K$, and conditional probabilities $p_{X|K}(x \mid k)$ make sense and are known. I understand that the penalty function, which does not make sense in a given application, is missing for the formulation of the Bayesian task. But why should I have to formulate the task as a non-Bayesian one in this case? Is it not better to remain within the Bayesian framework? Would the task not be formulated as a risk minimisation, but as the minimisation of the probability of the wrong decision? Such a criterion is fully understandable to the user and even more than is the risk. It seems to me that the criterion minimising the probability of the wrong decision has the right to exist fully and not only when it is derived from a specific penalty function.*

It is not possible to answer your question unambiguously. Of course, you can use strategies which give a small probability of the error. But you must be sure that something comprehensible actually corresponds to the criterion in your application. We only warn you not to think that the criterion of the smallest probability of the error is always usable.

Let us assume that your application is the diagnosis of an oncological illness in its early stage. You are aware that this is an important task, because if the illness is discovered as it emerges then the patient can be still saved. Unfortunately the task is also a quite difficult one. Imagine that we would come to you and insist that we had an algorithm which will surely not indicate a wrong diagnosis more often than in 1% of cases. You will be pleased and buy our algorithm immediately. You are content because you assume (watch out, the error follows!) that you will save 99% of people, who otherwise should die without your help. But your enthusiasm evaporates quickly as you learn that our algorithm evaluates all patients as healthy, and yields 1% wrong decisions only because sick people constitute not more than 1% of population.

You feel that we as merchants made you pay through the nose. We succeeded in cheating you even though you know quite well which requirements the algorithm has to satisfy, but you formulated these requirements using an inappropriate criterion. In the given case it is necessary to characterise the decision algorithm by two probabilities of the wrong decision and not merely by a single one. These two probabilities are formally expressed by means of numbers which can be added, averaged, etc.. However, these probabilities describe entirely different events and nothing corresponds to their sum that could be interpreted within the framework of your application. You have to formulate the task in the way that this non-interpretable operation should not be used. In this way only, you can achieve the formally proved strategy that will correspond to real demands of the application.

Instead of formulating these demands, you formulated another task in a hope that the optimal solution of this supplementary task will not be too bad for the actual task. Such an approach should be comprehensible if you had known only one statistical decision task. But you know more tasks of this kind and you need not keep the only one so convulsively. The most appropriate would be to arrange the task as Wald or Neyman–Pearson task in your case.

I was convinced in the lecture that the solution of the properly formulated task can significantly differ from the solutions which someone could guess only on the basis of mere intuition. In this respect Wald task seems to me to be the most distinctive one for $|K| > 2$, where the outcome is quite unexpected. And I resent it, since I understand the task only purely formally. If I am to use it I have to believe in formal methods too much as to my taste. I expected that the solution of Wald tasks will be much easier. Let me explain the algorithm which I considered to be the solution of the task. And then I would ask you to help me analyse it and tell me why it is wrong.

The formally proved algorithm can be expressed in the following form. The strategy depends on $2|K|$ numbers $\tau(k)$, $q(k)$, $k \in K$, and on basis of them the following values should be calculated

$$\gamma_k(x) = \frac{\tau(k)\, p_{X|K}(x \mid k)}{\sum\limits_{k \in K} \tau(k)\, p_{X|K}(x \mid k)}$$

which resemble very much the a posteriori probabilities of the state k under the condition x. Next, the largest number of these numbers is sought and is compared with the threshold $\theta(x)$. The threshold value is

$$\theta(x) = 1 - \frac{\sum\limits_{k \in K} q(k)\, p_{X|K}(x \mid k)}{\sum\limits_{k \in K} \tau(k)\, p_{X|K}(x \mid k)}\,,$$

and this value is different for each observation. It is very difficult to understand this step informally, because the threshold value was the same for all observations in the Bayesian variant of this task, i.e., independent of x. These considerations are not the proof, but I assumed that here in Wald task it should be the same case. This means that the strategy would have a form

$$k^* = \begin{cases} \operatorname*{argmax}\limits_{k \in K} \gamma_k(x)\,, & \text{if } \max\limits_{k \in K} \gamma_k(x) > (1 - \delta)\,, \\ 0\,, & \text{if } \max\limits_{k \in K} \gamma_k(x) < (1 - \delta)\,. \end{cases}$$

This strategy seems to me far more natural compared to the strategy used in the lecture. That is why I assumed that when the strategy was not the solution of the formulated task then it could be a useful strategy for some other task.

Let us try to make a retrospective analysis of your strategy and let us find tasks for which this strategy is optimal. The strategy suggested by you would be obtained if the variables $\tau(k)$, $q(k)$, $k \in K$, were constrained by the property that the ratio

$$\frac{\sum_{k \in K} q(k)\, p_{X|K}(x \mid k)}{\sum_{k \in K} \tau(k)\, p_{X|K}(x \mid k)} = \delta$$

is the same for all observations $x \in X$. The constraint can be expressed in a linear form

$$-\sum_{k \in K} q(k)\, p_{X|K}(x \mid k) + \delta \sum_{k \in K} \tau(k)\, p_{X|K}(x \mid k) = 0 \ .$$

These linear constraints will be embodied into the formulation of the dual task (2.62). This constitutes a new dual task.

Dual task:

$$\left.\begin{array}{ll}
\max \left(\sum_{x \in X} t(x) + (1-\varepsilon) \sum_{k \in K} \tau(k) \right) & \\
t(x) + \sum_{k \in K} \tau(k)\, p_{X|K}(x \mid k) - \sum_{k \in K} q(k)\, p_{X|K}(x \mid k) \leq 0\,, \quad x \in X, & \text{(c)} \\
t(x) + \tau(k)\, p_{X|K}(x \mid k) \leq 0\,, \quad x \in X\,, \quad k \in K\,, & \\
\delta \sum_{k \in K} \tau(k)\, p_{X|K}(x \mid k) - \sum_{k \in K} q(k)\, p_{X|K}(x \mid k) = 0\,, \quad x \in X, & \text{(a)} \\
\sum_{k \in K} q(k) = 1\,, & \text{(b)} \\
q(x) \geq 0\,, \quad \tau(k) \geq 0\,. &
\end{array}\right\} \quad (2.82)$$

This task differs from the previous task (2.62) which was composed in the lecture having in mind $|X|$ additional constraints (2.82a). The original task was somewhat deformed. We will build up the primal task, to which the task (2.82) is dual, rather formally in order to find out what happened after deformation. Before that we will perform several equivalent transformations of the task (2.82). First, on the basis of (2.82a) the constraint (2.82c) can have the form

$$t(x) + (1-\delta) \sum_{k \in K} \tau(k)\, p_{X|K}(x \mid k) \leq 0\,, \quad x \in X\,. \tag{2.83}$$

Furthermore, the pair (2.82a) and (2.82b) is equivalent to the pair

$$\left.\begin{aligned} \delta \sum_{k\in K} \tau(k)\, p_{X|K}(x\,|\,k) - \sum_{k\in K} q(k)\, p_{X|K}(x\,|\,k) = 0\,, \quad k \in K \\ \sum_{k\in K} \tau(k) = \frac{1}{\delta}\,. \end{aligned}\right\} \tag{2.84}$$

Indeed, from the constraint (2.82a) there follows

$$\delta \sum_{x\in X} \sum_{k\in K} \tau(k)\, p_{X|K}(x\,|\,k) = \sum_{x\in X} \sum_{k\in K} q(k)\, p_{X|K}(x\,|\,k)\,.$$

Based on (2.82b) and that $\sum_{x\in X} p_{X|K}(x\,|\,k) = 1$ we obtain

$$\delta \sum_{k\in K} \tau(k) = \sum_{k\in K} q(k) = 1\,.$$

This expression is substituted into the maximised function and we can write

$$\sum_{x\in X} t(x) + \frac{1-\varepsilon}{\delta}\,.$$

Maximisation of this function is equivalent to maximisation of the function

$$\sum_{x\in X} t(x)\,.$$

From what was said above the new shape of the task (2.82) follows:

$$\left.\begin{aligned} & \max\Big(\sum_{x\in X} t(x)\Big) \\ & t(x) + (1-\delta) \sum_{k\in K} \tau(k)\, p_{X|K}(x\,|\,k) \le 0\,, \quad x \in X\,, \\ & t(x) + \tau(k)\, p_{X|K}(x\,|\,k) \le 0\,, \quad x \in X\,, \quad k \in K\,, \\ & \delta \sum_{k\in K} \tau(k)\, p_{X|K}(x\,|\,k) - \sum_{k\in K} q(k)\, p(x\,|\,k) = 0\,, \quad x \in X\,, \quad \text{(a)} \\ & \sum_{k\in K} \tau(k) = \tfrac{1}{\delta}\,, \\ & \tau(k) \ge 0\,, \quad q(k) \ge 0\,, \quad k \in K\,. \end{aligned}\right\} \tag{2.85}$$

From such a form of the task it can be seen that the constraints (2.85a) are redundant. Indeed, if the variables $t(x)$, $x \in X$, $q(k)$, $k \in K$, and $\tau(k)$, $k \in K$, conform to all constraints except the constraint (2.85a) then only the variables $q(k)$, $k \in K$, can be changed so that the constraint (2.85a) will be satisfied. For example, the variable $q(k)$ can be selected to be equal to $\delta\,\tau(k)$. Thus the task (2.85) can be expressed as

$$\left.\begin{array}{r|l}
 & \max\left(\sum\limits_{x\in X} t(x)\right) \\
\alpha(x,0) & t(x) + (1-\delta) \sum\limits_{k\in K} \tau(k)\, p_{X|K}(x\,|\,k) \le 0\,, \quad x \in X\,, \\
\alpha(x,k) & t(x) + \tau(k)\, p_{X|K}(x\,|\,k) \le 0\,, \quad x\in X\,, \quad k\in K\,, \\
c & \sum\limits_{k\in K} \tau(k) = \frac{1}{\delta}\,, \\
 & \tau(k) \ge 0\,.
\end{array}\right\} \tag{2.86}$$

It should not surprise you that variables $q(k)$, $k \in K$, disappeared from the task. Well, you yourself wanted that the threshold with which '*a posteriori*' quantities $\gamma(k)$, $k \in K$, should be compared would not depend on the observation x. This dependence was realised in the former task through coefficients $q(k)$, $k \in K$. You will probably be surprised that as a result of a task deformation the parameter ε disappeared, which had determined an acceptable probability error limit. So we can affirm now that the algorithm, which seems natural to you, does not secure any more that the error will not be larger than ε. This is because the algorithm simply ignores the value ε. We will proceed further to find out in which sense is the algorithm proposed by you optimal.

The task (2.86) is dual to the following task

$$\left.\begin{array}{r|l}
 & \min\left(\frac{c}{\delta}\right) \\
\tau(k) & c + \sum\limits_{x\in X} \alpha(x,k)\, p_{X|K}(x\,|\,k) + (1-\delta) \sum\limits_{x\in X} \alpha(x,0)\, p_{X|K}(x\,|\,k) \ge 0,\ k\in K\,, \\
t(x) & \alpha(x,0) + \sum\limits_{k\in K} \alpha(x,k) = 1\,, \quad x\in X\,, \\
 & \alpha(x,0) \ge 0\,, \quad \alpha(x,k)\ge 0\,, \quad x\in X\,, \quad k\in K\,,
\end{array}\right\} \tag{2.87}$$

which can be interpreted, e.g., as: The algorithm proposed by you minimises the value $\max_{k\in K}\left(\omega(k) + \delta\,\chi(k)\right)$, where

$$\omega(k) = \sum_{x\in X} \left(\sum_{k'\neq k;\ k'\neq 0} \alpha(x,k')\right) p_{X|K}(x\,|\,k)$$

is the probability of the wrong decision provided that the object is in the state k and

$$\chi(k) = \sum_{x\in X} \alpha(x,0)\, p_{X|K}(x\,|\,k)$$

is the probability of the answer `not known` under condition that the object is in the state k. The parameter δ in your algorithm can be interpreted as the penalty given to the answer `not known` where the wrong decision is penalised by one. Your task formulation is admissible as well. You can now answer yourself whether the formulation is just what you intuitively wanted it to be when you tried to guess the algorithm.

After the task had been transformed into a unified dual tasks formalism, we obtained a finally intelligible task formulation. But we do not think that such a way of algorithm creation should be used often. The designer of the algorithm should behave like a person, who sees a locked door in front of him and who tries to make a key that will open the door. He should not proceed just the other way round, i.e., to seek an inspiration in the person, who tries to make the key himself first and then tries to find a door that the key opens.

I cannot definitely agree with your last remark although I acknowledge that it is distinctive and expressed in rich colours.

You may be right too.

I have one more question which is not very short. It was said in Lectures 1 and 2 that the strategy solving both Bayesian and non-Bayesian tasks could be expressed in the space of probabilities using convex cones. For all that, a significant difference remains between Bayesian and non-Bayesian strategies. The Bayesian strategy divides the whole space of probabilities into convex cones in such a way that each point from the space belongs just to a single cone including the points lying on the cone borders. This was in fact stated by the basic theorem about the deterministic property of Bayesian strategies, cf., Theorem 1.1.

It is somewhat different in non-Bayesian strategies. It was proved in the lecture that all points inside the cones have to be classified into a certain class in a deterministic way. Nothing was said, on purpose or maybe accidentally, about what is to be done with the points which lie exactly on the cone's borders. Non-Bayesian strategies are deterministic in almost all points of the probability space but not entirely everywhere. It is clear to me that decisions corresponding to observations fitting exactly to the borders of convex cones not only can but even must be random. This random decision can be better than any deterministic one. Naturally, I do not believe that miraculous results can be achieved with the help of this randomisation. Random strategies cannot be much better than deterministic strategies because the randomisation is useful only in a very small subset of points. I am more concerned that the deterministic character of strategies is no longer a peremptory imperative for non-Bayesian tasks as it was for Bayesian tasks. On this basis I am starting to believe that there might exist another broader area of statistical tasks in which randomised strategies will have decisive predominance over deterministic ones perhaps similarly as what happens in antagonistic games in the game theory.

It seems to me that such a situation could arise if the basic concepts were enriched by the penalty function that would not be fixed but would be dependent on a value of a certain non-random parameter, i.e., the intervention. In contradiction to interventions, which we met in testing of complex hypotheses (random or non-random), these interventions do not influence the observed object but merely, how the appropriate decision will be penalised. The task could be defined as seeking such a strategy which is good enough for any intervention. In other words, I would like to study also such strategies for which the penalty function is not defined uniquely but on the other hand it is not entirely unknown. Such a partial knowledge and partial ignorance is expressed by the help of the class of penalty functions and a strategy has to be found, which would be admissible for each penalty function from this class.

The given questions are not easy to answer. These issues will be treated neither at this moment nor during our entire course since we do not know the answer. It might be an open problem which is worth being examined. You can see that the answer to your question is far shorter than the question itself.

January 1997.

2.7 Bibliographical notes

The tool in this lecture was the pair of dual tasks of linear programming which was carefully studied in mathematics [Kuhn and Tucker, 1950; Zuchovickij and Avdejeva, 1967].

We have not found such a general view on non-Bayesian tasks anywhere. Actually, this was the main motivation for us to write the lecture. A mathematician can observe tasks from the height of great generality, e.g., Wald [Wald, 1950] stated that the finite nature of the observation space is such a severe constraint that it is seldom satisfied in statistical decision tasks. Nevertheless, in the statistical decision making theory, situations are sharply distinguished—when the estimated parameter is either random or non-random [Neyman, 1962].

A practitioner solves non-Bayesian tasks often subconsciously when she or he starts tuning parameters of the decision rule that was derived in a Bayesian manner with the aim of recognising all classes roughly in the same way. By doing this she or he actually solves the non-Bayesian task, in the given case a minimax one. Two articles [Schlesinger, 1979b; Schlesinger, 1979a] are devoted to formalisation of the practitioner's approach and they served as a starting material for writing the lecture.

The references to original sources will be mentioned for individual non-Bayesian tasks. References relevant to Neyman–Pearson task [Neyman and Pearson, 1928; Neyman and Pearson, 1933] and for deeper understanding the textbook of statistics [Lehmann, 1959] can be recommended. A minimax task is described in [Wald, 1950]. Wald task, as it was understood in the lecture, is a special case of Wald sequential analysis [Wald, 1947; Wald and Wolfowitz, 1948]. Statistical decision tasks with non-random interventions, also called tasks testing complex hypotheses, were formulated by Linnik [Linnik, 1966].

Lecture 3

Two statistical models of the recognised object

A distribution $p_{X|K} : X \times K \to \mathbb{R}$ of conditional probabilities of observations $x \in X$, under the condition that the object is in a state $k \in K$, is the central concept on which various task in pattern recognition are based. Now is an appropriate time to introduce examples of conditional probabilities of observations with the help of which we can elucidate the previous as well as the following theoretical construction. In this lecture we will stop progressing in the main direction of our course for a while to introduce the two simplest functions $p_{X|K}$ which are the most often used models of the recognised object.

3.1 Conditional independence of features

Let an observation x consist of n certain measured object features $x_1, x_2, \ldots, x_n$. Each feature x_i, $i \in I$, assumes values from a finite set X_i which in the general case is different for each feature. The set of observations X is a Cartesian product $X_1 \times X_2 \times \ldots \times X_n$. It is assumed that the probabilities $p_{X|K}(x \mid k)$ have the form

$$p_{X|K}(x \mid k) = \prod_{i=1}^{n} p_{X_i|K}(x_i \mid k) . \tag{3.1}$$

This means that at the fixed state the features become mutually independent. But this does not mean that the features are also *a priori* mutually independent. In general,

$$p_X(x) \neq \prod_{i=1}^{n} p_{X_i}(x_i) . \tag{3.2}$$

The object's features are dependent on each other but all the dependence is realised via the dependence on the state of the object. If the state is fixed then the mutual dependence among the features disappears. Such a relation is called *conditional independence of features*. The case mentioned describes the simplest model of the conditional independence.

It is easy to prove that in the case in which each of the features assumes only two values {0,1} and the number of states is 2, then the strategy solving any Bayesian and non-Bayesian task can be implemented as a decomposition of the set of vertices on an n-dimensional hypercube by means of a hyperplane. We do already know that the strategy solving any such task has the following form: to each decision d an interval of values of likelihood ratio corresponds, i.e., the decision d is taken when

$$\theta^d_{\min} < \frac{p_{X|K}(x \mid k=1)}{p_{X|K}(x \mid k=2)} \le \theta^d_{\max} , \tag{3.3}$$

where $\theta^d_{\min}$ and $\theta^d_{\max}$ are threshold values. The expression (3.3) is evidently equivalent to the relation

$$\theta^d_{\min} < \log \frac{p_{X|K}(x \mid k=1)}{p_{X|K}(x \mid k=2)} \le \theta^d_{\max} , \tag{3.4}$$

where the threshold values are different from those in the relation (3.3). If each feature x_i, $i \in I$, assumes only two values 0 or 1 then the following derivation in the form of several equations will bring us to the interesting property of the logarithm of the likelihood ratio

$$\begin{aligned} \log \frac{p_{X|K}(x \mid k=1)}{p_{X|K}(x \mid k=2)} &= \sum_{i=1}^{n} \log \frac{p_{X_i|K}(x_i \mid k=1)}{p_{X_i|K}(x_i \mid k=2)} \\ &= \sum_{i=1}^{n} x_i \log \frac{p_{X_i|K}(1 \mid k=1)\, p_{X_i|K}(0 \mid k=2)}{p_{X_i|K}(1 \mid k=2)\, p_{X_i|K}(0 \mid k=1)} + \sum_{i=1}^{n} \log \frac{p_{X_i|K}(0 \mid k=1)}{p_{X_i|K}(0 \mid k=2)} \, . \end{aligned}$$

The transition from the next to last line to the last line in the previous derivation can be verified when two possible cases $x_i = 0$ and $x_i = 1$ are considered separately. It can be seen that the *logarithm of the likelihood ratio is a linear function* of variables x_i. Because of this we can rewrite the expression (3.4) in the following way

$$\theta^d_{\min} < \sum_{i=1}^{n} \alpha_i \, x_i \le \theta^d_{\max} . \tag{3.5}$$

If the tasks are expressed by a firmly chosen function $p_{X|K}$ then various strategies (3.5) differ each from the other only by a threshold value. If, in addition, the function $p_{X|K}$ varies then also the coefficients α_i start varying. At all these changes, it remains valid that all decision regions are regions, where values of a linear function belong to a contiguous interval.

In special cases in which a set of decisions consists of two decisions only, i.e., when the observation set X is to be divided into two subsets X_1 and X_2 then the decision function assumes the form

$$x \in \begin{cases} X_1 , & \text{if} \quad \sum\limits_{i=1}^{n} \alpha_i \, x_i \le \theta \, , \\ X_2 , & \text{if} \quad \sum\limits_{i=1}^{n} \alpha_i \, x_i > \theta \, . \end{cases} \tag{3.6}$$

This means that for objects characterised by binary and conditionally independent features, the search for the needed strategy is equal to searching for coefficients α_i and the threshold value θ. The entire Lecture 5 on linear decision rules will be devoted to the manner how to tune these coefficients and thresholds properly.

3.2 Gaussian probability distribution

Let a set of observations X be an n-dimensional linear space. Let us note that this assumption is in a discrepancy with the content of the previous lectures. There we have emphasised many times that X is a finite set. Nevertheless, the results derived earlier can be used in most situations even in this case. It is sufficient to mention that the number $p_{X|K}(x \mid k)$ does not mean a probability but a probability density.

We will assume that the function $p_{X|K} \colon X \times K \to \mathbb{R}$ has the form

$$p_{X|K}(x \mid k) = C(A^k) \exp\left(-\frac{1}{2} \sum_{i=1}^{n} \sum_{j=1}^{n} a_{ij}^k \left(x_i - \mu_i^k\right)\left(x_j - \mu_j^k\right) \right), \tag{3.7}$$

where k is a superscript index and not a power. Multi-dimensional random variable with the given probability density (3.7) is called the multi-dimensional Gaussian (normal) random variable. In the expression (3.7), the x_i is a value of the i-th feature of the object, μ_i^k is the conditional mathematical expectation of the i-th feature under the condition that the object is in the state k. The symbol A^k introduces the inverse covariance matrix, i.e., the matrix equal to $(B^k)^{-1}$. The element b_{ij}^k in the matrix B^k corresponds to the covariance between the i-th and the j-th features, i.e., the conditional mathematical expectation of the product $(x_i - \mu_i^k)(x_j - \mu_j^k)$ under the condition that the object is in the state k. At last, $C(A^k)$ is a coefficient which ensures that the integral over the whole domain of the function (3.7) is equal to 1.

It is well known (and it can be simply shown too) that, in the case of two states and two decisions, the optimal strategy is implemented using a quadratic discriminant function. This means that

$$x \in \begin{cases} X_1, & \text{if} \quad \sum\limits_i \sum\limits_j \alpha_{ij}\, x_i\, x_j + \sum\limits_i \beta_i\, x_i \le \gamma\,, \\ X_2, & \text{if} \quad \sum\limits_i \sum\limits_j \alpha_{ij}\, x_i\, x_j + \sum\limits_i \beta_i\, x_i > \gamma\,. \end{cases} \tag{3.8}$$

Coefficients α_{ij}, β_i, $i, j = 1, 2, \ldots, m$, and the threshold value γ depend on a statistical model of the object, i.e., on matrices A^1, A^2, vectors μ^1, μ^2 and also on the fact which Bayesian or non-Bayesian decision task is to be solved. Even in the two-dimensional case, the variability of geometrical forms, which the sets X_1 and X_2 assume, is quite large. We will show some of them.

1. The border between the sets X_1 and X_2 can be a straight line which is situated in the way that the set X_1 lies on one side of the line and the set X_2 on its other side.

2. The border between the sets X_1 and X_2 can be determined by a pair of parallel lines located in the way that X_1 is positioned between the lines and X_2 is constituted by the remaining part of the space lying outside the two lines.
3. The border between sets can be constituted by a pair of intersecting straight lines which decompose the plane X into four sectors. Two sectors represent the set X_1 and two are constituted by the set X_2.
4. The border can be given by an ellipse (or a circle in a particular case) in the way that, for instance, X_1 lies inside the ellipse and X_2 corresponds to the part of the plane outside the ellipse.
5. The border can be created by a hyperbola, i.e., it is two curves in the way that one of the classes lies between the curves and the other class is expressed as two convex sets. Either set is marked off by one of the continuous hyperbolae.

In three-dimensional and multi-dimensional cases, geometrical forms of sets X_1 and X_2 can be even much more varied. That is why it is quite useful to understand that, in a certain sense, all variety of forms can be summarised into a single form, namely into that form, when the border between classes is constituted only by a hyperplane, and not an ellipse, hyperbolae, etc.. Then the set of strategies of the form (3.8) is equivalent to the set of strategies of the form

$$x \in \begin{cases} X_1, & \text{if} \quad \sum_i \alpha_i \, x_i \le \gamma \,, \\ X_2, & \text{if} \quad \sum_i \alpha_i \, x_i > \gamma \,. \end{cases} \tag{3.9}$$

The equivalence of different decision strategies with the hyperplane is achieved by the method that in pattern recognition is called the *straightening of the feature space*, sometimes called a Φ-processor. The transformation deforms the original feature space in such a way that a set of curves is transformed into a set of planes. In our case we are interested in a set of nonlinear surfaces of the class (3.8). The original n-dimensional feature space is nonlinearly straightened (transformed) into the $\left(n + \frac{1}{2}n(n+1)\right)$-dimensional feature space. The vector $x = (x_1, x_2, \ldots, x_i, \ldots, x_n)$ is transformed into the $\left(n + \frac{1}{2}n(n+1)\right)$-dimensional vector

$$\begin{array}{rlllllll} y = (& x_1\,, & x_2\,, & \ldots, & x_i\,, & \ldots, & x_{n-1}\,, & x_n\,, \\ & x_1x_1\,, & x_1x_2\,, & \ldots, & x_1x_i\,, & \ldots, & x_1x_{n-1}\,, & x_1x_n\,, \\ & & x_2x_2\,, & \ldots, & x_2x_i\,, & \ldots, & x_2x_{n-1}\,, & x_2x_n\,, \\ & & & & & & \vdots & \\ & & & & x_ix_i\,, & \ldots, & x_ix_{n-1}\,, & x_i\,x_n\,, \\ & & & & & & \vdots & \\ & & & & & & x_{n-1}x_{n-1}\,, & x_{n-1}x_n\,, \\ & & & & & & & x_nx_n\,)\,. \end{array} \tag{3.10}$$

If we denote the coordinates of a newly created vector y by y_i, where $i = 1, 2, \ldots, n + \frac{1}{2}n(n+1)$, and by Y_1, Y_2 sets, into which the sets X_1, X_2 are transformed by means of (3.10), then the strategy (3.8) can be written in the form

$$y \in \begin{cases} Y_1, & \text{if} \quad \sum_i \alpha_i\, y_i \le \gamma\,, \\ Y_2, & \text{if} \quad \sum_i \alpha_i\, y_i > \gamma\,. \end{cases} \tag{3.11}$$

Let us show the straightening of the feature space on an example.

Example 3.1 Straightening of the feature space. *Assume x is a one-dimensional random variable with Gaussian distribution. Let us assume that the strategy for two classes X_1, X_2 is of the following form,*

$$x \in \begin{cases} X_1, & \textit{if} \quad (x - x_0)^2 < \delta\,, \\ X_2, & \textit{if} \quad (x - x_0)^2 \ge \delta\,, \end{cases}$$

Then the strategy can be expressed in the equivalent linear form

$$x \in \begin{cases} X_1, & \textit{if} \quad \alpha_1\, y_1 + \alpha_2\, y_2 > \theta\,, \\ X_2, & \textit{if} \quad \alpha_1\, y_1 + \alpha_2\, y_2 \le \theta\,, \end{cases}$$

where $y_1 = x^2$, $y_2 = x$, $\alpha_1 = -1$, $\alpha_2 = 2\,x_0$, $\theta = x_0^2 - \delta$. ▲

It is evident from the strategy (3.11) obtained that it can be performed as a linear decomposition of certain linear space. A mapping of initial feature space into new space (the space straightening) is of the standard form. It is the same for all situations in which an observation x of an object in k-th state is a multi-dimensional Gaussian random variable. The more specific knowledge about statistical parameters of an appropriate Gaussian distribution and the solved task itself are needed only to determine coefficients α_i and the threshold γ which already determine the strategy (3.11) uniquely. How this tuning of the decision algorithm has to be performed in a specific situation will be shown in Lecture 5 concerning linear discriminant functions.

3.3 Discussion

Every time the model with independent features is used in publications it seems incredible to me that it is possible to describe a real object in such a simple way. Why is the model used so often? My answer is that there is a lack of knowledge in practical tasks about being able to use a more complex model. Available experimental data are sufficient to evaluate how particular features depend on the object state and are not sufficient to evaluate the dependence of feature group on the state. However, the lack of knowledge about the mutual dependence of features does not entitle anybody to make the conclusion that the features are independent. A more thorough approach should be used here which explicitly considers the insufficient knowledge of the statistical model of the object. In common practice the insufficient piece of knowledge is wilfully replaced by a specific model which has the only advantage that its analysis is simple. An analogy comes to mind, of a person looking for a lost item not at the spot he lost it but under the lantern. I have tried to settle accounts with the indicated difficulties. I have reached partial results but it is difficult for me to continue. I will explain to you my results first and I will ask you to help me to make a step ahead.

Let X be the set of observations $x = (x_1, x_2, \ldots, x_n)$ which is $X = X_1 \times X_2 \times \ldots \times X_n$, where X_i is the set of values of the feature x_i. Let the set of states K consist of two states 1 and 2, i.e., $K = \{1, 2\}$. Let $p_{X|k}(x)$ be a conditional probability of observation x under the condition of the state k. The functions $p_{X|k}$, $k \in K$, are unknown but the marginal probabilities $p_{X_i|k}(x_i)$ are known and are expressed by the relation

$$\left.\begin{array}{l} p_{X_1|k}(x_1) = \displaystyle\sum_{x \in X(1,x_1)} p_{X|k}(x)\,, \\ \quad\vdots \\ p_{X_i|k}(x_i) = \displaystyle\sum_{x \in X(i,x_i)} p_{X|k}(x)\,, \\ \quad\vdots \\ p_{X_n|k}(x_n) = \displaystyle\sum_{x \in X(n,x_n)} p_{X|k}(x)\,, \end{array}\right\} k = 1, 2\,. \tag{3.12}$$

In the previous formula the notation $X(i, x_i)$ stands for the set $X_1 \times X_2 \times \ldots \times X_{i-1} \times \{x_i\} \times X_{i+1} \times \ldots \times X_n$, i.e., the set of those sequences $x = (x_1, x_2, \ldots, x_n)$ in which the i-th position is occupied by the fixed value x_i.

If it is known that the features under the condition of a fixed state k constitute an ensemble of independent random variables, it means, in fact, that also the function $p_{X|k}$ is known, because

$$p_{X|k}(x) = \prod_{i=1}^{n} p_{X_i|k}(x_i)\,, \quad k = 1, 2 \tag{3.13}$$

holds in this case.

Let us sincerely admit that such a model is quite simple. If it actually occurs then difficult questions can hardly arise. But I am interested in how I should create a recognition strategy, when I am not sure that the assumption (3.13) about the conditional independence of features is satisfied, and I know only the marginal probabilities $p_{X_i|k}$, $i = 1, \ldots, n$, $k = 1, 2$. In other words, how should I recognise the state k, when I only know that the functions $p_{X|k}$, $k = 1, 2$, satisfy the relations (3.12) with known left sides, and nothing else.

I am not amazed that the question is difficult for me because many other questions seem difficult to me too. On the other hand, I am surprised that no one has been attracted by this question so far. As soon as I ask someone how I should recognise an object, about which only marginal probabilities $p_{X_i|k}$ are known, I usually get a witty answer based on the implicit assumption that *the features are mutually independent. If I ask why the recognition strategy should be oriented just to the case of independent features, I obtain a less understandable explanation. This strategy is said to be suitable also for the case in which the features are mutually dependent.*

When I pursue the additional explanation, I am told that the entropy of observation was the greatest for independent features. It should mean that such observations can also appear at the input of the classifier, which would not occur, if there existed a dependence between the features. That is why the strategy, which is successful with a certain set of observations, cannot be less successful with the subset of these observations.

On the one hand, all these considerations seem to me admissible because they are in harmony with the informal behaviour of a human in a statistically uncertain situation. I consider as acceptable the behaviour in which a human in uncertain conditions makes preparations for the worst situation and behaves in such a way that losses might not be too high even in that case. When it proves that reality is better than the worst case, it is then even better. On the other hand, these considerations are based on implicit assumptions which seem self-evident, but in fact need not be always satisfied.

The main question is if for some set of statistical models (in our case these are the models which satisfy the relation (3.12)) there exists the exceptional worst model to which the recognition strategy should be tuned. At the same time, the strategy tuned to this exceptional model should also suit any other model. The recognition results should not be worse for any other model than those for the worst one. I know that such an exceptional model need not exist in every set. I can illustrate it by the following simple situation.

Let X be a two-dimensional set (a plane), $K = \{1, 2\}$, and $p_{X|1}(x)$ be two-dimensional Gaussian probability distribution with independent components, with variance 1 and mathematical expectation $\mu_1 = \mu_2 = 0$. Let $p_{X|2}(x)$ be unknown, but let it be known that it is only one of the two possible distributions: either $p'_{X|2}(x)$ or $p''_{X|2}(x)$ which differs from $p_{X|1}(x)$ only in the mathematical expectation. It is the point $\mu'_1 = 2$, $\mu'_2 = 0$ in the first case, and the point $\mu''_1 = 0$, $\mu''_2 = 4$ in the second case, see Fig. 3.1. Thus we have the set consisting of two statistical models. The first model is determined by the pair of functions $(p_{X|1}(x), p'_{X|2}(x))$ and the second by the pair $(p_{X|1}(x), p''_{X|2}(x))$.

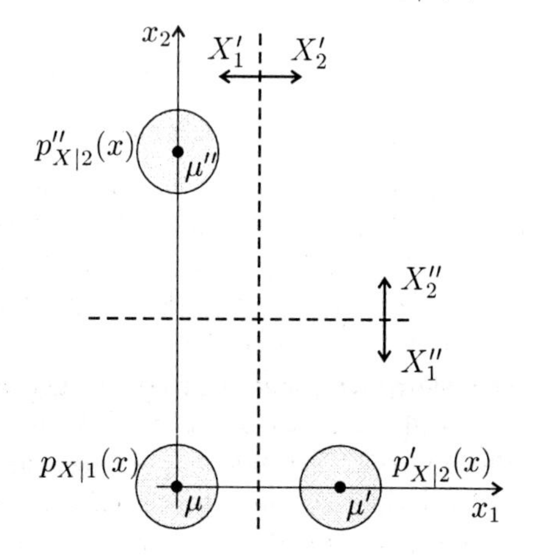

Figure 3.1 Difficulties in seeking for the 'worst statistical model' when attempting to decompose the plane into subsets.

The first model seems to me to be worse at the first glance. However, when the strategy tuned to this model will be used, i.e., the strategy that decomposes *the plane X into the subsets X_1' and X_2' in such a manner as is illustrated in Fig. 3.1, it is easy to find out that for this strategy the worst model will no longer be the first possibility $(p_{X|1}(x),\, p'_{X|2}(x))$, but the second possibility $(p_{X|1}(x),\, p''_{X|2}(x))$. It can be seen at the same time that when the plane X is decomposed into subsets X_1'' and X_2'', i.e., when the strategy is tuned for the second model $(p_{X|1}(x), p''_{X|2}(x\,|\,2))$, the first model $(p_{X|1}(x), p'_{X|2}(x))$ becomes the worst model. We see that the strategy q^*, which should cope simultaneously with both models, is not equal to any of two strategies that are tuned for each model separately. Thus the model, which I want to call the worst, does not exist in this group of models. I see that I should define the term 'the worst model' more precisely. Then I could ask more specifically if some specific set of the models contains the worst model.*

Let m be the statistical model that is determined by the pair of numbers $p_K(k)$, $k = 1,2$, which are the a priori probabilities of the state k, and the ensemble $p_{X|k}(x)$, $x \in X$, $k = 1,2$. Thus, $m = \big(p_K(k), p_{X|k}(x),\, k = 1,2;\, x \in X\big)$. Let M be the set of models. Let q denote the strategy of the form $X \to \{1,2\}$. Let Q denote the set of all possible strategies. Let $R(q,m)$ denote the risk obtained when the strategy q is used for the model m. For certainty, let us assume that $R(q,m)$ is the probability of the wrong estimate of the state k of the object. I would like to call a model $m^* \in M$ the worst model in the set M, if such a strategy q^* exists that:

1. *Any other strategy $q \in Q$ satisfies the inequality*

$$R(q,m^*) \geq R(q^*,m^*)\,, \quad q \in Q\,. \tag{3.14}$$

   This means that strategy q^* is the Bayesian strategy for the model m^*;
2. *When the strategy q^* is used with any other model $m \in M$ the risk is then reached which is not worse than that corresponding to the strategy q^* for the model m^*, i.e.,*

$$R(q^*,m^*) \geq R(q^*,m)\,, \quad m \in M\,. \tag{3.15}$$

   This means that the model m^* is the worst one for the strategy q^*.

Having introduced the concepts I can formulate questions that arise with respect to the set of models satisfying the equation (3.12). It concerns the set of models of the form $m = \big(p_K(1),\, p_K(2),\, p_{X|1}(x),\, p_{X|2}(x)\big)$, where the function $p_{X|1}$ satisfies the equation (3.12) for $k = 1$, and the function $p_{X|2}$ satisfies the equation (3.12) for $k = 2$.

Question 1: *Does a model m^* and a strategy q^* exist that satisfy conditions (3.14) and (3.15)?*

Question 2: *Assume a positive answer to Question 1. Does the model m^* include just those probability distributions $p_{X|1}(x)$ and $p_{X|2}(x)$ for which the equation (3.13) holds?*

I assume that the answer to Question 1 is related to the non-Bayesian statistical decision tasks with non-random interventions (see Subsection 2.2.5) which *require that the searched strategy q^* satisfies the inequality*

$$\max_{m \in M} R(q,m) \geq \max_{m \in M} R(q^*,m) \tag{3.16}$$

for any strategy q. This hypothesis comes from the property that if the pair (q^*, m^*) exists, which satisfies conditions (3.14) and (3.15), then q^* is the solution of the task (3.16). Indeed, the following equations and inequalities hold for the arbitrary strategy $q \in Q$

$$\max_{m \in M} R(q,m) \geq R(q,m^*) \geq R(q^*,m^*) = \max_{m \in M} R(q^*,m)$$

from which the inequality (3.16) follows. The first deduction step is self-evident. The second element of the deduction is the inequality (3.14) and the third one is just the inequality (3.15) in a different form.

I tried to solve the task (3.16) for a group of models of the form (3.12). I have found, as it seems to me, that there exists a model in this model group that satisfies the relations (3.14) and (3.15). I got to a positive answer to Question 1 on the basis of the following considerations.

Let $\mathcal{P}(1)$ be the set of functions $p_{X|1}$ which satisfy equation (3.12) for $k = 1$, and let $\mathcal{P}(2)$ be a similar set for $k = 2$. The solution of Linnik task, as I know from Lecture 2, is given by two functions $\alpha_1 \colon X \to \mathbb{R}$ and $\alpha_2 \colon X \to \mathbb{R}$ which solve this task by means of linear programming.

$$\left.\begin{array}{l}
\min c \\
c - \sum\limits_{x \in X} \alpha_1(x)\, p_{X|2}(x) \geq 0\,, \quad p_{X|2} \in \mathcal{P}(2)\,; \\
c - \sum\limits_{x \in X} \alpha_2(x)\, p_{X|1}(x) \geq 0\,, \quad p_{X|1} \in \mathcal{P}(1)\,; \\
\alpha_1(x) + \alpha_2(x) = 1\,, \quad x \in X\,; \\
\alpha_1(x) \geq 0\,, \quad \alpha_2(x) \geq 0\,, \quad x \in X\,.
\end{array}\right\} \tag{3.17}$$

This task has infinitely many constraints which are just $|\mathcal{P}(1)| + |\mathcal{P}(2)| + |X|$. I got rid of the unpleasant infinities like this: the set $\mathcal{P}(1)$ is the set of solutions of the system of linear equations (3.12). Having in mind that the solution $p_{X|1}$ of the system (3.12) has to satisfy the natural constraint $0 \leq p_{X|1}(x) \leq 1$ in any point $x \in X$, I come to conclusion that $\mathcal{P}(1)$ is a constrained convex set. As the solution of the finite system of linear equations is concerned, the set is a multi-dimensional polyhedron. A number of polyhedron vertices is quite large, but finite. I will denote the vertices by p_1^j, $j \in J(1)$, where $J(1)$ is a finite set of indices. It is obvious that when the inequality $c - \sum_{x \in X} \alpha_2(x)\, p_{X|1}(x) \geq 0$ holds for an arbitrary function $p_{X|1}$ from the set $\mathcal{P}(1)$ then the same inequality holds also for any function p_1^j, $j \in J(1)$, i.e.,

$$c - \sum_{x \in X} \alpha_2(x)\, p_1^j(x) \geq 0\,, \quad j \in J(1)\,, \tag{3.18}$$

because every vertex p_1^j, $j \in J(1)$, belongs to the set $\mathcal{P}(1)$. The opposite statement is correct too. This mean that from inequalities (3.18) also the *system of inequalities follows*

$$c - \sum_{x \in X} \alpha_2(x)\, p_{X|1}(x) \ge 0\,, \quad p_{X|1} \in \mathcal{P}(1)\,. \tag{3.19}$$

It is so because it is possible to express any function $p_{X|1}$ from the set $\mathcal{P}(1)$ as

$$p_{X|1} = \sum_{j \in J(1)} \gamma_j\, p_1^j\,,$$

where γ_j, $j \in J(1)$, are non-negative coefficients for which $\sum_{j \in J(1)} \gamma_j = 1$ holds. Thus, the conditions (3.18) and (3.19) are equivalent. The same is true for conditions

$$c - \sum_{x \in X} \alpha_1(x)\, p_{X|2}(x) \ge 0\,, \quad p_{X|2} \in \mathcal{P}(2)\,,$$

$$c - \sum_{x \in X} \alpha_1(x)\, p_2^j(x) \ge 0\,, \quad j \in J(2)\,,$$

where $\{p_2^j \mid j \in J(2)\}$ is the set of all vertices of the polyhedron $\mathcal{P}(2)$. The task (3.17) assumes the form

$$\left.\begin{array}{r|l}
 & \min c \\
\tau_{2j} & c - \sum\limits_{x \in X_1} \alpha_1(x)\, p_2^j(x) \ge 0\,, \quad j \in J(2)\,; \\
\tau_{1j} & c - \sum\limits_{x \in X_1} \alpha_2(x)\, p_1^j(x) \ge 0\,, \quad j \in J(1)\,; \\
t(x) & \alpha_1(x) + \alpha_2(x) = 1\,, \quad x \in X\,; \\
 & \alpha_1(x) \ge 0\,, \quad \alpha_2(x) \ge 0\,, \quad x \in X\,.
\end{array}\right\} \tag{3.20}$$

I remember from Lecture 2 that the task is solved by the strategy

$$\left.\begin{array}{ll}
 & \alpha_1^*(x) = 1\,,\ \alpha_2^*(x) = 0\,, \ \text{if} \ \sum\limits_{j \in J(1)} \tau_{1j}^*\, p_1^j(x) > \sum\limits_{j \in J(2)} \tau_{2j}^*\, p_2^j(x) \\
\text{or} & \alpha_1^*(x) = 0\,,\ \alpha_2^*(x) = 1\,, \ \text{if} \ \sum\limits_{j \in J(1)} \tau_{1j}^*\, p_1^j(x) < \sum\limits_{j \in J(2)} \tau_{2j}^*\, p_2^j(x)\,,
\end{array}\right\} \tag{3.21}$$

where $\tau_{1j}^*, j \in J(1)$, and $\tau_{2j}^*, j \in J(2)$, are Lagrange coefficients that solve the dual task

$$\left.\begin{array}{r|l}
 & \max \sum\limits_{x \in X} t(x)\,, \\
\alpha_1(x) & t(x) - \sum\limits_{j \in J(2)} \tau_{2j}\, p_2^j(x) \le 0\,, \quad x \in X\,; \\
\alpha_2(x) & t(x) - \sum\limits_{j \in J(1)} \tau_{1j}\, p_1^j(x) \le 0\,, \quad x \in X\,; \\
c & \sum\limits_{j \in J(1)} \tau_{1j} + \sum\limits_{j \in J(2)} \tau_{2j} = 1\,, \\
 & \tau_{1j} \ge 0\,, \quad j \in J(1)\,; \quad \tau_{2j} \ge 0\,, \quad j \in J(2)\,.
\end{array}\right\} \tag{3.22}$$

According to the form of the strategy (3.21), which is denoted by q^*, it is obvious that the strategy minimises the probability of the wrong decision for the statistical model, where the a priori probabilities of the states $k = 1, 2$ are

$$p_K^*(1) = \sum_{j \in J(1)} \tau_{1j}^* , \quad p_K^*(2) = \sum_{j \in J(2)} \tau_{2j}^* , \tag{3.23}$$

and for which the conditional probabilities of observed state x under the condition of states $k = 1, 2$ are

$$p_{X|1}^*(x) = \sum_{j \in J(1)} \frac{\tau_{1j}^*}{\sum_{i \in J(1)} \tau_{1i}^*} \, p_1^j(x) , \quad x \in X , \tag{3.24}$$

$$p_{X|2}^*(x) = \sum_{j \in J(2)} \frac{\tau_{2j}^*}{\sum_{i \in J(2)} \tau_{2i}^*} \, p_2^j(x) , \quad x \in X . \tag{3.25}$$

The statistical model (3.23), (3.24), (3.25) will be denoted by m^*. It is obvious that this model satisfies the condition (3.12) because both functions $p_{X|1}^*$ and $p_{X|2}^*$ represent the convex combination of functions p_1^j, $j \in J(1)$, and p_2^j, $j \in J(2)$, satisfying the condition (3.12). The strategy q^* is the solution of Linnik task that is formulated as the minimisation of the function $\max_{m \in M} R(q, m)$. The task is expressed by (3.17). I can write

$$\min c = \max_{m \in M} R(q^*, m) .$$

The coefficients τ_{1j}^*, $j \in J(1)$, and τ_{2j}^*, $j \in J(2)$, are the solution of the dual task (3.22) and thus

$$\max \sum_{x \in X} t(x) = \sum_{x \in X} \min \left(\sum_{j \in J(1)} \tau_{1j}^* \, p_1^j(x), \sum_{j \in J(2)} \tau_{2j}^* \, p_2^j(x) \right) , \tag{3.26}$$

where the expression on the right-hand side of (3.26) denotes the risk $R(q^*, m^*)$. By the first duality Theorem 2.1, I have $\min c = \max \sum_{x \in X} t(x)$, and consequently

$$R(q^*, m^*) = \max_{m \in M} R(q^*, m) . \tag{3.27}$$

I have proved that the set of models satisfying conditions (3.12) also comprises the worst model m^*, for which the following holds

$$R(q, m^*) \geq R(q^*, m^*) \geq R(q^*, m) , \quad q \in Q , \quad m \in M. \tag{3.28}$$

The first inequality in the expression (3.28) is correct because q^* is the Bayesian strategy for the model m^*. The second inequality is only the equation (3.27) rewritten in a different manner.

The answer to my Question 1 is therefore positive. Now an additional question can be formulated correctly, too. This question asks: What is the worst model from the ensemble of models that satisfy (3.12)? How must the strategy be chosen for the worst model? What should the recognition look like in a case in which only marginal probabilities on the left-hand side of (3.12) are known and nothing else?

I am helpless in an attempt to answer the abovementioned questions. When once I became convinced in such a complicated way that the questions were correct, I cannot help but get an impression that there are answers to these questions.

We are pleasantly surprised by the enormous work you have done. It seems that we should hardly distinguish who teaches whom here. It took a while until we found the correct answer to your question. It was worth the effort because the answer is entirely unexpected. We were quite pleased by your solution of the problems as well as that you have found a virgin field which seems to have been investigated in a criss-cross manner.

The worst model m^* the existence of which you proved in such an excellent way, cannot be described so transparently. The most interesting issue in your question is that the Bayesian strategy q^* can be found for the worst model m^* without the need to find the worst model. It is possible because it can be proved that the strategy q^* makes the decision about the state k only on the basis of a single feature and it must ignore all others.

It was not easy to prove this property well, and our explanation may not have been easily comprehensible for you. So we will discuss first, using simple examples, what this property means. Assume you have two features x_1 and x_2. You are to find out which of these two features lead to smaller probability of the wrong decision. You will make decision about the state k only on the basis of the better feature and you will not use the value of the other feature. Let us have a more detailed look at this situation and let us try to understand why we have to deal with it in just this way.

Say that the feature x_1 assumes two values 0, 1 only and its dependence on the state $k \in \{0, 1\}$ is determined by four conditional probabilities:

$$p(x_1 = 1 \,|\, k = 1) = 0.75\,, \quad p(x_1 = 0 \,|\, k = 1) = 0.25\,,$$
$$p(x_1 = 1 \,|\, k = 0) = 0.25\,, \quad p(x_1 = 0 \,|\, k = 0) = 0.75\,.$$

It is evident that you can create the strategy q^*, based on this feature, which will estimate the state k with the probability of the wrong decision 0.25. It is the strategy which chooses either the state $k = 1$ when $x_1 = 1$ or the state $k = 0$, when $x_1 = 0$. The probability of the wrong decision does not apparently rely on *a priori* probabilities of states. Be they of any kind, the wrong decision probability will be the same, that is 0.25.

Let us assume that this probability seems to you too large and that is why you would like to lower it by using one more feature. Let us assume that you have such a feature at your disposal. In our simplified example let it be the feature x_2 which also assumes two values only. Conditional probabilities corresponding to these values under the constraint of the fixed state k are

$$p(x_2 = 1 \,|\, k = 1) = 0.7\,, \quad p(x_2 = 0 \,|\, k = 1) = 0.3\,,$$
$$p(x_2 = 1 \,|\, k = 0) = 0.3\,, \quad p(x_2 = 0 \,|\, k = 0) = 0.7\,.$$

$k = 1$			
$p(k = 1) =?$			
	x_1	0	1
		$p(x_1 \mid k = 1)$	
x_2	$p(x_2 \mid k = 1)$	0.25	0.75
0	0.3	?	?
1	0.7	?	?

$k = 0$			
$p(k = 0) =?$			
	x_1	0	1
		$p(x_1 \mid k = 0)$	
x_2	$p(x_2 \mid k = 0)$	0.75	0.25
0	0.7	?	?
1	0.3	?	?

Table 3.1 The data of the example determining the probability of the wrong decision. The values in six table entries denoted by the question mark correspond to the unknown parameters of the statistical model.

All data having a relation to our example are given in a comprehensive way in Table 3.1. Furthermore the known values presented in the table, a space is reserved for unknown data. These are the *a priori* probabilities $p(k = 1)$ and $p(k = 0)$ and joint conditional probabilities $p(x_1, x_2 \mid k)$. There are question marks in the entries corresponding to unknown values in Table 3.1. The question marks can be replaced by arbitrary numbers that must only satisfy an obvious condition

$$p(k = 1) + p(k = 0) = 1$$

and also the condition (3.12) on marginal probabilities. This means that the sum of probabilities $p(x_1 = 1, x_2 = 1 \mid k = 1)$ and $p(x_1 = 1, x_2 = 0 \mid k = 1)$ has to be 0.75, etc.. The alternative, by which question marks are substituted by the numbers in Table 3.1, influences in the general case the probability of the wrong decision reached by means of a strategy. It will not be difficult for you to become convinced that the formerly introduced strategy q^*, which decides about the state only considering the feature x_1, secures the probability 0.25 of the wrong decision for an arbitrary substitution of the question marks by actual numbers.

Let us now have a look at whether the probability of the wrong decision can be lowered when both features x_1 and x_2 are used instead of a single feature x_1. It could be natural to use one of the two following strategies for this purpose. The first strategy decides for the state $k = 1$ if and only if $x_1 = 1$ and $x_2 = 1$. The second strategy selects $k = 0$ if and only if $x_1 = 0$ and $x_2 = 0$.

Let us analyse the first strategy first. Here the probability of the wrong decision, unlike that in the case of the strategy q^*, is dependent on which numbers substitute the question marks in Table 3.1. The numbers can be such that the probability of the wrong decision will be 0.55. Thus it will be worse than it would be if only the worse feature x_2 was used. In this case it would be 0.3. These numbers are displayed in Table 3.2(a) which shows a value only for $k = 1$. It is obvious that with $p(k = 0) = 0$ there is no longer influence whatever the probabilities $p(x_1, x_2 \mid k = 0)$ are.

When applying the second strategy, such numbers can substitute the question marks in Table 3.1 that the probability of the wrong decision will again be 0.55. These values are shown in Table 3.2(b).

	$k = 1$		
	$p(k = 1) = 1$		
	x_1	0	1
		$p(x_1 \mid k = 1)$	
x_2	$p(x_2 \mid k = 1)$	0.25	0.75
0	0.3	0	0.3
1	0.7	0.25	0.45

(a)

	$k = 0$		
	$p(k = 0) = 1$		
	x_1	0	1
		$p(x_1 \mid k = 0)$	
x_2	$p(x_2 \mid k = 0)$	0.75	0.25
0	0.7	0.45	0.25
1	0.3	0.3	0

(b)

Table 3.2 The decision with two features. Two most unfavourable cases with the probability of the wrong decision 0.55 are depicted which correspond to two different strategies.

If you liked, you could make sure that even any other strategy will not be better than the earlier presented strategy q^*. It is the strategy q^* which decides about the state k wrongly in 25% of cases for an arbitrary substitution of the question marks by actual numbers. For any other strategy there exists such a substitution of the question marks by numbers for which the probability of the wrong decision will be greater than 25%. We do not think that it would be difficult for you to try the remaining possibilities. There are only 16 possible strategies in our case and 3 of them have been already analysed. In spite of that, we think that you will not like trying it because this is only a simple example. We would like to study together with you the properties of the strategy found in the general case. We mean the case in which marginal probabilities are arbitrary, and the number of features and the number of values of each feature need not be just two but it can be arbitrary.

But before we start analyzing your task let us try to get used to a paradoxical fact, on which we will now concentrate. The fact is that the usage of a greater number of features cannot prove better, and it is usually worse compared to the use of only one single feature. When we manage to prove it (and we shall do so) then we will obtain a quite important constraint for procedures, of how to distinguish the object state on the basis of information available from various sources. Let us leave these quite serious problems aside in the meanwhile and let us deal with a less important but more instructive example, which will lead us to a clear idea, how to solve your task formulated by Question 2.

Imagine that you are a company director or a department head or someone else who has to make the decision 'yes' or 'no'. You will establish a board of advisors consisting of ten people for such a case. You will submit the question to be decided to the advisors and you will get ten answers $x_1, x_2, \ldots, x_{10}$ 'yes' or 'no'. After a certain time of your cooperation with the board of advisors, you will learn the quality of each expert which will be expressed with the help of probabilities $p_i(x_i|k)$, where k is the correct answer. You are faced with the question in the 'yes' or 'no' manner on the basis of ten answers of the advisors $x_1, x_2, \ldots, x_{10}$. This question would be easy if you were convinced that the experts from your advisory board are mutually independent. But you do know that such independence is not possible due to the complicated personal interrelations among the experts. But you do not know what the dependence is like.

The recommendation that follows from the analysis we performed according to your wish, says that you have to take into account only an opinion of a single expert in this case, namely that one who is the best in your consulting board. You can listen to the opinions of all other experts politely but you must ignore them or dissolve the advisory board. This recommendation is valid also in the case in which among the experts there are such that are not worse than the best one.

These conclusions seem to be too paradoxical. Therefore let us examine what consideration these results are owed to. For simplicity let us study the first situation in which the number features is 2 and then let us generalise the results obtained also for the case of the arbitrary number of features.

Let x and y be two features that assume values from the finite sets X and Y. Let the number of states be 2 that is k is either 1 or 2. The probability distribution $p_{XY|k}(x,y)$ is not known but the following probability distributions are known,

$$p_{X\,|\,k}(x) = \sum_{y \in Y} p_{XY\,|\,k}(x,y)\,, \quad x \in X\,, \quad k = 1,2\,, \tag{3.29}$$

$$p_{Y\,|\,k}(y) = \sum_{x \in X} p_{XY\,|\,k}(x,y)\,, \quad y \in Y\,, \quad k = 1,2\,. \tag{3.30}$$

Let M denote the set of statistical models of the form $m = \big(p_K(1),\ p_K(2),\ p_{XY|1},\ p_{XY|2}\big)$, where $p_{XY\,|\,1}$ and $p_{XY|2}$ satisfy the conditions (3.29) and (3.30). You have already proved that there is such a strategy $q^*\colon X \times Y \to \{1,2\}$ and such a model $m^* \in M$ for which there holds

$$R(q,m^*) \geq R(q^*,m^*) \geq R(q^*,m)\,, \quad q \in Q\,, \quad m \in M\,. \tag{3.31}$$

Let us analyse the properties of the strategy q^* that satisfies (3.31).

Let us denote as $XY^*(1)$ and $XY^*(2)$ the sets into which the strategy q^* divides the set $X \times Y$. Let the symbols $XY^+(1)$ and $XY^+(2)$ denote two such subsets of the sets $XY^*(1)$ and $XY^*(2)$ that $XY^+(1)$ contains just all points (x,y) for which there holds

$$XY^+(1) = \Big\{(x,y) \in X \times Y \;\Big|\; q^*(x,y) = 1\,,\ p^*_{XY|1}(x,y) > 0\Big\}\,,$$

and similarly

$$XY^+(2) = \Big\{(x,y) \in X \times Y \;\Big|\; q^*(x,y) = 2\,,\ p^*_{XY|2}(x,y) > 0\Big\}\,.$$

The subset $Z \subset X \times Y$ will be called a *rectangle* in the set $X \times Y$, when there are subsets $X' \subset X$ and $Y' \subset Y$ such that $Z = X' \times Y'$. The smallest rectangle containing the subset $Z \subset X \times Y$ will be called a *Cartesian hull* of the subset Z and will be denoted Z^c.

The sets $XY^+(1)$ and $XY^+(2)$ are fully determined by the pair (q^*,m^*). Let us prove an important property of these sets. If (q^*,m^*) satisfies the condition (3.31) then the Cartesian hulls of the set $XY^+(1)$ and $XY^+(2)$ do not intersect. This means that

$$\left(XY^{+}(1)\right)^{c} \cap \left(XY^{+}(2)\right)^{c} = \emptyset \, . \tag{3.32}$$

Assume that the equation (3.32) does not hold. Then a point (x^*, y^*) must exist which belongs both to $\left(XY^{+}(1)\right)^{c}$ and to $\left(XY^{+}(2)\right)^{c}$. The value $q^*(x^*, y^*)$ in that point is either 1 or 2. Let us choose $q^*(x^*, y^*) = 1$ for certainty. If $(x^*, y^*) \in \left(XY^{+}(2)\right)^{c}$ then there are two points (x^*, y) and (x, y^*) for which there holds

$$q(x^*, y) = 2 \, , \quad p^*_{XY|2}(x^*, y) > 0 \, , \quad q(x, y^*) = 2 \, , \quad p^*_{XY|2}(x, y^*) > 0 \, .$$

We will choose a positive quantity Δ which is not larger than $p^*_{XY|2}(x^*, y)$ and is not larger than $p^*_{XY|2}(x, y^*)$. It can be, for instance, the value

$$\Delta = \min \left(p^*_{XY|2}(x^*, y) \, , \; p^*_{XY|2}(x, y^*) \right) \, .$$

We create a new model $m = \left(p^*_K(1), \, p^*_K(2), \, p^*_{XY|1}, \, p_{XY|2}\right)$ in which only the function $p_{XY|2}$ changed when compared with the function $p^*_{XY|2}$ in the model m^*. Furthermore the function $p_{XY|2}$ differs from the function $p^*_{XY|2}$ only in four points (x^*, y^*), (x, y^*), (x^*, y) and (x, y) according to the following equations:

$$\left. \begin{aligned} p_{XY|2}(x, y^*) &= p^*_{XY|2}(x, y^*) - \Delta \, , \\ p_{XY|2}(x^*, y) &= p^*_{XY|2}(x^*, y) - \Delta \, , \\ p_{XY|2}(x^*, y^*) &= p^*_{XY|2}(x^*, y^*) + \Delta \, , \\ p_{XY|2}(x, y) &= p^*_{XY|2}(x, y) + \Delta \, . \end{aligned} \right\} \tag{3.33}$$

During such a transfer from m^* to the model m, the probability of the wrong decision when the actual state is 2 increases minimally by Δ. Indeed, the strategy q^* assigns the point (x^*, y^*) into the first class and the points (x, y^*) and (x^*, y) into the second class. The probability of just the point (x^*, y^*) in the second state increases by Δ and probabilities of both points (x, y^*) by Δ. Consequently and independently of the assignment of the point (x, y) the total probability of the points which actually belong to the second state and are assigned by the strategy q^* into the second state decreases minimally by Δ.

It is also obvious that when the function $p^*_{XY|2}$ satisfied the conditions (3.29) and (3.30) then also the new function $p_{XY|2}$ obtained by modifying the condition(3.33) satisfies these conditions too. It is proved by this that if (3.32) is not satisfied then there is the model $m \in M$ for which there holds $R(q^*, m) > R(q^*, m^*)$. This contradicts the assumption (3.31). That is why (3.32) follows from the assumption (3.31).

Based on the fact that the rectangles $\left(XY^{+}(1)\right)^{c}$ and $\left(XY^{+}(2)\right)^{c}$ do not intersect, the following assertion holds.

Assertion 3.1 *One of the following two possibilities holds at least:*

1. *Such a decomposition of the set X into two subsets $X(1)$ and $X(2)$ exists that*

$$\left(XY^+(1)\right)^c \subset X(1) \times Y\,, \quad \left(XY^+(2)\right)^c \subset X(2) \times Y\,. \tag{3.34}$$

2. *Such a decomposition of the set Y into two subsets $Y(1)$ and $Y(2)$ exists that*

$$\left(XY^+(1)\right)^c \subset X \times Y(1)\,, \quad \left(XY^+(2)\right)^c \subset X \times Y(2)\,. \tag{3.35}$$

▲

The proof of that at least one of the two assertions mentioned is valid is clear on the basis of purely geometrical considerations. For two non-intersecting rectangles with vertical and horizontal edges lying in a single plane, a vertical or horizontal line exists that separates the plane into two parts containing just one rectangle each. Let us remark that we are working here with generalised rectangles. Nevertheless, the generalisation of the given principle is easy and leads to the formulation of the previous assertion.

Assume for certainty that, for instance, the first part of Assertion 3.1 holds. Let us denote by q' the strategy which decomposes the set $X \times Y$ into classes $X(1) \times Y$ and $X(2) \times Y$. We can see that the strategy q' does not depend on the feature y. That is why the risk $R(q', m)$ does depend on the model m either, i.e.,

$$R(q', m^*) = R(q', m)\,, \quad m \in M\,. \tag{3.36}$$

Let us prove now that when the strategy q^* for the model m^* is a Bayesian strategy then also q' is the Bayesian strategy for m^*. Let it hold for some point (x, y) that $q'(x, y) \neq q^*(x, y)$. If such a point does not exist then the strategies q' and q^* are equal, and therefore also the strategy q' is Bayesian. If $q'(x, y) \neq q^*(x, y)$ then there holds

$$\text{either} \quad q'(x,y) = 1\,, \quad q^*(x,y) = 2\,, \qquad \text{or} \quad q'(x,y) = 2\,, \quad q^*(x,y) = 1\,.$$

If $q'(x, y) = 1$ and $q^*(x, y) = 2$ hold then

$$\begin{aligned}\left(q'(x,y) = 1\right) &\Rightarrow \left((x,y) \in X(1) \times Y\right) \Rightarrow \left((x,y) \notin XY^+(2)\right)^c \\ &\Rightarrow \left((x,y) \notin XY^+(2)\right) \Rightarrow \left(p^*_{XY|2}(x,y) = 0\right).\end{aligned}$$

It follows from the result of the derivation chain mentioned that the probability $p^*_{XY|1}(x, y)$ must equal to zero too. If the converse were true then the strategy q^*, which assigns (x, y) into the second class, would not be Bayesian. It can be proved in a similar way that for all points (x, y), for which $q'(x, y) = 2$, $q^*(x, y) = 1$ hold, it also holds that $p^*_{XY|1}(x, y) = p^*_{XY|2}(x, y) = 0$. The created strategy q', which depends on the feature x, differs from the strategy q^* only on the set whose probability in the model m^* is equal to zero. From the fact that q^* is the Bayesian strategy for m^* it therefore follows that q' is also the Bayesian strategy for m^*. In addition to the relation (3.36) obtained earlier we derived that for the strategy q' there holds

$$R(q, m^*) \geq R(q', m^*) \geq R(q', m)\,, \quad q \in Q\,, \quad m \in M\,.$$

We proved that in the case of two features the strategy sought depends on one of them only. We will explore a more general case now when the number of the features is arbitrary. The previous considerations are also valid for the general case except for the relation (3.32) which states that Cartesian hulls of certain sets do not intersect. We have to prove this property for a multi-dimensional case.

Let $x_1, x_2, \ldots, x_n$ be n features that assume values from the sets X_1, X_2, $\ldots$, X_n. Let x be an ensemble $(x_1, x_2, \ldots, x_n)$, $X = X_1 \times X_2 \times \ldots \times X_n$, m^* be the statistical model $\bigl(p^*_K(1), p^*_K(2), p^*_{X|1}, p^*_{X|2}\bigr)$, where the function $p^*_{X|1}$ satisfies the system of equations

$$p_{X_i|1}(x_i) = \sum_{x \in X(i,x_i)} p^*_{X|1}(x)\,, \quad x_i \in X_i\,, \quad i = 1, \ldots, n$$

and the function $p^*_{X|2}$ satisfies the system of equations

$$p_{X_i|2}(x_i) = \sum_{x \in X(i,x_i)} p^*_{X|2}(x)\,, \quad x_i \in X_i\,, \quad i = 1, \ldots, n\,.$$

Marginal probabilities are fixed in the left-hand sides of the abovementioned systems of equations. Let $q^*\colon X \to \{1,2\}$ be the Bayesian strategy for the model m^* for which there holds

$$R(q^*, m^*) \geq R(q^*, m)\,, \quad m \in M\,. \tag{3.37}$$

We will define the sets $X^+(1)$ and $X^+(2)$ as

$$X^+(1) = \Bigl\{ x \in X \mid q^*(x) = 1\,,\ p^*_{X|1}(x \,|\, 1) > 0 \Bigr\}\,,$$
$$X^+(2) = \Bigl\{ x \in X \mid q^*(x) = 2\,,\ p^*_{X|2}(x \,|\, 2) > 0 \Bigr\}\,,$$

and the sets $\bigl(X^+(1)\bigr)^c$ and $\bigl(X^+(2)\bigr)^c$ as Cartesian hulls of the sets $X^+(1)$ and $X^+(2)$. We will prove that from inequalities (3.37) there follows

$$\bigl(X^+(1)\bigr)^c \cap \bigl(X^+(2)\bigr)^c = \emptyset\,.$$

Assume that the previous expression does not hold, i.e., there is a point x^* which belongs both to $\bigl(X^+(1)\bigr)^c$ and to $\bigl(X^+(2)\bigr)^c$. Assume for uniqueness that $q^*(x^*) = 1$. Let the features $x_1, x_2, \ldots, x_n$ assume the value 0 at the point x^*. The point x^* is thus an ensemble of n zeros, i.e., $(0, 0, 0, \ldots, 0)$. If x^* belongs to $\bigl(X^+(2)\bigr)^c$, then such a set S of points $x^1, x^2, \ldots, x^t$, $t \leq n$, exists that for each $i = 1, 2, \ldots, n$ such a point x' exists in the set S that $x'_i = 0$. Furthermore, each point $x' \in S$ belongs to $X^+(2)$, i.e., for each of them $q^*(x') = 2$, $p^*_{X|2}(x') > 0$ holds.

We will show that in this case probabilities $p_{X|2}(x)$ can be decreased in the points $x^1, x^2, \ldots, x^t$ and probabilities can be increased in other points, including the point x^*, all that without changing marginal probabilities $p_{X_i|2}(x_i)$. In such case only probabilities of the points $x^1, x^2, \ldots \ldots, x^t$ are decreased, which are assigned by the strategy q^* to the second class. There is one point at least (it is the point x^*) which is assigned by the strategy q^* to the first class and the probability of which increases. As a consequence of a change of probabilities in deliberately selected points, the probability of the wrong decision about the object in the second state increases. As this contradicts to the requirement (3.37), it proves the relation (3.37) too.

We will prove that the possibility of expected change of probability follows from the inequality $\left(X^+(1)\right)^c \cap \left(X^+(2)\right)^c \neq \emptyset$. The proof will be based on two assertions.

Assertion 3.2 *If the set S contains two points x^1 and x^2 only then the model m exists for which the inequality $R(q^*, m^*) < R(q^*, m)$, $m \in M$, holds.* ▲

Proof. Select some point x'. Its coordinates x'_i, $i = 1, \dots, n$, are determined by the following rule. If $x^1_i = 0$ then $x'_i = x^2_i$. If $x^2_i = 0$ then $x'_i = x^1_i$. In other words, the i-th coordinate of the point x' is equal to the non-zero coordinate which is either x^1_i or x^2_i. If both these coordinates are equal to zero then the i-th coordinate of the point x' is equal to zero too. For the point x' determined in this way and for points x^1, x^2 and x^* it holds: How many times a certain value of the i-th coordinate occurs in a pair of points x^1 and x^2, exactly that many times this value occurs in the pair of points x' and x^*. Let us remind that all coordinates of the point x^* are zeros.

Example 3.2 *Let $x^1 = (0, 0, 0, 5, 6, 3)$ and $x^2 = (5, -2, 0, 0, 0, 0)$. As was said earlier, the point $x^* = (0, 0, 0, 0, 0, 0)$. In this case $x' = (5, -2, 0, 5, 6, 3)$.* ▲

Let $\Delta = \min\left(p^*_{X|2}(x^1), p^*_{X|2}(x^2)\right)$. Let the probability distribution $p_{X|2}$ be given by the equations:

$$\left.\begin{array}{rcl} p_{X|2}(x_1) &=& p^*_{X|2}(x^1) - \Delta\,; \\ p_{X|2}(x_2) &=& p^*_{X|2}(x^2) - \Delta\,; \\ p_{X|2}(x^*) &=& p^*_{X|2}(x^*) + \Delta\,; \\ p_{X|2}(x') &=& p^*_{X|2}(x') + \Delta\,; \\ p_{X|2}(x) &=& p^*_{X|2}(x)\,, \quad x \notin \{x^1, x^2, x^*, x'\}\,. \end{array}\right\} \tag{3.38}$$

If the probability $p^*_{X|2}$ changes to $p_{X|2}$ in this way then the marginal probability $p_{X_i|2}$ does not change. The reason is that as many summands in the right-hand side of (3.12) increased by Δ, as many summands decreased by Δ. At the same time the probability of the wrong decision about the object in the second state increases minimally by Δ. Indeed, the strategy q^* decides like this: $q^*(x^1) = 2$, $q^*(x^2) = 2$, $q^*(x^*) = 1$. If $q^*(x') = 1$ then the probability of the wrong decision increases by 2Δ. When $q^*(x') = 2$ then the probability of the wrong decision increases by Δ. ■

Assertion 3.3 *Let $S = \{x^1, x^2, \dots, x^t\}$, $t > 2$. In this case either such a model m exists for which $R(q^*, m^*) < R(q*, m)$ holds, or a set of points S' exists the number of points of which is smaller than the number of points in the set S, and its Cartesian hull contains the point x^* too.* ▲

Proof. We will denote by I the set of indices $\{1, 2, \dots, n\}$. For each point x we will denote by $I(x)$ the subset of those indices which correspond to the zero coordinate in the point x. Therefore $I(x^*) = \bigcup_{x\in S} I(x) = I$. The set S apparently contains two points. Let us denote them as x^1 and x^2 for which $I(x^1) \neq I(x^2)$. Let us create two new points x', x'', according to the points x^1, x^2, so that $I(x') = I(x^1) \cup I(x^2)$. This means that the i-th coordinate of the point x' is equal to zero if this coordinate is zero in one of two points x^1 or x^2 at least. All other coordinates will be determined in such a way that the same property holds for the quadruplet of points (x^1, x^2, x', x'') which we mentioned in the proof of Assertion 3.2. It is also valid here that the number of times a certain value of the i-th coordinate appears in the pair of points x^1 and x^2, it appears the same number of times in the pair of points x' and x''. It is easy to show that it is possible to find the pair of points with such properties.

Example 3.3

$$x^1 = (0, 0, 0, 0, 1, 2, 3, 4, 5)\,,$$
$$x^2 = (5, 4, 0, 0, 0, 0, 3, 2, 1)\,.$$

The pair of points x' and x'' can have, for example, the form

$$x' = (0, 0, 0, 0, 0, 0, 3, 4, 1)\,,$$
$$x'' = (5, 4, 0, 0, 1, 2, 3, 2, 5)\,.$$

▲

Let us denote the variable $\Delta = \min\left(p_{X|2}(x^1), p_{X|2}(x^2)\right)$ which is positive because $x^1 \in X^+(2)$ and $x^2 \in X^+(2)$. The new model is determined exactly according to the relation (3.38), where x^* is replaced by x''. Let the strategy q^* assign one of the points x', x'' to the first class at least. The probability of the wrong decision about the object in the second state will increase owing to the change of the model and thus the inequality $R(q^*, m^*) < R(q^*, m)$ will be satisfied. If $q^*(x') = q^*(x'') = 2$ then the point x' will belong to the set $X^+(2)$ in the new model because its probability is already positive. As $I(x') = I(x^1) \cup I(x^2)$ holds, the inequality $\cup_{x\in S'} I(x) = I$ holds too for the set $S' = \{x', x^3, \ldots, x^k\}$ which has one point less than the set S. The set S' is obtained so that the points x^1, x^2 are excluded from and the point x' is included into the set S. ■

And so we have proved in the more general case that the inequality $R(q^*, m^*) \geq R(q^*, m)$ does not hold for all models m when the Cartesian hulls of classes intersect. It follows from this property that the strategy q^*, for which the corresponding Cartesian hulls intersect, is not the strategy that we look for. But because you have already proved that the strategy sought really exists, this can be only a strategy whose Cartesian hulls do not intersect. All other considerations are the same as in the case in which the number of the features is 2. Therefore we can say that we have *viribus unitis* managed your task.

Well, strictly speaking we have not finished yet because it is not quite clear to me which of the considerations mentioned can be generalised easily for the case in which the number of states is larger than 2.

You have certainly noticed that your proof of the existence of the worst model can be generalised almost without any change also for the case of an arbitrary number of states. Furthermore, and it is not too difficult to prove it, if m^* is the worst model and q^* is the Bayesian strategy for this worst model then the classes, in which the strategy decomposes the set of observations $X = X_1 \times X_2 \times \ldots \times X_n$, will be rectangles again. It is similar to the case of two states.

But it does not follow from the abovesaid that this Bayesian strategy depends only on one feature. In this case the strategy depends on not more than $|K|-1$ features, where $|K|$ is the number of states. For example, if the number of states is 3 then it is decided, based on one single selected feature, whether it is a certain selected class, say the class k_1. If not then it is decided, based on another feature (and possibly the same one), whether the class is k_2 or k_3. We will not continue explaining these things to you. Perhaps, not because they are not interesting but rather because everyone would prefer to sing some nice songs to listening how his songs are sung by somebody else. There are a lot of such songs in the investigated area. We like to praise you and thank you for discovering this area.

Please, notice as well that even if it is sometimes difficult to formulate Linnik task of evaluating complex hypotheses (see Subsections 2.4.6 and 2.4.7), their solutions are surprisingly simple. It may not be by chance. Well, Linniks' tasks occur as soon the knowledge about the statistical model of the recognised object is uncertain or incomplete. That is why the strategy must not be too sensitive to the statistical model used in pattern recognition. The robustness is secured just by simple strategies.

Examine one more task and you will be surely convinced how simple the exact solution of the statistical task under the condition of incomplete knowledge of the statistical model of the object can be.

Assume X is a two-dimensional space (a plane), $K = \{1, 2\}$, and the random variable x to be, under the condition that $k = 1$, a two-dimensional Gaussian random variable with statistically independent components with unit variance. The conditional mathematical expectation μ_1 of the random variable x under the condition $k = 1$ is not known. It is only known that it is one of the vectors in the convex closed set M_1, as shown in Fig. 3.2.

Under the condition $k = 2$, x is the same random variable, but with different mathematical expectation which is also unknown. It is known only that it is one of the vectors from the set M_2 which is closed, convex and does not intersect with M_1, see Fig. 3.2.

You may have seen two proposals in literature how to recognise a given x under such uncertain conditions.

1. The *nearest neighbour classifier* calculates for each $x \in X$ the values

$$d_1 = \min_{\mu \in M_1} r(x, \mu) \quad \text{and} \quad d_2 = \min_{\mu \in M_2} r(x, \mu) ,$$

   where $r(x, \mu)$ is the Euclidean distance of points x and μ. Then, x is assigned into the first or second class, if $d_1 \leq d_2$ or $d_1 > d_2$, respectively.
2. The *classification according to the integral of the probability* is based on the assumption that μ_1 and μ_2 are random variables with the uniform proba-

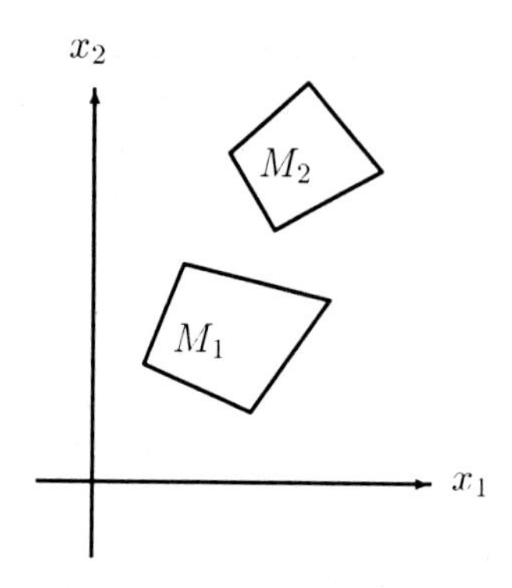

Figure 3.2 Two convex sets corresponding to $K = \{1, 2\}$.

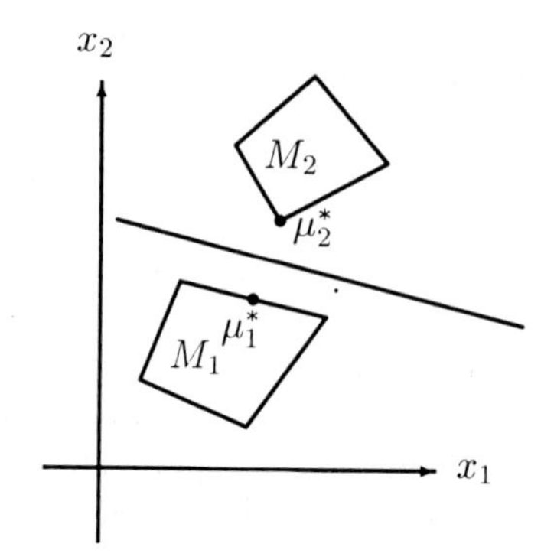

Figure 3.3 The decision as a separation of a plane by a straight line.

bility distribution on the sets M_1 and M_2. Two quantities are calculated

$$s_1 = \int_{M_1} f(x, \mu) \, \mathrm{d}\mu \text{ and } s_2 = \int_{M_2} f(x, \mu) \, \mathrm{d}\mu \,,$$

where $f(x, \mu)$ is the probability density of the Gaussian variable in the point x, the mathematical expectation being μ. The state k is then assessed according to the ratio s_1/s_2.

Your task is to find the strategy which solves Linnik task correctly in the given case.

It is needed to find the pair

$$(\mu_1^*, \mu_2^*) = \underset{(\mu_1, \mu_2) \in M_1 \times M_2}{\operatorname{argmin}} \; r(\mu_1, \mu_2) \,.$$

The next step is to determine the Bayesian strategy for the model m^*, in which the probabilities $p_K^*(1)$ and $p_K^*(2)$ are the same, and $p_{X|1}^*(x) = f(x, \mu_1^*)$ and $p_{X|2}^*(x) = f(x, \mu_2^*)$. This strategy decomposes the plane X into two classes by means of a straight line according to Fig. 3.3. I will denote this strategy as q^* and the probability of the error, which the strategy q^* assumes on the model m^*, will be denoted as ε^*. In view of q^* being the Bayesian strategy for m^*, $R(q, m^*) \geq \varepsilon^*$ holds for any strategy q. Furthermore, it is obvious from Fig. 3.3 that for any model $m = \big(p_K(1), p_K(2), f(x, \mu_1), f(x, \mu_2)\big)$, $\mu_1 \in M_1$, $\mu_2 \in M_2$, $p_K(1) + p_K(2) = 1$, $p_K(1) \geq 0$, $p_K(2) \geq 0$, the inequality $R(q^*, m) \leq \varepsilon^*$ holds. This means that the model m^* is the worst one and the strategy q^* is the solution of the task.

The strategy described is much simpler than the strategies usual in the literature, from which you mentioned only the nearest neighbour classification and the classification according to the integral of the probability. In the first proposal in which the observation x is assigned to a class according to the smallest distance from the exemplar, it is necessary to solve two quadratic programming *tasks. It is yet not so difficult in the two-dimensional case but problems can occur in a multi-dimensional case. In the second classification proposal according to the integral of the probability, it is necessary to calculate an integral of the Gaussian probability distribution on a polyhedron in recognising each observation. I do not know, even in a two-dimensional case, from where to begin my calculation. The simplicity of the exact solution of Linnik task is really surprising in comparison with the two mentioned strategies.*

You must add to your evaluation that the strategy q^* which you have found is the solution of a well defined task. Having any *a priori* probabilities of the states and any mathematical expectations μ_1 and μ_2, you can be sure that the probability of the wrong decision will not be greater than ε^*. The value ε^* serves as the guaranteed quality that is independent of the statistical model of the object. There is no other strategy about which the same can be said. It is not possible to express any similar assertion about the recognition based on the nearest neighbour and on the integral of the probability that would sound as a guarantee.

We would like to offer you a small implementation exercise that relates to the Gaussian random variables. It does not solve any basic question and it belongs more or less to mathematical or programming folklore. You will not lose anything when you look at such an exercise.

We have already mentioned in the lecture that in the case of two states and in the case that the observation x under the condition of each state is a multi-dimensional Gaussian random variable, in the search for a decision it is needed to calculate a value of a certain quadratic function in the observed point x and compare the obtained value with a threshold. When there are two features only and they are denoted by symbols x, y then the following discriminant function has to be calculated

$$f(x,y) = ax^2 + bxy + cy^2 + dx + ey + g\,. \tag{3.39}$$

Eight multiplications and five additions are needed for each observation x, y. The computational time can be saved for calculating these values in a recognition process. The function (3.39) can be calculated in advance into a table, say for 1000 integer values of x and y. Assume that it is needed in the application to make the quickest possible calculation of the table. A program can be written that requires only two additions for each point (x, y), $x = 0, 1, 2, \ldots, 999$, $y = 0, 1, 2, \ldots, 999$, when tabulating the function (3.39), and even no multiplication. Try to find out yourself how to organise such a calculation of values in the table.

I think I have found the procedure based on the deduction

$$f(x,y) = ax^2 + bxy + cy^2 + dx + ey + g\,,$$

$$f(x,y) = f(x-1,y) + 2ax + by + d + a\,, \tag{3.40}$$

$$f(0,y) = f(0,y-1) + 2cy + e - c\,. \tag{3.41}$$

The program, which I introduce later, will use the constants: $\mathtt{A} = 2a$, $\mathtt{B} = b$, $\mathtt{C} = 2c$, $\mathtt{D} = d + a$, $\mathtt{E} = e - c$, $\mathtt{G} = g$, *about which I assume that they were calculated in advance. By (3.40) and (3.41) I obtain the following formula*

$$f(x,y) = f(x-1,y) + \mathtt{A}x + \mathtt{B}y + \mathtt{D}\,, \tag{3.42}$$

$$f(0,y) = f(0,y-1) + \mathtt{C}y + \mathtt{E}\,, \tag{3.43}$$

$$f(0,0) = \mathtt{G}\,. \tag{3.44}$$

By (3.43) and (3.44), I can write the program in C programming language which tabulates the function f *for values* $x = 0$, $y = 0, 1, \ldots, 999$.

```
fCur = f[0][0] = G;  DeltaCur = E;
for (y=1;  y < 1000;  y++)
  f[0][y] = fCur += DeltaCur += C;                    /* L1 */
```

When each command `L1` *is satisfied then the value* `DeltaCur` *assumes the value* $\mathtt{C}y + \mathtt{E}$. *The value* `fCur` *is* $f(0,0)$ *before the command* `L1` *is performed for the first time and then, when performing the command again and again, it is increased by the value* `DeltaCur` $= \mathtt{C}y + \mathtt{E}$. *Each new value* `fCur`, *which is calculated correctly according to formula (3.43), is transferred into* `f(0,y)`.

The following program fragment creates the array `Delta(x)`, $x = 1, 2, \ldots,$ 999 *with values* `Delta(x)` $= \mathtt{A}x + \mathtt{D}$. *Furthermore, the variables* $f(x,y)$, *for* $x = 1, 2, \ldots, 999$ *and for* $y = 0$ *are filled up.*

```
fCur = G; DeltaCur = D
for (x=1; x<1000; x++)
  f[x][0] = fCur += Delta[x] = DeltaCur += A;     /* L2 */
```

After each command `L2` *is satisfied, the variable* `DeltaCur` *is filled by the value* $\mathtt{A}x + \mathtt{D}$. *This variable is stored in the element* `Delta(x)` *and it is added to the variable* `fCur`, *whose content is* `f(x-1,0)` *before the command* `L2` *is issued and it is stored in the element* `f(x,0)` *after the command is issued which is correct with respect to (3.42).*

Finally, let us show the last program section which fills up the rest of the array $f(x,y)$ *for variables* $x = 1, 2, \ldots, 999$ *and* $y = 1, 2, \ldots, 999$.

```
for (y=1; y < 1000; y++) {
  fcur = f[0][y];
  for (x=1; x < 1000; x++)
    f[x][y] = fcur += Delta[x] += B;                /* L3 */
}
```

The logic of this program is almost the same as that in the two previous program fragments. The variable `Delta(x)` *is filled up by the value* $\mathtt{A}x + \mathtt{B}y + \mathtt{D}$ *after the command* `L3` *is issued each time and therefore the variable* `f(x,y)` *is created according to the formula (3.42).*

In the program fragments mentioned the addition is performed only in commands with attached comments `L1` *to* `L3`. *It was* $2(n_y - 1)$ *in the first program,*

$2(n_x - 1)$ in the second program, and $2(n_x - 1)(n_y - 1)$ in the third program, where n_x, and n_y are the numbers of values of variables x, and y, respectively. The total number of additions is thus $2(n_x n_y - 1)$ which means that the average number of additions for every value $f(x, y)$ is less than 2. It is definitely less than 8 multiplications and 5 additions that would have to be performed if each value of the function $f(x, y)$ was calculated directly according to formula (3.39).

Assume for the moment that you have to tabulate a quadratic function that depends on three variables x_1, x_2, x_3 and not just on two variables x, y as in the previous case,

$$\begin{aligned} f(x_1, x_2, x_3) = {} & a_{11}\, x_1^2 + a_{22}\, x_2^2 + a_{33}\, x_3^2 \\ & + \; a_{12}\, x_1\, x_2 + a_{23}\, x_2\, x_3 + a_{13}\, x_1\, x_3 \\ & + \; b_1\, x_1 + b_2\, x_2 + b_3\, x_3 \\ & + \; c\,. \end{aligned}$$

In this case 15 multiplications and 9 additions are needed to calculate the value of a given function in one point (x_1, x_2, x_3). See how the calculation complexity increases for the best tabulation possible in comparison to the two-dimensional case.

It seems incredible, but the tabulation of the function of three variables needs again only 2 additions and no multiplication for each entry of the table. This property even does not depend on the number of variables being tabulated. When I encountered this property I said to myself again that the quadratic functions are really wonderful.

You are right, but do not think that other functions are worse. When you think a bit about how you should tabulate a cubic function, you will find out quite quickly that 3 additions and no multiplication are again needed to tabulate it. Again, this property does not depend on a number of variables. In the general case, when tabulating a polynomial function of degree k of an arbitrary large number of variables, only k additions and no multiplications are needed for each table entry.

It has been said in the lecture that if all features are binary, then the strategy (3.3) is implementable by a hyperplane. I have seen and heard this result many times, for example in the book by Duda and Hart [Duda and Hart, 1973]. It is a pleasant property, of course, that eases the analysis of these strategies and makes it more illustrative. That is why I am surprised why hardly anybody has noticed that a similarly pleasant property is valid not only in the case of binary features. I will try to make such a generalisation.

Let $k(i)$ be the number of values of the feature x_i. Without loss of generality I can assume that the set of values of the feature x_i is $X_i = \{0, 1, 2, \ldots, k(i)-1\}$. I shall express the observation $x = (x_1, x_2, \ldots, x_n)$ as a different observation y with binary features in the following way. I will number the elements of observations y using two indices $i = 1, 2, \ldots, n$ and $j = 0, 1, 2, \ldots, k(i) - 1$.

I shall express the feature y_{ij} in such a way that $y_{ij} = 1$, if the feature x_i assumes its j-th value, and $y_{ij} = 0$, if the feature x_i does not assumes its j-th value. The discriminant function $\sum_{i=1}^{n} \log \frac{p_{X_i|1}(x_i)}{p_{X_i|2}(x_i)}$ can be written in the form $\sum_{i=1}^{n} \sum_{j=0}^{k(i)-1} y_{ij}\,\alpha_{ij}$, where the coefficient α_{ij} is $\log \frac{p_{X_i|1}(j)}{p_{X_i|2}(j)}$, and the strategy (3.46) obtains the form

$$x \in \begin{cases} X_1\,, & \text{if } \sum\limits_{i=1}^{n} \sum\limits_{j=0}^{k(i)-1} \alpha_{ij}\, y_{ij} \geq \theta\,, \\ X_2\,, & \text{if } \sum\limits_{i=1}^{n} \sum\limits_{j=0}^{k(i)-1} \alpha_{ij}\, y_{ij} < \theta\,. \end{cases} \tag{3.45}$$

This result actually asserts that any strategy of the form

$$x \in \begin{cases} X_1\,, & \text{if } \sum\limits_{i=1}^{n} \log \frac{p_{X_i|1}(x_i)}{p_{X_i|2}(x_i)} \geq \theta\,, \\ X_2\,, & \text{if } \sum\limits_{i=1}^{n} \log \frac{p_{X_i|1}(x_i)}{p_{X_i|2}(x_i)} < \theta\,, \end{cases} \tag{3.46}$$

can be expressed using the linear discriminant function. But it is not yet the generalisation of the result given in the lecture. If all the features x_i are binary, i.e., if $k(i) = 2$, then from the relation (3.45) it follows that the strategy (3.46) can be implemented by a hyperplane in a $2n$-dimensional space. On the other hand, the result from the lecture asserts that the strategy (3.46) is implementable in this case by means of a linear discriminant function in an n-dimensional space. The form of the strategy (3.45) can be improved so that the number of binary features will not be $\sum_{i=1}^{n} k(i)$ but $\sum_{i=1}^{n} k(i) - n$. I shall introduce new variables $y'_{ij} = y_{ij} - y_{i0}$, $i = 1, 2, \ldots, n$, $j = 1, 2, \ldots, k(i)$, and the new threshold $\theta' = \theta - \sum_{i=1}^{n} \alpha_{i0}$. It is obvious that the strategy (3.45) is equivalent to the strategy

$$x \in \begin{cases} X_1\,, & \text{if } \sum\limits_{i=1}^{n} \sum\limits_{j=1}^{k(i)-1} \alpha_{ij}\, y'_{ij} \geq \theta'\,, \\ X_2\,, & \text{if } \sum\limits_{i=1}^{n} \sum\limits_{j=1}^{k(i)-1} \alpha_{ij}\, y'_{ij} < \theta'\,. \end{cases}$$

This result generalises the result given in the lecture for the case in which the features x_i are not binary.

Each strategy of the form (3.46) can be expressed using linear discriminant functions in $(\sum_{i=1}^{n} k(i) - n)$-dimensional space. In the particular case in which $k(i) = 2$ for every i, the dimension of this space is n.

I am sure that this more general result can simplify the analysis of strategies of the form of (3.46) in various theoretical considerations.

Very good. We look forward to it.

March 1997

3.4 Bibliographical notes

In this brief lecture we have not introduced new results. A more substantial contribution arose here in the new view on the conditional independence of features with the help of Jiří Pecha in the discussion after the lecture.

Two statistical models, i.e., the conditional independence of features and the Gaussian probability distribution are described in greater detail from the pattern recognition point of view in [Duda and Hart, 1973; Devijver and Kittler, 1982; Fukunaga, 1990]. The conditional independence of features was treated by Chow [Chow, 1965].

We adopted into the lecture from [Duda and Hart, 1973; Fukunaga, 1990] the proof concerning the property that for conditionally independent binary features the classifier is linear. The generalisation of this property for the case of non-binary features with a finite number of values was proposed by Jiří Pecha.

Properties of multi-dimensional Gaussian vectors were carefully analysed outside pattern recognition [Anderson, 1958].

Lecture 4

Learning in pattern recognition

4.1 Myths about learning in pattern recognition

The development of various areas of science and technology that substantially change human possibilities passes almost all the time through the following three stages.

In the first stage, as in fairy tales, a miraculous instrument is usually sought that would allow us to perform what has been impossible until now (for example to develop a flying carpet and float in the air). In the second stage, various models are created which imitate dreams in fairy tale stage (i.e., the models were in a way already flying) although they are too far from any practical exploitation because they are, after all, nothing more than mere toys. Something similar to a product (e.g., an airplane) appears only in the third stage and it fulfils the practical requirements, a little at the beginning, and more and more later.

There is not any doubt about the importance of the third stage. However, it is plausible to realise quite clearly that the two first stages have their own essential place too. Thinking in a fairy tale manner is clearly nothing more than an effort to perceive the result demanded. During the creation and examination of toys the principles are cleared up that check whether it is possible to realise this or that dream. It is checked whether some wishes happen to be unrealistic (even if they sound extremely urgent), namely, owing to a discrepancy with the laws of nature. In particular, a theoretical substance of a future product is created in a model construction stage because quite different problems are to be solved in the third stage. A quick and sloppy passing through the first two stages of fairy tales and toys can lead to deep, long term, and negative consequences.

Current pattern recognition is an uncanny mixture of fairy tales, toys and products. This can be said especially and to the highest degree about the part of pattern recognition that is called learning.

In the previous lectures we have seen that certain knowledge had been needed to construct recognition strategies, i.e., functions $q\colon X \to K$. This is a serious obstacle on its own since not every one possesses this knowledge. We will be convinced many times in the following lectures that there are, unfortunately, far more obstacles of this kind than we would wish. The reaction to such a situation usually occurs as a dream about a miraculous tool, e.g., in the 'Lay table, lay!' form. With its help it would be possible to avoid all the difficulties at once. This fairy tale has usually the following wording in pattern recognition:

'There is a system (a genetic, evolutionary, neural, or exotic in another way) which works in the following manner. The system learns first, i.e., the *training multi-set* $x_1, x_2, \ldots, x_l$ of observational examples is brought to its input. Simultaneously, each observation x_i from the training multi-set is accompanied by the information k_i representing the reaction to the observation which is considered correct. When the learning finishes after l steps, the normal exploitation stage of the system begins, during which the system reacts through the correct answer k to each observation x, and even to one which did not appear in the learning stage. Thanks to the information about the correct answer not having been provided explicitly, the system is able to solve any pattern recognition task.'

In such cases it is usually hopeless to try to find an understandable answer to the question of how the task is formulated, for which the solution is intended, and to learn more specifically how the system works. The expected results seem, at least to the authors of the proposals mentioned, to be so wonderful and easily accessible that they regret on losing time on trifles like those of the unambiguous task formulation and the exact derivation of the particular algorithm which should solve the task. The fairy tale is simply so wonderful that it is merely spoiled by down to earth questions.

The more realistic view of this fairy tale leads to current models of learning and their formulations that are going to be brought forward in the following section.

4.2 Three formulations of learning tasks in pattern recognition

We shall denote the conditional probability of observation x under the condition of the state k in two ways. The first notation $p_{X|K}(x \mid k)$ already has been used. Here the function $p_{X|K}$ has been considered as a function of two variables x and k, i.e., the function of the shape $X \times K \to \mathbb{R}$ with the domain given by the Cartesian product $X \times K$. The same function of two variables can be understood as an ensemble of functions of the form $X \to \mathbb{R}$ of one single variable x where each specific function from the ensemble is determined by the value of the state k. The function from this ensemble that corresponds to the state $k \in K$ is denoted by $p_{X|k}$. The conditional probability of observation x under the condition of the state k is thus the value of the function $p_{X|k}$ in the point x which will be denoted by $p_{X|k}(x)$.

When is learning necessary? It is at the time when knowledge about the recognised object is insufficient to solve a pattern recognition task without learning. Most often the knowledge about the probabilities $p_{X|K}(x \mid k)$ is insufficient, i.e., it is not known exactly enough how the observation x depends on the state k. The lack of knowledge can be expressed in such a way that the function $p_{X|K}$ is known to belong to a class $\mathcal{P}$ of functions but it is not known which specific function from the class $\mathcal{P}$ actually describes the object. Expressed differently, knowledge can be determined by the ensemble of sets $\mathcal{P}(k)$, $k \in K$. Each of the sets comprises the actual function $p_{X|k}$; however, which one is not known. The set $\mathcal{P}$ or, what is the same, the ensemble of sets $\mathcal{P}(k)$, $k \in K$, can quite often (roughly speaking almost always) be parameterised in such a way that the function $f(x, a)$ of two variables x, a is known and determines the function $f(a)\colon X \to \mathbb{R}$ of one single variable for each fixed value of the parameter a. At present it is not necessary to specify more exactly what is meant by the parameter a and to constrain the task prematurely and unnecessarily. The parameter a can be a number, vector, graph, etc.. The set $\mathcal{P}(k)$ is thus $\{f(a) \mid a \in A\}$, where A is the set of values of the unknown parameter a. Our knowledge about the probabilities $p_{X|K}(x \mid k)$ which is given by the relation $p_{X|K} \in \mathcal{P}$ means that the value a^* of the parameter a is known to exist for which $p_{X|k} = f(a^*)$.

Example 4.1 Parametrisation of $\mathcal{P}(k)$. *Let $\mathcal{P}$ be a set consisting of a probability distributions of n-dimensional Gaussian random variables with mutually independent components and unit variances. Then the set $\mathcal{P}(k)$ in a parameterised form is the set $\{f(\mu) \mid \mu \in \mathbb{R}^n\}$ of the functions $f(\mu)\colon X \to \mathbb{R}$ of the form*

$$f(\mu)(x) = \prod_{i=1}^{n} \frac{1}{\sqrt{2\pi}} \exp\left(\frac{-(x_i - \mu_i)^2}{2}\right) .$$

▲

Based on knowledge of the functions $p_{X|k}$, $k \in K$, defined up to values of the unknown parameters $a_1, a_2, \ldots, a_n$, the function $q(x, a_1, a_2, \ldots, a_n)$ can be created which will be understood as a strategy given up to the values of unknown parameters. The function $q(x, a_1, a_2, \ldots, a_n)$ illustrates how the observation x would be assessed if the parameters a_k, $k = 1, 2, \ldots, n$ determining the distribution $p_{X|k}$, were known. In other words the parametric set of strategies can be created

$$Q = \{q(a_1, a_2, \ldots, a_n) \mid a_1 \in A, a_2 \in A, \ldots, a_n \in A\}$$

into which the strategy sought surely belongs.

We have already learned from the previous lectures that the statistical tasks in pattern recognition can be formulated and solved not only in the case in which the statistical model of an object is determined uniquely, but also when it is known to belong to a certain set of models. A set of unknown parameters can be considered as a non-random intervention which influences a statistical model of the object, and the task is formulated as a non-Bayesian statistical estimate with non-random intervention.

Nevertheless, it can happen that in such an approach the guaranteed level of risk will be insufficient. It happens when an *a priori* known set of models is too extensive (as it happened in Example 4.1). Because of that the admissible set of strategies is so rich that it cannot be substituted by a single strategy without incurring essential losses. In such situations it is necessary to narrow the set of models or the possible strategies set by using additional information. This additional piece of information is obtained from the teacher in a process of learning. The information has the form of an ensemble $T = \big((x_1, k_1), (x_2, k_2), \ldots, (x_l, k_l)\big)$ in which $x_i \in X$ and $k_i \in K$. If the processing of the additional information mentioned depends on how many times an element (x, k) occurred in the ensemble T then the ensemble is treated as a multi-set and it is called the *training multi-set*. If the learning outcome depends only on whether an element occurred in the ensemble at least once and does not depend on how many times it occurred then the ensemble T is treated as a *training set*.

Supervised learning (i.e., learning with a teacher) must choose a single strategy being chosen in one or other convincing way from the set of *a priori* known strategies, namely on the basis of information provided in a learning process.

The most natural criterion for the strategy choice is, of course, the risk

$$\sum_{x\in X} \sum_{k\in K} p_{XK}(x, k)\, W\big(k, q(x)\big) \,, \tag{4.1}$$

which will be obtained in using the strategy the wrong decisions of which are quantified by the penalty W. But the *criterion cannot be computed because the function $p_{XK}(x, k)$ is not known.* The lack of knowledge about the function $p_{XK}(x, k)$ is substituted to a certain degree by the training set or multi-set.

Various formulations of the learning task differ in how the most natural criterion is replaced by the substitute criteria which can be calculated on the basis of information obtained during the learning. Nevertheless, a gap always remains between the criterion that should, but cannot, be calculated, and the substitute criterion which can be computed. This gap can be based on conscientiousness (intuition or experience) of the learning algorithm's designer or can be estimated in some way. We will first introduce the most famous substitute criteria on which the approaches to learning, popular today, are based. Later we will introduce the basic concepts of the statistical learning theory, main task of which is just the evaluation of how large the gap we spoke of can be.

4.2.1 Learning according to the maximal likelihood

Let $p_{X|K}(x \mid k, a_k)$ be a conditional probability of the observation x under the condition of the state k which is known up to an unknown value of the parameter a_k. Let the *training multi-set*

$$T = \big((x_1, k_1), (x_2, k_2), \ldots, (x_l, k_l)\big)\,, \quad x_i \in X\,, \quad k_i \in K\,,$$

be available. The selection is treated similarly as it is common in statistics. The most important assumption about the elements of the training multi-set T is that they are understood as mutually independent random variables with probability distribution

$$p_{XK}(x, k) = p_K(k)\, p_{X|K}(x \mid k, a_k)\,.$$

In this case the probability of the training multi-set T can be computed for each ensemble of unknown parameters $a = (a_k, k \in K)$ as

$$L(T, a) = \prod_{i=1}^{l} p_K(k_i)\, p_{X|K}(x_i \mid k_i, a_{k_i}) \tag{4.2}$$

under the condition that the statistical model of an object is represented by the mentioned values.

In learning according to the maximal likelihood, such values a_k^*, $k \in K$, are found that maximise the probability (4.2),

$$a^* = (a_k^*, k \in K) = \operatorname*{argmax}_{(a_k, k \in K)} \prod_{i=1}^{l} p_K(k_i)\, p_{X|K}(x_i \mid k_i, a_{k_i})\,. \tag{4.3}$$

Then the ensemble a^* of values $(a_k^*, k \in K)$ found is treated in the same way as if the values were real. This means that the ensemble $(a_k^*, k \in K)$ is substituted into the general expression $q(x, a_1, a_2, \ldots, a_n)$ and the recognition is performed according to the strategy $q(x, a_1^*, a_2^*, \ldots, a_n^*)$.

The expression (4.3) can be expressed in a different but equivalent form which will be useful in the coming analysis. Let $\alpha(x, k)$ be a number that indicates how many times the pair (x, k) occurred in the training multi-set. We can write under the condition of non-zero probabilities $p_{X|K}(x \mid k, a_k)$

$$\begin{aligned}
a^* &= \operatorname*{argmax}_{(a_k, k \in K)} \prod_{x \in X} \prod_{k \in K} \left(p_K(k)\, p_{X|K}(x \mid k, a_k)\right)^{\alpha(x,k)} \\
&= \operatorname*{argmax}_{(a_k, k \in K)} \sum_{k \in K} \sum_{x \in X} \alpha(x, k) \log p_K(k)\, p_{X|K}(x \mid k, a_k) \\
&= \operatorname*{argmax}_{(a_k, k \in K)} \sum_{k \in K} \sum_{x \in X} \alpha(x, k) \log p_{X|K}(x \mid k, a_k)\,.
\end{aligned} \tag{4.4}$$

The expression (4.4) maximised according to the values $(a_k, k \in K)$ constitutes the sum in which each term of addition depends only on one single element of this set. The maximisation task (4.4) decomposes into $|K|$ independent maximisation tasks that search for a_k^* according to the requirement

$$a_k^* = \operatorname*{argmax}_{a_k} \sum_{x \in X} \alpha(x, k) \log p_{X|K}(x \mid k, a_k)\,. \tag{4.5}$$

The previous Equation (4.5) shows that it is not needed to know *a priori* probabilities $p_K(k)$ when determining a_k^*.

4.2.2 Learning according to a non-random training set

The strategy obtained for the solution of the learning task according to the maximal likelihood (4.3), (4.4), (4.5) depends on the training multi-set T very substantially. Learning according to the maximal likelihood demands that the training multi-set is composed of mutually independent examples of random pairs with certain statistical properties. The second approach to learning (which is common mainly in the recognition of images) does not tune the recognition algorithm using random examples because it is not easy to get them. Instead, a carefully selected patterns are used for learning which from the designer point of view:

1. represent well the whole set of images which are to be recognised, and
2. any of the images chosen for learning is good enough, of a satisfying quality, not damaged, so the recognition algorithm should evaluate it as a very probable representative of its class.

These considerations, up to now informal and inaccurate, are formalised in the following way. Let $X(k)$, $k \in K$, be the ensemble of examples each of them consisting of representatives reliably selected by the teacher. The recognition algorithm should be tuned using that ensemble of examples in which each $x \in X(k)$ was regarded as a quite probable representative of k-th class. The parameter a_k^* which determines the probability distribution $p_{X|K}$ is to be chosen in such a way that

$$a_k^* = \operatorname*{argmax}_{a_k \in A} \min_{x \in X(k)} p_{X|K}(x \mid k, a_k) .$$

It is seen from the previous requirement that in such an approach to learning the information from the teacher is expressed by means of a training set and not a multi-set. The solution of the task no longer depends on how many times this or that observation has occurred. It is significant that it has occurred at least once.

Example 4.2 Comparison of two learning methods for multi-dimensional Gaussian distributions. *If $\mathcal{P}(k)$ is a set of functions of the form*

$$p(x|k, \mu_k) = \prod_{i=1}^{m} \frac{1}{\sqrt{2\pi}} \exp\left(\frac{-(x_i - \mu_{ik})^2}{2}\right)$$

then in the first formulation (learning according to the maximal likelihood) the μ_k^* is estimated as the mean value $(1/l)\sum_{i=1}^{l} x_i$ of observations of the object in the k-th state. If the learning task is solved in its second formulation (based on the non-random training set) then the μ_k^* is estimated as the centre of the smallest circle containing all vectors which were selected by the teacher as rather good representatives of objects in the k-th state. ▲

4.2.3 Learning by minimisation of empirical risk

Let $W(k, d)$ be a penalty function and $Q = \{q(a) \mid a \in A\}$ be a parameterised set of strategies expressed as the strategy $q(a): X \to D$ defined up to values of certain parameters a which are unknown. The quality of each strategy $q(a)$ is measured by the risk R which is achieved when this strategy is used,

$$R(a) = \sum_{k \in K} \sum_{x \in X} p_{XK}(x, k)\, W\big(k, q(a)(x)\big) . \tag{4.6}$$

The risk $R(a)$ should be minimised by an appropriate selection of the value a. However, the risk cannot be measured exactly because the statistical model $p_{XK}(x,k)$ is not known. Fortunately, based on the training multi-set $T = ((x_1,k_1), (x_2,k_2), \ldots, (x_l,k_l))$ the empirical risk can be defined,

$$\widehat{R}(a) = \frac{1}{l} \sum_{i=1}^{l} W\big(k_i, q(a)(x_i)\big), \tag{4.7}$$

which can be measured and seems to be a close substitute of the actual risk (4.6).

The third approach to learning in pattern recognition tries to create parametric a set of strategies on the basis of partial knowledge about the statistical model of the object. From this parametric set, such a strategy is next chosen which secures the minimal empirical risk (4.7) on the submitted training multi-set.

Example 4.3 Learning by minimisation of empirical risk for multi-dimensional Gaussian distributions. *Let us have a look at what the third approach just discussed means in a special case, the same as that in which we have recently illustrated the dissimilarity between the learning according to the maximal likelihood and the learning according to the non-random training set, see Example 4.2.*

If the number of states and number of decisions is equal to two and the observation is a multi-dimensional Gaussian random variable with mutually independent components and unit variance then the set of strategies contains strategies separating classes by the hyperplane. The third approach to learning aims at finding the hyperplane which secures the minimal value of the empirical risk (or the minimal number of errors in the particular case) on the training multi-set. ▲

A variety of approaches to learning in pattern recognition (and we have not mentioned all of them, by far) does not at all mean that we would prefer one approach to the other. We think that it will be cleared up in the future that each approach will have its advantages with respect to certain additional requirements, whose importance has not yet been fully comprehended. Such a clarification has not yet appeared which in exaggeration means that not everything needed and possible has been dug out from the fairy tale and toys-creating stage. Nevertheless, it already is possible to state, though based on an imperfectly theoretically analysed model, that some fairy tales have undergone substantial modifications. The most important observation is that learning has already and surely lost the meaning of a magic wand for idle people which would allow one to avoid laborious and careful work during the construction of a recognition algorithm. Such an idea, hoping that the algorithm would be found in the learning process on its own, has disappeared too.

We see that there is a *difference between recognition itself and recognition with learning.* The recognition itself is used for a single recognition task, whereas the recognition with learning is used for an unambiguously defined class of tasks. The learning is nothing else than the recognition of what task has to be solved and, subsequently, the choice of the right algorithm for this task. To be able to make such a choice a designer of a learning algorithm has himself to solve all the tasks that can occur. In other words, he has to find a general solution for the whole class of the tasks and present this general solution as the set of parametric strategies. When this is done this general solution is then to be incorporated into the body of the learning algorithm. Such a deformed fairy tale about pattern recognition with learning has totally lost its gracefulness and charm, no doubt, but it has gained a prosaic solidity and reliability because it has stopped being a miracle.

4.3 Basic concepts and questions of the statistical theory of learning

In spite of all the varieties of approaches to learning in pattern recognition (and we have not introduced all of them, by far), there exists a group of questions which arises in the framework of any known approaches. The formulation of questions and, mainly, the effort to give a comprehensible answer to them constitutes the contents of statistical learning theory in pattern recognition. These questions will be described informally first and then in the form of explicit statements.

4.3.1 Informal description of learning in pattern recognition

Basic problems related to learning have been analysed in the familiar works of Chervonenkis and Vapnik. We will introduce the main results of these studies. Before doing so we will show in an informal example what these problems are.

Imagine someone, whom we will call a customer, coming to someone else, whom we will call a producer. The customer has ordered a pattern recognition algorithm from the producer. After, perhaps, a long dialogue the customer and the producer are convinced that they have come to an agreement on what the recognition task is about. The customer has submitted as an appendix to the agreement the experimental material that consists of the set of images $x_1, x_2, \ldots, x_l$ and the corresponding sequence $k_1, k_2, \ldots, k_l$ of answers (states) which the algorithm ordered should give for the images provided. It is agreed that the algorithm will be checked principally by this experimental material. Assume that the experimental material is fairly extensive, say $l = 10000$. The outcome that the producer should deliver to the customer is the recognition strategy.

The customer in this respect does not care at all how the strategy has been found, and, in addition to that, if something called learning has been used in creating the algorithm. The customer is only interested in the quality of the created strategy. Assume, to ensure uniqueness, that the quality is given by the probability of a wrong decision. Because this probability cannot be measured directly, the customer and the producer have agreed that the number of errors which the strategy makes on the experimental material will be used as a substitute for the quality. Both sides agreed at the same time that such a substitution can be justified (pay attention, the error follows!) by the law of large numbers which, roughly speaking, claims that in a large number of experiments the relative frequency of an even differs only a little from its probability.

The question to which the answer is sought is far more complicated in reality to be smoothed away by a mere, and not quite well thought out reference to the law of large numbers. Let us express more exactly what this complexity pivots on.

Let Q be a set of strategies and q a strategy from this set, i.e., $q \in Q$. Let the ensemble T be a training multi-set $(x_1, k_1), (x_2, k_2), \ldots, (x_l, k_l)$ and T^* be the set of all possible training multi-sets. Let $\widehat{R}(T, q)$ denote the relative frequency of wrong decisions that the strategy q makes on the multi-set T. Let us denote by $R(q)$ the probability of the wrong decision that is achieved when the strategy q is used. And, finally, let us denote by $V \colon T^* \rightarrow Q$ the learning algorithm, i.e., the algorithm which for each selected multi-set $T \in T^*$ determines the strategy $V(T) \in Q$. The number $\widehat{R}\big(T, V(T)\big)$ thus represents the quality achieved on the training multi-set T using the strategy which was created based on the same multi-set T. By the law of large numbers it is possible to state, in a slightly vulgarised manner for the time being, that for any strategy q the random number $\widehat{R}(T, q)$ converges to the probability $R(q)$ provided the length of the multi-set approaches infinity. The length of the training multi-set is understood as the number of its elements. This not very exact, but basically correct, statement does not say anything about the relation between two random numbers; the first of them is the number $\widehat{R}\big(T, V(T)\big)$ and the second is the number $R\big(V(T)\big)$.

If we assumed, with reference to the law of large numbers, that these two numbers coincide for the large lengths of the multi-sets T then it testifies that the concept of the 'law of large numbers' is used as a mere magic formula without clear understanding of what it relates to. The law does not say anything about the relation between the two numbers mentioned.

In reality the convergence of the random numbers $\widehat{R}\big(T, V(T)\big)$ and $R\big(V(T)\big)$ to the same limits is not secured. In some cases this pair of random numbers converges to the same limits and in other cases the numbers $\widehat{R}\big(T, V(T)\big)$ and $R\big(V(T)\big)$ remain different whatever the length of the training multi-set T is. We will show the example of the second mentioned situation.

Example 4.4 The estimate of the risk and the actual risk can also differ for infinitely long training multi-sets T. *Let the set X of observation be a one-dimensional continuum, for example, an interval of real numbers. Let two functions $p_{X|1}(x)$ and $p_{X|2}(x)$ define two probability distributions of the random variable x on the set X under the condition that the object is in the first or in the second state. Let it be known that densities $p_{X|1}(x)$ and $p_{X|2}(x)$ are not infinitely large in any point x which means that the probability of each value x, as well as of any finite number of values, is equal to zero. Let $V(T)$ be the following strategy: it is analysed for each $x \in X$ if the value x occurs in the training multi-set T. If it does, i.e., if some x_i is equal to x then the decision*

k_i is given for observation x. If the observation x does not occur in the training multi-set T then $k = 1$ is decided.

Two assertions are valid for this learning algorithm V. The probability of the wrong decision $R\big(V(T)\big)$ is an a priori probability of the second state $p_K(2)$ because the strategy $V(T)$ assigns practically all observations into the first class independently on the training multi-set T. Indeed, the probability that the random x appears in the finite multi-set T is equal to zero. This means that the probability of the answer $k = 2$ for random observation is equal to zero too.

On the other hand, the number $\widehat{R}\big(T, V(T)\big)$ is equal, of course, to zero with the probability 1. Indeed, $\widehat{R}\big(T, V(T)\big) = 0$ for an arbitrary multi-set T, in which any element (x, k) does not occur more than once and the total probability of all other multi-sets is equal to 0.

Consequently we have two random variables. The first is $p_K(2)$ with the probability 1 and the second is equal to zero with probability one. This fact holds for an arbitrary length of the training set T. Therefore it cannot happen for any lengths of the multi-set T that random variables $\widehat{R}\big(T, V(T)\big)$ and $R\big(V(T)\big)$ approach each other. It does not contradict the law of large numbers since it does not have anything in common with it. ▲

The learning algorithm presented is apparently a deception because it is based on remembering the whole training multi-set and the following correct recognition only of those observations that occurred in the training multi-set. The deception was made possible because of the assumption that the value of empirical risk $\widehat{R}\big(T, V(T)\big)$ can serve as a decent approximation of the actual risk $R\big(V(T)\big)$. The notion of the learning algorithm should be narrowed in such a way that not all algorithms of the form $V : T^* \to Q$ could have a right to exist. Not only deceptions such as the one mentioned should be excluded but also all less obvious incorrectness in which it is manifested that the empirical risk has nothing in common with the actual risk.

The set of learning algorithms which remain after this exclusion is not entirely homogeneous and can be separated into groups of better and worse algorithms. Let us illustrate again the sense of such a classification of learning algorithms on our informal example.

Let us assume that after the imaginary contact between the customer and the producer (as introduced at the beginning of Subsection 4.3.1) the customer has made the following correct conclusion. He has noticed that he can be cheated by the producer and he cannot escape from this situation even with the help of increasing the amount of experimental material. None, even any arbitrarily large number, of the test images can guarantee that, in the practical phase of the recognition algorithm application, the same recognition quality will be secured as was found on the test images. The customer notices that these troubles have two reasons. First, the failure is made possible because the producer knows the experimental material in advance and has enough possibilities to adapt to any experimental data. Second, the customer has noticed that he must not rely on the law of large numbers without clear understanding of what the law concerns. So far he has understood the law in the not very accurate formulation that a variability of a large number of independent (or conditionally dependent) random variables is compensated for so much that its sum is relatively constant (for instance with respect to the mean value). The customer substituted an insufficient accuracy of this formulation by the following examples.

Example 4.5 Law of large numbers and the pressure of gas. *The pressure of gas on the surface of a container is almost constant even though each molecule hits the surface at random instants.* ▲

Example 4.6 Law of large numbers and chemical reactions. *The progress of chemical reactions can be predicted with the help of differential equations even if it concerns the resultant of random behaviour of single molecules. There is a large number of participating molecules and their mutual dependence is small.* ▲

Now he concludes that he should know more exactly what the law concerns. The original formulation of the law of large numbers was made by Bernoulli. Let q be the *a priori* chosen strategy. One of several possible, and for our explanation suitable, expressions of the law of large numbers is

$$P\left\{\left|\nu_l(q) - p(q)\right| > \varepsilon\right\} \leq \exp\left(-2\varepsilon^2 l\right) , \tag{4.8}$$

where l is the length of the training multi-set $\nu_l(q)$ is a relative frequency of errors which the strategy q makes on the training multi-set of length l, $p(q)$ is the probability of the wrong decision achieved when the strategy q is used, $P\{\}$ is the probability of the event that is expressed within the brackets. Equation (4.8) illustrates that the experiment about indirect measurement of the actual quality $p(q)$ of the strategy q is characterised by means of three parameters. The first parameter is the length l of the experiment. The second parameter gives the accuracy ε of the strategy with help of which the probability of error of the strategy $p(q)$ is appreciated by the sentence 'the probability $p(q)$ is not larger than $\nu_l(q) + \varepsilon$ and it is not smaller than $\nu_l(q) - \varepsilon$' or, as is usually expressed in the handbooks of practitioners:

$$p(q) = \nu_l(q) \pm \varepsilon \, . \tag{4.9}$$

And finally, the third parameter is the reliability η of the assertion (4.9) which stresses that the assertion (4.9) can be erroneous. Thus the reliability is the probability of the event that (4.9) is an erroneous statement, in other words, that the wrong strategy will pass the test (called also false positive). The law of large numbers (4.8) claims that these three parameters of experiments contradict each other. Consequently in a short experiment such an estimate of the probability $p(q)$ cannot be achieved that is exact and reliable at the same time. The law (4.8) shows at the same time that no pair of three parameters mentioned is in contradiction. For instance, for an arbitrarily short length l of the experiment the arbitrarily small probability $\eta > 0$ can be achieved that expresses the reliability of the assertion (4.9). However, this can occur only for a rather large value ε that expresses the precision of $p(q)$ estimate in Equation (4.9). What is important for us is that the experiment's length can be planned so that arbitrarily small precision $\varepsilon > 0$ and arbitrary reliability $\eta > 0$

can be achieved. Actually, the length l of the experiment can be quite large, that is,

$$l \geq \frac{-\ln \eta}{2\varepsilon^2} . \tag{4.10}$$

Example 4.7 Accuracy, reliability, and length of the experiment shown on specific numbers. *Having in mind the previous relation (4.10), the customer can determine more exactly, at least for himself, what result of an experiment will be considered positive. The customer realises that any experiment has the restricted accuracy ε given by the highest admissible probability of the wrong decision and the reliability η. In his particular case he chooses $\varepsilon = 2\%$ and $\eta = 0.1\%$ and formulates the rule according to which he accepts or rejects the proposed strategy q. If the strategy q recognises about 9000 observations without error, then he accepts it and deduces that the probability of the wrong decision for the accepted strategy does not exceed 2%. Such a rule can be justified on the basis of the correctly understood law of large numbers. He substitutes $\varepsilon = 2\%$ and $\eta = 0.1\%$ into the inequality (4.10) and writes*

$$l \geq \frac{-\ln 0.001}{2\,(0.02)^2} = \frac{-(-6.9077)}{0.0008} \doteq 8635 .$$

▲

The customer equipped with this knowledge enters a shop selling programs for pattern recognition and chooses a program which does not make any single mistake on the testing multi-set prepared in advance. He is convinced that he already has what he needs this time. Indeed, the purchased program has been created not considering the experimental material. That is why he concludes, that the possibility of a direct swindle has been excluded. The customer makes a mistake here again.

Despite all illusory cogency, the rules used do not protect the customer from choosing a wrong pattern recognition strategy, because it is not taken into account from how extensive a set the strategy is chosen. The extent of this set has a substantial significance. Indeed, the customer makes a wrong choice if one single strategy out of the bad ones passes the test. Imagine the counterexample in which a choice is made from the set of wrong examples only, even in the case of an extremely strict test being used. The probability that some of the wrong strategies will pass the test can actually be quite large if the set of examined strategies is quite extensive.

The customer, having acquired this important but not very pleasant experience, comes to the conclusion that it is not enough for reliable choice of the strategy that the length of the experimental material satisfies the condition

$$P\left\{\left|\nu_l(q) - p(q)\right| > \varepsilon\right\} < \eta , \tag{4.11}$$

which is too weak. Instead, it has to fulfil the condition

$$P\left\{\sup_{q\in Q}\left|\nu_l(q) - p(q)\right| > \varepsilon\right\} < \eta ,$$

which is: (a) much more strict compared to the condition (4.11); (b) it requires that the probability that a single wrong algorithm from the set Q passes the test is low; and (c) this condition depends significantly on the cardinality of the set of recognition algorithms Q, from which the choice is made. Our illusory customer starts understanding the merit of the questions that the statistical theory of learning (to be explained in the next subsection) tries to answer.

4.3.2 Foundations of the statistical learning theory according to Chervonenkis and Vapnik

Let Q be the set of strategies of the form $q: X \to K$, $p_{XK}: X \times K \to \mathbb{R}$ be a statistical model of the recognised object that is not known. The probability of the wrong decision $p(q)$ corresponding to the strategy q is given by the formula

$$p(q) = \sum_{x \in X} \sum_{k \in K} p_{XK}(x, k)\, W\big(k, q(x)\big) ,$$

in which it for the penalty W

$$W(k, k^*) = \begin{cases} 1, & \text{if} \quad k \neq k^* , \\ 0, & \text{if} \quad k = k^* , \end{cases} \tag{4.12}$$

holds and the symbol k^* denotes an estimate of the actual state k using the strategy q.

Let $\nu_l(q)$ be a random variable represented by the frequency of a wrong decision which the strategy q assumes on the random training multi-set $T = \big((x_1, k_1), (x_2, k_2), \ldots, (x_l, k_l)\big)$ of length l,

$$\nu_l(q) = \frac{1}{l} \sum_{i=1}^{l} W\big(k_i, q(x_i)\big) .$$

It is known by the law of large numbers that the relation between the value $p(q)$ and the random variable $\nu_l(q)$ can be expressed by the following inequality for an arbitrary $\varepsilon > 0$

$$P\left\{\big|\nu_l(q) - p(q)\big| > \varepsilon\right\} \leq \exp\left(-2\varepsilon^2 l\right) . \tag{4.13}$$

The strategies can be divided into correct and wrong ones. The strategy q is considered correct if $p(q) \leq p^*$ and wrong if $p(q) > p^*$. It is not possible to decide immediately about the correctness of a strategy. However, on the basis of the relation (4.13), the following test can be performed. If $\nu_l(q) \leq p^* - \varepsilon$ then the strategy has passed the test (it is likely to be correct); otherwise it has not passed the test (it is likely to be wrong). It is possible to calculate the reliability of the test, which means the probability of the event that the wrong strategy passes the test. By (4.13) this probability is not larger than $\exp\left(-2\varepsilon^2 l\right)$.

In the case in which learning is used for finding this strategy, not only a single strategy q but a set Q of strategies are put through the test. A strategy

$q^* \in Q$ is chosen in this or that manner and it is checked whether the variable $\nu_l(q^*)$ is greater or less than $p^* - \varepsilon$. The strategy found is definitely accepted or rejected. The test described can fail if there is one wrong strategy at least in the set Q which passes the test. The probability of this situation is small if the probability

$$P\left\{ \exists q \in Q \left(\left|\nu_l(q) - p(q)\right| > \varepsilon \right) \right\}$$

is small, or equivalently, if the probability

$$P\left\{ \max_{q \in Q} \left|\nu_l(q) - p(q)\right| > \varepsilon \right\} \tag{4.14}$$

is small.

The reliability of the whole learning process is influenced by the probability (4.14), not by the probability (4.13). The probabilities (4.13) and (4.14) are substantially different. The main property of the probability (4.13) is that it can assume an arbitrary small value for the arbitrary $\varepsilon > 0$ when the proper length l is chosen. Owing to this property the relation (4.13) is one of the basic formulæ of classical mathematical statistics.

A similar property is not guaranteed for the probability (4.14). The probability (4.14) can no always be arbitrarily decreased by increasing the length l. In other words, the probability (4.13) converges always to zero for $l \to \infty$, whereas the probability (4.14) may or may not converge to zero for $l \to \infty$ in dependence on the set of strategies Q. This fact expresses the *central issue of learning in pattern recognition*, which cannot be solved by mere reference to the law of large numbers.

Let us show the most important properties of the probability (4.14). We will start from the simplest case in which the set Q consists of a finite number N of strategies. In this case

$$\begin{aligned} P\left\{ \max_{q \in Q} \left|\nu_l(q) - p(q)\right| > \varepsilon \right\} &\leq \sum_{q \in Q} P\left\{ \left|\nu_l(q) - p(q)\right| > \varepsilon \right\} \\ &\leq N \exp(-2\varepsilon^2 l)\,. \end{aligned} \tag{4.15}$$

We will show how this simply derived relation can be interpreted in learning within pattern recognition.

1. Q is a set which consists of N strategies in the form $X \to K$;
2. T is a random multi-set $(x_1, k_1), (x_2, k_2), \ldots, (x_l, k_l)$ of the length l with the probability $\prod_{i=1}^{l} p_{XK}(x_i, k_i)$; $p_{XK}(x, k)$ is a joint probability of the observation $x \in X$ and the state $k \in K$;
3. We will determine two subsets of strategies for two certain numbers p^* and ε. The strategy belongs to the set of wrong strategies if $p(q) > p^*$. The strategy belongs to the subset of strategies that passed the test if $\nu_l(q) < p^* - \varepsilon$.
4. An arbitrary strategy that passed the test is selected from the set Q.
5. It follows from the relation (4.15) that the probability of the wrong strategy selection is not larger than $N \exp(-2\varepsilon^2 l)$.

A generality of the interpretation given above is important. Its validity depends neither on the set Q nor on the statistical model p_{XK} and not on the learning algorithm either.

This fact was expressed by Chervonenkis and Vapnik in the following, though not very thoroughly formulated, theorem which is presented here in nearly original wording.

Theorem 4.1 Chervonenkis and Vapnik. The estimate of the training multi-set length. *If from the set consisting of N strategies a strategy is chosen that has the smallest relative frequency ν of errors on the training multi-set of length l, then with a probability $1 - \eta$ it can be stated that applying this strategy the probability of the wrong decision will be smaller than $\nu + \varepsilon$ provided that*

$$l = \frac{\ln N - \ln \eta}{2\varepsilon^2} . \tag{4.16}$$

▲

This theorem correctly expresses the *most substantial property of learning:* the broader the class of strategies is, i.e., the less the specific pattern recognition task was investigated in advance, the longer the learning must last to become reliable enough (which is always needed).

From a practical point of view, Equation (4.16) defines the demands for the length of learning too roughly, that is with a too big reserve. For instance, when the set Q is infinite, and as a matter of fact only such cases occur in practice, then the recommendation (4.16) does not yield anything because it requires endlessly long learning. This contradicts our intuition and, as we will see later, the intuition is correct. This means that the relation (4.16) can be substantially improved. The length of learning will not depend on such a rough characteristic of the set Q as the number of its elements is, but on other more gentle properties of this set. These properties are the entropy, the growth function, and the capacity of the class. Let us introduce definitions of these concepts.

Let Q be a set of strategies and $x_1, x_2, \ldots, x_l$ be a sequence of observations. Two strategies $q_1 \in Q$ and $q_2 \in Q$ are called *equivalent* with respect to the sequence $x_1, x_2, \ldots, x_m$, if for any i the equality $q_1(x_i) = q_2(x_i)$ holds. Thus each sequence of observations induces the equivalence relation on the set Q. Let denote number of equivalence classes corresponding to this relation by $\Delta(Q, x_1, x_2, \ldots, x_l)$. In other words, the $\Delta(Q, x_1, x_2, \ldots, x_l)$ corresponds to the number of different decompositions of the sequence $x_1, x_2, \ldots, x_l$ by means of strategies from the set Q.

Example 4.8 Decomposition of real numbers through a threshold. *Let $x_1, x_2, \ldots, x_l$ be real numbers and q be a strategy of the following form: each strategy is characterised by the threshold value θ and maps an observation x into the first class if $x < \theta$, and into the second class if $x \geq \theta$. It is obvious that the number $\Delta(Q, x_1, x_2, \ldots, x_l)$ is greater by one than the number of different numbers in the sequence $x_1, x_2, \ldots, x_l$. Well, $\Delta(Q, x_1, x_2, \ldots, x_l) = l + 1$ happens almost always.*

▲

Because the sequence of observation is random then the number $\Delta(Q, x_1, x_2, \ldots, x_l)$ is also random. The mathematical expectation of the logarithm of this number can be defined

$$\sum_{x_1 \in X} \cdots \sum_{x_l \in X} \sum_{k_1 \in K} \cdots \sum_{k_l \in K} \prod_{i=1}^{l} p_{XK}(x_i, k_i) \log \Delta(Q, x_1, x_2, \ldots, x_l) . \tag{4.17}$$

We will denote it $H_l(Q)$ and call it the *entropy* of the set of strategies Q on the sequences of the length l.

Our main goal is to show how large the length of learning l should be in order to obtain a fairly accurate and reliable result of learning. This means that

$$P\left\{ \max_{q \in Q} \left| \nu_l(q) - p(q) \right| > \varepsilon \right\}$$

should be fairly small for a quite small ε. Before doing so we must exclude from consideration all situations in which this probability does not converge to zero at all. In this case learning does not make any sense because the frequency of errors on the learning sequence does not have anything in common with the probability of error for a learning sequence of arbitrarily large length. The complete description of all such hopeless situations is given by the following theorem.

Theorem 4.2 Chervonenkis and Vapnik. The necessary and sufficient condition of a uniform convergence of empirical risk convergence to the real risk. *The probability*

$$P\left\{ \max_{q \in Q} \left| \nu_l(q) - p(q) \right| > \varepsilon \right\} \tag{4.18}$$

converges to zero for $l \to \infty$ and for any $\varepsilon > 0$, if and only if the relative entropy $H_l(Q)/l$ *converges to zero for $l \to \infty$.* ▲

Proof. The proof on the Theorem 4.2 is rather complicated and long [Vapnik and Chervonenkis, 1974]. ■

Theorem 4.2 provides an exhaustive answer to that difficult question. Of course, this theorem, like any exact and general statement, can be used only with difficulties in particular cases. The theorem says that the difficult question about the convergence of the probability

$$P\left\{ \max_{q \in Q} \left| \nu_l(q) - p(q) \right| > \varepsilon \right\}$$

to zero is equivalent to the convergence of the relative entropy $H_l(Q)/l$ to zero. This second issue is not easy either. It suffices to look at the formula (4.17) and realise at the same time that the function $p_{XK}(x, k)$ is not known either. That is why the two following steps are so important and lead to rougher but more constructive conditions.

The first step is based on the term of the growth function. Let $\Delta_l(Q, x_1, x_2, \ldots, x_l)$ be the number of possible decompositions of the sequence $x_1, x_2, \ldots, x_l$ by strategies of the set Q. Let us introduce the number $m_l(Q)$ by

$$m_l(Q) = \max_{x_1, \ldots, x_l} \Delta_l(Q, x_1, x_2, \ldots, x_l) .$$

The sequence of numbers $m_l(Q)$, $l = 1, 2, \ldots, \infty$, is called the *growth function*. The number $\log m_l(Q)$ is tied up with the entropy $H_l(Q)$ by the simple expression $\log m_l(Q) \geq H_l(Q)$. Thus if

$$\lim_{l \to \infty} \frac{\log m_l(Q)}{l} = 0 , \tag{4.19}$$

then $\lim_{l \to \infty} \frac{H_l(Q)}{l} = 0$ and the expression (4.19) can be used as a sufficient condition (but not a necessary one) to assure convergence of the probability (4.18) to zero. Equation (4.19) can be checked in an easier manner because the probability distribution $p_{XK}(x, k)$ need not be known in order to calculate the growth function. With the help of the growth function it is possible not only to prove the convergence of the probability (4.18) to zero but also to find the upper bound of the empirical risk deviation from the actual risk.

Theorem 4.3 On the upper bound of the empirical risk deviation from the actual risk.

$$P\left\{ \max_{q \in Q} \left| \nu(\overline{q}) - p(q) \right| > \varepsilon \right\} < 3\, m_{2l}(Q) \mathrm{e}^{-\frac{1}{4}\varepsilon^2 (l-1)} . \tag{4.20}$$

▲

It can be seen that Equation (4.20) assessing the reliability of learning is similar to Equation (4.15) that holds for the case of the finite set Q. The growth function plays in Equation (4.20) the same role as does the number of strategies in Equation (4.15). This means that the growth function can be considered the measure of the complexity of the set Q which is analogous to the number of strategies for the case of the finite sets Q. Certainly, if the growth function can be calculated for the finite set Q then exactly the growth function should be used, and not the mere number N of strategies in the set Q. The growth function describes the structure of the set of strategies more expressively and more precisely than the simple number of strategies in that set because it considers the diversity of strategies. A mere number of strategies simply ignores this diversity.

The second step towards the simplified assessment of learning reliability is based on the concept of the capacity of the set of strategies. The concept of the *capacity of the set of strategies*, informally speaking, is the smallest possible number of observations which cannot be classified in an arbitrary way by the strategies from the appropriate set. The name *VC dimension* it is also used in the literature for the capacity of the set of strategies according to the first letters of the surnames of the original authors. We use the name introduced by Chervonenkis and Vapnik in their original publications.

First, let us have a look at how the capacity of the set of strategies is defined in simple examples. The exact definition for the general case will be introduced later.

Example 4.9 Capacity of the set of strategies for classification of real numbers according to the threshold. *Let X be the set of real numbers and Q be the set of strategies of the abovementioned form for classification into two classes only: any strategy is characterised by the threshold θ and assigns the number $x \in X$ to the first class if $x < \theta$, and to the second class if $x \geq \theta$. For any $x \in X$ there is such a strategy q' in the set Q which assigns the observation x to the first class, and another strategy q'' which assigns x to the second class. Let x_1 and x_2 be two different points on the coordinate axis X. For these two points either $x_1 < x_2$ or $x_2 < x_1$ holds. Let us choose $x_1 < x_2$ to assure certainty. This pair of points cannot already be classified arbitrarily into two classes by means of strategies from the set Q. Indeed, there is no strategy in the set Q that assigns x_2 into the first class and x_1 into the second class because $x_2 > x_1$. Thus the given set Q of strategies is such that there is a point on the straight line X that can be classified in an arbitrary manner using different strategies from the given set. But for any pair of points such classification of these points exists, which can be made with no strategy from the given set. In this case the capacity of the set of strategies Q is equal to two.* ▲

Example 4.10 Capacity of the richer set of strategies. *Let us extend the set Q and illustrate how the capacity of the set of strategies is defined in this richer case. Let Q be the set of strategies each of which being determined by the pair of numbers α, θ. The observation x is assigned in the first class if $\alpha x < \theta$, and in the second class if $\alpha x \geq \theta$. Let x_1, x_2 be two distinct points on the straight line X such that $x_1 \neq x_2$. There are 2^2 possible decompositions of this pair of points in two classes and each of them can be implemented with some strategy from the set Q. Let us analyse any triplet of distinct points x_1, x_2, x_3 and let us assume that $x_1 < x_2 < x_3$. There are 2^3 decompositions of this triplet in two classes but not all of them are implementable by means of the strategies from the set Q. No strategy from Q can assign x_1 and x_3 in the first class and the x_2 into the second class. The given set Q is such that some pair of points x_1, x_2 can be decomposed in an arbitrary way in two classes and then no triplet of points can be already decomposed in an arbitrary manner into two classes. Such a set of strategies has the capacity 3.* ▲

Having presented two simple particular cases, we can proceed to the general *definition of the capacity of the set Q of strategies $q\colon X \to \{1,2\}$*. Let $x_1, x_2, \ldots, x_l$ be the sequence of observations, $C_l\colon \{1,2,\ldots,l\} \to \{1,2\}$ be the decomposition (classification) of this sequence into two classes, C_l^* be the set of all possible decompositions in the form$\{1,2,\ldots,l\} \to \{1,2\}$ which is a set consisting of 2^l decompositions.

The number r defines the capacity of the set Q of strategies of the form $q\colon X \to \{1,2\}$ iff

1. There exists a sequence $x_1, x_2, \ldots, x_{r-1}$ of the length $r-1$ such that for any classification $C_{r-1} \in C_{r-1}^*$ a strategy $q \in Q$ exists such that $q(x_i) = C_{r-1}(i)$, $i = 1,2,\ldots,r-1$;

2. For any sequence $x_1, x_2, \ldots, x_r$ of the length r there exists a classification $C_r \in C_r^*$ such that no strategy $q \in Q$ satisfies the equalities $q(x_i) = C_r(i)$, $i = 1, 2, \ldots, r-1$.

The *definition of the capacity of the set* Q can be *formulated in an equivalent way* by means of the growth function. Let

$$m_1(Q), m_2(Q), \ldots, m_{r-1}(Q), m_r(Q), \ldots, m_l(Q), \ldots \tag{4.21}$$

be the growth function for a set Q. In this sequence $m_1(Q)$ is not larger than 2^1, $m_2(Q)$ is not larger than 2^2 and in the general case the l-th element $m_l(Q)$ is not larger than 2^l. If an element, say the l-th element $m_l(Q)$ has value 2^l, then the preceding $(l-1)$-th element is 2^{l-1} too. The reason is that if some sequence of the length l can be decomposed in all possible manners, then it naturally holds also for any its subsequence of the length $l-1$. It follows from the abovesaid that the sequence (4.21) consists of two contiguous parts. The initial part, whose length can even be zero, is composed of sequence elements, which has the value 2^l, where l is an ordinal number of the element in the sequence (4.21). The elements in the second part of the sequence are lesser than 2^l.

The capacity of the set Q is an ordinate number by which the start of the second part is indexed, i.e., the minimal l, for which $m_l(Q) < 2^l$ holds.

It is immediately obvious from the definition that if r is the capacity of the set of strategies Q, then $m_l(Q) = 2^l$ holds for any $l < r$. Much less expected is that the values $m_l(Q)$ for $l \geq r$ are also influenced by the capacity and cannot assume arbitrary values. It follows from the next theorem.

Theorem 4.4 Upper limit of the growth function. *If r is the capacity of the set Q then for all lengths of sequences $l \geq r$*

$$m_l(Q) \leq \frac{1,5\, l^{r-1}}{(r-1)!} \tag{4.22}$$

holds. ▲

Proof. We refer the interested reader to [Vapnik and Chervonenkis, 1974]. ■

By Theorem 4.4, the inequality (4.20) can be rewritten in the form

$$\eta < 4.5 \frac{(2\,l)^{r-1}}{(r-1)!}\, \mathrm{e}^{-\frac{1}{4}\varepsilon^2(l-1)} \tag{4.23}$$

which expresses the explicit relation between all three parameters describing learning: the accuracy ε, the reliability η and the sufficient length l of the training multi-set. The set Q in the formula (4.23) is represented by its only one parameter, i.e., by the capacity r.

Let us summarise the main supporting points that lead to the sought result.

1. For the analysis of reliability of learning in pattern recognition, the knowledge how the probability

$$P\{\,|\nu_l(q) - p(q)| > \varepsilon\} \tag{4.24}$$

behaves when $l \to \infty$ does not suffice. It is necessary to analyse a more complicated probability

$$P\{ \max_{q \in Q} |\nu_l(q) - p(q)| > \varepsilon \} . \tag{4.25}$$

2. The fundamental difference between (4.24) and (4.25) is that at $l \to \infty$ the probability (4.24) converges to zero for any strategy q and any $\varepsilon > 0$, whereas the probability (4.25) in some cases converges to zero and in some cases it does not, according to the complexity of the set Q.
3. The exhaustive description of the set Q from the point of view of the convergence of (4.25) to zero is the entropy of the set of strategies Q (see Theorem 4.2) which can be defined uniquely, but cannot be constructively calculated if the statistical model is not known.
4. The growth function of the set Q can be constructively calculated because it does not depend on the statistical model of the object. If the growth function is known then the upper bound of the probability (4.25) can be estimated as well as the speed at which this probability converges to zero, see the formula (4.20).
5. The simplest description of the set Q is its capacity which explicitly influences the length of the training multi-set which suffices for required accuracy and reliability of learning.

4.4 Critical view of the statistical learning theory

The fundamentals explained of the statistical learning theory deserve all respect and thanks to its mathematical justification they do not require any indulgence. That is why they withstand any critical remarks given, particularly, in this section.

The results of the statistical learning theory, as well as the other theoretical results, have the form of the implication 'if A is valid then B is valid'. Let us pry into the rightfulness of recommendations following from the statistical theory of learning, for instance, from the elegant formula of (4.23). Naturally, we will not manage it if we start from the same assumptions from which the relation (4.23) was derived. If we want cast a doubt on the practical applicability of the relation (4.23) we have to pose the question first of all of whether the basic assumptions are transparent enough (the statement A in our implication) to be able to answer definitely whether the assumptions are satisfied. At the same time it has to be known if the theoretical results (that is the assertion B) can be experimentally verified.

The central assumption on which the statistical learning theory stands is the assumption about randomness of the training multi-set (x_1, k_1), (x_2, k_2), $\ldots, (x_l, k_l)$. It is assumed that the elements of training multi-set are mutually independent and their probability distribution $p_{X|K}(x \mid k)$ is the same as that during recognition. The problem is not that sometimes the assumption is satisfied and sometimes not. It can happen with any supposition. The crucial problem is that it is not known how it should be investigated if the assumption is satisfied in the real task and not merely in its artificial substitute. The problem of the relation between theoretical models and reality is difficult in general because it cannot be solved in the framework of any theoretical construction. But particularly difficult problems are those where the relation between statistics and real problem is sought.

This vexed nature, even in the exaggerated form, is expressed by the following tale, which is known in the pattern recognition folklore.

Example 4.11 The customer is a geologist. [Zagorujko, 1999] *Imagine a geologist being a customer and coming to a supplier demanding a solution of a pattern recognition task. It is to be found out by measuring physical properties of a piece of rock whether it contains iron or whether it is dead. The customer can take the responsibility that the decision strategy solving this task is implementable with the help of linear functions of the chosen measurable physical properties of the rock. The supplier is required to find out professionally the parameters of the appropriate linear decision function. It appears that this is a situation ideally suited to pattern recognition based on learning.*

When the supplier requests that he needs a training multi-set to fulfil the job, it does not embarrass the customer. The customer is already prepared for this situation and takes two pieces of rock out his rucksack. He is sure that one of them contains the iron and the other is dead. Such a training multi-set appears to be too short to the supplier who calculates, very quickly applying known formulæ, that he needs at least 200 samples of the rock containing iron and at least 200 dead pieces of rock to assure quite reliable and accurate learning. Neither this demand embarrasses the customer: he takes out of his rucksack a geological hammer and crushes each piece of rock into 200 pieces. The supplier clearly understands that he has obtained something quite different from what he needs but he is not able to express in an understandable way the recommendation which the customer should follow when he prepares the training multi-set. ▲

The first serious objection to the practical applicability of recommendations that follows from the statistical learning theory is this: recommendations follow from assumptions that cannot be constructively proved. In another words, in the recommendation of the form 'if A is valid then B is valid', the statement A is formulated in a way about which it is not possible to say whether it is satisfied. But this is not the only imperfection. Let us have a look at, and we will be extremely critical again if the statement B can be constructively proved.

Let us draw our attention to the crucial difference between the two following statements. The first is 'the probability of the wrong decision is not larger than ε'. The second statement reads 'the probability of the fact that the probability of the wrong decision will be larger than ε is not larger than η'. The first statement characterises the specific strategy. Each specific strategy, including the one that will be obtained as a result of learning, can be analysed and it can be found out if the first statement is true.

The second statement does not characterise a specific strategy but a population of strategies. We do not even consider that it can be quite difficult to find out empirically the validity of this statement. What is important is that it is entirely impossible to find out the correctness of a statement on the basis of one single strategy from the population in case the statement relates to the whole population. We want to draw attention to a serious drawback now, i.e., that a single strategy is considered to be the result of learning but the statistical theory of learning actually concerns properties of a population of strategies. This means that the theory speaks about something quite different. Such a cogitation naturally cannot serve as a guarantee to a demanding customer.

This is a very serious discrepancy between the producer of a pattern recognition algorithm and its user. Imagine that a producer supplied us with a product. When it is discovered after some time that the product did not work, the producer starts to insist, that it is a mere coincidence and that such cases do not occur in his business more than in one in ten thousand cases. The worst thing we can do in this situation is to start discussing the topic with him. On the basis of a single product, which is at disposal at the moment, the supplier cannot prove to us that he is right as well as we cannot prove to him that he is wrong. In this case we should kindly interrupt his speech and tell him that he is talking about something in which we have no interest at all. We are indifferent to whether the products he supplied to other customers work correctly. We are interested only in the functionality of the single product he has delivered to us.

Of course, the criticism presented of the statistical view on learning is very strict. It testifies much to the maturity of the statistical theory of learning. Its inadequacy appears only if such a strict view is used. These critical considerations would probably not be introduced if there was not another, deterministic, view of learning, which is indeed not as developed as the statistical one but is deprived only of inadequacies we now point to.

4.5 Outlines of deterministic learning

To avoid the inadequacies mentioned of statistical learning, we have to get on with our comprehension more deeply and challenge some assumptions which have been accepted as self-evident up to now. We will even repudiate some assumptions.

In a widely accepted view of learning in pattern recognition, it is taken as self-evident that the result of learning is a recognition strategy. The goal already formulated in such a way hides in itself an incorrectness because the initial information usable for learning is not sufficient to determine the learning strategy unambiguously. In a simplified formulation, the stated goal of learning does not differ substantially from such a nonsense as the desire 'on the basis that a number q satisfies the inequality $3 \leq q \leq 4$ it should be found out what the number q is equal to'.

The initial information about the strategy sought consists of two parts. The first is an *a priori* information about the set of strategies which includes the actual but unknown strategy, too. The second part is the training set or multi-set. The information contained in both parts usually does not determine the actual strategy in an unambiguous way. The training set or multi-set only allows to exclude such strategies from the *a priori* possible strategies that are in contradiction to it. As a result of this narrowing, a narrower set of possible strategies is obtained. In the general case it is still a set containing more than a single strategy. In such a case the demand to find such a strategy about which it is known only that it belongs to a certain set is similar to the abovementioned nonsense.

In order not to come across such nonsense, it is necessary to explicitly say good bye to the idea that the result of learning is a strategy. The strategy $q^*: X \to K$ cannot be considered as the goal of learning because q^* cannot be determined uniquely. The goal can only be to find out what result is provided by a correct but unknown strategy $q^*(x)$ for some given observation x which is to be recognised. The construction that we call *taught in recognition* is based[1] on this principal idea: *even in spite of the unambiguity of the strategy q^*, its value for some observations $x \in X$ can be determined uniquely.* It is natural that the unambiguity cannot be reached for all observations $x \in X$. In the case in which such an ambiguous observation is recognised, the learning algorithm does not provide any answer and merely says that learning was insufficient to assess correctly the observation. We will show more precisely how such a taught in recognition can work.

Let X be a set of observations and $q^*: X \to D$ be a strategy that will be called the correct strategy. The strategy q^* is unknown but the set of strategies Q, into which q^* belongs, is known.

Let illustrate the described construction by an example.

Example 4.12 Decomposition of a plane by a straight line. *Assume X be a two-dimensional space (a plane), Q is the set of strategies, each of which separates the plane X into two parts by means of a straight line.* ▲

Let XD be a training set of the form

$$XD = \big((x_1, d_1), (x_2, d_2), \ldots, (x_l, d_l)\big)\,, \tag{4.26}$$

where $d_i = q^*(x_i)$ is the decision, $i = 1, 2, \ldots, l$. The sequence (4.26) differs considerably from the training multi-set T on which statistical learning is based, and which has the form

$$T = \big((x_1, k_1), (x_2, k_2), \ldots, (x_l, k_l)\big)\,, \tag{4.27}$$

where k_i is the state in which the object was when the observation x_i was observed.

Special conditions are necessary for obtaining the sequence (training multi-set) T under which the state of the object becomes directly observable. Sometimes it is not possible to provide such conditions. We will not insist that one of these two approaches is always preferred to the other. But sometimes, in the case of image recognition, it is much easier to obtain the sequence (training set) XD than the sequence T.

[1] V.H. The term 'taught in recognition' was selected for the translation from Czech into English even when 'supervised recognition' seems to match the Czech or Russian equivalent better. The latter term is likely to be confused with 'supervised learning'.

There is another important difference. Statistical learning requires that the sequence T has certain statistical features, but it can be impossible to check them practically. The construction to be created in the sequel does not require that the sequence XD has such statistical properties. To obtain the sequence XD it is necessary that there is a device at disposal, possibly even a quite sophisticated one, let us call it *teacher*, that is able to point out the required decision for any arbitrary observation. The device which is being taught will eventually replace the teacher and recognise the new unknown observation on its own. The device will be called *taught in classifier*. We have chosen the unusual animate term because we like to stress the active role of the taught subject (as will be explained soon).

The taught in recognition uses the knowledge that the correct strategy satisfies the relation

$$q^* \in Q\,, \quad q^*(x_i) = d_i\,, \quad i = 1,\ldots,l\,. \tag{4.28}$$

In the previous relation, the set of strategies Q, the observation x_i and the decision d_i, $i = 1,\ldots,l$ are considered as known. The strategy q^* is unknown. It has to be determined in the taught in recognition task about each pair $x \in X$, $d \in D$, whether it follows from Equation (4.28) that $q^*(x) = d$. Let us denote by $Q(XD)$ the set of strategies which satisfy Equation (4.28). It is to be determined in the task whether the set $Q(XD)$ is not empty, and in addition for chosen $x \in X$ to verify

$$\exists d \in D\ \left[\forall\, q \in Q\,(XD)\ (q(x) = d)\right]\,. \tag{4.29}$$

The previous formula says that all the strategies which satisfy Equation (4.28) (consequently, the correct strategy q^* too) assign x to the same class d. If this statement is correct then the correct value $q^*(x)$ be determined unambiguously. If the statement (4.29) is not satisfied then the only possible answer is `not known`, because the information obtained from the teacher does not suffice for a justified conclusion about the value $q^*(x)$.

Example 4.13 Decomposition of the plane by a straight line. *(Continuation of Example 4.12) Figure 4.1 shows the training set represented by white and black circles. The white circles illustrate points for which $q^*(x) = 1$ holds and the black circles display points for which $q^*(x) = 2$ holds. The set $Q(XD)$ is not shown in the figure but it is the set of straight lines that correctly separate the training set. Two convex sets $\widetilde{X}_1$, $\widetilde{X}_2$ are shown in the figure. Both of them are bound by the polyline of infinite length. The set $\widetilde{X}_1 \cup \widetilde{X}_2$ consists*

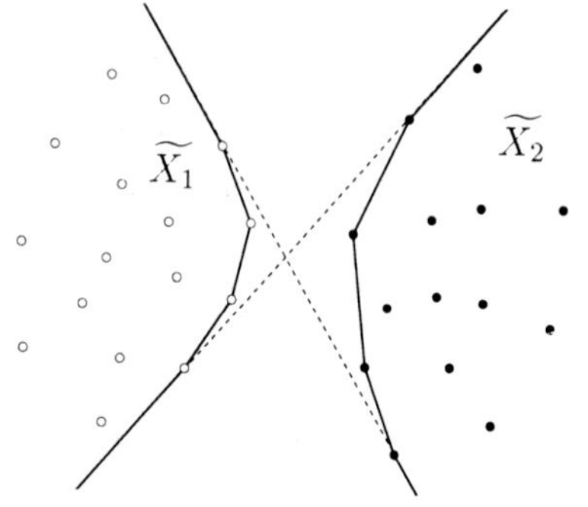

Figure 4.1 Training set consisting of white and black circles. $\widetilde{X}_1$ and $\widetilde{X}_2$ denote regions where unambiguous decision is possible.

of those points $x \in X$ for which the statement (4.29) holds. For each $x \in \widetilde{X}_1$ the $q^*(x) = 1$ holds, and for each $x \in \widetilde{X}_2$ the $q^*(x) = 2$ holds. ▲

To check the validity of the statement (4.29) it is not needed to represent in some particular manner the set $Q(XD)$ of all possible strategies which satisfy Equation (4.28). In fact, it can be seen that the original Equation (4.28) is already the most suitable for direct verification of the statement (4.29). We will show how such a verification can be done.

For each decision $d \in D$ and observation $x \in X$ which has to be recognised, we will write a relation similar to Equation (4.28),

$$\left.\begin{array}{l} q^* \in Q\,, \\ q^*(x_i) = d_i\,, \quad i = 1, 2, \ldots, l\,, \\ q^*(x) = d\,, \end{array}\right\} d \in D\,. \tag{4.30}$$

Equation (4.30) does not express a single expression but an ensemble consisting of $|D|$ relations and each of them corresponds to one decision $d \in D$. There has to be checked for each relation from the ensemble (4.30), i.e., for each value d, whether it is contradictive. Thus the statement (4.29) is equivalent to the statement that there is just a single relation in the ensemble (4.30) which is not contradictive.

We will show at the end of our example how the equivalence of these two statements can be used to recognise a specific observation.

Example 4.14 Recognition of the single observation based on the training set. *Two auxiliary sets $XD_1 = \big(XD \cup \{(x,1)\}\big)$ and $XD_2 = \big(XD \cup \{(x,2)\}\big)$ have to be created based on the training set XD. It is to be checked whether there is a straight line for both sets which classifies them correctly. The result of the analysis can be just one of the four following possibilities.*

1. *If the set XD_1 can be correctly classified with the help of the straight line and the set XD_2 cannot be classified in a similar way this means that the answer $q^*(x) = 1$ is sure to be correct.*
2. *If the set XD_1 cannot be classified with the help of the straight line and it can be done for the set XD_2, then the answer $q^*(x) = 2$ is sure to be correct.*
3. *If each of the sets can be classified with the help of the straight line then it means that the device did not make enough progress in teaching to be able to recognise correctly the submitted observation. That is why it must give the answer* `not known`. *It is important that the taught in classifier detects its inadequacy by itself in the learning process. In this case it can address its teacher with a question of how the given observation is to be classified correctly. When it receives the answer then it can incorporate it into the training set and use it later to recognise next observations better.*
4. *If none of the sets can be classified with the help of the straight line then it means that the initial information which the device had learned from the teacher was contradictive. The taught in classifier with a good sense of humour would be able to give the answer 'you do not know' in this case, and present to the teacher the smallest possible part of the training set provided*

earlier by the teacher that contains the discovered contradiction. In this case the teacher could modify the answers which he had earlier considered to be correct or change the set Q. As is seen, it is not easy to distinguish in this case who actually learns from whom. ▲

In the next lecture, which will be devoted to the linear discriminant functions, we will see that the contradiction in sequences can be discovered constructively.

In the general case the analysis of the contradiction in Equation (4.30) provides more useful information, from a pragmatic point of view, than a mere statement about the validity of Equation (4.29). Let $D(x) \subset D$ be the set of those decisions $d \in D$ whose incorporation into Equation (4.30) does not lead to contradiction. In this case, even if $|D(x)| \neq 1$ but $D(x) \neq \emptyset$, $D(x) \neq D$, i.e., when it is not possible to determine the correct answer uniquely, those decisions $d \notin D(x)$ can be determined which cannot be correct for the observation x. The practical useful result of recognition may not be only the single correct answer $q^*(x)$ but the whole set $D(x)$ of decisions which do not contradict observations.

In the proposed taught in classifier hardly anything has remained from what has been earlier considered as learning in pattern recognition. What has remained from the earlier case is only that the result of recognition is influenced by the training set. The phase of actual learning disappeared entirely in the taught in classifier created. However, the taught in classifier has not lost features of intelligent behaviour; moreover, it seems to have even improved them. Indeed, an approach in which teaching precedes recognising, and later, the recognition starts without being taught, is very far from the mutually fruitful relations between the teacher and the student. Such an approach much more resembles the drill of correct behaviour than education. As we have seen, hard solvable problems occur as a result of separating learning and recognition in time. Well, it is difficult, or even impossible, to mediate during the teaching phase all that has to be sufficient in any future case in which learning will be no longer possible.

The arrangement is entirely different with the suggested procedure. The *taught in classifier* is ready to be recognising in each stage of its activity. Knowledge obtained so far from the teacher suffices to solve some tasks and does not suffice for others. The first group of tasks is solved by the taught in classifier correctly, no doubt. For the other group of tasks the classifier approaches the teacher and enlarges its knowledge. In each stage of such a behaviour, i.e., in the taught in recognition, the taught in classifier has the possibility of processing the information obtained from the teacher. It can detect contradiction in it, or redundancy on the other hand. Redundancy means that some elements in the training set follow from other elements. Finally, the taught in classifier need not wait until an observation occurs which it is not yet able to recognise correctly. It can create such an observation artificially and approach the teacher with it. By doing so it can influence in an active way the content of the training set, i.e., the knowledge received from the teacher.

It is only now that we can see how much we miss in the current statistical theory of learning, to be able to call it the theory of intelligent behaviour. With a certain degree of exaggeration it can be said that ingenious analysis was applied to algorithms resembling more drill than intellectual behaviour.

4.6 Discussion

I would like to ask several questions concerning learning. The first of them, I am afraid, is not very concrete. I feel a deference to subtle mathematical considerations with help of which the main asymptotic properties of learning algorithms were formulated. I can only presume how back breakingly difficult the proofs of theorems, presented in the lecture, can be. It may be natural that after an enthusiastic declaration a sentence starting with 'but' usually follows. I became rather firmly convinced that the scientific significance of the all theory about learning discussed lies hidden within the theory and that there are considerably fewer recommendations resulting from the theory and addressing the world outside it. When reading the part of the lecture about the relation between the length of the training multi-set, the accuracy and reliability of recognition, I have spontaneously recalled the various oriental souvenirs. For instance, a bottle with a narrow neck and a ship built inside it, or several Chinese balls carved from a single piece of ivory and hidden one inside another. When I see such objects I admire the craftsmen's mastery, and mainly the patience needed to create it. At the same time, I cannot get rid of an unpleasant impression from the unanswered question: for what, in fact, could the sailing boat inside a bottle or Chinese balls be useful? I would not like to formulate my questions more precisely because they would not sound polite enough.

The question is indeed somewhat philosophical. That is why we would like to remark first that the most important knowledge, substantially changing the human mind about the world, does not have the form of recommendations but rather of prohibitions. They do not answer the question 'how to do' but rather 'what is impossible'. Recall the energy conservation law. The knowledge of the law will hardly be useful when you would like to lift a piano to the apartment on the tenth floor. On the other hand, the knowledge of the energy conservation law will surely help you if someone comes to you with a project of a self-acting machine which can lift the piano without consuming external energy. Without getting deeply into, how the machine is invented, without looking out for specific erroneous suppositions on which the machine's design is based (and such an analysis can be quite difficult), you can save time and effort and be quite sure that the person bringing you the design is either cheating or is simply an ignoramus. It is not so long, about three hundred years, since mankind did not have as strong and quite general weapon at its disposal as the energy conservation law is. That is why an enormous intellectual effort was wasted first in designing a *perpetuum mobile* and again later in trying to understand why it did not work. Well, it seemed so obvious that a *perpetuum mobile* should work, and moreover, it would be so excellent if it worked.

The scientific and practical value of Chervonenkis–Vapnik theory is that it clearly prohibits some trends in designing learning algorithms in pattern recognition. Let us have a look at the fundamental Theorem 4.2 about the necessary and sufficient condition allowing the recognition device to learn on its own. The condition says that the relative entropy $H(l)/l$ has to converge to zero when the length l of the training multi-set increases to infinity. Even though the entropy $H(l)$ can be almost never calculated, the theorem has, in spite of it's lack of constructiveness, a strong prohibiting power. The entropy $H(l)$ can be easily calculated for a universal recognition device which can implement an arbitrary decomposition of the space of observations. The entropy value $H(l)$ in this case is equal to the lengths l of the learning sequence. The relative entropy $H(l)/l$ then is equal to one for an arbitrary l. That is why the necessary condition that the device can learn is not satisfied. The strict restriction consequently follows from the statistical theory of learning: learning in a universal pattern recognition device is impossible.

Knowledge of this restriction can save you a lot of time and stress. You surely have heard lectures several times or you have read articles, in which it is proved in the first paragraph that the proposed recognition algorithm is universal, and in the second paragraph the learning procedure of the same algorithm is described. Usually it is quite difficult to find some counterexample which would prove that the algorithm is not universal. It is even more difficult to prove that there are classifications which cannot be achieved by learning. Without analysing it in a complicated way, you can be quite sure that the author has made a blunder in at least one of the two paragraphs mentioned. You can also require the author to explain how his results agree with the Chervonenkis–Vapnik theorem. If he does not know anything about the theorem you can stop discussing it without any hesitation, because his professional level can be compared to that of the mechanics 300 years ago. He has not yet known, after all, what everyone should know in these days.

I did not exactly expect such an answer to my question. It is perhaps my fault because I did not ask it precisely enough. I assumed you would stress more the practical usefulness of relations specifying the necessary length of the training data for learning. And neither do I understand everything even I would wish to. I assume that I will have problems in determining the capacity of the set of strategies. Are there some recommendations on how to calculate the capacity in particular cases?

Unfortunately, we can add hardly anything to the definition of the capacity of the set of strategies given in the lecture. Only experience is needed to learn how to estimate the capacity quickly. Practice the calculation of capacities of the following two sets of strategies.

Let X be a two-dimensional linear space in the first case and let the set of strategies contain all strategies which decompose the space X into two parts

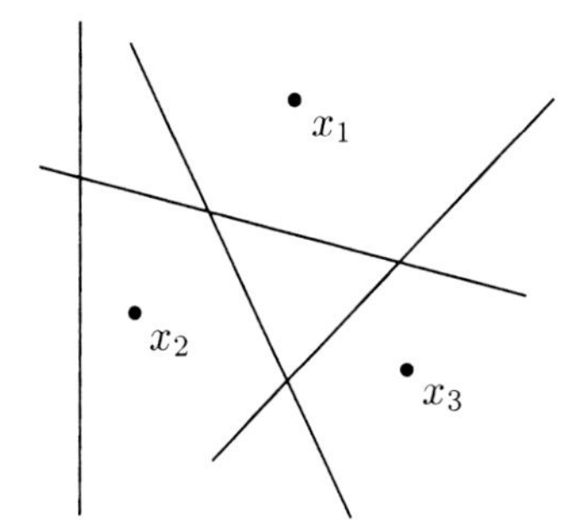

Figure 4.2 Four possible decompositions of three points in the plane with the help of straight lines.

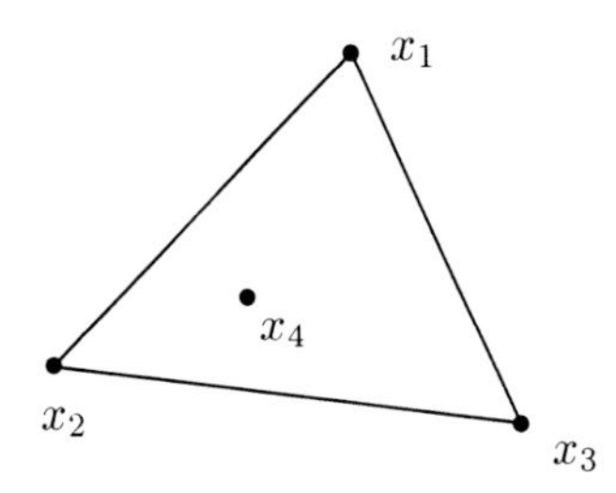

Figure 4.3 The convex hull of four points can be a triangle.

with the help of a straight line. The second case resembles the first one but X is a three-dimensional linear space in which each strategy decomposes the space X into two parts with the help of a plane.

I tried to get used to the concept of the capacity of a set in the two-dimensional example. I have found out that the capacity of the set is $CAP = 4$ on the basis of the following purely geometrical thoughts. There is such a triplet of points which can be decomposed by a straight line into two classes in an arbitrary way. For example, it can be the triplet of points x_1, x_2, and x_3 shown in Fig. 4.2, where all four possible decompositions using four straight lines are illustrated.

It is possible to imagine that no quadruplet of points x_1, x_2, x_3, x_4 can be decomposed by a straight line into two classes in an arbitrary manner. It is obvious because if three out of four points lie on a single straight line then not all decompositions are realisable. Actually, the point in the middle cannot be separated by any straight line from the other two points. If no triplet of points is collinear then the convex hull of the quadruplet of points x_1, x_2, x_3, x_4 can constitute just one of the two following configurations: either a triangle (Fig. 4.3); or a quadrilateral (Fig. 4.4).

In the first case it is not possible to separate the point which is located inside the triangle from the other three vertices of the triangle. In the second case it is not possible to separate one pair of the opposite quadrangle vertices by means of a straight line. That is why the capacity of the class mentioned is equal to four.

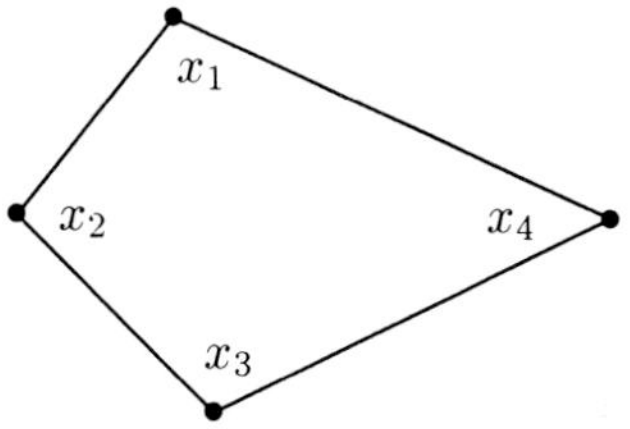

Figure 4.4 *The convex hull of four points can be a quadrilateral.*

Purely geometrical considerations do not suffice in the three-dimensional case. I have analysed the case analytically and the result suits not only the three-dimensional case but also the general k-dimensional case.

Let X be a k-dimensional space and x_i, $i = 0, 1, \ldots, k+1$, be an ensemble of $k+2$ points in this space. Certainly the vectors $x_i - x_0$, $i = 1, \ldots, k+1$, are linearly dependent. This means that the coefficients α_i, $i = 1, \ldots, k+1$, exist, *not all of them being equal to zero and satisfying the equation*

$$\sum_{i=1}^{k+1} \alpha_i \, (x_i - x_0) = 0 \,,$$

or the equivalent equation

$$x_0 \sum_{i=1}^{k+1} \alpha_i = \sum_{i=1}^{k+1} \alpha_i \, x_i \,.$$

The sum $\sum_{i=1}^{k+1} \alpha_i$ is denoted as $-\alpha_0$ and the equation $\sum_{i=0}^{k+1} \alpha_i \, x_i = 0$ is obtained in which the sum $\sum_{i=0}^{k+1} \alpha_i$ is equal to zero too. As not all coefficients α_i equal zero, some of them are positive and others are negative. The symbol I^+ denotes the set of positive coefficients and I^- denotes the set of negative coefficients, i.e.,

$$I^+ = \{i \,|\, \alpha_i > 0\} \,, \quad I^- = \{i \,|\, \alpha_i < 0\} \,.$$

As $\sum_{i=0}^{k+1} \alpha_i = 0$ holds, the equation

$$\sum_{i \in I^-} \alpha_i = - \sum_{i \in I^+} \alpha_i \neq 0$$

holds too. Let new variables β_i, $i = 0, 1, \dots, k+1$, be introduced such that

$$\beta_i = \begin{cases} -\dfrac{\alpha_i}{\sum_{i \in I^-} \alpha_i} \,, & \text{if} \quad i \in I^- \,, \\[2ex] \dfrac{\alpha_i}{\sum_{i \in I^+} \alpha_i} \,, & \text{if} \quad i \in I^+ \,. \end{cases}$$

There follows from the equation $\sum_{i=0}^{k+1} \alpha_i \, x_i = 0$ that

$$\left. \begin{array}{l} \displaystyle\sum_{i \in I^-} \alpha_i \, x_i + \sum_{i \in I^+} \alpha_i \, x_i = 0 \\ \displaystyle\qquad \Rightarrow \left(\sum_{i \in I^-} \alpha_i \, x_i = - \sum_{i \in I^+} \alpha_i \, x_i \right) \\ \displaystyle\qquad \Rightarrow \left(\sum_{i \in I^-} \beta_i \, x_i = \sum_{i \in I^+} \beta_i \, x_i \right) , \end{array} \right\} \tag{4.31}$$

where

$$\sum_{i \in I^-} \beta_i = \sum_{i \in I^+} \beta_i = 1 \,. \tag{4.32}$$

I can assert now that there is no such hyperplane for which the ensemble of points x_i, $i \in I^-$, lies on one side of it and the ensemble of points x_i, $i \in I^+$, lies *on its other. Indeed, if such a hyperplane existed then also the vector $\alpha \in X$ and a number θ would exist that would satisfy the system of inequalities*

$$\left.\begin{array}{ll} \langle \alpha, x_i \rangle \geq \theta\,, & i \in I^+\,, \\ \langle \alpha, x_i \rangle < \theta\,, & i \in I^-\,, \end{array}\right\} \tag{4.33}$$

where $\langle \alpha, x_i \rangle$ denotes the scalar product of vectors α and x_i. Because the sum of all coefficients β_i, $i \in I^+$, is equal to 1 and all coefficients are positive, it follows from the first group of inequalities in the system (4.33) that

$$\sum_{i \in I^+} \beta_i \langle \alpha, x_i \rangle \geq \theta\,.$$

and it follows from the second group of inequalities that

$$\sum_{i \in I^-} \beta_i \langle \alpha, x_i \rangle < \theta\,.$$

This means that the inequality

$$\sum_{i \in I^+} \beta_i \langle \alpha, x_i \rangle > \sum_{i \in I^-} \beta_i \langle \alpha, x_i \rangle \tag{4.34}$$

holds. But from the derivation (4.31) the equation

$$\sum_{i \in I^-} \beta_i \langle \alpha, x_i \rangle = \sum_{i \in I^+} \beta_i \langle \alpha, x_i \rangle$$

follows which contradicts the inequality (4.34). I have proved that in the k-dimensional space no ensemble of $k+2$ points can be decomposed in an arbitrary manner into two classes by means of a hyperplane.

It is not difficult to show that an ensemble of $k+1$ points exists that can be decomposed into two classes in an arbitrary manner using a hyperplane. For instance, it could be the ensemble: the point x_0 has all coordinates equal to zero, and the point x_i, $i = 1, \ldots, k$, has all coordinates equal to zero except i-th coordinate which is non-zero. I can thus consider the following statement as proved.

Let X be the k-dimensional space and Q be a set of strategies of the form $X \to \{1, 2\}$. Each strategy decomposes the space X into two classes using a hyperplane. The capacity of the set Q is $k+2$.

Indeed, the capacities of the set of strategies are not as terrible as it might look at first glance.

We are glad to hear that. Have a look at the following set of strategies in a two-dimensional space to be more certain about capacities: each strategy is defined by a circle which decomposes a plane into two parts: the inner and the outer part of the circle. We will prompt you a little: apply the straightening of the feature space.

I have managed the circle case with your hint quite quickly. First, there is a quadruplet of points in the plane which can be decomposed by means of circles into two classes in an arbitrary manner. For instance, it can be the quadruplet illustrated in Fig. 4.5, where also all possible decompositions using eight circles are shown. Consequently the capacity of given set of strategies is not less than 5.

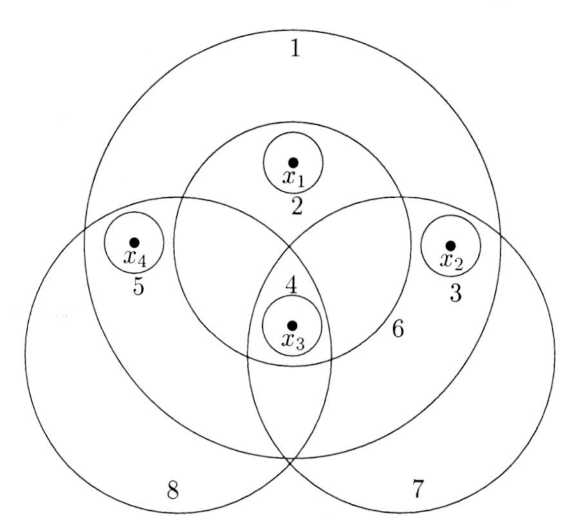

Figure 4.5 Eight possible decompositions of four points in the plane by means of eight circles.

Second, I will prove that already no ensemble of five points can be decomposed into arbitrary two classes with the help of circles. Thus the capacity of the introduced set of strategies is equal to 5.

Let x and y be the coordinates of an arbitrary point in the plane. The given set of strategies contains strategies q in the form:

$$q(x,y) = \begin{cases} 1\,, & \text{if } (x-x_0)^2 + (y-y_0)^2 \le r^2\,, \\ 2\,, & \text{if } (x-x_0)^2 + (y-y_0)^2 > r^2\,, \end{cases} \tag{4.35}$$

or

$$q(x,y) = \begin{cases} 1\,, & \text{if } (x-x_0)^2 + (y-y_0)^2 > r^2\,, \\ 2\,, & \text{if } (x-x_0)^2 + (y-y_0)^2 \le r^2\,. \end{cases} \tag{4.36}$$

Each strategy in the form (4.35) or (4.36) can be expressed in the form

$$q(x,y,z) = \begin{cases} 1\,, & \text{if } \alpha\,x + \beta\,y + \gamma\,z \ge \theta\,, \\ 2\,, & \text{if } \alpha\,x + \beta\,y + \gamma\,z < \theta\,, \end{cases} \tag{4.37}$$

where $z = x^2 + y^2$. The converse holds, i.e., any strategy in the form (4.37) on the set of points satisfying the constraint $z = x^2 + y^2$ can be expressed either in the form (4.35) or in the form (4.36). A direct statement is sufficient for me from which it follows that the capacity of the class (4.35), (4.36) cannot be *greater than the capacity of the class (4.37). The class (4.37) is the class of linear decision functions in the three-dimensional space. I had proved before that a capacity of the class (4.37) was 5. In such a way I have proved that the capacity of the class (4.35), (4.36) is equal to 5.*

Now I can see that the capacity can be also determined exactly for the set of strategies given by the quadratic discriminant functions in a general form. Those strategies in question are those optimal in the case in which the observation x under the condition of each fixed state k is an n-dimensional Gaussian random variable in a general form. The strategy of this kind then has the form

$$\left.\begin{aligned} q(x) &= 1\,, \quad \text{if} \quad f(x) \geq \theta\,, \\ q(x) &= 2\,, \quad \text{if} \quad f(x) < \theta\,, \end{aligned}\right\} \tag{4.38}$$

for a threshold value θ and for a quadratic function f which has the form

$$f(x) = \sum_{i=1}^{n} \alpha_i\, x_i^2 + \sum_{i=1}^{n} \sum_{j=i+1}^{n} \beta_{ij}\, x_i\, x_j + \sum_{i=1}^{n} \gamma_i\, x_i\,. \tag{4.39}$$

The designation x_i in Equation (4.39) denotes the i-th coordinate of the point x. I will show that the set X^* exists which consists of $2n + \frac{1}{2}n(n-1) + 1$ points and can be decomposed into two classes in an arbitrary manner. From this it will immediately follow that for the capacity CAP of the set of strategies studied

$$CAP > 2n + \frac{1}{2}n(n-1) + 1 \tag{4.40}$$

holds. The ensemble X^* of points is defined in the following way. The ensemble will consist of the sets X_1^- and X_1^+ introduced in addition (each of them consisting of n points), and also of the set X_2 (consisting of $\frac{1}{2}n(n-1)$ points) and the set X^0 containing a single point. The points in the set X_1^- will be numbered by the index $i = 1, 2, \ldots, n$. The point with the index i will be denoted by $(x^i)^-$ and will be defined as the point in which all the coordinates equal zero but the i-th one the value of which is -1. In a similar way the i-th point in the ensemble X_1^+ will be denoted by $(x^i)^+$. It will be defined that all coordinates in this point equal zero except the i-th one which assumes the value $+1$. The points in the set X_2 will be numbered by two indices $i = 1, 2, \ldots, n$ and $j = i+1, i+2, \ldots, n$. The (ij)-th point will be denoted as x^{ij} and will be determined as the point in which all the coordinates equal zero except the i-th and j-th the value of which is 1. The single point of which the set X_0 consists will be denoted by x^0 and will be defined as the origin of the coordinate system.

Let X_1^* and X_2^* be an arbitrary decomposition of the ensemble $X^* = X_1^- \cup X_1^+ \cup X_2 \cup X_0$ into two classes. Let us prove that such coefficients α_i, β_{ij}, γ_i exist for the given decomposition and such a threshold value θ exists in the strategy (4.38), (4.39) that it satisfies the system of inequalities

$$\left.\begin{aligned} f(x) &\geq \theta\,, \quad x \in X_1^*\,, \\ f(x) &< \theta\,, \quad x \in X_2^*\,. \end{aligned}\right\} \tag{4.41}$$

Because $f(x^0) = 0$ the threshold value θ cannot be positive if $x^0 \in X_1^*$. On the other hand, the θ must be positive if $x^0 \in X_2^*$. I will choose $\theta = \frac{1}{2}$ if $x^0 \in X_2^*$, and $\theta = -\frac{1}{2}$ if $x^0 \in X_1^*$. I will analyse only the first case because the analysis of the second case is almost the same. I will show how the coefficients α_i, β_{ij}, γ_i can be determined which satisfy the constraints

$$\left.\begin{aligned} f(x) = 1\,, \quad x \in X_1^*\,, \\ f(x) = 0\,, \quad x \in X_2^*\,, \end{aligned}\right\} \tag{4.42}$$

and thus satisfy the constraints (4.41) too. We will introduce the auxiliary notation, i.e., the numbers $(k^i)^-$, $(k^i)^+$, k^{ij}, $i = 1, 2, \ldots, n$, $j = i+1, i+2, \ldots, n$ such that

$$\begin{aligned} (k^i)^- &= 0\,, \text{ if } (x^i)^- \in X_2^*\,, \\ (k^i)^- &= 1\,, \text{ if } (x^i)^- \in X_1^*\,, \\ (k^i)^+ &= 0\,, \text{ if } (x^i)^+ \in X_2^*\,, \\ (k^i)^+ &= 1\,, \text{ if } (x^i)^+ \in X_1^*\,, \\ k^{ij} &= 0\,, \text{ if } \quad x^{ij} \in X_2^*\,, \\ k^{ij} &= 1\,, \text{ if } \quad x^{ij} \in X_1^*\,. \end{aligned}$$

If this notation is used the system of equations (4.42) assumes the form

$$\left.\begin{aligned} f((x^i)^-) &= (k^i)^-\,, \\ f((x^i)^+) &= (k^i)^+\,, \\ f(x^{ij}) &= k^{ij}\,, \end{aligned}\right\} \quad \begin{aligned} &i = 1, 2, \ldots, n, \\ &j = i+1, i+2, \ldots, n, \end{aligned} \tag{4.43}$$

or, which is equivalent,

$$\left.\begin{aligned} \alpha_i - \gamma_i &= (k^i)^-\,, \\ \alpha_i + \gamma_i &= (k^i)^+\,, \\ \alpha_i + \alpha_j + \gamma_i + \gamma_j + \beta_{ij} &= k^{ij}\,, \end{aligned}\right\} \quad \begin{aligned} &i = 1, 2, \ldots, n, \\ &j = i+1, i+2, \ldots, n. \end{aligned} \tag{4.44}$$

The system of equations (4.44) has an obvious solution

$$\alpha_i = \frac{(k^i)^- + (k^i)^+}{2}\,, \quad \gamma_i = \frac{(k^i)^+ - (k^i)^-}{2}\,, \quad \beta_{ij} = k^{ij} - (k^i)^+ - (k^j)^+\,,$$

which proves the inequality (4.40).

Furthermore, I will prove relatively easily that the capacity of the set of strategies analysed cannot be greater than $2n + \frac{1}{2}n(n-1) + 2$. It was shown in the lecture that the n-dimensional linear space X can be mapped into the $2n + \frac{1}{2}n(n-1)$-dimensional space Y in such a way that an arbitrary strategy in the form (4.38) to (4.39) in the space X corresponds to a decomposition of the space Y using the hyperplane. I have already proved earlier that the capacity of the set of hyperplanes in an m-dimensional space is equal to $m + 2$. From what has been said it follows that the capacity of the set of hyperplanes in the space Y is $2n + \frac{1}{2}n(n-1) + 2$. The capacity of the set of strategies in the form (4.38) to (4.39) cannot be greater and thus

$$CAP \leq 2n + \frac{1}{2}n(n-1) + 2\,.$$

From this, with regard to the inequality (4.40) proved earlier, the equation

$$CAP = 2n + \frac{1}{2}n(n-1) + 2$$

follows.

When you have so thoroughly investigated the model with Gaussian random variables, could you do, for completeness, the same also for the model with conditional independence? Try to determine the capacity of the set of strategies in the form

$$x \in \begin{cases} X_1\,, & \text{if } \sum_{i=1}^{n} \log \dfrac{p_{X_i|1}(x_i)}{p_{X_i|2}(x_i)} \geq \theta\,, \\ X_2\,, & \text{if } \sum_{i=1}^{n} \log \dfrac{p_{X_i|1}(x_i)}{p_{X_i|2}(x_i)} < \theta\,, \end{cases} \tag{4.45}$$

which are optimal in the case in which the observation x under the condition of a fixed state is the random variable $x = (x_1, x_2, \ldots, x_n)$ with independent components.

In the discussion after Lecture 3 I proved that any strategy of the form (4.45) could be expressed as a linear discriminant function of the dimension

$$\sum_{i=1}^{n} k(i) - n\,,$$

where $k(i)$ was a number of values of the variable x_i. From this result there immediately follows

$$CAP \leq \sum_{i=1}^{n} k(i) - n + 2\,, \tag{4.46}$$

because I have just proved that the capacity of the set of hyperplanes in an m-dimensional space is $m + 2$.

I will now show that in the inequality (4.46) the relation $\leq$ can be substituted by an equality. I need to prove that an ensemble of $\sum_{i=1}^{n} k(i) - n + 1$ of points in the space X exists which can be decomposed into two classes by means of strategies of the form (4.45).

The ensemble sought may be constructed in the following way. Let me choose an arbitrary point $x^0 = (x_1^0, x_2^0, \ldots, x_n^0)$ and include it in the ensemble. Then every point x' differing from x^0 in only one component is included in the ensemble too. So the number of points in the ensemble will be $\sum_{i=1}^{n} k(i) - n + 1$. Have I to prove that such a set of observation can be decomposed in an arbitrary way?

It is not necessary. We think that it is quite clear. We thank you for cooperation and this will be enough for today.

April 1997.

4.7 Bibliographical notes

Three formulations of learning tasks have been introduced in this lecture. The first formulation with respect to the maximal likelihood is a direct transfer of known statistical methods into pattern recognition [Nilsson, 1965]. Let us mention only Gauss and Fisher [Fisher, 1936] from the statistical sources related to the most likely estimates. If someone is interested in the matter we recommend the textbook by Waerden [Waerden, 1957]. A theoretical analysis of the properties of learning algorithms in pattern recognition according to the first formulation is represented by [Raudys and Pikelis, 1980].

The second minimax formulation of learning according to a non-random ensemble was suggested in [Schlesinger, 1989], who was inspired by practical tasks [Schlesinger and Svjatogor, 1967]. The theoretical analysis of the approach will be given in this monograph in Lecture 8.

The third formulation seeks a strategy which correctly recognises the training sequence [Rosenblatt, 1962; Ajzerman et al., 1970]. Many other publications stem from these works. The third formulation was analysed by Chervonenkis, Vapnik [Vapnik and Chervonenkis, 1974; Vapnik, 1995], [Vapnik, 1998] and has been developed into a deep theory. The first work mentioned established the basis of our explanation in this lecture.

Let us compare Raudis' theory analysing the learning in the first formulation with respect to the maximal likelihood [Raudys and Pikelis, 1980] with the conclusions by Vapnik with respect to the third formulation. The first case yields less general assumptions and thus estimates a shorter training sequence. The second approach is more general and thus more pessimistic in its estimates.

Another interesting view of statistical learning theory is given in [Vidyasagar, 1996].

We have adapted Zagorujko's example with geologist [Zagorujko, 1999]. This book is of interest on its own, as it gives insight into the research of one Russian group strong in clustering, and lists several tasks solved practically.

Lecture 5

Linear discriminant function

5.1 Introductory notes on linear decomposition

In the previous lectures we have pointed out several times that linear discriminant functions deserve some special attention. First, some statistical models are known to have the Bayesian or non-Bayesian strategy implemented, namely, by means of linear discriminant functions.

Second, some nonlinear discriminant functions can be expressed as linear functions through straightening of the feature space which has been discussed in Section 3.2. This is possible in the case in which it is known that the nonlinear discriminant function $f: X \to \mathbb{R}$ can be expressed in the form

$$f(x) = \sum_{j \in J} \alpha_j \, f_j(x)$$

with known functions $f_j: X \to \mathbb{R}$, $j \in J$ and unknown coefficients α_j. In this case the searching for a discriminant function, which in the input space X is nonlinear, is reduced to searching for a linear discriminant function in the straightened space Y of the dimension $|J|$. The space X is mapped into the space Y so that the point $x \in X$ corresponds to the point $y \in Y$ the j-th coordinate of which is $f_j(x)$.

Third, from the theoretical standpoint it is important that there exists a universal way of representing the initial observation space in the space of probabilities where the solution of any known statistical task is implemented by means of linear discriminant functions, i.e., by means of the decomposition of the probability space into convex cones.

Fourth, the capacity of linear strategies in an n-dimensional space is known to be $n + 2$ and the learning task to be correct. From that it follows that the strategy tuned to a finite training multi-set does not differ much from the correct strategy tuned to the statistical model. Therefore the tuning strategy for a concrete statistical model can be replaced by tuning for the given training multi-set. We point out once more that the replacement is possible thanks to the finite capacity of the linear discriminant function class. In the preceding lecture, for this case an explicit relation between the necessary length of the training multi-set, the accuracy, and reliability of learning was mentioned, expressed by a practically applicable formula.

All these advantages would naturally not be of great importance if there were not at our disposal procedures for finding linear discriminant functions. They are the main topic of this lecture. We will see that different linear discriminant function tasks, which seem not to be similar to each other at first glance, are acting together (are collaborating, in fact). We will see that the properties of one type of tasks are helpful for solving other tasks the properties of which are not so evident.

5.2 Guide through the topic of the lecture

We start with a survey which will provide an overview of linear discriminant functions and so make easier the understanding of basic outcomes from the very beginning. Further explanation in the lecture will provide a deeper insight in them and prove them.

Let X be a multi-dimensional linear space. The result of the observation of an object is a point x in this space. Let k be the state of the object inaccessible to observation and let it assume two values $\{1, 2\}$ only. Let it be known that the distribution of conditional probabilities $p_{X|K}(x \mid k)$, $x \in X$, $k \in K$, is a multi-dimensional Gaussian distribution. The mathematical expectation μ_k and the covariance matrix σ_k, $k = 1, 2$, of these probability distributions are not known. However, it is known that the parameters (μ_1, σ_1) belong to a certain finite set of parameters $\{(\mu^j, \sigma^j) \mid j \in J_1\}$. Similarly (μ_2, σ_2) are also unknown parameters belonging to the finite set $\{(\mu^j, \sigma^j) \mid j \in J_2\}$. We used both the superscript and subscript indices. For example, μ_1 and σ_1 mean real, but unknown, parameters of an object that is in the first state. Parameters (μ^j, σ^j) for some of the superscripts j are one of the possible value pairs which the parameter can assume.

This case can be illustrated with Fig. 5.1. Ellipses in the figure show five random Gaussian quantities which assume values in a two-dimensional space (on a plane). For the time being let us ignore the separating straight line q. Let, e.g., $J_1 = \{1, 2, 3\}$ and $J_2 = \{4, 5\}$. It would mean that the object is characterised in the first state by a random vector, which has the first, second, or third probability distribution, but it would not be known which of them it was. It is similar for the second state and the fourth and fifth probability distribution.

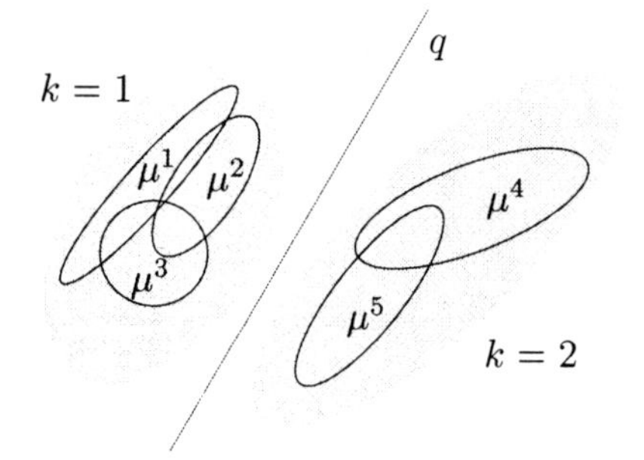

Figure 5.1 Generalised Anderson task in a two-dimensional space.

Thus we have two object classes and each of them is described by a mixture of Gaussian distributions. The components of each mixture are known, and unknown are only their weights. If the state of the object k is to be found from the mentioned incomplete *a priori* knowledge of the statistical model and the known observation x then the task is to be formulated as a task of statistical decision making with non-random interventions (it has been described in a more general manner in Subsection 2.2.5). If we used the results of the analysis in our case, we would be seeking the strategy $q\colon X \to \{1,2\}$ which minimises the value

$$\max_{j \in J_1 \cup J_2} \varepsilon(j, \mu^j, \sigma^j, q)\,, \tag{5.1}$$

where $\varepsilon(j, \mu^j, \sigma^j, q)$ is the probability that the Gaussian random vector x with mathematical expectation μ^j and the covariance matrix σ^j satisfies either the relation $q(x) = 1$ for $j \in J_2$ or $q(x) = 2$ for $j \in J_1$.

In other words, the minimal value ε and the strategy q^* are sought that satisfy two conditions:

1. The probability of a wrong estimate of the state under the condition that the actual state is 1 is not greater than ε, which is valid independently of the values of the mathematical expectation μ_1 and the covariance matrix σ_1, but only when $(\mu_1, \sigma_1) \in \{(\mu^j, \sigma^j), j \in J_1\}$.
2. The probability of a wrong evaluation of a state under the condition that the actual state is 2, is not greater than ε, either, which is valid independently of the values of mathematical expectation μ_2 and the covariance matrix σ_2, but only when $(\mu_2, \sigma_2) \in \{(\mu^j, \sigma^j), j \in J_2\}$.

From the results presented in Subsection 2.2.5 it follows that the statistical decision making task with non-random interventions is reduced to searching for a minimax solution in the weight space of mixture components.

We are interested in the task (5.1) with an additional constraint on the strategy q. We require the discriminant function to be linear, i.e., to be the hyperplane $\langle \alpha, x\rangle = \theta$ and

$$q(x) = \begin{cases} 1\,, & \text{if} \quad \langle \alpha, x\rangle > \theta\,, \\ 2\,, & \text{if} \quad \langle \alpha, x\rangle < \theta\,, \end{cases} \tag{5.2}$$

at a certain vector $\alpha \in X$ and the threshold value θ. Recall that $\langle \alpha, x\rangle$ denotes the scalar product of the vectors α, x. For the two-dimensional case illustrated above in Fig. 5.1, the observation plane should be divided into two half-planes so that the first half-plane should contain the majority of random realisations from the first, second, and third probability distributions, and the second half-plane contains those from the fourth and fifth probability distribution. This distribution is represented in Fig. 5.1 by the separating straight line q.

The task (5.1) satisfying the requirement (5.2) is a generalisation of the known task by Anderson and Bahadur [Anderson and Bahadur, 1962], who formulated and solved the task for the case $|J_1| = |J_2| = 1$. Our more general case will be called *generalised Anderson task*.

The abovementioned formulation of generalised Anderson task includes another particular case worth attention, i.e., when all covariance matrices σ^j,

$j \in J_1 \cup J_2$ are unit. This task is used even in the case of pattern recognition algorithms which are determined by a training set. It is referred to as the *optimal separation of a finite sets of points.* Let $\widetilde{X}$ be a finite set of points $x_1, x_2, \ldots, x_n$ from the space X which is decomposed into two subsets $\widetilde{X}_1$ and $\widetilde{X}_2$. A separating hyperplane is sought which will let the subset $\widetilde{X}_1$ remain in one half-space, and the subset $\widetilde{X}_2$ in the other half-space. And moreover, the hyperplane is as distant from the both divided subsets as possible. More precisely speaking, a vector α and the threshold value θ are sought in order that:

1. all $x \in \widetilde{X}_1$ satisfy the inequality

$$\langle \alpha, x \rangle > \theta \, ; \tag{5.3}$$

2. all $x \in \widetilde{X}_2$ satisfy the inequality

$$\langle \alpha, x \rangle < \theta \, ; \tag{5.4}$$

3. under the conditions (5.3) and (5.4) the number

$$\min \left(\min_{x \in \widetilde{X}_1} \frac{\langle \alpha, x \rangle - \theta}{|\alpha|} \, , \; \min_{x \in \widetilde{X}_2} \frac{\theta - \langle \alpha, x \rangle}{|\alpha|} \right) \tag{5.5}$$

   reaches its maximal value.

Another simplification of the preceding task of optimal separation of two finite sets of points is a case in which an arbitrary hyperplane separating the sets $\widetilde{X}_1$ and $\widetilde{X}_2$ is sought. This task is referred to as the *simple separation of finite sets of points.* This means that the solution must satisfy the conditions (5.3), (5.4) and does not take into account the requirement (5.5).

We will begin this lecture by thoroughly analysing Anderson task. The minimised optimisation criterion will appear to be unimodal. This is a positive statement since for the optimisation easy steepest descent methods can be used with which the optimum can be found without being stuck in local extremes. A disadvantageous statement is that the minimised unimodal criterion is neither convex, nor differentiable. Therefore neither method calculating the gradient, nor that the gradient in the point corresponding to the minimum is equal to zero can be applied. The minimum will occur in the point where no gradient exists, and therefore other conditions for the minimum are to be found which are not based on the concept of the gradient. Such necessary and sufficient conditions for the minimum will be formulated and the steepest-decreasing algorithm will be presented which can be applicable even in this case.

For a particular case of Anderson task, which is the optimal separation of the finite sets of points, we will prove that the optimisation conditions will be simplified to the minimisation of the quadratic function on a convex polyhedron. Such a task can be easily solved by elaborate methods of convex optimisation.

For an even more specified task of a simple separation of sets of points we will first remember perceptron algorithms and quote some less known algorithms proposed by the Russian mathematician Kozinec.

At the close of the lecture we will present results which will cover all the tasks studied here in a single frame. This will reveal less obvious relations between Anderson general task, the optimal separation of the sets of points, and the most specified simple separation of the sets of points. Understanding these relations, we can properly modify perceptron and Kozinec algorithms so that they will be applicable even for solving Anderson general task.

5.3 Anderson tasks

5.3.1 Equivalent formulation of generalised Anderson task

We will be dealing with a generalisation of Anderson task in which the numbers of elements of the classes J_1 and J_2 need not be equal to 1. Without loss of generality we can assume that the recognition strategy based on comparing the value of the linear function $\langle\alpha, x\rangle$ with the threshold value θ can be replaced by an equivalent strategy making decision according to the sign of the linear function $\langle\alpha, x\rangle$. It can be obtained in the following standard way.

Let us map the original n-dimensional space X into $(n+1)$-dimensional space X' so that the point x in the space X is mapped to the point x' in the space X'. The first n coordinates of the point x' are the same as the n corresponding coordinates in the point x and the coordinate number $(n+1)$ is always $+1$. We will denote the threshold value θ as $-\alpha_{n+1}$. We find out that the strategy based on examining the inequality $\langle\alpha, x\rangle > \theta$ can be replaced by an equivalent strategy based on examining the inequality $\langle\alpha', x'\rangle > 0$. In this case the first n components of the vector α' are the same as those with the vector α and the component number $(n+1)$ is $-\theta$. With respect to this modification we formulate generalised Anderson task once more.

Let X be the multi-dimensional linear space as before, and J be a set of indices of a certain ensemble of Gaussian random variables which assume their values in this space. For each random variable from the ensemble determined by the index $j \in J$ the mathematical expectation μ^j and the covariance matrix σ^j are known. The group J is divided into two classes J_1 and J_2. Let the decision making strategy have the following form given by the vector α. At $\langle\alpha, x\rangle > 0$ a decision is made that the observation x is a realisation of the random variable from the first class and the $\langle\alpha, x\rangle < 0$ is considered to be a realisation of the random variable from the second class. We will denote by the symbol $\varepsilon^j(\alpha)$ the probability of the event that the realisation of the j-th random variable will not be included into the class where it actually belongs. This means that for $j \in J_1$ the symbol $\varepsilon^j(\alpha)$ denotes the probability that the random Gaussian vector x with the mathematical expectation μ^j and the covariance matrix σ^j will satisfy the inequality $\langle\alpha, x\rangle \leq 0$. Similarly for $j \in J_2$ the symbol $\varepsilon^j(\alpha)$ denotes the probability of the inequality $\langle\alpha, x\rangle \geq 0$ for the random Gaussian vector x, whose mathematical expectation is μ^j and the covariance matrix is σ^j.

In generalised Anderson task, a vector $\alpha \neq 0$ is sought which minimises the criterion $\max_{j\in J} \varepsilon^j(\alpha)$. Thus the following vector has to be calculated

$$\alpha = \underset{\alpha}{\operatorname{argmin}} \max_{j\in J} \varepsilon^j(\alpha) . \tag{5.6}$$

To make further analysis more convenient, we will present the task (5.6) in a slightly different equivalent formulation. Let us introduce vectors μ'^j in this form

$$\mu'^j = \begin{cases} \mu^j , & \text{for} \quad j \in J_1 , \\ -\mu^j , & \text{for} \quad j \in J_2 . \end{cases}$$

Figure 5.2 illustrates the transformation for the case in which $J_1 = \{1, 2\}$ and $J_2 = \{3, 4\}$. For any vector α there holds that the probability of the inequality $\langle \alpha, x \rangle \geq 0$ for the random Gaussian vector x with the mathematical expectation μ^j and the covariance matrix σ^j is the same as the probability of the inequality $\langle \alpha, x \rangle \leq 0$ for the random Gaussian vector x with the mathematical expectation $-\mu^j$ and the covariance matrix σ^j. Thus generalised Anderson task (5.6) can be expressed in the following equivalent formulation.

For the ensemble $\big((\mu^j, \sigma^j), j \in J\big)$, a non-zero vector α has to be sought which minimises the number $\max_j \varepsilon^j(\alpha)$,

$$\alpha = \underset{\alpha}{\operatorname{argmin}} \; \max_j \; \varepsilon^j(\alpha) , \tag{5.7}$$

where $\varepsilon^j(\alpha)$ is the probability that the random Gaussian vector x with the mathematical expectation μ^j and the covariance matrix σ^j will satisfy the inequality $\langle \alpha, x \rangle \leq 0$.

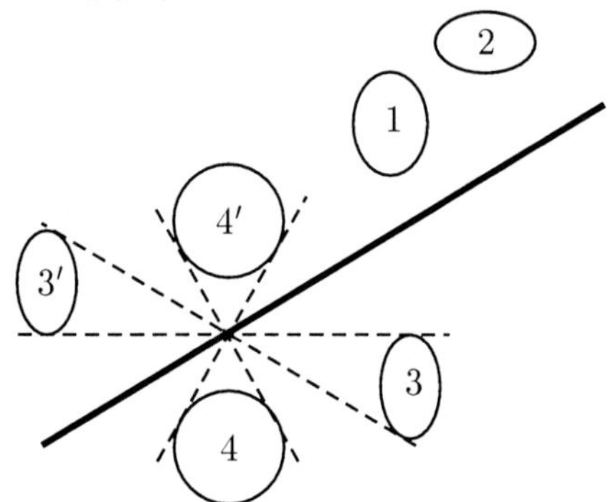

Figure 5.2 The straight line separating the ellipses 1, 2 from the ellipses 3, 4 is equivalent to the straight line leaving the ellipses 1, 2 and 3′, 4′ along one side.

For better illustration let us go back to geometrical considerations. In the original task there were two sets corresponding to Gaussian random variables and they were separated by a hyperplane into two parts. In the present formulation we have one set and want to achieve to have it in one half-space (i.e., on one side of the separating hyperplane). That very formulation of the task is convenient since in further examination it will not be necessary to keep in memory for each j whether it belongs to the set J_1 or to the set J_2. For these two cases, therefore, different formulæ will not have to be used.

Before starting a formal analysis of the task (5.7) we will examine it informally.

5.3.2 Informal analysis of generalised Anderson task

The input data for generalised Anderson task are formed by an ensemble of pairs $\big((\mu^j, \sigma^j),\ j \in J\big)$. The ensemble characterises a certain group of multidimensional Gaussian random variables that assume their values in the linear space X. For the given vector μ, the positive-definite matrix σ and for the number r a set of points x will be introduced which satisfy the condition

$$\big\langle (x - \mu) \, , \, \sigma^{-1} \cdot (x - \mu) \big\rangle \leq r^2 , \tag{5.8}$$

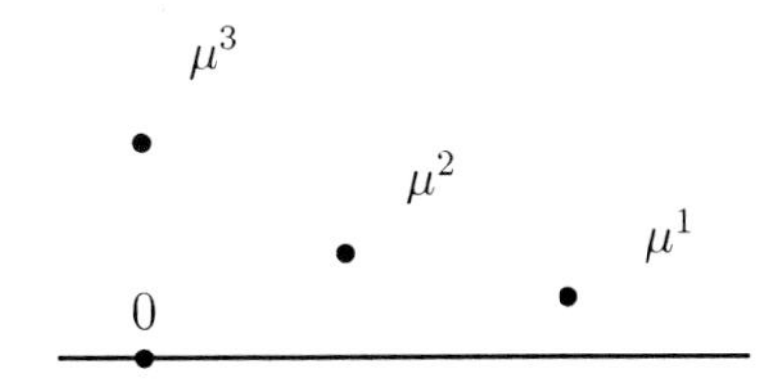

Figure 5.3 Straight line passing through the origin, leaving the points μ_1, μ_2 and μ_3 along one side.

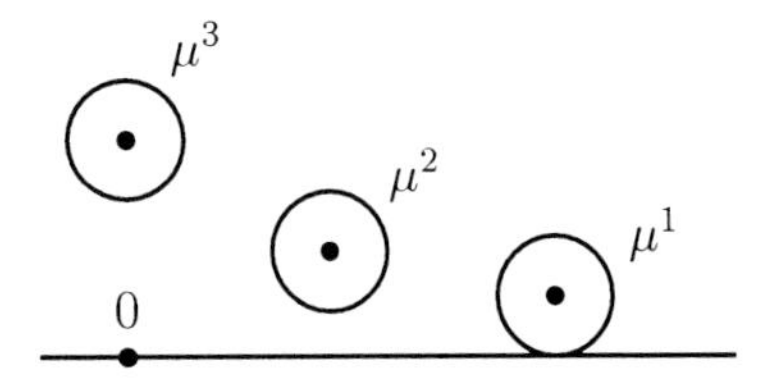

Figure 5.4 Contact of one ellipse with the straight line.

where $\cdot$ denotes a matrix product, in our case the product of a matrix and a vector.

The set of points defined by the preceding inequality will be denoted $E(r, \mu, \sigma)$ and will be referred to as an ellipse of the size r, even in which we have in mind a multi-dimensional body. The concept of ellipse will be used also for a multi-dimensional case where from the geometrical point of view it would be called an ellipsoid.

Let us express generalised Anderson task in the following equivalent form (on the basis of common sense and without any proof, for the time being). For the given ensemble of pairs $\big((\mu^j, \sigma^j),\ j \in J\big)$, such a vector α is to be found for the half-space $\{x \in X\,,\ |\ \langle \alpha, x\rangle \geq 0\}$ to contain the union of the ellipses $\bigcup_{j\in J} E(r, \mu^j, \sigma^j)$ with their largest possible size r.

If the formulation presented was really equivalent to the requirement (5.7) (and we will see later that it really is so) then the hyperplane, which is the solution of the task (5.7), could be sought by means of a procedure which will be presented first in the simplest two-dimensional case.

Let μ^1, μ^2 and μ^3 be the mathematical expectations of three random variables, as it is shown in Fig. 5.3. First, a straight line is to be drawn that passes through the origin of the coordinates and leaves all three points μ^1, μ^2 and μ^3 in the same half-space.

If such a straight line were not exist, it would mean that for each linear discriminant function at least one random variable existed for which the probability of the wrong decision was greater than 0.5. In such a case it would not be necessary to solve the task because even the best result would not be practically applicable. In Fig. 5.3 we can see that such a straight line does exist, e.g., as a horizontal straight line. Around the points μ^1, μ^2 and μ^3 the ellipses begin to grow whose sizes are the same at each instant and whose orientation depends on the matrices σ^1, σ^2 and σ^3. At the same time, with growing sizes of the ellipses, the position of the straight line changes so that all three ellipses should lie, all the time, in one half-plane defined by the straight line. The growth of the ellipse sizes continues till some ellipses (it may be even one ellipse) force the straight line into the only one possible position. Here a further growth of the ellipse sizes is no longer possible, since there is no such straight line to allow all three ellipses to lie in one half-plane.

Let us see what the growth of ellipse sizes will look like in the case in Fig. 5.3. At the beginning the ellipse sizes grow without forcing the straight line to be

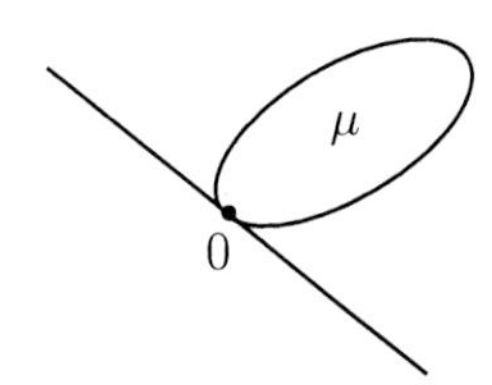

Figure 5.5 A particular case in which the straight line is closely contacted with one ellipse.

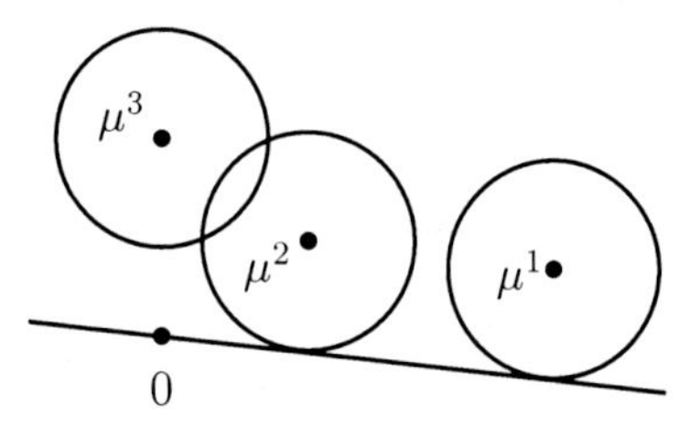

Figure 5.6 The straight line contacts another ellipse.

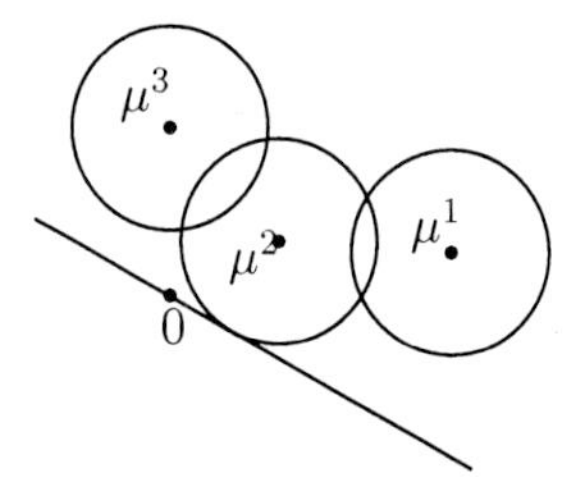

Figure 5.7 Turning the straight line clockwise further.

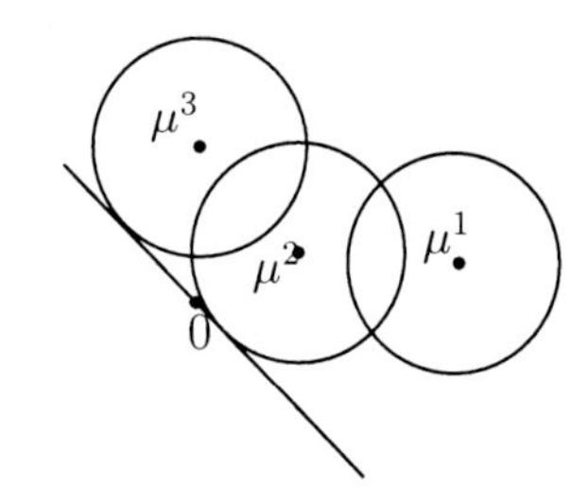

Figure 5.8 The straight line has contacted another ellipse. The growth of the ellipses ends.

rotated. The initial growth will last only till one of the ellipses touches the straight line. In our case (when the matrices σ_1, σ_2, σ_3 are the same) it is the ellipse 1, see Fig. 5.4. If the contact point were to fall exactly in the coordinate origin, further growth of the ellipse sizes would not be possible. This particular case of determining the straight line by one single ellipse is presented in Fig. 5.5. But in our case the contact point is not a coordinate origin, and thus the growth of the ellipse sizes continues and the straight line continues turning clockwise till it touches another ellipse. In our case it is the ellipse 2 in Fig. 5.6. If the contact points of the first and second ellipses were to lie along different sides with respect to the coordinate origin no further growth of the ellipse sizes would be possible and thus the growth of the ellipses would end. In our case such a situation has not occurred and ellipse sizes can grow further and the straight line is turning clockwise at the same time. The first ellipse stops to touch the straight line and the turning up now depends on the second ellipse only, Fig. 5.7.

The growth of ellipses continues either until the contact point does not reaches the origin of coordinates or until the straight line touches some other ellipse. In our our case it is the ellipse 3, see Fig. 5.8. The contact points of the second and third ellipses lie on opposite half-lines with respect to the coordinate origin and therefore with the growing size of the second ellipse the straight line would have to turn in one direction, and the growing size of the third ellipse would force the straight line to turn in the opposite direction. The growth of sizes of both the ellipses at the same time is no longer possible, and thus the found out position represents the solution of our task.

With a certain amount of imagination we can obtain some idea of what the growth of ellipse sizes might look like in a three-dimensional space (in terms of geometry they would be ellipsoids). Also the ellipse sizes here are growing till some ellipses force the separating plane into one possible position. This can happen either when the contact point gets to the coordinate origin, or when two contact points and the coordinate origin appear to be on one straight line, or, finally, when the triangle formed by three contact points incorporates the coordinate origin as well.

On the basis of such an informally understood task we can consider the following necessary and sufficient condition for the optimal position of the hyperplane in the task (5.7).

Let H be a hyperplane and μ^j, σ^j, $j \in J$, be parameters of $|J|$ random Gaussian vectors. The variable r^j is a positive real number. Let x_0^j, $j \in J$, represent a point in which the ellipse

$$\left\langle (x - \mu^j) \, , \, (\sigma^j)^{-1} \cdot (x - \mu^j) \right\rangle = r^j$$

touches the hyperplane H. Let J^0 be a subset of those $j \in J$ for which

$$r^j = \min_{j \in J} r^j \, .$$

For the optimal position of the hyperplane H with respect to the task (5.7) it is necessary and sufficient that the coordinate origin should lie inside the polyhedron the vertices of which are the contact points x_0^j, $j \in J^0$.

This statement will be formulated more elaborately and will be proved. For a more accurate analysis of Anderson tasks both the original and generalised tasks, we will state more precisely the concepts of the ellipse and the contact point which we introduced when referring to intuitive understanding.

5.3.3 Definition of auxiliary concepts for Anderson tasks

Let X be n-dimensional linear space, $\mu \in X$ be n-dimensional vector and σ be symmetrical positive-definite matrix of the dimension $(n \times n)$. Furthermore, let the vector $\alpha \in X$ and the number θ decompose the space X into three subsets:

$$\begin{array}{lcl} \text{the positive half space } X^+ &=& \{x \in X \mid \langle \alpha, x \rangle > \theta\} \, , \\ \text{the negative half space } X^- &=& \{x \in X \mid \langle \alpha, x \rangle < \theta\} \, , \\ \text{and the hyperplane } \quad X^0 &=& \{x \in X \mid \langle \alpha, x \rangle = \theta\} \, . \end{array}$$

Let us assume $\mu \in X^+$. Let us denote by F the quadratic function

$$F(x) = \left\langle (x - \mu) \, , \, \sigma^{-1} \cdot (x - \mu) \right\rangle \tag{5.9}$$

and for a certain non-negative number r the set $E(r, \mu, \sigma) = \{x \in X \,|\, F(x) \leq r^2\}$ is to be referred to as the *ellipse of the size* r. The highest value of r, at which the ellipse $E(r, \mu, \sigma)$ is a subset of the set $X^+ \cup X^0$, will be denoted by r^* and referred to as the *distance* of the pair (μ, σ) from the hyperplane X^0. It is obvious that there exists one single point which belongs to both the hyperplane

X^0 and the ellipse $E(r^*, \mu, \sigma)$. It will be denoted by x_0 and referred to as the *contact point.* It is also obvious that in the contact point the minimal value of the function F in the hyperplane X^0 is reached and the value of the function F in the contact point is $(r^*)^2$, i.e., the square power of the distance (μ, σ) from X^0. Explicit expressions for the distance and the contact point can be derived.

To the respective optimisation task the Lagrange function corresponds

$$\Phi(x, \lambda) = \big\langle (x - \mu)\,,\, \sigma^{-1} \cdot (x - \mu) \big\rangle + \lambda \cdot \langle \alpha, x \rangle$$

and the point x_0 sought is the solution of the pair of equations

$$\left. \begin{aligned} \operatorname{grad} \Phi(x, \lambda) &= 0\,, \\ \langle \alpha, x \rangle &= \theta \end{aligned} \right\} \tag{5.10}$$

with respect to the variables x and λ. In particular, the first equation in the system (5.10) is

$$2\,\sigma^{-1} \cdot (x - \mu) + \lambda \cdot \alpha = 0\,,$$

from which it follows that

$$x_0 = \mu - \frac{\lambda}{2}\,\sigma \cdot \alpha\,, \tag{5.11}$$

where the Lagrange coefficient λ is to assume such a value at which the second equation in the system (5.10) is satisfied, i.e., the equation

$$\left\langle \alpha\,,\, \left(\mu - \frac{\lambda}{2}\,\sigma \cdot \alpha \right) \right\rangle = \theta\,.$$

Its solution with respect to the coefficient λ is

$$\lambda = 2\,\frac{\langle \alpha, \mu \rangle - \theta}{\langle \alpha,\, \sigma \cdot \alpha \rangle}\,. \tag{5.12}$$

Let us substitute the expression (5.12) for the value λ into the formula for the contact point (5.11) to obtain an explicit expression for the contact point x_0. It no longer contains the undetermined coefficient λ,

$$x_0 = \mu - \frac{\langle \alpha, \mu \rangle - \theta}{\langle \alpha,\, \sigma \cdot \alpha \rangle}\,\sigma \cdot \alpha\,. \tag{5.13}$$

When the expression for x_0 is substituted into the formula (5.9) we obtain the size $(r^*)^2$ of the ellipse $\langle (x_0 - \mu)\,,\, \sigma^{-1} \cdot (x_0 - \mu) \rangle = (r^*)^2$ at which the ellipse touches the hyperplane $\langle \alpha, x \rangle = \theta$. Let us do it.

$$\begin{aligned}
(r^*)^2 &= \big\langle (x_0 - \mu)\,,\, \sigma^{-1} \cdot (x_0 - \mu) \big\rangle \\
&= \left\langle \left(\mu - \frac{\lambda}{2}\sigma \cdot \alpha - \mu \right),\, \sigma^{-1} \cdot \left(\mu - \frac{\lambda}{2}\,\sigma \cdot \alpha - \mu \right) \right\rangle \\
&= \left(\frac{\lambda}{2} \right)^2 \big\langle (\sigma \cdot \alpha)\,,\, \sigma^{-1} \cdot (\sigma \cdot \alpha) \big\rangle = \left(\frac{\lambda}{2} \right)^2 \big\langle (\sigma \cdot \alpha)\,,\, \alpha \big\rangle \\
&= \left(\frac{\lambda}{2} \right)^2 \langle \alpha,\, \sigma \cdot \alpha \rangle = \left(\frac{\langle \alpha, \mu \rangle - \theta}{\langle \alpha,\, \sigma \cdot \alpha \rangle} \right)^2 \langle \alpha,\, \sigma \cdot \alpha \rangle = \left(\frac{\langle \alpha, \mu \rangle - \theta}{\sqrt{\langle \alpha,\, \sigma \cdot \alpha \rangle}} \right)^2 .
\end{aligned}$$

This means that the dimension r^* of the ellipse contacting the hyperplane $\langle\alpha, x\rangle = \theta$ is

$$r^* = \left| \frac{\langle\alpha, \mu\rangle - \theta}{\sqrt{\langle\alpha,\, \sigma \cdot \alpha\rangle}} \right| . \tag{5.14}$$

If we take into consideration that the vector μ belongs to the positive half-space X^+ then we will obtain an expression without the sign for the absolute value

$$r^* = \frac{\langle\alpha, \mu\rangle - \theta}{\sqrt{\langle\alpha,\, \sigma \cdot \alpha\rangle}} . \tag{5.15}$$

In case the vector μ belonged to the negative half-space X^-, the corresponding expression would be

$$r^* = \frac{\theta - \langle\alpha, \mu\rangle}{\sqrt{\langle\alpha,\, \sigma \cdot \alpha\rangle}} . \tag{5.16}$$

We will continue concentrating our attention on the formulæ (5.15) and (5.16) obtained above. The numerator in (5.15) is the mathematical expectation of the random variable $\langle\alpha, x\rangle - \theta$ for the random vector x with mathematical expectation μ. The denominator is the mean square deviation of the random variable $\langle\alpha, x\rangle - \theta$ for the random vector x with the covariance matrix σ. From this it directly follows that the size of the ellipse contacting the hyperplane is a strictly monotonically decreasing function of the probability that the random variable x will get to the half-space defined by the hyperplane X^0 and different from the half-space in which the mathematical expectation μ occurs. In this way we have proved the following lemma.

Lemma 5.1 *Let x be a multi-dimensional random Gaussian variable with the mathematical expectation μ and the covariance matrix σ, which assumes the values in a linear space X. Let the vector $\alpha \in X$ and the number θ decompose the space X into three subsets:*

$$\begin{aligned} &\text{positive half-space } X^+ &= \{x \in X \mid \langle\alpha, x\rangle > \theta\} , \\ &\text{negative half-space } X^- &= \{x \in X \mid \langle\alpha, x\rangle < \theta\} , \\ &\text{and the hyperplane } X^0 &= \{x \in X \mid \langle\alpha, x\rangle = \theta\} . \end{aligned}$$

1. *If $\mu \in X^+$ then the probability of the event $x \in X^-$ is a strictly decreasing function of the distance of the pair (μ, σ) from the hyperplane X^0;*
2. *If $\mu \in X^-$ then the probability of the event $x \in X^+$ is a strictly decreasing function of the distance of the pair (μ, σ) from the hyperplane X^0.* ▲

Proof. Lemma 5.1 is an obvious consequence of the relations (5.15) and (5.16). ■

5.3.4 Solution of Anderson original task

Now we will formulate Anderson original task using concepts presented in the previous section. Let μ_1 and μ_2 be two vectors and σ_1 and σ_2 be two matrices. A vector α and a number θ have to be found that are to decompose the space X into half-spaces X^+ and X^- by the hyperplane X^0 so that $\mu_1 \in X^+$, $\mu_2 \in X^-$ and the distances of the pairs (μ_1, σ_1) and (μ_2, σ_2) from the hyperplane X^0 should be as large as possible.

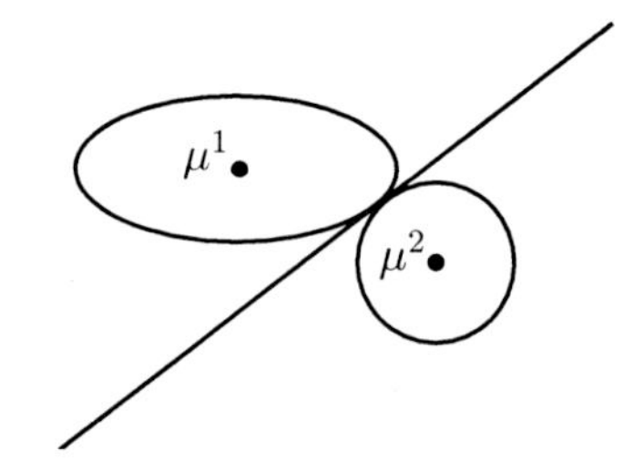

Figure 5.9 Both ellipses touch the hyperplane in the same point.

Referring to the intuitive understanding, we claim that the hyperplane $X^0 = \{x \in X \mid \langle\alpha, x\rangle = 0\}$ is (a) a tangent hyperplane common with the two ellipses of the same size; (b) both ellipses touch the hyperplane in the same point, as can be seen in Fig. 5.9.

We will use explicit formulæ (5.15), (5.16) for the maximal ellipse sizes and the formula (5.11) for the contact point. The statement will be proved by means of the system

$$\left.\begin{aligned}
\mu_1 - \tfrac{1}{2}\lambda_1\,\sigma_1\cdot\alpha &= \mu_2 - \tfrac{1}{2}\lambda_2\,\sigma_2\cdot\alpha\,,\\
\left|\frac{\langle\alpha,\mu_1\rangle - \theta}{\sqrt{\langle\alpha,\,\sigma_1\cdot\alpha\rangle}}\right| &= \left|\frac{\langle\alpha,\mu_2\rangle - \theta}{\sqrt{\langle\alpha,\,\sigma_2\cdot\alpha\rangle}}\right|,\\
\lambda_1 &= 2\,\frac{\langle\alpha,\mu_1\rangle - \theta}{\langle\alpha,\,\sigma_1\cdot\alpha\rangle}\,,\\
\lambda_2 &= 2\,\frac{\langle\alpha,\mu_2\rangle - \theta}{\langle\alpha,\,\sigma_2\cdot\alpha\rangle}\,,
\end{aligned}\right\} \tag{5.17}$$

which by means of the newly introduced variables $\lambda'_1 = \frac{1}{2}\lambda_1$, $\lambda'_2 = -\frac{1}{2}\lambda_2$ will be presented in the form

$$\left.\begin{aligned}
\alpha &= (\lambda_1\cdot\sigma_1 + \lambda_2\cdot\sigma_2)^{-1}\cdot(\mu_1 - \mu_2)\,,\\
\frac{\langle\alpha,\mu_1\rangle - \theta}{\sqrt{\langle\alpha,\,\sigma_1\cdot\alpha\rangle}} &= -\frac{\langle\alpha,\mu_2\rangle - \theta}{\sqrt{\langle\alpha,\,\sigma_2\cdot\alpha\rangle}}\,,\\
\lambda_1 &= \frac{\alpha\cdot\mu_1 - \theta}{\alpha\cdot\sigma_1\cdot\alpha}\,,\\
\lambda_2 &= -\frac{\langle\alpha,\mu_2\rangle - \theta}{\langle\alpha,\,\sigma_2\cdot\alpha\rangle}\,.
\end{aligned}\right\} \tag{5.18}$$

The second equation in the relation (5.17) has been rewritten to the form (5.18) with respect to the requirement $\mu_1 \in X^+$, $\mu_2 \in X^-$. Both coefficients λ_1 and λ_2 are positive, and therefore their sum $\lambda_1 + \lambda_2$ is also positive. Note that the vector α need not be determined precisely, but only up to a multiple by a positive coefficient. The solution of the task, therefore, does not depend on precise values of the coefficients λ_1 and λ_2, but only on their ratio. Both coefficients can be, e.g., tied together by the following relation

$$\lambda_1 + \lambda_2 = 1\,, \tag{5.19}$$

since any ratio between λ_1 and λ_2 can be achieved even on the condition given by the equation (5.19). Their ratio is

$$\begin{aligned}\frac{\lambda_1}{\lambda_2} &= \frac{\langle\alpha,\mu_1\rangle-\theta}{\langle\alpha,\,\sigma_1\cdot\alpha\rangle}\cdot\left(-\frac{\langle\alpha,\,\sigma_2\cdot\alpha\rangle}{\langle\alpha,\mu_2\rangle-\theta}\right)\\ &= \frac{\langle\alpha,\mu_1\rangle-\theta}{\sqrt{\langle\alpha,\,\sigma_1\cdot\alpha\rangle}}\cdot\left(-\frac{\sqrt{\langle\alpha,\,\sigma_2\cdot\alpha\rangle}}{\langle\alpha,\mu_2\rangle-\theta}\right)\cdot\sqrt{\frac{\langle\alpha,\,\sigma_2\cdot\alpha\rangle}{\langle\alpha,\,\sigma_1\cdot\alpha\rangle}}\,.\end{aligned} \tag{5.20}$$

Thanks to the second equality in the system (5.18) the product of the two first coefficients in the right-hand part of (5.20) is unit, and thus

$$\frac{\lambda_1}{\lambda_2}=\sqrt{\frac{\langle\alpha,\,\sigma_2\cdot\alpha\rangle}{\langle\alpha,\,\sigma_1\cdot\alpha\rangle}}\,.$$

In this way we have arrived to Anderson smart original solution of the task. Note the condition (5.19) and write down the first formula of the system (5.18) in the form

$$\alpha=\big((1-\lambda)\cdot\sigma_1+\lambda\cdot\sigma_2\big)^{-1}\cdot(\mu_1-\mu_2)\,, \tag{5.21}$$

which explicitly states the dependence of the vector α on the input data μ_1, μ_2, σ_1 and σ_2 up to the value of the coefficient λ. It is to be chosen such as to satisfy the condition

$$\frac{1-\lambda}{\lambda}=\sqrt{\frac{\langle\alpha,\,\sigma_2\cdot\alpha\rangle}{\langle\alpha,\,\sigma_1\cdot\alpha\rangle}}\,.$$

The obtained result is elegant since a complex minimax task with $n+1$ variables (n can be rather large) has been successfully simplified to search for a single number λ. As soon as the number λ is found the remaining n unknowns are obtained that are elements of the vector α, see (5.21). The value of the coefficient λ sought can be iteratively calculated. An arbitrary initial value is taken, e.g., $\lambda=0.5$. Then by means of the formula (5.21) the vector α is calculated as well as the ratio

$$\gamma=\sqrt{\frac{\langle\alpha,\,\sigma_2\cdot\alpha\rangle}{\langle\alpha,\,\sigma_1\cdot\alpha\rangle}}\,,$$

and it is to find out if the already obtained ratio is equal to $(1-\lambda)/\lambda$. If it is so then the task has been solved. If not then a new value of the coefficient λ', is stated that already satisfies the condition $(1-\lambda')\lambda'=\gamma$, rewritten as $\lambda'=1(1+\gamma)$, and the iteration continues.

Without intending to diminish the ingenuity of the procedure, we want to remark that from a computational standpoint the smartness was achieved in such a way that all the clumsiness was hidden in the relation (5.21) in the operation of matrix inversion. We would like to find an algorithm which would do without inverting, and, moreover, we would like to solve a more general task than that by Anderson. To achieve the result desired a more accurate and elaborate analysis will be needed compared to the current informal analysis.

5.3.5 Formal analysis of generalised Anderson task

First, let us recall the main concepts of generalised Anderson task which has been referred to in Subsection 5.3.1. Recall that X is an n-dimensional linear space and J is a finite set of indices. For each index j from the set J an n-dimensional vector μ^j and a symmetrical positive-definite $(n \times n)$-dimensional matrix σ^j are defined. Further on α is an n-dimensional vector. For each triplet α, μ and σ a number $\varepsilon(\alpha, \mu, \sigma)$ is defined which means the probability that a random Gaussian vector x with the mathematical expectation μ and the covariance matrix σ will satisfy the inequality $\langle \alpha, x \rangle \leq 0$.

In the task a vector α is sought that minimises the number $\max_{j \in J} \varepsilon(\alpha, \mu^j, \sigma^j)$ for the known μ^j, σ^j, $j \in J$. We write

$$\alpha = \operatorname*{argmin}_{\alpha} \max_{j \in J} \varepsilon(\alpha, \mu^j, \sigma^j)\,.$$

Let us denote the minimised function $\max_{j \in J} \varepsilon(\alpha, \mu^j, \sigma^j)$ by the symbol $f(\alpha)$. The given data suffice for proving the theorem which states that the function $f(\alpha)$ is unimodal and thus its minimisation can be achieved.

Theorem 5.1 Convexity of the set of vectors α. *The set of vectors α satisfying the inequality $f(\alpha) \leq b$ is convex for each number $b < 0.5$.* ▲

Proof. Lemma (5.1) states that the probability $\varepsilon(\alpha, \mu^j, \sigma^j)$ strictly decreases with the growth of

$$\frac{\langle \alpha, \mu^j \rangle}{\sqrt{\langle \alpha,\ \sigma^j \cdot \alpha \rangle}}\,.$$

This means that for each real b there exists such real c, that when $\varepsilon(\alpha, \mu^j, \sigma^j) \leq b$ is valid then the following expression is also satisfied

$$\frac{\langle \alpha, \mu^j \rangle}{\sqrt{\langle \alpha,\ \sigma^j \cdot \alpha \rangle}} \geq c\,.$$

If $\varepsilon(\alpha, \mu^j, \sigma^j) = 0.5$ then there holds

$$\frac{\langle \alpha, \mu^j \rangle}{\sqrt{\langle \alpha,\ \sigma^j \cdot \alpha \rangle}} = 0\,.$$

The condition $f(\alpha) \leq b$ is equivalent to the following system of equations

$$\varepsilon(\alpha, \mu^j, \sigma^j) \leq b\,, \quad j \in J\,.$$

Thanks to the monotonicity the preceding system can be replaced by an equivalent system of inequalities

$$\frac{\langle \alpha, \mu^j \rangle}{\sqrt{\langle \alpha,\ \sigma^j \cdot \alpha \rangle}} \geq c\,, \quad j \in J\,, \tag{5.22}$$

where, in a way, the number c depends on the number b. The system (5.22) can be written in the form

$$\langle \alpha, \mu^j \rangle - c \cdot \sqrt{\langle \alpha,\ \sigma^j \cdot \alpha \rangle} \geq 0\,, \quad j \in J\,. \tag{5.23}$$

The functions in the left-hand side of each inequality of the system (5.23) consist of two summands. The first of them is a linear function of the vector α, and as such it is concave. The function $\sqrt{\langle \alpha, \sigma^j \cdot \alpha \rangle}$ is a convex function of the vector α. Thus the function $-c\sqrt{\langle \alpha, \sigma^j \cdot \alpha \rangle}$ is concave since the number c is strictly positive by the assumption $b < 0.5$. The left-hand side in each inequality in the system (5.23) is a sum of two concave functions, and thus it is also concave. Therefore for each j the set of vectors satisfying the j-th inequality is a convex set. The set of vectors satisfying the system (5.23) is an intersection of convex sets, and thus it is also convex. ∎

From the theorem proved it directly follows that in the domain where $f(\alpha) < 0.5$ no strict local minimum can exist which would not be identical to the global minimum. The strict local minimum is here the point α_0, for which a δ-neighbourhood of the point α_0 exists in which for each $\alpha \neq \alpha_0$ the strict inequality $f(\alpha_0) < f(\alpha)$ is satisfied. Let us assume the opposite, let α' and α'' be two strict local minima. Without loss of generality, let us assume that $f(\alpha'') \leq f(\alpha') = c < 0.5$. We will connect the points α'' and α' with a straight line segment. Since the point α' is the local minimum then on this line segment there is a point α (and it can be quite near the point α') for which $f(\alpha) > f(\alpha') = c$ must hold. This would, however, mean that the set of vectors α, for which $f(\alpha) \leq c < 0.5$ holds is not convex. From Theorem 5.1 it follows that if the point α is reached in which the value $f(\alpha) \leq 0.5$ then from this point we can get to the global minimum α^* directly along the straight line α which connects the current point with the point α^*. When moving along the straight line, the function $f(\alpha)$ will not be rising in any position.

Actually even stronger statements are valid than those presented here. From each point α for which $f(\alpha) < 0.5$ holds it is possible to pass along the straight line to the point α^* in which the global minimum of the function f is reached. When moving along the straight line from the point α to the point α^* the function f will be continuously decreasing.

This property of the function f could be used for organising the procedure of its minimisation. But this procedure cannot be based on the motion in the direction of the gradient getting towards a zero-gradient point, as usually happens, since the function being maximised is neither convex, nor differentiable. Necessary and sufficient conditions for the existence of a maximum are to be stated that are not based on the concept of the gradient.

Further on a lemma is proved which deals with the necessary and sufficient conditions for the minimum of the number $\varepsilon(\alpha, \mu, \sigma)$. The formulation of the lemma is based on the concept of the contact point of an ellipse $\{x \in X \mid \langle x - \mu, \sigma \cdot (x - \mu) \rangle \leq r^2\}$ and a hyperplane $\{x \in X \mid \langle \alpha, x \rangle = 0\}$. The contact point is marked $x_0(\alpha, \mu, \sigma)$. On the basis of (5.13) referring to $\theta = 0$, the expression for $x_0(\alpha, \mu, \sigma)$ is

$$x_0(\alpha, \mu, \sigma) = \mu - \frac{\langle \alpha, \mu \rangle}{\langle \alpha, \sigma \cdot \alpha \rangle} \sigma \cdot \alpha \,. \tag{5.24}$$

The proof is based on the concept of distance of the pair (μ, σ) from the hyperplane $\{x \in X \mid \langle \alpha, x \rangle = 0\}$. The distance is marked $r^*(\alpha, \mu, \sigma)$. Based on (5.15) and remembering that $\theta = 0$ the expression for $r^*(\alpha, \mu, \sigma)$ can be written as follows

$$r^*(\alpha, \mu, \sigma) = \frac{\langle \alpha, \mu \rangle}{\sqrt{\langle \alpha, \sigma \cdot \alpha \rangle}} . \tag{5.25}$$

Lemma 5.2 Necessary and sufficient condition for optimality of α for one distribution. *Let for a triplet (α, μ, σ) hold that $\langle \alpha, \mu \rangle > 0$. Let $x_0(\alpha, \mu, \sigma)$ be the contact point and $\Delta\alpha$ any vector which is not collinear with the vector α. For this case two implications are valid:*

1. *If the following condition is satisfied*

$$\big\langle \Delta\alpha,\ x_0(\alpha, \mu, \sigma) \big\rangle > 0 \tag{5.26}$$

then a positive number T exists such that for any t, $0 < t \leq T$, the following inequality is satisfied

$$\varepsilon\,(\alpha + t \cdot \Delta\alpha, \mu, \sigma) < \varepsilon\,(\alpha, \mu, \sigma) . \tag{5.27}$$

2. *If the following condition is satisfied*

$$\big\langle \Delta\alpha,\ x_0(\alpha, \mu, \sigma) \big\rangle \leq 0 \tag{5.28}$$

then there holds

$$\varepsilon\,(\alpha + \Delta\alpha, \mu, \sigma) > \varepsilon\,(\alpha, \mu, \sigma) . \tag{5.29}$$

▲

Remark 5.1 *Lemma 5.2 states virtually that the necessary and sufficient condition for the optimality of the vector α is that the contact point $x_0(\alpha, \mu, \sigma)$ is identical with the coordinate origin, which was intuitively understood in informally examining Anderson task.*

And in fact, if $x_0(\alpha, \mu, \sigma) = 0$ then the condition (5.28) is satisfied for any vector $\Delta\alpha$, and thus for any vector $\Delta\alpha$, that is not collinear with the vector α. The inequality (5.29) states that $\varepsilon(\alpha, \mu, \sigma) < \varepsilon(\alpha', \mu, \sigma)$ holds for any vector α', that is not collinear with the vector α, which means that the vector α ensures the least possible value of the probability $\varepsilon(\alpha, \mu, \sigma)$.

On the other hand, if $x_0(\alpha, \mu, \sigma) \neq 0$ then a vector $\Delta\alpha$ exists for which (5.26) holds. It can be, e.g., the vector $\Delta\alpha = x_0(\alpha, \mu, \sigma)$. Thus there exists a point $\alpha' = \alpha + t \cdot \Delta\alpha$ having the value $\varepsilon(\alpha', \mu, \sigma)$ which is less than $\varepsilon(\alpha, \mu, \sigma)$.

Lemma 5.2 was stated in a less lucid form since just in the presented form it is helpful for the proof of the next theorem. Our objective is not to minimise the probability $\varepsilon(\alpha, \mu, \sigma)$ for one single pair (μ, σ), but to minimise the number $\max_{j \in J} \varepsilon(\alpha, \mu^j, \sigma^j)$ for a certain ensemble of pairs. Lemma 5.2 is only an auxiliary result for our final aim. ▲

Proof. (Lemma 5.2) We will examine a case where the condition (5.26) is satisfied, i.e., $\big\langle \Delta\alpha\,,\ x_0(\alpha, \mu, \sigma) \big\rangle > 0$. We will consider the function $r^*(\alpha + t \cdot \Delta\alpha, \mu, \sigma)$ of the variable t. According to (5.25) it is the function

$$\frac{\langle \alpha + t \cdot \Delta\alpha,\ \mu \rangle}{\sqrt{\langle \alpha + t \cdot \Delta\alpha,\ \sigma \cdot (\alpha + t \cdot \Delta\alpha) \rangle}} .$$

Its derivative in the point $t = 0$ is

$$\left.\frac{\mathrm{d}r^*(\alpha + t \cdot \Delta\alpha, \mu, \sigma)}{\mathrm{d}t}\right|_{t=0} = \frac{\left\langle \left(\mu - \frac{\langle \alpha, \mu \rangle}{\langle \alpha\,,\, \sigma \cdot \alpha \rangle}(\sigma \cdot \alpha)\right),\ \Delta\alpha \right\rangle}{\sqrt{\langle \alpha\,,\, \sigma \cdot \alpha \rangle}}\,. \tag{5.30}$$

On the basis of the expression (5.24) for $x_0(\alpha, \mu, \sigma)$ it is clear that in the fraction on the right-hand side of (5.30) the numerator is the scalar product $\langle x_0(\alpha, \mu, \sigma)\,,\ \Delta\alpha \rangle$, which is positive, as assumed. The examined derivative is, therefore, also positive. It follows from it that there exists such a positive T that for each t, $0 < t \leq T$, the following inequality is satisfied

$$r^*(\alpha + t \cdot \Delta\alpha, \mu, \sigma) > r^*(\alpha, \mu, \sigma)\,,$$

from which, thanks to Lemma 5.1 the inequality (5.27) follows which was to prove. In this way the first statement of Lemma 5.2 is proved.

We will now prove the second statement of Lemma 5.2, where the following condition is assumed

$$\big\langle \Delta\alpha\,, x_0(\alpha, \mu, \sigma) \big\rangle \leq 0\,.$$

Here the behaviour of the function $r^*(\alpha, \mu, \sigma)$ is to be examined not only in the neighbourhood of the vector α, as was in the previous case, but in a global sense, and therefore for an analysis of such a behaviour the knowledge of the derivatives of this function in the point α will not do. Thus additional considerations, but not very complicated ones, are needed.

To be brief, we will denote $\alpha + \Delta\alpha$ as α' and $x_0(\alpha, \mu, \sigma)$ as x_0. For the vector α' three cases can appear:

$$\langle \alpha', \mu \rangle \leq 0\,; \tag{5.31}$$

$$\langle \alpha', \mu \rangle > 0\,, \quad \langle \alpha', x_0 \rangle < 0\,; \tag{5.32}$$

$$\langle \alpha', \mu \rangle > 0\,, \quad \langle \alpha', x_0 \rangle = 0\,. \tag{5.33}$$

The case in which $\langle \alpha', x_0 \rangle > 0$ is excluded since $\langle \alpha', x_0 \rangle = \big\langle (\alpha + \Delta\alpha), x_0 \big\rangle = \langle \alpha, x_0 \rangle + \langle \Delta\alpha, x_0 \rangle \leq 0$. And actually the summand $\langle \alpha, x_0 \rangle$ is zero since according to the definition the contact point x_0 belongs to the hyperplane $X^0(\alpha) = \{x \in X \mid \langle \alpha, x \rangle = 0\}$, and $\langle \Delta\alpha, x_0 \rangle$ is not positive according to the assumption (5.28).

When the condition (5.31) is satisfied then the statement of Lemma 5.2 is obviously valid, since α satisfies the inequality $\langle \alpha, \mu \rangle > 0$, and thus $\varepsilon(\alpha, \mu, \sigma) < 0.5$. The inequality (5.31) means that $\varepsilon(\alpha', \mu, \sigma) \geq 0.5$.

Let us examine the cases (5.32) and (5.33). The symbol $F(x)$ will denote a quadratic function $\big\langle (x - \mu), \sigma^{-1} \cdot (x - \mu) \big\rangle$ and we will prove that in both cases (5.32) and (5.33) such a point x^* in the hyperplane $X^0(\alpha') = \{x \in X \mid \langle \alpha', x \rangle = 0\}$ exists that

$$F(x^*) < F(x_0)\,. \tag{5.34}$$

Thus the inequality (5.29) will be proved, and so will the entire Lemma 5.2. If the inequality (5.34) is valid (and we will prove its validity) then there holds

$$\left(r^*(\alpha', \mu, \sigma)\right)^2 = \min_{x \in X^0(\alpha')} F(x) \leq F(x^*) < F(x_0) = \left(r^*(\alpha, \mu, \sigma)\right)^2 . \quad (5.35)$$

By Lemma 5.1 it leads to the inequality (5.29) which was to be proved.

We will prove first that for the constraint (5.32) the statement (5.34) is valid. The scalar product $\langle \alpha', x \rangle$ depends continuously on the vector x, and so from the inequalities (5.32) it follows that there exists a number $0 < k < 1$ such that the point $x^* = x_0 \cdot (1-k) + k \cdot \mu$ lies in the hyperplane $X^0(\alpha')$. If μ lies on one side from the hyperplane $X^0(\alpha')$ and x_0 lies on the other side of it then some intermediate point x^* must lie just on the hyperplane. The value of the function F in the point x^* is $F(x^*) = F\big(x_0 \cdot (1-k) + k \cdot \mu\big) = (1-k)^2 \cdot F(x_0) < F(x_0)$, which proves (5.34).

Now we will prove that the statement (5.34) follows from the condition (5.33) as well. Since the vectors α and α' are not collinear, neither of the hyperplanes $X^0(\alpha)$ and $X^0(\alpha')$ are identical. Therefore a point x' exists which belongs to the hyperplane $X^0(\alpha')$ and does not belong to the hyperplane $X^0(\alpha)$. This point is not identical with the point x_0 because x_0 lies in the hyperplane $X^0(\alpha)$. The point x_0 also lies in the hyperplane $X^0(\alpha')$, since the assumption (5.33) states that $\langle \alpha', x_0 \rangle = 0$. Let us draw a straight line through the points x_0 and x' which also lies in the hyperplane $X^0(\alpha')$ as stated before. Let us examine the behaviour of the function F along this straight line.

$$\begin{aligned} & F\big(x_0 + k \cdot (x' - x_0)\big) && (5.36) \\ & = \big\langle \big(\mu - x_0 - k \cdot (x' - x_0)\big),\, \sigma^{-1} \cdot \big(\mu - x_0 - k \cdot (x' - x_0)\big) \big\rangle \\ & = F(x_0) - 2k \cdot \big\langle (x' - x_0),\, \sigma^{-1}(\mu - x_0) \big\rangle + k^2 \cdot \big\langle (x' - x_0),\, \sigma^{-1} \cdot (x' - x_0) \big\rangle . \end{aligned}$$

If for the number k the following expression is supplied

$$k = \frac{\big\langle (x' - x_0),\, \sigma^{-1} \cdot (\mu - x_0) \big\rangle}{\big\langle (x' - x_0),\, \sigma^{-1} \cdot (x' - x_0) \big\rangle} ,$$

we can continue in modifying (5.36)

$$\begin{aligned} & F\big(x_0 + k \cdot (x' - x_0)\big) \\ & = F(x_0) - 2k \cdot \big\langle (x' - x_0),\, \sigma^{-1} \cdot (\mu - x_0) \big\rangle \\ & \quad + k \cdot \frac{\big\langle (x' - x_0),\, \sigma^{-1} \cdot (\mu - x_0) \big\rangle}{\big\langle (x' - x_0),\, \sigma^{-1} \cdot (x' - x_0) \big\rangle} \cdot \langle (x' - x_0),\, \sigma^{-1} \cdot (x' - x_0) \rangle \\ & = F(x_0) - k \cdot \big\langle (x' - x_0),\, \sigma^{-1} \cdot (\mu - x_0) \big\rangle \\ & = F(x_0) - \frac{\Big(\big\langle (x' - x_0),\, \sigma^{-1} \cdot (\mu - x_0) \big\rangle\Big)^2}{\big\langle (x' - x_0),\, \sigma^{-1} \cdot (x' - x_0) \big\rangle} . \end{aligned} \quad (5.37)$$

If we use the expression (5.11) for x_0 then we obtain $\sigma^{-1} \cdot (\mu - x_0) = -\frac{1}{2}\lambda\alpha$ which simplifies the formula (5.37) to

$$F\big(x_0 + k \cdot (x' - x_0)\big) = F(x_0) - \frac{\big(\frac{1}{2}\lambda\langle\alpha, (x' - x_0)\rangle\big)^2}{\big\langle(x' - x_0), \sigma^{-1} \cdot (x' - x_0)\big\rangle}.$$

Because $x' \notin X^0(\alpha)$ and $x_0 \in X^0(\alpha)$, the relations $\langle\alpha, x'\rangle \neq 0$ and $\langle\alpha, x_0\rangle = 0$ are valid. Consequently the scalar product $\big\langle\alpha, (x' - x_0)\big\rangle$ is not zero. According (5.12) λ is a nonzero number as well as $\frac{1}{2}\lambda\big\langle\alpha, (x' - x_0)\big\rangle$. Thus

$$F\big(x_0 + k \cdot (x' - x_0)\big) < F(x_0).$$

From this it follows that on the straight line passing points x_0 and x', there exists a point $x^* \in X^0(\alpha')$ for which (5.34) holds. ■

Now we have had sufficient knowledge to state the necessary and sufficient conditions for the vector α to minimise the value $\max_{j\in J} \varepsilon(\alpha, \mu^j, \sigma^j)$. Let us note that (μ^j, σ^j), $j \in J$, are vectors and matrices. Let α be a vector that satisfies the inequalities $\langle\alpha, \mu^j\rangle > 0$, $j \in J$. Further on we will need the numbers

$$(r^j)^2 = \min_{x\in H(\alpha)} \big\langle x - \mu^j, \sigma^j \cdot (x - \mu^j)\big\rangle, \quad j \in J,$$

the subset J^0 of indices $j \in J$, for which $r^j = \min_{j\in J} r^j$ holds. We will need the set of corresponding contact points x_0^j, $j \in J^0$, too.

Theorem 5.2 Necessary and sufficient conditions for the solution of generalised Anderson task. *If the convex hull of the set of contact points x_0^j, $j \in J^0$, includes the coordinate origin then for any vector α' which is not collinear with the vector α the following inequality holds:*

$$\max_{j\in J} \varepsilon(\alpha', \mu^j, \sigma^j) > \max_{j\in J} \varepsilon(\alpha, \mu^j, \sigma^j).$$

If the abovementioned convex hull does not include the coordinate origin then a vector $\Delta\alpha$ and a positive number T exist so that for any t, $0 < t \leq T$, the following inequality is satisfied:

$$\max_{j\in J} \varepsilon(\alpha + t \cdot \Delta\alpha, \mu^j, \sigma^j) < \max_{j\in J} \varepsilon(\alpha, \mu^j, \sigma^j).$$ ▲

Proof. First, we will prove the first statement of Theorem 5.2. It is assumed that such numbers γ^j, $j \in J^0$, exist which satisfy the conditions

$$\gamma^j \geq 0, \; j \in J^0, \qquad \sum_{j\in J^0} \gamma^j = 1, \qquad \sum_{j\in J^0} \gamma^j \cdot x_0^j = 0.$$

This means that any vector α' satisfies the equality

$$\sum_{j\in J^0} \gamma^j \cdot \langle\alpha', x_0^j\rangle = 0.$$

The equality is certainly valid for some nonzero vector α' which is not collinear with α. This sum can be zero only when at least for one $j^* \in J^0$ the following inequality is satisfied

$$\langle \alpha', x_0^{j^*} \rangle \leq 0 \,.$$

The equation $\langle \alpha, x_0^j \rangle = 0$ is satisfied for each $j \in J$ and thus also for j^*. This means that the vector $\Delta\alpha = \alpha' - \alpha$ satisfies the inequality $\langle \Delta\alpha \,, x_0^{j^*} \rangle \leq 0$. With respect to Lemma 5.2 we write $\varepsilon(\alpha', \mu^{j^*}, \sigma^{j^*}) > \varepsilon(\alpha, \mu^{j^*}, \sigma^{j^*})$. The number $\max_{j \in J} \varepsilon(\alpha, \mu^j, \sigma^j)$ is evidently $\varepsilon(\alpha, \mu^{j^*}, \sigma^{j^*})$, since $j^* \in J^0$, and the number $\max_{j \in J} \varepsilon(\alpha', \mu^j, \sigma^j)$ is not less than $\varepsilon(\alpha', \mu^{j^*}, \sigma^{j^*})$. Thus $\max_{j \in J} \varepsilon(\alpha', \mu^j, \sigma^j) > \max_{j \in J} \varepsilon(\alpha, \mu^j, \sigma^j)$. In this way the first statement of Theorem 5.2 is proved.

We will denote by X_0 the convex hull of the set of contact points x_0^j and prove the second statement of Theorem 5.2 in which $0 \notin X_0$ is assumed. Then a vector $\Delta\alpha$ exists for which the inequality $\langle \Delta\alpha \,, x_0^j \rangle > 0$ for each $j \in J^0$ holds. It can be, e.g., the point $\arg\min_{x \in X_0} |x|$. As a result of the first statement of Lemma 5.2 it follows that there exist such positive numbers T^j, $j \in J^0$ that

$$\forall (j \in J^0) \, \forall (t \mid 0 \leq T^j) \colon \varepsilon(\alpha + \Delta\alpha \cdot t, \, \mu^j, \sigma^j) < \varepsilon(\alpha, \mu^j, \sigma^j) \,.$$

The preceding statement remains valid when all numbers T^j are substituted by the number $T' = \min_{j \in J_0} T^j$, and after this substitution the order of quantifiers is changed. In this way we obtain the relation

$$\forall (t \mid 0 < t \leq T') \, \forall (j \in J^0) \colon \varepsilon(\alpha + \Delta\alpha \cdot t, \, \mu^j, \sigma^j) < \varepsilon(\alpha, \mu^j, \sigma^j) \,. \tag{5.38}$$

According to the definition of the set J^0 each value $\varepsilon(\alpha, \mu^j, \sigma^j)$, $j \in J^0$, is equal to $\max_{j \in J} \varepsilon(\alpha, \mu^j, \sigma^j)$ and the expression (5.38) can be modified as follows:

$$\forall (t \mid 0 < t \leq T') \, \forall (j \in J^0) \colon \varepsilon(\alpha + \Delta\alpha \cdot t, \, \mu^j, \sigma^j) < \max_{j \in J} \varepsilon(\alpha, \mu^j, \sigma^j) \,,$$

$$\forall (t \mid 0 < t \leq T') \colon \max_{j \in J^0} \varepsilon(\alpha + \Delta\alpha \cdot t, \, \mu^j, \sigma^j) < \max_{j \in J} \varepsilon(\alpha, \mu^j, \sigma^j) \,. \tag{5.39}$$

The dependence of $\varepsilon(\alpha, \mu^j, \sigma^j)$ on the vector α is continuous. Therefore when for an index j' the inequality

$$\varepsilon(\alpha, \mu^{j'}, \sigma^{j'}) \neq \max_{j \in J} \varepsilon(\alpha, \mu^j, \sigma^j)$$

is satisfied it then remains valid at least for small values of t too,

$$\varepsilon(\alpha + \Delta\alpha \cdot t, \mu^{j'}, \sigma^{j'}) \neq \max_{j \in J} \varepsilon(\alpha + \Delta\alpha \cdot t, \mu^j, \sigma^j) \,.$$

Thus a positive number T exists (it may be less than T') for which the inequality

$$\max_{j \in J^0} \varepsilon(\alpha + \Delta\alpha \cdot t, \mu^j, \sigma^j) = \max_{j \in J} \varepsilon(\alpha + \Delta\alpha \cdot t, \mu^j, \sigma^j)$$

is valid for any t, $0 < t \le T$. Based on this we will rewrite the statement (5.39) in the form

$$\forall(t \mid 0 < t \le T): \max_{j \in J} \varepsilon(\alpha + \Delta\alpha \cdot t, \mu^j, \sigma^j) < \max_{j \in J} \varepsilon(\alpha, \mu^j, \sigma^j)\,,$$

and in this way also the second statement of Theorem 5.2 is proved. ■

The Theorem 5.2 proved shows the procedure of minimisation of the number $\max_{j \in J} \varepsilon(\alpha, \mu^j, \sigma^j)$. The algorithm solving this optimisation task is to reach a state in which the polyhedron circumscribing the contact points $x_0(\alpha, \mu^j, \sigma^j)$, $j \in J_0$, includes the coordinate origin. In this state the algorithm can be finished. If such a state has not occurred then a direction $\Delta\alpha$ is sure to exist such that when moving in that direction then the number $\max_{j \in J} \varepsilon(\alpha, \mu^j, \sigma^j)$ is decreasing. In the proof of Theorem 5.2 it can be seen how to find the direction $\Delta\alpha$. The direction sought is a vector having non-negative scalar products with the vectors corresponding to the contact points. The task of searching for the direction $\Delta\alpha$ appears to be identical with the task of the simple separation of a finite set of points. We can see that for solving generalised Anderson task we must have an algorithm for a simple separation of the sets of points and use it iteratively for each task minimising $\max_{j \in J} \varepsilon(\alpha, \mu^j, \sigma^j)$.

5.3.6 Outline of a procedure for solving generalised Anderson task

On the basis of the analysis used up to now, we can outline a framework of an algorithm for solving generalised Anderson task.

1. First, a vector α is to be found such that all scalar products $\langle \alpha, \mu^j \rangle$, $j \in J$, should be positive. Finding such a vector appears to be identical with the task of the simple separation of finite sets of points.
   Another alternative would be to make sure that such a vector α does not exist. Then the task cannot be solved because the theory built so far, and even the present informal considerations, hold only for the domain where $\langle \alpha, \mu^j \rangle > 0$, $j \in J$. This resignation does not cost us much. Even if the task for this case was solved for some $j \in J$ the probability of an error would be greater than 0.5. But the same error is produced by the decision making rule which takes no regard to the observation x, and decides on inclusion to the first or second class in a random way.
2. After finding the vector α which satisfies the condition

$$\langle \alpha, \mu^j \rangle > 0\,, \quad j \in J\,, \tag{5.40}$$

   it is necessary to calculate the contact points x_0^j, $j \in J$, numbers r^{*j}, to select the set J^0 of those j for which there holds $r^{*j} = \min_{j \in J} r^{*j}$ and to find such a direction $\Delta\alpha$, which satisfies the conditions

$$\langle \Delta\alpha\,,\, x_0^j \rangle > 0\,, \quad j \in J^0\,. \tag{5.41}$$

   We come again to the task of the simple separation of finite sets of points.

Another alternative would be to make sure that such a direction $\Delta\alpha$ does not exist. In this other case a result is obtained that the vector α, which has been found, solves the task.
If a vector $\Delta\alpha$ exists which satisfies (5.41) then it is not unique, but there is a whole set of vectors $\Delta\alpha$ satisfying the condition. Among them the vector $\Delta\alpha$ could be sought in whose direction, in a certain sense, all errors $\varepsilon(\alpha, \mu^j, \sigma^j)$, $j \in J$, will be best reduced. It is natural that the vector $\Delta\alpha$ will be chosen so that the derivatives of all functions $-\varepsilon(\alpha, \mu^j, \sigma^j)$, $j \in J^0$, should be as great as possible, i.e.,

$$\Delta\alpha = \operatorname*{argmax}_{\Delta\alpha} \min_j \left(- \frac{\partial \varepsilon \left(\alpha + t \cdot \frac{\Delta\alpha}{|\Delta\alpha|} \right)}{\partial t} \right)$$

or, which is the same,

$$\Delta\alpha = \operatorname*{argmax}_{\{\Delta\alpha \,\big|\, |\Delta\alpha|=1\}} \min_j \left\langle \Delta\alpha \,, \frac{x_0^j}{\sqrt{\langle \alpha, \sigma^j \cdot \alpha \rangle}} \right\rangle .$$

If we denote $x_0^j / \sqrt{\langle \alpha, \sigma^j \cdot \alpha \rangle}$ by the symbol y^j then we will obtain

$$\Delta\alpha = \operatorname*{argmax}_{\Delta\alpha} \min_j \frac{\langle \Delta\alpha, y^j \rangle}{|\Delta\alpha|} . \tag{5.42}$$

We can see that searching for such a vector is identical with the task of the best separation of the sets of points.
3. After finding the vector $\Delta\alpha$, which is characterised by the condition (5.42), we must find

$$t = \operatorname*{argmax}_{t} \min_{j \in J} \varepsilon(\alpha + \cdot \Delta\alpha, \mu^j, \sigma^j) \tag{5.43}$$

and find a new vector α, as $\alpha ::= \alpha + t \cdot \Delta\alpha$. This vector is also sure to satisfy the condition (5.40).
4. Go to the step 2 of the procedure.

Let us see how far the outlined procedure is apt to be a basis for writing a practically applicable program. From the theoretical standpoint, the most important drawback is that the iterative cycle need not end. From the practical standpoint, it is not as inconvenient. The value

$$\max_{j \in J} \varepsilon(\alpha, \mu^j, \sigma^j)$$

decreases monotonically during iterations. This will usually do for a practical application: the user observes how the preceding value changes and informally decides whether to continue with the iterations hoping to obtain even substantially better results, or whether to stop the iterations at that moment.

The computation of the number t according to the formula (5.43) represents the optimisation of a function, which is not very simply ordered, but it is, at least, an unimodal function of one variable. There is a number of methods for its optimisation. For the realisation of the formula (5.43) they are suitable rather equally. In spite of that, the formula (5.43) is treacherous for programmers, so that in careless programming the program can run about ten to twenty times longer than is needed.

Let us examine the auxiliary tasks (5.40), (5.41) and (5.42). The tasks (5.40) and (5.41) are the same. The task (5.42) includes the above two tasks. First, the task (5.42) is a particular case of generalised Anderson task, to whose solving the whole procedure sought is intended. Second, the task (5.42) is a particular case of generalised Anderson task, where all matrices σ^j, $j \in J$, are unitary matrices. We might seem to be stacked in a logical loop: to solve Anderson task it is necessary to know how to solve the task (5.42), which can be solved only through the algorithm for solving generalised Anderson task. Actually, there is no logical loop since the particular case (5.42) has additional positive features, thanks to which its solving is much easier than that with the general task.

Furthermore the property that the solution of the particular case (5.42) contributes to the solution of generalised Anderson task, this task itself has a further importance for the separation of the finite sets of points through linear discriminant functions. Such a kind of task is favourite in pattern recognition as one of the methods of learning. At the beginning of the lecture we made a note that it was worth being a favourite one.

Now we will part with Anderson task for some time. First we will study the task of the linear separation of finite sets of points and then within the scope of this lecture we will again return to Anderson task.

5.4 Linear separation of finite sets of points

5.4.1 Formulation of tasks and their analysis

Let J be a finite set of indices, which is decomposed into two subsets J_1 and J_2, and $X = \{x^j, \, j \in J\}$ be a finite set of points in a linear space. A vector α is sought which satisfies the system of inequalities

$$\left.\begin{aligned} \langle \alpha, x^j \rangle > 0\,, \quad j \in J_1\,, \\ \langle \alpha, x^j \rangle < 0\,, \quad j \in J_2\,. \end{aligned}\right\} \tag{5.44}$$

This system is referred to as a *simple separation of finite sets of points.* If this task has a solution then also the task

$$\left.\begin{aligned} \left\langle \frac{\alpha}{|\alpha|}, x^j \right\rangle > 0\,, \quad j \in J_1\,, \\ \left\langle \frac{\alpha}{|\alpha|}, x^j \right\rangle < 0\,, \quad j \in J_2\,, \end{aligned}\right\} \tag{5.45}$$

has a solution. If the system (5.45) has one solution then the same system has an infinite number of solutions. Therefore, let us make (5.45) stricter through the requirement to seek a vector α and the greatest positive value of r satisfying the system

$$\left.\begin{aligned} \left\langle \frac{\alpha}{|\alpha|}, x^j \right\rangle &\geq r\,, \quad j \in J_1\,, \\ \left\langle \frac{\alpha}{|\alpha|}, x^j \right\rangle &\leq -r\,, \quad j \in J_2\,. \end{aligned}\right\} \tag{5.46}$$

In other words, a vector α is to be found that maximises the number r when satisfying the system (5.46). An identical procedure is to seek the vector

$$\alpha = \underset{\alpha}{\operatorname{argmax}} \min \left(\min_{j \in J_1} \frac{\langle \alpha, x^j \rangle}{|\alpha|}, \min_{j \in J_2} \frac{-\langle \alpha, x^j \rangle}{|\alpha|} \right). \tag{5.47}$$

The task (5.47) is referred to as the *optimal separation of finite sets of points.* This task is a particular case of generalised Anderson task in which for all $j \in J$ the matrices σ^j are unit matrices.

In this particular case the task has an illustrative *geometrical interpretation* which will later be several times our basis in formal as well as in informal considerations. The task (5.45) requires a hyperplane to be found separating the set of points $\{x^j,\ j \in J_1\}$ from the set of points $\{x^j,\ j \in J_2\}$. The left-hand sides of inequalities in (5.45) represent the distance of the points from the hyperplane. The tasks (5.46) and (5.47) require us to find a hyperplane among all possible hyperplanes satisfying (5.45) which is most distant from the given points.

From the analysis of generalised Anderson task we can see that the tasks (5.46), (5.47) can have even a different geometrical interpretation. An arbitrary vector satisfying the system (5.45) separates not only the points x^j, $j \in J_1$, from the points x^j, $j \in J_2$, but it also separates a certain r-neighbourhoods of these points. The size of the neighbourhood, i.e., the number r, depends on the vector α. The task (5.46), (5.47) requires to find such a vector α which separates together with separating one set of points from the other even their largest possible neighbourhoods.

We will denote the vectors x'^j, $j \in J$, so that $x'^j = x^j$, $j \in J_1$, and $x'^j = -x^j$, $j \in J_2$. The objective of (5.45) is to find a vector α for which there holds

$$\langle \alpha, x'^j \rangle > 0\,, \quad j \in J\,, \tag{5.48}$$

and the requirement (5.47) assumes the form

$$\alpha = \underset{\alpha}{\operatorname{argmax}} \min_{j \in J} \left\langle \frac{\alpha}{|\alpha|}, x'^j \right\rangle. \tag{5.49}$$

Our tasks will be analysed in both formulations. The first formulation (5.48) has its origin in the task of simple separation of sets of points. Now a hyperplane is to be found that will get all points into one half-space. The second task (5.49) originates in the task of optimal separation of the set of points. Now a hyperplane is to be found that, in addition to satisfying the conditions (5.48) of the first formulation, is most possibly distant from the set of points.

The distance is meant in the usual Euclidian sense. The formulation of the necessary and sufficient conditions for the optimal position of the hyperplane, which in the general case is provided by Theorem 5.2, can be expressed in the following nearly geometrical form. Assume we already operate with transformed vectors x'^j and we will simply write them as x^j.

Theorem 5.3 Geometrical interpretation of conditions for optimal hyperplane. *Let $\overline{X}$ be a convex hull of the set of points $\{x^j,\ j \in J\}$ and α^* be a point from $\overline{X}$, which lies nearest the coordinate origin,*

$$\alpha^* = \operatorname*{argmin}_{x \in \overline{X}} |x| \,. \tag{5.50}$$

When $\alpha^* \neq 0$ then α^* is the solution of the task (5.49). ▲

Proof. For each $j \in J$ we will consider a triangle whose three vertices are the coordinate origin, the point α^*, and the point x^j. We will denote by the symbol $\overline{X}^j$ the side of a triangle that connects the vertices α^* and x^j. The relation $\overline{X}^j \subset \overline{X}$ is valid since $\alpha^* \in \overline{X}$, $x^j \in \overline{X}$ and $\overline{X}$ is convex. On these conditions, from the assumption (5.50) there follows that

$$\alpha^* = \operatorname*{argmin}_{x \in \overline{X}^j} |x| \,.$$

This means that in the side $\overline{X}^j$ of the triangle, it is the vertex α^* which is the nearest point to the coordinate origin. Thus the angle at the vertex α^* cannot be acute. The result is that the scalar product of the vectors $-\alpha^*$ and $x^j - \alpha^*$ cannot be positive,

$$\langle -\alpha^*,\, x^j - \alpha^* \rangle \leq 0 \,.$$

The same can be expressed as an equivalent statement

$$|\alpha^*|^2 \leq \langle \alpha^*, x^j \rangle \,. \tag{5.51}$$

The vector α^* belongs to $\overline{X}$, and therefore a set of non-negative coefficients γ^j exists the sum of which is 1, and there holds

$$\alpha^* = \sum_{j \in J} \gamma^j \cdot x^j \,. \tag{5.52}$$

From what has been said the equality $\langle \alpha^*, \alpha^* \rangle = \sum_{j \in J} \gamma^j \cdot \langle \alpha^*, x^j \rangle$ follows which will be written in a somewhat different form

$$\sum_{j \in J} \gamma^j \cdot \left(\langle \alpha^*, x^j \rangle - |\alpha^*|^2 \right) = 0 \,.$$

We can see that the sum of non-negative numbers (see (5.51)) is zero. This, however, can occur only when all summands equal to zero, i.e.,

$$\gamma^j \cdot \left(\langle \alpha^*, x^j \rangle - |\alpha^*|^2 \right) = 0 \,, \quad j \in J \,.$$

Some coefficients γ^j, $j \in J$, must not be zero since their sum must be 1. We will denote by J^0 the set of indices j for which $\gamma^j \neq 0$ holds. For each such $j \in J^0$ the equation $\langle \alpha^*, x^j \rangle = |\alpha^*|^2$ must hold, which together with the inequality (5.51) means that for each $j \in J^0$, and for an arbitrary $j' \in J$ there holds

$$\langle \alpha^*, x^j \rangle \le \langle \alpha^*, x'^j \rangle \,.$$

From the above there follows that

$$\left\langle \frac{\alpha^*}{|\alpha^*|}, x^j \right\rangle = \min_{j' \in J} \left(\left\langle \frac{\alpha^*}{|\alpha^*|}, x'^j \right\rangle \right), \quad j \in J^0 \,.$$

We have proved that in the expression (5.52) the non-zero coefficients are only the coefficients γ^j by which the vectors nearest to the hyperplane $\langle \alpha, x \rangle = 0$ are multiplied. So the expression (5.52) assumes the form

$$\alpha^* = \sum_{j \in J^0} \gamma^j \cdot x^j \,.$$

We will start from the previous expression and prove that the convex hull of the contact points includes the coordinate origin, and this with respect to Theorem 5.2 will prove that α^* is the solution of the formulated task. And in fact, if we use the formula (5.15) for the contact point and if we take into account that $\sigma^j = 1$ then we can write

$$x_0^j = x^j - \alpha^* \cdot \frac{\langle \alpha^*, x^j \rangle}{|\alpha^*|^2} \,,$$

and

$$\begin{aligned} \sum_{j \in J_0} \gamma^j \cdot x_0^j &= \sum_{j \in J_0} \gamma^j \cdot x^j - \alpha^* \cdot \frac{\left\langle \alpha^*, \sum_{j \in J_0} \gamma^j \cdot x^j \right\rangle}{|\alpha^*|^2} \\ &= \alpha^* - \alpha^* \cdot \frac{\langle \alpha^*, \alpha^* \rangle}{|\alpha^*|^2} = \alpha^* - \alpha^* = 0 \,. \end{aligned}$$

This means that the convex hull of the contact points includes the coordinate origin. ■

The theorem proved is already the solution of the task of optimal separation of the finite sets of points, since it reduces the task to minimisation of the quadratic function the domain of definition of which is a multi-dimensional convex polyhedron. Special features of the task allow to use even simpler and more illustrative algorithms. They will be quoted later. The theorem proved and the ideas of the proof can be summarised in relations which are valid for an arbitrary vector α and an arbitrary vector $x \in \overline{X}$, i.e.,

$$\min_{j \in J} \left\langle \frac{\alpha}{|\alpha|}, x^j \right\rangle \le \min_{j \in J} \left\langle \frac{\alpha^*}{|\alpha^*|}, x^j \right\rangle = |\alpha^*| \le |x| \,. \tag{5.53}$$

The previous relations will be helpful in analysing algorithms solving the tasks of simple and optimal separations of finite sets of points.

5.4.2 Algorithms for linear separation of finite sets of points

The task of optimal separation of finite sets of points might seem, at first glance, to be more important than the task of simple separation. Naturally, if an algorithm for solving the task (5.49) is at our disposal then it can also be used for solving the task (5.48). In spite of the indubitable truth of the preceding statement we realise that the task of the simple separation is valuable in itself. This is not only because the solution of the task (5.48) can be simpler compared with the problem (5.49) which is again quite natural. The fact is, as we will see later, that on the basis of thoroughly understanding the algorithm for a simple separation not only an algorithm for an optimal separation can be easily created, but also it may concern even the algorithm for solving generalised Anderson task.

Kozinec algorithm linearly separating finite sets of points

Let $\{x^j,\, j \in J\}$ be a finite set of points for which an unknown vector $\overline{\alpha}$ exists which satisfies the system of inequalities

$$\langle \overline{\alpha}, x^j \rangle > 0\,, \quad j \in J\,. \tag{5.54}$$

We will show a procedure known as Kozinec algorithm, which can find a vector α, that satisfies the condition (5.54), even if it may, naturally, be different from $\overline{\alpha}$.

We will create a sequence of vectors $\alpha_1, \alpha_2, \ldots, \alpha_t, \alpha_{t+1}, \ldots$ according to the following algorithm.

Algorithm 5.1 Kozinec algorithm for simple separation of sets

The vector α_1 can be an arbitrary vector from the set $\overline{X}$, i.e., from the convex closure of the set $\{x^j \mid j \in J\}$. For example, it can be one of the vectors x^j, $j \in J$. Let us admit that the vector α_t has already been calculated. The vector α_{t+1} will be found according to the following rules:

1. Such a vector x^j, $j \in J$, is sought that satisfies the condition

$$\langle \alpha_t, x^j \rangle \le 0\,. \tag{5.55}$$

2. If such a vector x^j does not exist then it means that the solution of the task has already been found and α_t is the vector sought.
3. If the vector x^j exists then we will denote it as x_t. The vector α_{t+1} is determined in such a way that on a straight line connecting the points α_t and x_t a point is sought which is nearest the coordinate origin. This means that

$$\alpha_{t+1} = (1-k)\cdot \alpha_t + k \cdot x_t\,, \quad k \in \mathbb{R}\,, \tag{5.56}$$

where

$$k = \operatorname*{argmin}_k |(1-k)\cdot \alpha_t + k \cdot x_t|\,. \tag{5.57}$$

It is proved of Algorithm 5.1 that the vector α_t is sure to occur in one of the steps which satisfies (5.54). This is stated in the following theorem.

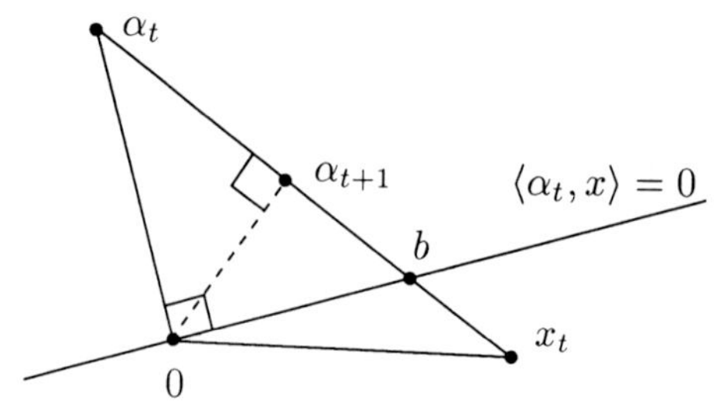

Figure 5.10 Geometrical interpretation of properties of the points α_t, α_{t+1} and x_t.

Theorem 5.4 Convergence of the Kozinec algorithm. *For the sequence α_1, α_2, ..., α_t, α_{t+1}, ... obtained using the Kozinec algorithm, such t^* exists for which*

$$\langle \alpha_{t^*}, x^j \rangle > 0 \,, \quad j \in J \,,$$

is valid. ▲

Proof. The proof is based on the geometrical interpretation of properties of the points α_t, α_{t+1} and x_t, which the algorithm defines. The vector α_{t+1} is the foot of a perpendicular that goes through the coordinate origin and is perpendicular to the straight line passing through the points α_t and x_t, as can be seen in Fig. 5.10. In addition, it can be seen that the vector α_{t+1} is a convex linear combination of the vectors α_t and x_t.

Since α_1 is a member of a convex set $\overline{X}$ and α_{t+1} is a convex linear combination of points α_t and x_t, $x_t \in \overline{X}$, it follows that the vectors $\alpha_2, \alpha_3, \ldots, \alpha_t, \ldots$ are members of the $\overline{X}$. For each of these vectors, therefore, the inequality

$$|\alpha_t| \geq \varepsilon \,, \text{ where } \varepsilon = \min_{x \in \overline{X}} |x| \tag{5.58}$$

is valid. It follows from the strict inequality in condition (5.54) that the set $\overline{X}$ does not include the coordinate origin, which means that the length of the vector α_t cannot converge to zero.

On the basis of the geometrical interpretation of the relations between the vectors α_t, α_{t+1} and x_t we will evaluate the ratio $|\alpha_{t+1}|/|\alpha_t|$ for $\alpha_{t+1} \neq \alpha_t$. The point b is an intersection of the straight line interlaced with the points α_t, x_t with the hyperplane $\langle \alpha_t, x \rangle = 0$. We can see from Fig. 5.10 that

$$\frac{|\alpha_{t+1}|}{|\alpha_t|} = \frac{|b|}{\sqrt{|\alpha_t|^2 + |b|^2}} = \frac{1}{\sqrt{1 + (|\alpha_t|^2/|b|^2)}} \,.$$

We will denote $D = \max_{j \in J} |x_j|$. Thanks to $b \in \overline{X}$ the inequality $|b| \leq D$ holds. Using (5.58) we can write

$$\frac{|\alpha_t|^2}{|b|^2} \geq \frac{\varepsilon^2}{D^2} \,, \quad \sqrt{1 + \frac{|\alpha_t|^2}{|b|^2}} \geq \sqrt{1 + \frac{\varepsilon^2}{D^2}}$$

and

$$\frac{|\alpha_{t+1}|}{|\alpha_t|} \leq \frac{1}{\sqrt{1+\varepsilon^2/D^2}} < 1\,.$$

It can be seen that the sequence of values $|\alpha_1|, \ldots, |\alpha_t|, \ldots$ is decreasing faster than a decreasing geometrical sequence. If the sequence $\alpha_1, \ldots, \alpha_t, \ldots$ was infinite the number $|\alpha_t|$ could be less than any arbitrary positive number. Thanks to (5.58) the number $|\alpha_t|$ cannot be less than ε. Therefore for some t^* the vector α_{t^*} must cease changing. The theorem has been proved. ■

For completeness we will indicate that the number t^* can be estimated by means of the inequality

$$\frac{|\alpha_{t+1}|}{|\alpha_1|} \leq \left(\frac{1}{\sqrt{1+\varepsilon^2/D^2}}\right)^{t^*}, \qquad \frac{\varepsilon}{D} \leq \left(\frac{1}{\sqrt{1+\varepsilon^2/D^2}}\right)^{t^*},$$

$$-t^*\,\frac{1}{2}\ln\left(1+\frac{\varepsilon^2}{D^2}\right) \geq \ln\frac{\varepsilon}{D}\,; \qquad t^* \leq \frac{\ln(D^2/\varepsilon^2)}{\ln\left(1+\varepsilon^2/D^2\right)}\,.$$

At sufficiently small values of ε^2/D^2 the property $\ln(1+x) \approx x$ can be used for a simplified estimate

$$t^* \approx \frac{D^2}{\varepsilon^2}\ln\frac{D^2}{\varepsilon^2}\,.$$

Perceptron and Novikoff theorem

The Kozinec algorithm for a linear classifier provides a smart and simple relation $\alpha_{t+1} = (k-1)\,\alpha_t + k\,x_t$. At first glance it might seem that only with difficulty could something simpler be found. It appears that such a simpler algorithm is used by the perceptron, i.e., $\alpha_{t+1} = \alpha_t + x_t$. We will formulate the perceptron algorithm more precisely and introduce Novikoff theorem which proves that the perceptron algorithm solves the task of simple separation of finite sets.

Let $X = \{x^j \mid j \in J\}$ be a finite set of vectors, $\overline{X}$ be a convex hull of this set, and let

$$\varepsilon = \min_{x\in\overline{X}} |x| > 0\,, \quad D = \max_{j\in J} |x^j|\,.$$

We will create the sequence of vectors $\alpha_1, \alpha_2, \ldots, \alpha_t, \alpha_{t+1}, \ldots$ in the following way.

Algorithm 5.2 Separation of finite sets of points by means of the perceptron

1. The vector α_1 is zero. When the vector α_t, $t = 1, 2, \ldots$ is known then the vector α_{t+1} is determined according to the rules:
2. If for all $j \in J$ the inequality $\langle \alpha_t, x^j \rangle > 0$ is valid then the algorithm finishes.
3. If x_t is one of the vectors x^j, $j \in J$, for which the inequality $\langle \alpha_t, x^j \rangle \leq 0$ is satisfied then

$$\alpha_{t+1} = \alpha_t + x_t\,.$$

The American mathematician Novikoff proved the following famous theorem.

Theorem 5.5 Novikoff theorem on perceptron convergence. *There exists a number $t^* \leq D^2/\varepsilon^2$, such that the vector α_{t^*} satisfies the inequality*

$$\langle \alpha_{t^*}, x^j \rangle > 0$$

for each $j \in J$. ▲

Proof. Let us see what follows from the property that for some t the conditions $\langle \alpha_t, x^j \rangle > 0$, $j \in J$, are not satisfied and $\alpha_{t+1} \neq \alpha_t$ occurs. First, it follows that at each $t' \leq t$ also $\alpha_{t'+1} \neq \alpha_{t'}$ occurs. In addition, for each $t' \leq t$ an $x_{t'}$ was found such that $\langle \alpha_{t'}, x_{t'} \rangle \leq 0$. Therefore there holds

$$|\alpha_{t+1}|^2 = |\alpha_t + x_t|^2 = |\alpha_t|^2 + 2 \cdot \langle \alpha_t, x_t \rangle + |x_t|^2 \leq |\alpha_t|^2 + |x_t|^2 \leq |\alpha_t|^2 + D^2 ,$$

from which it follows that

$$|\alpha_{t+1}|^2 \leq t \cdot D^2 , \tag{5.59}$$

since $\alpha_1 = 0$. We will denote $\alpha^* = \operatorname{argmin}_{x \in \overline{X}} |x|$. The number $|\alpha^*|$ is then ε. According to (5.53) we obtain

$$\left\langle \frac{\alpha^*}{|\alpha^*|}, x^j \right\rangle \geq |\alpha^*| = \varepsilon .$$

There holds for the scalar product $\langle \alpha^*/|\alpha^*|, \alpha_{t+1} \rangle$,

$$\left\langle \frac{\alpha^*}{|\alpha^*|}, \alpha_{t+1} \right\rangle = \left\langle \frac{\alpha^*}{|\alpha^*|}, (\alpha_t + x_t) \right\rangle = \left\langle \frac{\alpha^*}{|\alpha^*|}, \alpha_t \right\rangle + \left\langle \frac{\alpha^*}{|\alpha^*|}, x_t \right\rangle \geq \left\langle \frac{\alpha^*}{|\alpha^*|}, \alpha_t \right\rangle + \varepsilon .$$

From that there immediately follows

$$\left\langle \frac{\alpha^*}{|\alpha^*|}, \alpha_{t+1} \right\rangle \geq t \cdot \varepsilon ,$$

since $\alpha_1 = 0$. By the triangular inequality the scalar product of vectors is not greater than the product of their absolute values. Therefore

$$|\alpha_{t+1}| \geq |\alpha_{t+1}| \cdot \left| \frac{\alpha^*}{|\alpha^*|} \right| \geq \left\langle \frac{\alpha^*}{|\alpha^*|}, \alpha_{t+1} \right\rangle \geq t \cdot \varepsilon .$$

The result can be expressed in a more concise manner as an inequality

$$|\alpha_{t+1}| \geq t \cdot \varepsilon .$$

If we divide the inequality (5.59) by the inequality $|\alpha_{t+1}|^2 \geq t^2 \cdot \varepsilon^2$ then we obtain $t \leq D^2/\varepsilon^2$. From this it follows that in the perceptron the vector α can be changed only if $t \leq D^2/\varepsilon^2$. Thus, not later than in the step number $(D^2/\varepsilon^2) + 1$ the inequality $\langle \alpha_t, x^j \rangle > 0$ is satisfied for each $j \in J$. ■

If we compare the perceptron algorithm with the Kozinec algorithm it might seem that the former is worth being preferred, since the upper limit for the number of iterations in the Kozinec algorithm is approximately

$$\frac{D^2}{\varepsilon^2} \ln \frac{D^2}{\varepsilon^2}$$

and in the perceptron algorithm it is D^2/ε^2. Such a conclusion, however, would be too hasty because both the former and the latter evaluations are too rough. Even when it is not proved theoretically which of these two algorithms converges better, our practical experience allows us to claim that the convergence of the Kozinec algorithm is significantly better. Even despite the empirical experience, we do not intend to claim that the perceptron algorithm is worse. It is just the matter of empirical experience. Both the algorithms are simple and rich of ideas. We mean by it that they can be easily modified even for other tasks for which they may not have been originally intended. We will show such modifications and their unusual application later on.

5.4.3 Algorithm for ε-optimal separation of finite sets of points by means of the hyperplane

Let $\{x^j,\ j \in J\}$ be a finite set of points the convex hull $\overline{X}$ of which does not include the coordinate origin. Further on, let

$$r^* = \min_{x \in \overline{X}} |x| > 0\,, \quad D = \max_{x \in \overline{X}} |x| = \max_{j \in J} |x^j|\,, \quad \alpha^* = \operatorname*{argmin}_{x \in \overline{X}} |x|\,.$$

We have already proved before that the vector α^* maximises the number

$$\min_{j \in J} \left\langle \frac{\alpha}{|\alpha|}\,,\ x^j \right\rangle .$$

This means that the vector α^* is the solution of the task of optimal separation of finite sets of points.

The vector α (it can be different from α^*) is defined as the *ε-optimal solution of the task* of finite sets of points separation, when for the positive value ε the following relation will be satisfied

$$\min_{j \in J} \left\langle \frac{\alpha^*}{|\alpha^*|}\,,\ x^j \right\rangle - \min_{j \in J} \left\langle \frac{\alpha}{|\alpha|}\,,\ x^j \right\rangle \le \varepsilon\,.$$

The Kozinec algorithm for a simple separation of finite sets of points has a favourable feature, i.e., after a slight modification it becomes the algorithm for the ε-optimal solution of the task. The algorithm creates a sequence of vectors $\alpha_1, \alpha_2, \ldots, \alpha_t, \alpha_{t+1}, \ldots$ in the following way.

Algorithm 5.3 ε-optimal separation of finite sets of points

1. The vector α_1 can be any vector from the set $\overline{X}$, such as one of the vectors x^j, $j \in J$. Assume that the vector α_t has been found. The vector α_{t+1} is created in the following way:

2. The satisfaction of the following condition is checked

$$|\alpha_t| - \min_{j \in J} \left\langle \frac{\alpha_t}{|\alpha_t|}, x^j \right\rangle \leq \varepsilon, \quad j \in J. \tag{5.60}$$

3. When the preceding condition is satisfied then the algorithm finishes.
4. When the condition (5.60) is not satisfied then a point x_t in the set $\{x^j, j \in J\}$ is to be found the scalar product of which with the vector $\alpha_t/|\alpha_t|$ is the least.
5. The vector α_{t+1} is determined as a point on the straight line segment connecting the points α_t and x_t distance of which from the coordinate origin is the least one.

It can be seen that this algorithm hardly differs from the procedure quoted above for the simple separation given by the relations (5.55)–(5.56). The difference is only in the stopping condition. In the algorithm for the simple separation, the condition (5.55) finishes the algorithm when all scalar products $\langle \alpha_t/|\alpha_t|, x^j \rangle$ are positive. In the algorithm for the ε-optimal solution another condition, (5.60) is used which is stricter with small ε. According to this condition, the algorithm ends its operation only when all scalar products are not less than $|\alpha_t| - \varepsilon$.

For such a modified algorithm the condition (5.60) is surely satisfied in a certain step, since the lengths $|\alpha_t|$ is decreasing faster than does the geometrical series with a quotient less than 1. In this way, in an infinite continuation, the length $|\alpha_t|$ would converge to zero. This is, however, not possible because the vector α_t at any step t does not get over the limit of the convex set $\overline{X}$. Thus its length cannot converge to zero.

If the algorithm ended after creating the vector α_t then this vector is the solution of the ε-optimal task. From the condition (5.60) for the algorithm stop and from the inequality

$$\min_{j \in J} \left\langle \frac{\alpha^*}{|\alpha^*|}, x^j \right\rangle \leq |\alpha_t|,$$

which was many times referred to (see 5.53), we obtain the inequality

$$\min_{j \in J} \left\langle \frac{\alpha^*}{|\alpha^*|}, x^j \right\rangle - \min_{j \in J} \left\langle \frac{\alpha_t}{|\alpha_t|}, x^j \right\rangle \leq \varepsilon$$

stating that α_t is the solution of the ε-optimal task.

When once the modified algorithm is so easy, a question may arise of how it would behave at $\varepsilon = 0$, i.e., when it is not to be tuned to searching for an ε-optimal solution, but directly to the optimal solution. In this case the algorithm will usually operate for an infinitely long time (under the assumption, of course, that it is implemented on an ideal computer without numerical errors). But then it could hardly be called an algorithm. Despite this, it can be proved for this 'algorithm' that an infinite sequence $\alpha_1, \alpha_2, \ldots, \alpha_t, \ldots$ converges to the vector α^*, which is the optimal solution of the task. But the sequence of lengths cannot be said to be upper bound by the decreasing geometrical sequence, and therefore the analysis of convergence with such a sequence requires much finer considerations. Let us leave this analysis to future generations, since the theoretical incompleteness of the present analysis is no obstacle for practical application of the given 'algorithm' at $\varepsilon = 0$. The user lets the algorithm run and observes at each iterative step the number $|\alpha_t|$ and the number

$$\max_{t' \le t} \min_{j \in J} \left\langle \frac{\alpha_{t'}}{|\alpha'_t|}, x^j \right\rangle = f_t .$$

The latter of the numbers shows that among the vectors calculated before a vector α occurred quality of which was f_t. The former number $|\alpha_t|$ shows that even at an infinite continuation of the 'algorithm' a better quality will not be attained than that of $|\alpha_t|$. The following of the sequence development of the two above numbers is usually sufficient for the user to decide whether the operation of the 'algorithm' is to be interrupted or if it is to be let running in a hope that a substantially better solution may be attained.

5.4.4 Construction of Fisher classifiers by modifying Kozinec and perceptron algorithms

Let us now examine the tasks being solved and the particular algorithms not from the pattern recognition standpoint, but from a slightly different side. We will see that the tasks we are dealing with are nothing else than a solution of a special system of strict linear inequalities. Specific features of such tasks lie only in an *a priori* assumption that the system of equations is not contradictory. If we look at the tasks being solved from such a point of view then we can see that the given algorithms can be applied not only to the separation for which purpose they were originally created, but for any task that can be reduced to solving linear inequalities. Such tasks need not be sought outside the domain of pattern recognition; within the scope of pattern recognition we can find plenty of them.

As one of them, the *Fisher classifier* [Fisher, 1936] will be studied, which will be introduced now. Let X be, as before, an n-dimensional linear space, K be an integer number, and α_k, $k = 1, \ldots, K$, be K vectors which determine the decomposition of the space X into K convex cones X_k, $k = 1, \ldots, K$, so that the point $x \in X$ lies in the set X_k, if

$$\langle \alpha_k, x \rangle > \langle \alpha_j, x \rangle , \quad j = 1, 2, \ldots, K , \quad j \ne k$$

is satisfied.

The decomposition of the space X into convex cones of the above properties is the *Fisher classifier*.

Let $\widetilde{X}$ be a finite sets of points in the space X which is decomposed into K subsets $\widetilde{X}_1, \widetilde{X}_2, \ldots, \widetilde{X}_K$. In the task, which we will refer to as *Fisher task*, such vectors $\alpha_1, \alpha_2, \ldots, \alpha_K$ are to be found for the inequality

$$\langle \alpha_k, x \rangle > \langle \alpha_j, x \rangle \tag{5.61}$$

to be valid for any triplet (x, k, j) satisfying the condition $x \in \widetilde{X}_k$, $j \neq k$. The system (5.61) thus consists of a finite number of inequalities. They are just $|\widetilde{X}|(K-1)$. Our objective is to solve the system (5.61) under the condition that the solution of such a task is previously known to exist. It is obvious that the task of a linear separation of two finite sets of points is a special case of Fisher task at $K = 2$. The linear separation is achieved by means of the vector $\alpha_2 - \alpha_1$. An unexpected result is that any Fisher task can be reduced into its particular case. And now we will show how to do it.

Let Y be a space of the dimension nK. We will map into it the set $\widetilde{X}$ and the set of vectors $\alpha_1, \alpha_2, \ldots, \alpha_K$. The set of coordinates of the space Y will be decomposed into K subsets. Each of them consists n coordinates. Thus we can use the expression the 'first n-tuplet of coordinates', 'second n-tuplet of coordinates', 'n-tuplet of coordinates with the ordinal number k'.

The ensemble of vectors $\alpha_1, \alpha_2, \ldots, \alpha_K$ will be represented as a (nK)-dimensional vector α the k-th n-tuplet of coordinates of which is the vector α_k. Simply speaking, the sequence of (nK) coordinates of vectors α is created so that the coordinates of the vectors $\alpha_1, \alpha_2, \ldots, \alpha_K$ are written into one sequence one after another.

For each $x \in \widetilde{X}$ a set $\widetilde{Y}(x) \subset Y$ will be created which contains $K-1$ vectors. It will be done in the following way. Let k be the ordinal number of the subset $\widetilde{X}(k)$ to which x belongs. We will enumerate the vectors from the set $\widetilde{Y}(x)$ with numbers $j = 1, 2, \ldots, K$, $j \neq k$. The symbol $y(j, x)$ will denote j-th vector from the set $\widetilde{Y}(x)$. It will be created so that its j-th n-tuplet of coordinates is $-x$, k-th n-tuplet is x, and all other coordinates are equal to zero. We will introduce $\widetilde{Y}$ as the set

$$\widetilde{Y} = \bigcup_{x \in \widetilde{X}} \widetilde{Y}(x) .$$

Let k and j be different numbers. Let x be a point from the subset $\widetilde{X}_k$. In the manner of creating vectors α and the set $\widetilde{Y}$ presented there holds $\langle \alpha_k, x\rangle - \langle \alpha_j, x\rangle = \big\langle \alpha, y(j, x)\big\rangle$, and therefore the inequality $\langle \alpha_k, x\rangle > \langle \alpha_j, x\rangle$ is equivalent to the inequality $\big\langle \alpha, y(j, x)\big\rangle > 0$. The system of inequalities (5.61) will become equivalent to the system

$$\langle \alpha, y\rangle > 0 \,, \quad y \in \widetilde{Y} . \tag{5.62}$$

The system (5.62) can be solved by means of a perceptron or a Kozinec algorithm. We will obtain an (nK)-dimensional vector α, which satisfies the condition (5.62), and apparently expresses the ensemble that consists of K vectors α_1, α_2, ..., α_K, each from which is n-dimensional. This ensemble satisfies the conditions (5.61).

The procedure for solving the task serves as an explanation only, and not as immediate programming hint. In writing the piece of software that is to construct the Fisher classifier practically, the set $\widetilde{Y}$ consisting of an (nK)-dimensional vectors need not be explicitly present. This set is expressed by the structure of the program in which an algorithm for solving the system (5.62) of strict inequalities is modified. We will take into account that the system

(5.62) originates from the system (5.61). We will demonstrate that, e.g., the modification of a perceptron algorithm will appear incredibly simple.

Let, in the step t of the algorithm, the vectors α_k^t, $k = 1, \ldots, K$, be calculated. These vectors are to be verified and the existence of the point x in the set $\widetilde{X}$ is to be found which will be wrongly recognised by these vectors. The vectors $x \in \widetilde{X}$ are to be examined one after another so that for each vector the number

$$b = \max_k \langle \alpha_k^t, x \rangle$$

is calculated and then it is checked whether the equality $\langle \alpha_j^t, x \rangle = b$ is satisfied for some $j \neq k$. Let us mention that k is the number of the subset $\widetilde{X}_k$ to which the point x belongs. As soon as such a point x occurs, it means that it will not be correctly classified. The vectors α_j and α_k for the next iteration will be changed so that

$$\alpha_k^{t+1} = \alpha_k^t + x, \quad \alpha_j^{t+1} = \alpha_j^t - x.$$

When such a point does not occur then this means that the task has been solved.

We could hardly find an algorithm the programming of which is simpler. The modification of the Kozinec algorithm leading to the Fisher classifier is slightly more complicated, but it is also rather simple.

5.4.5 Further modification of Kozinec algorithms

Let $\widetilde{X}_1$ and $\widetilde{X}_2$ be two finite sets of points in a linear space X the convex hulls $\overline{X}_1$ and $\overline{X}_2$ of which are disjunctive. Let the vector α and the number θ decompose the space X into three subsets $X^+(\alpha, \theta)$, $X^-(\alpha, \theta)$ and $X^0(\alpha, \theta)$ so that

$$\begin{aligned} X^+(\alpha, \theta) &= \{x \in X \mid \langle \alpha, x \rangle > \theta\}, \\ X^-(\alpha, \theta) &= \{x \in X \mid \langle \alpha, x \rangle < \theta\}, \\ X^0(\alpha, \theta) &= \{x \in X \mid \langle \alpha, x \rangle = \theta\}. \end{aligned}$$

The task of a simple separation of the sets $\widetilde{X}_1$ and $\widetilde{X}_2$ lies in finding such a vector α and the number θ, so that $\widetilde{X}_1 \subset X^+(\alpha, \theta)$ and $\widetilde{X}_2 \subset X^-(\alpha, \theta)$, or, which is the same, that the following strict inequalities should be satisfied

$$\left. \begin{aligned} \langle \alpha, x \rangle > \theta, \quad x \in \widetilde{X}_1, \\ \langle \alpha, x \rangle < \theta, \quad x \in \widetilde{X}_2, \end{aligned} \right\} \tag{5.63}$$

with respect to the vector α and the number θ at the known sets $\widetilde{X}_1$ and $\widetilde{X}_2$.

We have already demonstrated how this task can be reduced to the task

$$\begin{aligned} \langle \alpha, x \rangle > 0, \quad x \in \widetilde{X}_1, \\ \langle \alpha, x \rangle < 0, \quad x \in \widetilde{X}_2, \end{aligned}$$

and which was further reduced to become

$$\langle \alpha, x \rangle > 0, \quad x \in \widetilde{X}.$$

Now we formulate the tasks of the optimal and ε-optimal set separations and we will introduce Kozinec solution of the tasks without their equivalent transformations. We will see that such a direct solution has certain advantages.

Let α and θ be the solution of the system (5.63). The distance of the point $x \in \widetilde{X}_1$ from the hyperplane $X^0(\alpha, \theta)$ is $(\langle \alpha, x \rangle - \theta)/|\alpha|$ and the distance of the point $x \in \widetilde{X}_2$ from the hyperplane $X^0(\alpha, \theta)$ is $(\theta - \langle \alpha, x \rangle)/|\alpha|$. The task of the optimal separation of the sets $\widetilde{X}_1$ and $\widetilde{X}_2$ is defined as finding such a solution of the system (5.63), that maximises the number

$$f(\alpha, \theta) = \min \left(\min_{x \in \widetilde{X}_1} \frac{\langle \alpha, x \rangle - \theta}{|\alpha|}, \min_{x \in \widetilde{X}_2} \frac{\theta - \langle \alpha, x \rangle}{|\alpha|} \right) .$$

For $r^* = \max_{\alpha, \theta} f(\alpha, \theta)$ the task of the ε-optimal separation of the sets $\widetilde{X}_1$ and $\widetilde{X}_2$ is defined as finding the vector α and the number θ for which

$$r^* - f(\alpha, \theta) \leq \varepsilon .$$

Let us have a brief look at the main considerations leading to the solution of these tasks. The key idea is that the sets $\widetilde{X}_1$ and $\widetilde{X}_2$ can be optimally separated by a hyperplane which is perpendicular to the vector $\alpha_1^* - \alpha_2^*$, and passes through the centre of the straight line connecting the points α_1^* and α_2^*. The points α_1^* and α_2^* belong to the convex hulls $\overline{X}_1$ and $\overline{X}_2$. These two points determine the shortest distance between the two convex hulls. Algorithms for the optimal and ε-optimal separations, as well as the algorithm for creating a sequence that converges to the optimal separation are based on the minimisation of the distance between the points α_1 and α_2 on the condition that $\alpha_1 \in \overline{X}_1$ and $\alpha_2 \in \overline{X}_2$. If we wanted to derive these algorithms we would recall, with slight changes, the way of deriving Kozinec algorithms for the previous case, when it was assumed that $\theta = 0$. We will present the algorithm for solving the tasks formulated here without deriving or proving them. For the reader who is eager to learn more they are left as a kind of individual exercise.

We will create a sequence of points $\alpha_1^1, \alpha_1^2, \ldots, \alpha_1^t, \alpha_1^{t+1}, \ldots$ and a sequence of points $\alpha_2^1, \alpha_2^2, \ldots \ldots, \alpha_2^t, \alpha_2^{t+1}, \ldots$ according to the following algorithm.

Algorithm 5.4 Modification of Kozinec algorithm

1. The point α_1^1 is any point from the set $\widetilde{X}_1$ and the point α_2^1 is any point from the set $\widetilde{X}_2$.
2. Assume that the pair α_1^t and α_2^t has already been created. For these pairs either the point $x^t \in \widetilde{X}_1$ is to be found for which the following condition holds

$$\left\langle x^t - \alpha_2^t, \frac{\alpha_1^t - \alpha_2^t}{|\alpha_1^t - \alpha_2^t|} \right\rangle \leq |\alpha_1^t - \alpha_2^t| - \frac{\varepsilon}{2}, \tag{5.64}$$

or the point $x^t \in \widetilde{X}_2$ for which there holds

$$\left\langle x^t - \alpha_1^t, \frac{\alpha_2^t - \alpha_1^t}{|\alpha_2^t - \alpha_1^t|} \right\rangle \leq |\alpha_2^t - \alpha_1^t| - \frac{\varepsilon}{2}. \tag{5.65}$$

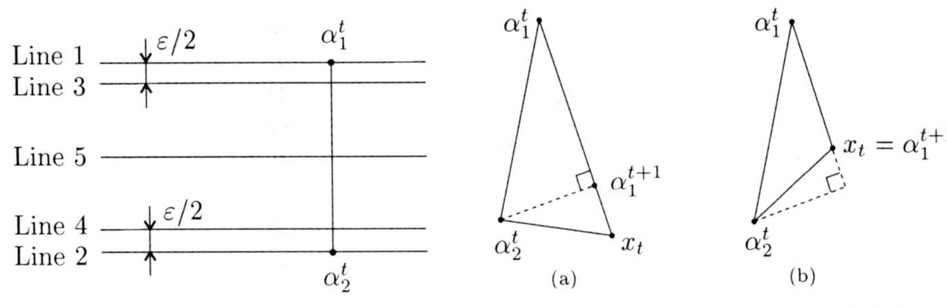

Figure 5.11 Geometrical interpretation of conditions as a relation of two points and five possible straight lines.

Figure 5.12 Two possible geometrical interpretations of the change of point α_1 positions.

3. If neither of these points exists then the algorithm stops the operation and provides the vectors α_1^t and α_2^t at the output.
4. If the vector $x^t \in \widetilde{X}_1$ exists which satisfies the relation (5.64) then the vector α_2 does not change, i.e., $\alpha_2^{t+1} = \alpha_2^t$, and the vector α_1 is changed according to the rule

$$\left.\begin{aligned} \alpha_1^{t+1} &= \alpha_1^t(1-k) + x^t \cdot k\,, \\ \text{where}\quad k &= \min\left(1, \frac{\langle \alpha_1^t - \alpha_2^t\,,\, \alpha_1^t - x^t\rangle}{|\alpha_1^t - x^t|^2}\right) . \end{aligned}\right\} \tag{5.66}$$

The above rule means that the vector α_1^{t+1} is determined as a point on the abscissa connecting the points α_1^t and x^t that is nearest to the point α_2^t.

5. When the vector $x^t \in \widetilde{X}_2$ exists satisfying (5.65) then $\alpha_1^{t+1} = \alpha_1^t$ and

$$\left.\begin{aligned} \alpha_2^{t+1} &= \alpha_2^t(1-k) + x^t \cdot k\,, \\ \text{where}\quad k &= \min\left(1, \frac{\langle \alpha_2^t - \alpha_1^t\,,\, \alpha_2^t - x^t\rangle}{|\alpha_2^t - x^t|^2}\right) . \end{aligned}\right\} \tag{5.67}$$

The above rule says that the vector α_2^{t+1} is determined as a point on the abscissa connecting the points α_2^t and x^t that is nearest to the point α_1^t.

According to the quantity ε in the expressions (5.64) and (5.65) the expressions (5.64)–(5.67) provide three different algorithms.

1. When ε is a positive constant then this is an algorithm for the ε-optimal separation of the sets.
2. When ε is a variable $\frac{1}{2}\,|\alpha_1^t - \alpha_2^t|$ then this is an algorithm for the simple separation of the sets.
3. When $\varepsilon = 0$ then this is an algorithm for creating an infinite sequence of vectors α_1^t and α_2^t, which converges to the vectors

$$(\alpha_1^*, \alpha_2^*) = \operatorname*{argmin}_{(\alpha_1,\alpha_2)\in \overline{X}_1 \times \overline{X}_2} |\alpha_1 - \alpha_2|\ .$$

For illustration, let us recall previous considerations with respect to the *geometrical interpretation*. The conditions (5.64) and (5.65) can be visualised by means of Fig. 5.11, where two points α_1^t, α_2^t and five straight lines are shown, which are perpendicular to an abscissa connecting the points α_1^t and α_2^t. Line 1 passes through the point α_1^t. Line 2 passes through the point α_2^t. Line 3 and Line 4 lie between Line 1 and Line 2 so that the distance between Line 3 and Line 1 is $\frac{1}{2}\varepsilon$. Similarly, Line 4 is in the distance $\frac{1}{2}\varepsilon$ from the Line 2. Finally, Line 5 lies half way between the points α_1^t and α_2^t.

The conditions (5.64) and (5.65) have the following geometrical interpretation. When $\varepsilon = |\alpha_1^t - \alpha_2^t|$ then the conditions (5.64) or (5.65) state that either one point from the set X_1 gets below the Line 5 or a point from the set X_2 gets above this straight line, i.e., one of the points is not correctly classified. In this case one of the points α_1 or α_2 changes its position. If such a point does not exist it means that Line 5 separates the sets $\widetilde{X}_1$ and $\widetilde{X}_2$. When ε is a positive constant then the condition (5.64) means that a point from the set $\widetilde{X}_1$ occurs below Line 3. The condition (5.65) states that a point from the set $\widetilde{X}_2$ comes above Line 4. In that case either of the points α_1 or α_2 changes its position and Line 5 can already classify the set $\widetilde{X}_1 \bigcup \widetilde{X}_2$ correctly. When no such a point exists then it means that Line 5 separates ε-optimally the set $\widetilde{X}_1 \bigcup \widetilde{X}_2$. And finally, when no point from $\widetilde{X}_1$ occurs below Line 1 and no point from $\widetilde{X}_2$ lies above Line 2 then Line 5 optimally separates sets $\widetilde{X}_1$ and $\widetilde{X}_2$.

The algorithm changing the position of points α_1 and α_2, which is expressed by relations (5.66) and (5.67), is also easy to comprehend in its geometrical interpretation. The case in which the vector α_1 changes and the vector α_2 does not change is illustrated in Fig. 5.12.

The point α_1^{t+1} is a point on the abscissa connecting the point α_1^t to the point x_t which is the closest to the point α_2^t. It is either the bottom of the perpendicular drawn from the point α_2^t towards the straight line passing through points α_1^t and x_t (Fig. 5.12a), or the point x_t (Fig. 5.12b). The first case occurs when the bottom of the perpendicular fits inside the abscissa and the second when the bottom lies outside the abscissa.

We have made sure in an informal, and hopefully easy to understand, way that the described algorithm in the steady state solves one of three tasks depending on the value of the ε.

1. The task of simple separation of the point sets if the algorithm used changing variable ε,

$$\varepsilon = \frac{1}{2}\,\left|\alpha_1^t - \alpha_2^t\right| .$$

2. The task of ε-optimal separation of the point sets if the algorithm used the constant value ε.

3. The task of optimal separation of the point sets if $\varepsilon = 0$.

What remains now is to check whether the described algorithm converges to the stable state for sure. It is possible to prove using simple reasoning similar to that one applied in analysis of the Kozinec algorithm from Subsection 5.4.3 that the algorithm definitely ends up in the stable state provided that $\varepsilon \neq 0$. The reason is that the sequence of lengths $|\alpha_1^t - \alpha_2^t|$ monotonically decreases and decreases faster than geometric sequences with quotient

$$\frac{1}{\sqrt{1+\varepsilon^2/D^2}} < 1 .$$

The variable D here is no longer

$$\max_{x \in \widetilde{X}} |x|$$

as was the case in the Kozinec algorithm separating finite sets of points, but it is the value

$$\max\left(\max_{x,y \in \widetilde{X}_1} |x-y| , \max_{x,y \in \widetilde{X}_2} |x-y| \right),$$

which can be much smaller. The algorithm can converge faster and typically it is the case.

5.5 Solution of the generalised Anderson task

The comprehension of the task separating linearly finite sets of points allows us to express the solution of the generalised Anderson task completely. The basic thoughts aimed at the solution were sketched in Subsection 5.3.6. The piece of information which was missing to the complete solution of the tasks (5.40), (5.41) and (5.42) has just been talked through.

However, the algorithms separating finite point sets are of importance not only because they are part of the solution of the Anderson task. We will show in this final chord of the lecture that Anderson tasks can be slightly modified and reduced to the simplest task which separates two point sets. This modification is almost negligible from the practical point of view. However, the point sets are not finite any more. Nevertheless the algorithms for simple separation which we already know can be reasonably modified to separate infinite sets too.

5.5.1 ε-solution of Anderson task

Recall the generalised Anderson task which has already been formulated in Subsection 5.3.1. Let $\{\mu^j, \, j \in J\}$ is a finite set of n-dimensional vectors and $\{\sigma^j, \, j \in J\}$ is a finite set of positive-definite symmetrical matrices of dimension $n \times n$. The set J is divided into two subsets J_1 and J_2. For $j \in J_1$ the vector α and the number θ, the number $\mathrm{er}(j, \alpha, \theta)$ represents the probability that the Gaussian vector x with mathematical expectation μ^j and covariance matrix σ^j satisfies the inequality $\langle \alpha, x \rangle \leq \theta$. The designation 'er' is introduced on the place of the function ε used before to prevent conflict with the symbol ε denoting a small value here. Similarly for $j \in J_2$ the number $\mathrm{er}(j, \alpha, \theta)$ is the probability of the inequality $\langle \alpha, x \rangle \geq \theta$. The task was defined earlier (see the criterion (5.6)) as seeking

$$(\alpha^*, \theta^*) = \operatorname*{argmin}_{\alpha, \theta} \max_j \; \mathrm{er}(j, \alpha, \theta) . \tag{5.68}$$

Now let us break away from the purely mathematical content of generalised Anderson task, even though it is quite rich in itself. Let us see the task from the practical point of view. In the great majority of practical applications the main interest of the project designer is not to have an optimal recognition procedure, but very often it is sufficient to choose just a good one. If the optimal discrimination rule is found and errors occur, e.g., in 30% cases, the task has not been solved from the practical point of view. Neither the procedure being optimal does not help. But in another situation in which recognition is correct in 99.9% cases, one can hardly imagine in practice that a procedure would be rejected only because of not being optimal. Simply speaking, optimal and well applicable are two different concepts. An optimal procedure may not be applicable and in another situation, a not optimal procedure can be acceptable.

The previous informal, but still reasonable, considerations make us replace the task (5.68) by seeking the vector α and the numbers θ which satisfy the inequality

$$\max_j \operatorname{er}(j, \alpha, \theta) < \varepsilon\,, \tag{5.69}$$

where ε is the probability of the wrong decision which in a given application must not be exceeded. Further on we shall formally analyse the task (5.69) under the usual assumption that it has a solution.

Let $E(r, \mu, \sigma)$ be an ellipse, i.e., a closed set of points x, which satisfy the inequality

$$\big\langle (\mu - x)\,,\, \sigma^{-1} \cdot (\mu - x) \big\rangle \le r^2\,.$$

Let us point out two closed sets

$$X_1(r) = \bigcup_{j \in J_1} E\left(r, \mu^j, \sigma^j\right) \quad \text{and} \quad X_2(r) = \bigcup_{j \in J_2} E\left(r, \mu^j, \sigma^j\right)\,.$$

The task (5.69) can be reduced to a task of the simple separation of infinite sets $X_1(r)$ and $X_2(r)$ at a certain value r. It results from the following theorem.

Theorem 5.6 On ε-solution of generalised Anderson task. *Let the number r be the solution of the following equation for the given positive number $\varepsilon < 0.5$,*

$$\varepsilon = \int_r^\infty \frac{1}{\sqrt{2\pi}} \mathrm{e}^{-\frac{1}{2}x^2}\, \mathrm{d}x\,.$$

The vector α and the number θ satisfy the requirement

$$\max_j \mathit{er}(j, \alpha, \theta) < \varepsilon \tag{5.70}$$

if and only if α and θ satisfy the infinite system of linear inequalities

$$\left.\begin{aligned} \langle \alpha, x \rangle &> \theta\,, \quad x \in X_1(r)\,, \\ \langle \alpha, x \rangle &< \theta\,, \quad x \in X_2(r)\,. \end{aligned}\right\} \tag{5.71}$$

▲

Proof. In the proof an explicit expression is needed for the maximal and minimal values of the scalar product

$$f(x) = \langle \alpha, x \rangle \tag{5.72}$$

under the condition

$$F(x) = \big\langle \mu - x\,,\, \sigma^{-1} \cdot (\mu - x) \big\rangle \le r^2\,. \tag{5.73}$$

The points sought which are the solution of this task satisfy the equation

$$\operatorname{grad}\big(f(x) + \lambda \cdot F(x)\big) = 0 \tag{5.74}$$

at some value of the coefficient λ. On the basis of (5.74) we can write

$$\alpha + \lambda \cdot \sigma^{-1} \cdot (\mu - x) = 0\,,$$

from which it follows that

$$x = \mu + \frac{1}{\lambda} \sigma \cdot \alpha\,.$$

It is obvious that the extreme of the linear function on the convex set (5.73) will be achieved at its limit, i.e., when $F(x) = r^2$. We will substitute the expression derived for x to the equation $F(x) = r^2$ and so ensure the value λ,

$$\begin{aligned}
&\left\langle \left(\mu - \mu - \frac{1}{\lambda}\sigma \cdot \alpha\right),\, \sigma^{-1} \cdot \left(\mu - \mu - \frac{1}{\lambda}\sigma \cdot \alpha\right) \right\rangle \\
&= \frac{1}{\lambda^2} \big\langle (\sigma \cdot \alpha),\, \sigma^{-1} \cdot (\sigma \cdot \alpha) \big\rangle \\
&= \frac{1}{\lambda^2} \big\langle (\sigma \cdot \alpha),\, (\sigma^{-1} \cdot \sigma) \cdot \alpha \big\rangle = \frac{1}{\lambda^2} \langle \alpha,\, \sigma \cdot \alpha \rangle = r^2\,.
\end{aligned}$$

From the above there follows in seeking the minimum that

$$\lambda = \frac{\sqrt{\langle \alpha\,,\, \sigma \cdot \alpha \rangle}}{r}\,,$$

since the number r is positive at $\varepsilon < 0.5$. Similarly, in seeking the maximum,

$$\lambda = -\frac{\sqrt{\langle \alpha\,,\, \sigma \cdot \alpha \rangle}}{r}\,.$$

The position vectors of the extremes sought are

$$\left.\begin{aligned}
x_{\min} &= \mu - \frac{r}{\sqrt{\langle \alpha\,,\, \sigma \cdot \alpha \rangle}} \cdot \sigma \cdot \alpha\,,\\
x_{\max} &= \mu + \frac{r}{\sqrt{\langle \alpha\,,\, \sigma \cdot \alpha \rangle}} \cdot \sigma \cdot \alpha\,.
\end{aligned}\right\} \tag{5.75}$$

The minimum and maximum (5.72) under the condition (5.73) are

$$\left.\begin{aligned}
\min f(x) &= \langle \alpha, \mu \rangle - r \cdot \sqrt{\langle \alpha\,,\, \sigma \cdot \alpha \rangle}\,,\\
\max f(x) &= \langle \alpha, \mu \rangle + r \cdot \sqrt{\langle \alpha\,,\, \sigma \cdot \alpha \rangle}\,.
\end{aligned}\right\}$$

We will prove now that any solution of the system (5.71) satisfies the condition (5.70). When (5.71) is satisfied then the inequality $\langle\alpha, x\rangle > \theta$ is satisfied for any point of the ellipse $E\left(r, \mu^j, \sigma^j\right)$, $j \in J_1$. Since this ellipse is a closed set the system of inequalities

$$\langle\alpha, x\rangle > \theta\,, \quad x \in E\left(r, \mu^j, \sigma^j\right)\,, \quad j \in J_1,$$

is a different written form of the expression

$$\min_{x \in E(r,\mu^j,\sigma^j)} \langle\alpha, x\rangle > \theta\,, \quad j \in J_1,$$

or with respect to the expressions (5.75) already proved

$$\langle\alpha, \mu^j\rangle - r \cdot \sqrt{\langle\alpha\,,\, \sigma^j \cdot \alpha\rangle} > \theta\,, \quad j \in J_1\,,$$

and

$$\frac{\langle\alpha, \mu^j\rangle - \theta}{\sqrt{\langle\alpha\,,\, \sigma^j \cdot \alpha\rangle}} > r\,, \quad j \in J_1\,. \tag{5.76}$$

Similarly, thanks to the second inequality in (5.71), we have

$$\langle\alpha, x\rangle < \theta\,, \quad x \in E\left(r, \mu^j, \sigma^j\right)\,, \quad j \in J_2\,,$$

$$\max_{x \in E(r,\mu^j,\sigma^j)} \langle\alpha, x\rangle < \theta\,, \quad j \in J_2\,,$$

$$\langle\alpha, \mu^j\rangle + r \cdot \sqrt{\langle\alpha\,,\, \sigma^j \cdot \alpha\rangle} < \theta\,, \quad j \in J_2\,,$$

$$\frac{\theta - \langle\alpha, \mu^j\rangle}{\sqrt{\langle\alpha\,,\, \sigma^j \cdot \alpha\rangle}} > r\,, \quad j \in J_2\,. \tag{5.77}$$

The numerator in the left-hand part of (5.76) is the mathematical expectation of a random number $\langle\alpha, x\rangle - \theta$ and the denominator is the mean square deviation of the same number under the condition that x is a random vector with the mathematical expectation μ^j and the covariance matrix σ^j. The inequality (5.76) means that the probability of the event that the number $\langle\alpha, x\rangle - \theta$ will be negative is less than

$$\int_r^\infty \frac{1}{\sqrt{2\pi r}} \mathrm{e}^{-\frac{1}{2}x^2}\, \mathrm{d}x\,.$$

This means that $\mathrm{er}(j, \alpha, \theta) < \varepsilon$ for $j \in J_1$. Similarly, from the inequality (5.77) there follows $\mathrm{er}(j, \alpha, \theta) < \varepsilon$ for $j \in J_2$. In this way we have proved that the arbitrary solution of the system (5.71) satisfies (5.70) as well.

Now we will prove that when the pair (α, θ) does not satisfy the system (5.71) then it is not satisfied by the inequality (5.70) either. Assume that the inequality expressed in the first line of (5.71) is not satisfied. Then $j \in J_1$ and $x \in E(r, \mu^j, \sigma^j)$ exist such that $\langle\alpha, x\rangle \le \theta$. From this there immediately follows that

$$\min_{x \in E(j,\mu^j,\sigma^j)} \langle\alpha, x\rangle \le \theta\,, \quad \langle\alpha, \mu^j\rangle - r\,\sqrt{\langle\alpha\,,\, \sigma^j \cdot \alpha\rangle} \le \theta\,, \quad \frac{\langle\alpha, \mu^j\rangle - \theta}{\sqrt{\langle\alpha\,,\, \sigma^j \cdot \alpha\rangle}} \le r\,.$$

Eventually $\text{er}(j, \alpha, \theta) \geq \varepsilon$ is true. Similarly, when an inequality of the second line in the system (5.71) is not satisfied then $j \in J_2$ exists such that $\text{er}(j, \alpha, \theta) \geq \varepsilon$. ■

Thanks to Theorem 5.6 the ε-solution of Anderson task is reduced to seeking the number θ and the vector α which satisfy the infinite system of inequalities

$$\left.\begin{array}{lll} \langle \alpha, x \rangle > \theta\,, & x \in \left\{ x \mid \langle\, \mu^j - x\,,\, (\sigma^j)^{-1} \cdot (\mu^j - x) \rangle \leq r^2 \right\}, & j \in J_1\,, \\ \langle \alpha, x \rangle < \theta\,, & x \in \left\{ x \mid \langle\, \mu^j - x\,,\, (\sigma^j)^{-1} \cdot (\mu^j - x) \rangle \leq r^2 \right\}, & j \in J_2\,, \end{array}\right\} \qquad (5.78)$$

at a certain r, which depends on ε.

With this simplification it is convenient that Anderson task has been transformed to the already well explored task of the simple linear separation. But at the same time, some rather serious apprehensions arise in connection with this simplification. First, the system (5.78) consists of an infinitely large number of inequalities. The perceptron algorithms explored and the algorithms by Kozinec were expressed only for finite sets of linear inequalities. Second, the expression of the system (5.78) contains an inverted covariance matrix. Remember that from the very beginning we have tried to avoid that. In spite of these apprehensions we continue analysing the task as expressed in (5.78). We will see that in spite of the existence of the inverse matrix in (5.78) we will be able to avoid inverting the matrix in solving the mentioned task.

5.5.2 Linear separation of infinite sets of points

To implement the Kozinec algorithm for simple linear separation of the sets $\widetilde{X}_1$ and $\widetilde{X}_2$ all points of the set $\widetilde{X}_1 \cup \widetilde{X}_2$ need not be examined, but in a way the point $x \in \widetilde{X}_1 \cup \widetilde{X}_2$, is to be found, which by means of current values of α^t and θ^t is not classified correctly. When the sets $\widetilde{X}_1$ and $\widetilde{X}_2$ are properly expressed this wrong point can be then found without examining all points of the sets $\widetilde{X}_1$ and $\widetilde{X}_2$. The Kozinec algorithm can be implemented even in the case of the system consisting of an infinite number of inequalities, which is, e.g., in our system (5.78). Further, for the validity of the Kozinec algorithm (for unambiguity it will be referred to as the algorithm of Subsection 5.4.4) the sets $\widetilde{X}_1$ and $\widetilde{X}_2$ need not be finite. It is sufficient for the 'diameters' of these sets expressed as

$$\max_{x,y \in \widetilde{X}_1} |x - y| \quad \text{and} \quad \max_{x,y \in \widetilde{X}_2} |x - y|$$

not to be infinitely large and for their convex hulls to be disjunctive. We will show that the diameters of the sets $\widetilde{X}_1$ and $\widetilde{X}_2$ are always finite. For $\widetilde{X}_1$ and $\widetilde{X}_2$ there hold

$$\widetilde{X}_1 = \bigcup_{j \in J_1} \left\{ x \mid \langle\, \mu^j - x,\, (\sigma^j)^{-1} \cdot (\mu^j - x) \rangle \leq r^2 \right\},$$

$$\widetilde{X}_2 = \bigcup_{j \in J_2} \left\{ x \mid \langle\, \mu^j - x,\, (\sigma^j)^{-1} \cdot (\mu^j - x) \rangle \leq r^2 \right\}.$$

The matrix σ in the relations (5.78) are positive-definite, and thus the sets $\widetilde{X}_1$ and $\widetilde{X}_2$ are bound. The disjunctive character of $\widetilde{X}_1$ and $\widetilde{X}_2$ cannot be ensured in the general case, and so the algorithm quoted later can solve the ε-task, if such a solution exists.

Let us study the problem how to look for an inequality in the system (5.76), which for the vectors α and the number θ is not satisfied. Thus the index $j \in J$ and the vector x are to be found which satisfy the inequality $\langle x-\mu^j\,,(\sigma^j)^{-1}\cdot(x-\mu^j)\rangle \le r^2$ and $\langle\alpha, x\rangle \le \theta$ for $j \in J_1$ or the inequality $\langle x-\mu^j\,,(\sigma^j)^{-1}\cdot(x-\mu^j)\rangle \le r^2$ and $\langle\alpha, x\rangle \ge \theta$ for $j \in J_2$. On the basis of our considerations used in the proof of Theorem 5.6, we claim that the first condition is equivalent to the statement that the minimal value of the scalar product $\langle\alpha, x\rangle$ on the set $\{x \mid \langle(x-\mu^j)\,,\,(\sigma^j)^{-1}\cdot(x-\mu^j)\rangle \le r^2\}$ is $\le \theta$ for some $j \in J_1$. The second condition is equivalent to the statement that for some $j \in J_2$ the maximal value of the scalar product $\langle\alpha, x\rangle$ is $\ge \theta$ on a similar set, i.e.,

$$\left(\exists j \in J_1 \;\Big|\; \min_{x\in E(j,\mu^j,\sigma^j)} \langle\alpha, x\rangle \le \theta\right) \;\vee\; \left(\exists j \in J_2 \;\Big|\; \max_{x\in E(j,\mu^j,\sigma^j)} \langle\alpha, x\rangle \ge \theta\right).$$

Recall that the symbol $\vee$ expresses the disjunction of the two conditions.

Using the expressions (5.76) and (5.77) we will transpose the statement that the pair (α,θ) does not satisfy the infinite system of inequalities (5.78) into an equivalent statement that the same pair (α,θ) satisfies some of the inequalities

$$\begin{aligned} &\frac{\langle\alpha,\mu^j\rangle-\theta}{\sqrt{\langle\alpha\,,\,\sigma^j\cdot\alpha\rangle}} \le r\,, \quad j\in J_1\,,\\ &\frac{\theta-\langle\alpha,\mu^j\rangle}{\sqrt{\langle\alpha\,,\,\sigma^j\cdot\alpha\rangle}} \le r\,, \quad j\in J_2\,, \end{aligned} \tag{5.79}$$

which though their character is non-linear, their number is finite. In addition, each inequality can be easily verified because there is no need for inverting the covariance matrix σ^j.

Assume that one inequality from the system (5.79) is satisfied. Let it be, for example, the inequality of the first line of (5.79). Then we can find a linear inequality from the infinite system (5.78) which is not satisfied and learn that it corresponds to the point x determined by the formula

$$x = \mu^j - \frac{r}{\sqrt{\langle\alpha\,,\,\sigma^j\cdot\alpha\rangle}}\cdot\sigma^j\cdot\alpha\,. \tag{5.80}$$

When the pair (α,θ) satisfies some inequality from the second line of the system (5.79), i.e., for $j \in J_2$, then the inequality from the system (5.78) which will not be satisfied corresponds to the point

$$x = \mu^j + \frac{r}{\sqrt{\langle\alpha\,,\,\sigma^j\cdot\alpha\rangle}}\cdot\sigma^j\cdot\alpha\,. \tag{5.81}$$

Both expressions (5.80) and (5.81) can be calculated fast and easily, since they do not require the inversion of the covariance matrix σ^j either. We see that the validity of the system (5.78) containing an infinite number of linear inequalities can be constructively verified. In addition, a concrete inequality can be found that is not satisfied when the system (5.78) is not valid. This property can be summarised in the following algorithm seeking the vector α and the number θ, which satisfy the conditions (5.78) on assumption that the values sought actually exist.

Algorithm 5.5 ε-solution of generalised Anderson task

1. For the given ε the number r is to be found which is the solution of the equation

$$\varepsilon = \int_r^\infty \frac{1}{\sqrt{2\pi}} \mathrm{e}^{-\frac{1}{2}x^2} \, \mathrm{d}x \, .$$

2. The algorithm creates two sequences of vectors $\alpha_1^1, \alpha_1^2, \ldots, \alpha_1^t, \ldots$ and $\alpha_2^1, \alpha_2^2, \ldots$ $\ldots, \alpha_2^t, \ldots$, to which the searched vector α^t and the number θ^t correspond

$$\alpha^t = \alpha_1^t - \alpha_2^t \, , \quad \theta^t = \frac{1}{2} \left(|\alpha_1^t|^2 - |\alpha_2^t|^2 \right) \, .$$

   The vector α_1^1 is an arbitrarily selected vector from the set $\widetilde{X}_1$, for example, one of the vectors μ^j, $j \in J_1$, and the vector α_2^1 is, for example, one of the vectors μ^j, $j \in J_2$.

3. Assume the vectors α_1^t and α_2^t have already been created. For them the vector α^t, the number θ^t are to be calculated and the following conditions are to be checked

$$\frac{\langle \alpha^t, \mu^j \rangle - \theta^t}{\sqrt{\langle \alpha^t \, , \, \sigma^j \cdot \alpha^t \rangle}} > r \, , \quad j \in J_1 \, , \tag{5.82}$$

   as well as the conditions

$$\frac{\theta^t - \langle \alpha^t, \mu^j \rangle}{\sqrt{\langle \alpha^t \, , \, \sigma^j \cdot \alpha^t \rangle}} > r \, , \quad j \in J_2 \, . \tag{5.83}$$

4. When all these conditions have been satisfied then it means that the vector α^t and the number θ^t are the ε-solution of the task.

5. When for $j \in J_1$ some of the inequalities (5.82) is not satisfied then a vector

$$x^t = \mu^j - \frac{r}{\sqrt{\langle \alpha^t \, , \, \sigma^j \cdot \alpha^t \rangle}} \cdot \sigma^j \cdot \alpha^t$$

   is to be calculated as well as the new vector α_1^{t+1} according to the formula

$$\alpha_1^{t+1} = \alpha_1^t \cdot (1 - k) + x^t \cdot k \, ,$$

   where

$$k = \min \left(1 \, , \, \frac{\langle \alpha_1^t - \alpha_2^t \, , \, \alpha_1^t - x^t \rangle}{|\alpha_1^t - x^t|^2} \right) \, .$$

   The vector α_2^t does not change in this case, i.e., $\alpha_2^{t+1} = \alpha_2^t$.

6. If for $j \in J_2$ some the inequalities (5.83) is not satisfied then a vector

$$x^t = \mu^j + \frac{r}{\sqrt{\langle \alpha^t \, , \, \sigma^j \cdot \alpha^t \rangle}} \cdot \sigma^j \cdot \alpha^t$$

is to be calculated as well as the new vector α_2^{t+1} according to the formula

$$\alpha_2^{t+1} = \alpha_2^t \cdot (1-k) + x^t \cdot k \,,$$

where

$$k = \min\left(1\,,\ \frac{\langle \alpha_2^t - \alpha_1^t\,,\ \alpha_2^t - x^t\rangle}{|\alpha_2^t - x^t|^2}\right).$$

The vector α_1^t does not change in this case, i.e., $\alpha_1^{t+1} = \alpha_1^t$.

If Anderson task has an ε-solution then the above algorithm is sure to arrive at a state in which both conditions (5.82) and (5.83) are satisfied, and therefore the algorithm stops.

5.6 Discussion

I have noticed an important discrepancy which erodes the analysis of Anderson task. On the one hand, you kept assuming during the analysis that covariance matrices σ^j were positive-definite. This assumption was used by you several times in the proofs. On the other hand, you claimed from the very beginning that without loss of generality the separating hyperplane sought went through the coordinate origin. But you achieved that by introducing an additional constant coordinate. The variance of the additional coordinate is zero, and just for this reason, the covariance matrix cannot be positive-definite but only positive semi-definite.

I have waited to see how this discrepancy will be settled in the lecture, but in vain. Now I am sure that you have made a blunder in the lecture. I believe that you did this teacher's trick on purpose to check if I had read the lecture properly. There was no use doing that. Your lectures are very interesting to me, even if they are not easy reading matter.

The teacher's trick you are speaking about is used by us from time to time, but not now. The discrepancy mentioned was made neither on purpose, nor through an oversight. It really is a discrepancy, but it has no negative effect on the final results. Have another glance at the complete procedure of Anderson task and its proof and you will see that everywhere where the assumption of positive-definiteness of matrices is made use of we could do without it. But we would only have to mention the case in which the covariance matrix is degenerate. It would not be difficult but it would interrupt the continuity of the argument. Let us now examine the most important moments of deriving the procedure.

In deriving the procedure for solving Anderson task, covariance matrices are used in two situations. First, it is in calculating the values

$$f^j(\alpha) = \frac{\langle \mu^j, \alpha\rangle}{\sqrt{\langle \alpha, \sigma^j \cdot \alpha\rangle}}\,, \quad j \in J\,, \tag{5.84}$$

and for the second time in searching for the contact points

$$x_0^j(\alpha) = \mu^j - \frac{\langle \mu^j, \alpha \rangle}{\langle \alpha, \sigma^j \cdot \alpha \rangle} \cdot (\sigma^j \cdot \alpha) , \tag{5.85}$$

which are calculated for j for which there holds

$$f^j(\alpha) = \min_{j \in J} f^j(\alpha) .$$

In the algorithm for the ε-solution of Anderson task the covariance matrices are again helpful in calculating the points

$$x^j = \mu^j - \frac{r}{\sqrt{\langle \alpha, \sigma^j \cdot \alpha \rangle}} \cdot (\sigma^j \cdot \alpha) \tag{5.86}$$

which minimise the scalar product $\langle \alpha, x \rangle$ on the set of vectors x which satisfy the inequality

$$\big\langle (x - \mu^j), (\sigma^j)^{-1} \cdot (x - \mu^j) \big\rangle \le r^2 . \tag{5.87}$$

The algorithms use only the formulæ (5.84), (5.85) and (5.86). The formula (5.87) only illustrates the meaning of the relation (5.86) and is practically not used by the algorithm. Formally speaking, for the calculation of the formulæ (5.84), (5.85) and (5.86) the matrices σ^j, $j \in J$, are to be positive-definite. The matrices involve the quadratic function $\langle \alpha, \sigma^j \cdot \alpha \rangle$, whose values form the denominators of fractions and thus the value $\langle \alpha, \sigma^j \cdot \alpha \rangle$ must be greater than zero for any $\alpha \neq 0$. However, based on the way in which the algorithm uses the value $f^j(\alpha)$ in further calculation, (see formula (5.84)) and according to the meaning of the vectors x^j (see formula (5.86)), the algorithm can be defined even for the case of a zero value of the quadratic function $\langle \alpha, \sigma^j \cdot \alpha \rangle$. The values $f^j(\alpha)$ are calculated because the least value is to be chosen from them. The value $f^j(\alpha)$ for $\langle \alpha, \sigma^j \cdot \alpha \rangle = 0$ can be understood as a rather great number which is definitely not less than $f^j(\alpha)$ for the indices j, where $\langle \alpha, \sigma^j \cdot \alpha \rangle \neq 0$. The contact points x_0^j are thus to be calculated only for the indices j, for which $\langle \alpha, \sigma^j \cdot \alpha \rangle \neq 0$ holds. When such points do not exist it means that there is a zero probability of wrong classification of the j-th random Gaussian vector for any index j. In this case the algorithm can stop the operation, since no better recognition quality can be achieved. We can see that such an augmented algorithm holds even for the case of degenerate matrices σ^j.

Let us see how the vector x^j is calculated according to the formula (5.86) when $\langle \alpha, \sigma^j \cdot \alpha \rangle = 0$. Recall that x^j is a point of an ellipse and maximises the scalar product $\langle \alpha, x \rangle$. Formally speaking, the formula (5.87) defines the given ellipse only in the case in which the matrix σ^j is positive-definite, i.e., if all eigenvalues of the matrix are positive. Only then the matrix σ^j can be inverted. The ellipse, however, can be defined even in the case in which some eigenvalues are zero. For any size r it will be an ellipse whose points will lie in the hyperplane whose dimension is equal to the number of non-zero eigenvalues. In all cases, irrespective of whether the matrix σ^j can be inverted, the point x^j sought in such a created ellipse is given by the formula (5.86). An exception is the case in which $\langle \alpha, \sigma^j \cdot \alpha \rangle$ assumes a zero value. In this case the scalar products of all the points of the ellipse with the vector α are the same. This is because the whole ellipse lies in a hyperplane which is parallel with the hyperplane $\langle \alpha, x \rangle = 0$. Thus any arbitrary point of the ellipse can be chosen for the point x^j. The simplest way is to choose the point μ^j.

As you can see, the final results can be stated for the case of degenerate, i.e., positively semi-definite, matrices. It is not even difficult, but only painstaking. If you feel like it then examine all the places in the analysis where the inversion of matrices is assumed and make sure that the assumption is not necessary for proving the above statements. But we do not think that you would come across significant results from the pattern recognition standpoint during this examination. In the minimal case it would not be a bad exercise in linear algebra and the theory of matrices for you.

Could I, perhaps, ask you now to examine, together with me, a case which is part of the solution of Anderson task, i.e., seeking the vector α which maximises the function

$$f(\alpha) = \min_{j \in J} \frac{\langle \alpha, \mu^j \rangle}{\sqrt{\langle \alpha, \sigma^j \cdot \alpha \rangle}} .$$

Assume we have found the direction $\Delta\alpha$ in which the function $f(\alpha)$ is growing. And now we have to find a specific point $\alpha + t \cdot \Delta\alpha$ in this direction for which $f(\alpha + t \cdot \Delta\alpha) > f(\alpha)$ holds. In another way, and better stated,

$$t = \operatorname*{argmax}_{t} \min_{j \in J} \frac{\langle\, \alpha + t\Delta\alpha,\ \mu^j \rangle}{\sqrt{\langle\, \alpha + t\Delta\alpha,\ \sigma^j \cdot (\alpha + t\Delta\alpha) \rangle}} . \tag{5.88}$$

The solving of this task was not mentioned at all in the lecture, and so I do not know what to think of it. On the one hand, it is an optimisation of a one-dimensional function which has one extreme in addition. It might seem that this is a simple task. On the other hand, I know that even such tasks are objects of serious research.

Let us try it. The task of one-dimensional optimisation is only seemingly simple. If sufficient attention is not paid to it, it can be even troublesome. Moreover, the task (5.88) is a suitable probing field for optimisation and programmer's trifles. None of them is very significant in itself but if the application programmer knows about one hundred of them, it indicates that he/she is an expert. Even for this reason these trifles are worth knowing. Now we are at a loss which trifles to explain first. It would be better to explain to us how would you handle the task (5.88) yourself.

First, I am to select a finite number of points on the straight line on which the point $\alpha + t \cdot \Delta\alpha$ lies. To the points a finite number of values of the parameters $t_0, t_1, \ldots, t_L$ correspond. I will get a little bit ahead and say that the number of points $L+1$ is to be odd and equal to $2^h + 1$ for me to find the best point by

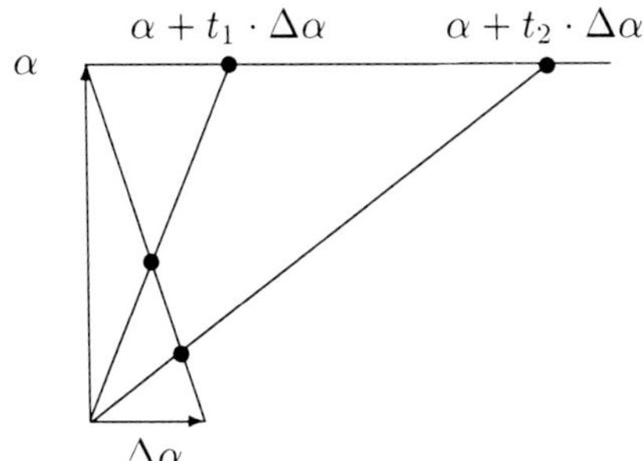

Figure 5.13 A half-line mapped to an abscissa.

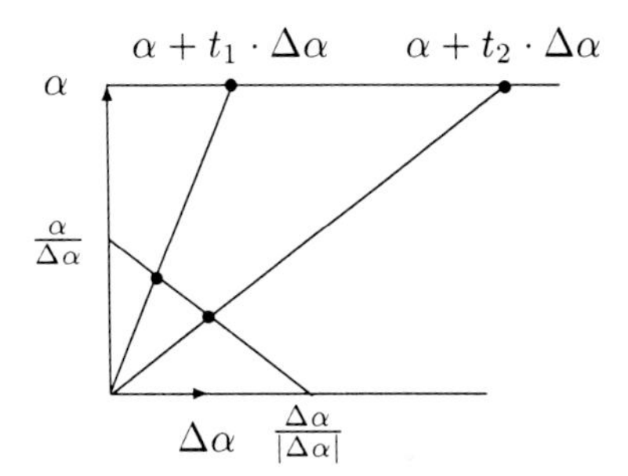

Figure 5.14 Another mapping of a half-line onto abscissa.

dividing the sequence into two parts having the same number of points. If it lay in a previously determined interval T then I would not bother much about the selection of the values $t_0, \ldots, t_L$. I would simply divide the interval into L equal segments. But we know only that $0 < t < \infty$ in our task. The sequence $t_0, t_1, \ldots, t_L$ can thus be selected in several different ways which seem nearly the same to me. The basic motivation for the selection is due to the property that for the function f the relation $f(\alpha) = f(k \cdot \alpha)$ for an arbitrary positive $k \in \mathbb{R}$ is valid. Therefore, instead in the points $\alpha + t \cdot \Delta\alpha$, $0 < t < \infty$, I can examine this function in the points

$$\frac{1}{1+t} \cdot \alpha + \frac{t}{1+t} \cdot \Delta\alpha\,, \text{ or, which is the same, in the points } \alpha \cdot (1-\tau) + \Delta\alpha \cdot \tau\,,$$

where τ already lies in the finite interval $0 < \tau \leq 1$. The finite sequence of points can be selected in a natural way by dividing the interval into L segments of equal length.

The abovementioned way means mapping a set of points of the form $\alpha + t \cdot \Delta\alpha$, $0 < t < \infty$, on a set of points of the form $\alpha \cdot (1-\tau) + \Delta\alpha \cdot \tau$. The half-line is mapped on the finite abscissa which connects the points α and $\Delta\alpha$, as it is shown in Fig. 5.13.

The half-line can be mapped on a finite set in a lot of reasonable ways. In addition to the matching already given, even later matching seems to be natural. In Fig. 5.14, a mapping of a half-line is shown on an abscissa, which connects the points $\alpha/|\alpha|$ and $\Delta\alpha/|\Delta\alpha|$. They seem to be better than the previous ones because the vectors α and $\Delta\alpha$ are becoming 'of equal rights'. A natural way of matching is the matching of a half-line on a set of points of the form

$$\alpha \cdot \cos\varphi + \frac{|\alpha| \cdot \Delta\alpha}{|\Delta\alpha|} \cdot \sin\varphi\,, \quad 0 \leq \varphi < \frac{\pi}{2}\,,$$

i.e., on a quarter-circle as can be seen in Fig. 5.15, or on the sides of a square as shown in Fig. 5.16. I do not think that it would be of much importance to explore which of these ways, and perhaps of other ways too, is the best. The matter is that they are nearly the same. In any of the ways a finite number

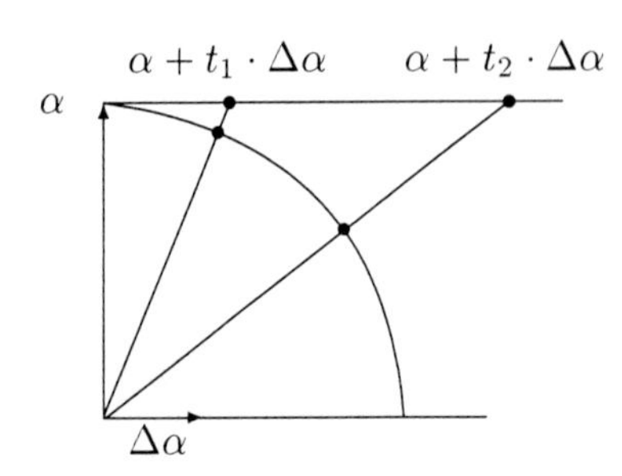

Figure 5.15 Mapping of a half-line on a quarter-circle.

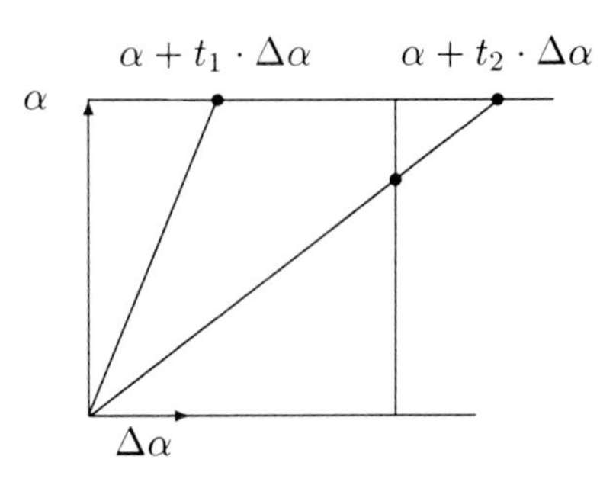

Figure 5.16 Mapping of a half-line on two sides of a square.

of $L + 1$ points is to be selected, which are in a sense uniformly spaced on the straight line, in the circle, or the square. From these points we have to find the best. Since we know that our function has one single extreme, the best point can be found in $\log_2 L$ steps, because a method of gradual division of the sequence into two equal parts can be applied (usually referred to as the method of interval halving, or the bisection method). That is, in fact, all I find interesting with this task.

You have omitted a number of interesting items since you took some trends in solving the task for granted and for the only possible ones. From the very beginning you have decided that you must inevitably match the half-line to a finite number of points and so replace the maximisation of the function of one real variable by searching for the greatest number from the finite set of numbers. Even though you noticed that there were many possibilities of such a replacement and none of them seemed convincing enough to you, it did not occur to you that such a replacement could be avoided.

Is something like that possible?

In the general case not, but in our particular case it is. But we will leave it to some later time. But now let us have a more profound look at the procedure which you have, rather carelessly, denoted as a method of interval halving, and even more so, in a strictly unimodal sequence. You said that by means of this method you would achieve the best point in $\log_2 L$ steps. We suspect that you do not see clearly enough the difference between searching for the zero point in a strictly decreasing number sequence and searching for the greatest number in a strictly unimodal sequence. Understand this difference well, and then we will continue.

They are two different tasks, indeed. I looked at either of them and saw that even when their respective calculations were similar they were still different. First, I will present the procedure for the simpler task of searching for zero, even when we evidently do not need it in our task.

Let $f(l)$, $l = 0, 1, \ldots, L$, $L = 2^h$, be a strictly decreasing series of numbers in which $f(0) > 0$, $f(L) < 0$, and let there be an index l^* for which $f(l^*) = 0$ holds. To find the index l^* in the fastest way, the following procedure is to be used. It is based on the idea that the difference between certain indices l^t_{beg} and l^t_{end} is stepwise diminished, so that in each step t the inequality $l^t_{\text{beg}} < l^* < l^t_{\text{end}}$. is fulfilled. At the beginning $l^0_{\text{beg}} = 0$ and $l^0_{\text{end}} = L$ is substituted. Let the values l^t_{beg} and l^t_{end} after the step t be known. The new values l^{t+1}_{beg} and l^{t+1}_{end} will be found in the following way. For the index $l_{\text{mid}} = \frac{1}{2}(l^t_{\text{beg}} + l^t_{\text{end}})$ the number $f(l_{\text{mid}})$ is determined and then

1. *if $f(l_{\text{mid}}) = 0$ then $l^* = l_{\text{mid}}$ and the algorithm stops;*
2. *if $f(l_{\text{mid}}) > 0$ then $l^{t+1}_{\text{beg}} = l_{\text{mid}}$ and $l^{t+1}_{\text{end}} = l^t_{\text{end}}$;*
3. *if $f(l_{\text{mid}}) < 0$ then $l^{t+1}_{\text{beg}} = l^t_{\text{beg}}$ and $l^{t+1}_{\text{end}} = l_{\text{mid}}$.*

This simple procedure is evidently to end when the difference $l^t_{\text{end}} - l^t_{\text{beg}}$ is 2 and it is decided that $l^* = \frac{1}{2}(l^t_{\text{beg}} + l^t_{\text{end}})$. Naturally, the procedure can end even sooner if the condition quoted in the first item is satisfied. The procedure ends no later than before the step number $\log_2 L$, because the difference $l^0_{\text{end}} - l^0_{\text{beg}}$ is L before the first step and is twice diminished with each step.

If I now wanted to use this procedure for seeking the greatest number in a strictly unimodal sequence then I would immediately find out that it is not suitable for this new task. The difference is that on the basis of a mere number $f(l_{\text{mid}})$ it can be stated whether the index l^* sought lies to the right or to the left of the index l_{mid}. In seeking the greatest number the knowledge of the number $f(l_{\text{mid}})$ is not sufficient for a such a conclusion. The algorithm seeking the greatest number in a strictly unimodal sequence is based on successive changes of a triplet of indices, not of the pair of them, as was the case in the previous task. I will give the algorithm in a more concrete form.

Let $f(l)$, $l = 0, 1, \ldots, L$, $L = 2^h$, be a series of numbers, which I call unimodal in the sense that there is such an index l^* that for any index $l < l^*$ the inequality $f(l) < f(l+1)$ is valid; and for any index $l > l^*$ the inequality $f(l) < f(l-1)$ is valid. The index l^* is sought in such a way that a triplet of indices $(l^t_{\text{beg}}, l^t_{\text{mid}}, l^t_{\text{end}})$, is changed stepwise so as to satisfy the inequalities $f(l^t_{\text{beg}}) < f(l^t_{\text{mid}})$ and $f(l^t_{\text{mid}}) > l^t_{\text{end}}$ at every step.

This triplet of indices has to be $l^0_{\text{beg}} = 0$, $l^0_{\text{mid}} = \frac{1}{2}L$, $l^0_{\text{end}} = L$ at the beginning. Let the triplet of indices l^t_{beg}, l^t_{mid} and l^t_{end} be obtained after the step t. Then the new triplet l^{t+1}_{beg}, l^{t+1}_{mid} and l^{t+1}_{end} is created in the following manner.

1. *An index l' is determined that divides the greater of the two intervals $(l^t_{\text{beg}}, l^t_{\text{mid}})$ and $(l^t_{\text{mid}}, l^t_{\text{end}})$ into two subintervals of the same lengths. In concrete terms, if $(l^t_{\text{mid}} - l^t_{\text{beg}}) < (l^t_{\text{end}} - l^t_{\text{mid}})$ then $l' = \frac{1}{2}(l^t_{\text{mid}} + l^t_{\text{end}})$. If $(l^t_{\text{mid}} - l^t_{\text{beg}}) \geq (l^t_{\text{end}} - l^t_{\text{mid}})$ holds then $l' = \frac{1}{2}(l^t_{\text{mid}} + l^t_{\text{beg}})$.*
2. *The quadruplet l^t_{beg}, l', l^t_{mid} and l^t_{end} obtained is ordered in an ascendant way as l_1, l_2, l_3, l_4. This means that $l_1 = l^t_{\text{beg}}$, $l_2 = \min(l', l^t_{\text{mid}})$, $l_3 = \max(l', l^t_{\text{mid}})$, $l_4 = l^t_{\text{end}}$.*

3. *In the quadruplet* l_1, l_2, l_3, l_4 *an index is found which corresponds to the greatest of the four values* $f(l_1), f(l_2), f(l_3), f(l_4)$. *It can be only an index* l_2 *or* l_3.
4. *If* $f(l_2) < f(l_3)$ *then* $l^{t+1}_{\text{beg}} = l_2$, $l^{t+1}_{\text{mid}} = l_3$ *and* $l^{t+1}_{\text{end}} = l_4 = l^t_{\text{end}}$.
5. *If* $f(l_2) > f(l_3)$ *then* $l^{t+1}_{\text{beg}} = l_1 = l^t_{\text{beg}}$, $l^{t+1}_{\text{mid}} = l_2$ *and* $l^{t+1}_{\text{end}} = l_3$.
6. *The algorithm ends when* $l^t_{\text{mid}} - l^t_{\text{beg}} = l^t_{\text{end}} - l^t_{\text{mid}} = 1$ *and decides that the index* l^* *sought is* l^t_{mid}.

From the above algorithm there follows that at least at every two steps the length of the longer of the two intervals $(l_{\text{mid}} - l_{\text{beg}})$ *and* $(l_{\text{end}} - l_{\text{mid}})$ *is shortened twice. Since prior to the first step the length was* $\frac{1}{2}L$, *the length becomes unitary prior to the step* $2 \log L$ *at the latest, and the algorithm ends. Thus I have arrived at the main conclusion that searching for the greatest number in a strictly unimodal sequence takes about twice as much time than searching for a known number in a strictly unimodal sequence. They are, as a matter of fact, two different tasks which are solved by applying different procedures.*

We thank you for your rather transparent explanation. But tell us now why you stated, without any further thought, that the length of the longer of the two intervals should be divided in every step into two equal parts. Why should an interval not be divided in another ratio.

Well, this is quite clear! Only in such a way can the greatest number in a unimodal sequence be found at the smallest number of queries for the value of a number. I do not know how to account for it except for saying that everybody does so.

We do not think that everybody does so. Perhaps all your acquaintances do so, and even that is doubtful. We would prefer you not to do so any longer and use the method that was proposed in the early 13th century by the Italian mathematician Leonardo Fibonacci. Fibonacci may have learned about the method in Central Asia, where science, including mathematics, was flourishing at that time. Fibonacci himself got to that part of the world as a merchant commissioned by his father. He surely, as you and we do, devoted his time also to interests not at all connected with his main commercial duties.

Let $l(i)$, $i = 1, 2, \ldots$, be a series of numbers for which $l(1) = l(2) = 1$ holds, and for each $i > 2$ the equality $l(i) = l(i-1) + l(i-2)$ holds. The numbers of this series are called Fibonacci numbers. Every Fibonacci number can be expressed as a sum of two other Fibonacci numbers and it can only be done in an unique way.

Two intervals are changed in each step in the procedure you created to find the greatest number. The lengths of intervals are $l_{\text{mid}} - l_{\text{beg}}$ and $l_{\text{end}} - l_{\text{mid}}$. The lengths are either equal, and in that case they are 2^h for the integer h, or they are different, and in that case they are 2^h and 2^{h-1}. Create a new algorithm which differs from what you proposed earlier. The length $l_{\text{end}} - l_{\text{beg}}$ must be one of Fibonacci numbers in each step, say $l(i)$. The index l_{mid} is to be such that the lengths $l_{\text{end}} - l_{\text{mid}}$ and $l_{\text{mid}} - l_{\text{beg}}$ are Fibonacci numbers as well. Thus, one length is $l(i-1)$ and the other is $l(i-2)$. We will call this modification the Fibonacci algorithm. Do you see it?

Yes, I do.

Then try both algorithms at the same strictly unimodal sequence and verify that the greatest number can be sought by means of the Fibonacci algorithm faster than by means of the algorithm you proposed before and which you without any reason claimed to be the fastest.

Well, this is surprising! The method of halving an interval is really not the fastest. The Fibonacci algorithm works faster, but only a little bit.

We have not expected anything else.

Why then does everybody say that the method of halving an interval is the fastest?

We are repeating once more that it is not said by everybody. Fibonacci did not say it.

But now a question arises. I know now that the method of halving an interval is not optimal because the Fibonacci method is better. But I cannot so far say that the Fibonacci method is an optimal one. What is, then, the ratio for halving an interval that allows to achieve the highest speed in finding the greatest number?

We are sure that this is not a difficult question for you. It would be the worst for you to think that you had already known the optimal algorithm and would not ask such questions. Now, when you have come across that question you will quickly find the right answer.

It was not very quick, but I have found an answer. Though I have not formulated the optimisation task very precisely, I understand it like this.

I have a certain class of algorithms the two representatives of which are already known to me. One is the algorithm seeking the greatest number in a strictly unimodal sequence, which I quoted before and which is based on halving the longer of two intervals. The latter algorithm for the same purpose has the same form as the former except that the longer interval is divided into unequal parts corresponding to Fibonacci numbers. A general algorithm is formulated by me as an algorithm in which the longer interval is divided into parts proportional to the values α *and* $1-\alpha$, *where* $0 < \alpha < 1$. *The number* α *is not known and it is necessary to determine it in a certain sense optimally. I am not going to formulate the criterion of optimality precisely now. I will make a not very complex analysis of an algorithm for the fixed value* α *and the analysis will show in what sense the value should be optimal.*

I assume that, before some of the iterations of the algorithm, I had numbers l_{beg}, l_{mid} and l_{end} at my disposal. Without loss of generality I can assume that they are the numbers $l_{\text{beg}} = 0$, $l_{\text{mid}} = 1 - \alpha$ and $l_{\text{end}} = 1$ and I accept that they are no longer integers. Furthermore, I assume that $\alpha \leq 0.5$. In accordance with the algorithm I am expected to divide the greater interval, i.e., $l_{\text{beg}}, l_{\text{mid}}$ into two parts that will be, thanks to their fixed value α of the lengths $(1-\alpha)^2$ and $(1-\alpha)\cdot\alpha$. This means that the index l' will be $(1-\alpha)^2$. I am again ignoring that it will be not an integer.

The new values of the numbers l_{beg} and l_{end} will be either 0 and $(1-\alpha)$ or $(1-\alpha)^2$ and 1 depending on the values $f(l')$ and $f(l_{\text{mid}})$. The lengths of the new interval $l_{\text{end}} - l_{\text{beg}}$ will then be either $1-\alpha$ or $1-(1-\alpha)^2$. Only now I can formulate the requirement that the parameter of the algorithm α is to be chosen so that the length of the longer of the two intervals should be the shortest possible one,

$$\alpha^* = \underset{\alpha}{\operatorname{argmin}} \max\left(1-(1-\alpha)^2, 1-\alpha\right).$$

The solution of this simple optimisation task is the number

$$\alpha^* = \frac{3-\sqrt{5}}{2}.$$

With the computed value of the parameter α, the sequence of the lengths of the interval $l_{\text{beg}} - l_{\text{end}}$ is during the algorithm upper bound by a decreasing geometrical series with the quotient $(\sqrt{5}-1)/2$ which is approximately 0.618. The optimality of that particular value of the parameter α lies in that at another arbitrary value of the parameter α the sequence of the lengths of the intervals will be bound by a geometrical series that is decreasing at a slower rate. For example, in the algorithm I proposed before only one thing was guaranteed, i.e., that in every pair of steps the lengths of the interval is shortened twice. This means, roughly speaking, that the sequence of the lengths is decreasing with the rate of the geometrical series with the quotient $\sqrt{2}/2 \sim 0.707$.

I was greatly pleased that I have managed to solve the task up to such a concrete results, up to a number, as the mathematicians say.

You are not the first to have been pleased by this task. Since several thousands of years, since the time of ancient Greece, the ratio $(\sqrt{5}-1)/2$ has been attracting attention and is referred to as the golden cut.

The number did remind me of something. The analysis was very interesting, but from a pragmatic standpoint it does not offer very much. You said in your lecture that a carelessly written program for this optimisation worked 100 times slower than one which was well thought out. Did you not have in mind anything else than the analysis just completed?

You are right. We had in mind something quite different, but we are glad that you have dragged us into the problem already dealt with.

Let us go back to our task in which for a given group of vectors μ^j and matrices σ^j, $j \in J$, and for given vectors α and $\Delta\alpha$ the parameter t is to be found which maximises the function $f(t)$,

$$f(t) = \min_{j \in J} \frac{\langle\, \alpha + t\Delta\alpha,\, \mu^j \rangle}{\sqrt{\langle\, \alpha + t\Delta\alpha,\, \sigma^j \cdot (\alpha + t\Delta\alpha) \rangle}} \,. \tag{5.89}$$

It is maximisation of a function of one single variable t. The proposed algorithm looks for a maximum so that it calculates the set of values of function (5.89) for some finite set of values of parameter t. As we can see, you have proposed to calculate the values of the vector $\alpha + t\,\Delta\alpha$ for the chosen set of values of t. You are then going to calculate values of $f(t)$ using the formula (5.89). This means you go to Paris via a neighbouring village. Throw away everything redundant from the proposed procedure.

I hope I have understood your idea. I will denote by $f(j,\alpha)$ the function $\langle \alpha, \mu^j \rangle / \sqrt{\langle \alpha, \sigma^j \cdot \alpha \rangle}$. For the given sequence of numbers $t_0, t_1, \dots, t_L$, the index

$$l^* = \operatorname*{argmax}_{l} \min_{j \in J} f\big(j, \alpha \cdot (1 - t_l) + \Delta\alpha \cdot t_l\big)$$

is to be calculated. The following numbers will be calculated

$$\left.\begin{aligned} s^j &= \langle \alpha, \mu^j \rangle \,, \\ s'^j &= \langle \Delta\alpha, \mu^j \rangle \,, \\ \Delta s^j &= s'^j - s^j \,, \\ \sigma^j_\alpha &= \langle \alpha,\, \sigma^j \cdot \alpha \rangle \,, \\ \sigma^j_\Delta &= \langle \Delta\alpha,\, \sigma^j \cdot \Delta\alpha \rangle \,, \\ \sigma^j_{\alpha\Delta} &= \langle \Delta\alpha,\, \sigma^j \cdot \alpha \rangle \,. \end{aligned}\right\} \tag{5.90}$$

Having done that, the numbers $f\big(j, \alpha(1 - t_l) + \Delta\alpha \cdot t_l\big)$ can be calculated as

$$\frac{s^j + t \cdot \Delta s^j}{\sqrt{(1 - t_l)^2 \cdot \sigma^j_\alpha + 2t \cdot (1 - t_l) \cdot \sigma^j_{\alpha\Delta} + t_l^2 \cdot \sigma^j_\Delta}} \,. \tag{5.91}$$

Looking at it, we can see that all multi-dimensional operations are performed only in the calculation of $6\,|J|$ numbers according to the formula (5.90). This calculation does not depend on the index l, and thus it is performed outside the cycle in which different indices l are tested. In testing the indices l everything is calculated according to the formula (5.91) which does not contain any multi-dimensional operations any longer.

Where the hundred fold acceleration is achieved?

It is so, for example, when the calculation of a quadratic function

$$\big\langle (1 - t_l) \cdot \alpha + t_l \cdot \Delta\alpha, \sigma^j \cdot \big((1 - t_l) \cdot \alpha + t_l \cdot \Delta\alpha\big) \big\rangle \,,$$

where σ^j is a matrix of dimension 32×32. The equivalent calculation of the quadratic function

$$(1-t_l)^2 \cdot \sigma_\alpha^j + 2t \cdot (1-t_l) \cdot \sigma_{\alpha\Delta}^j + t_l^2 \cdot \sigma_\Delta^j$$

is used in which σ_α^j, $\sigma_{\alpha\Delta}^j$ and σ_Δ^j are numbers.

At the very beginning of our discussion you stated that a function of one variable could be numerically maximised without substituting an infinite straight line by a finite number of values. But I do not know the way in which it could be done.

Once we have become so deeply absorbed in the analysis of a one-dimensional task we would like you to add one more method to the methods you have already mastered. It is different from those we have studied so far. So far we have assumed without any doubt that the domain of a maximised function must be substituted by a finite set of points, i.e., simply speaking, that the argument of a maximised function is to be made discrete. Quite often, however, it is not easy to find a reasonable degree of discretisation.

In fact, you had already come across these difficulties at the beginning of our discussion, when you suggested even several ways of discretisation you thought to be equally reasonable. We will also find out that all discretisations you have suggested are also equally unsubstantiated. They are based only on intuitive understanding of the nature of a function and on the belief that if an interval from 0 to 1 is substituted, e.g., by 1025 uniformly placed points, then the value in one of the points will be sure to differ only by a bit from the maximal value in the whole interval from 0 to 1. In the general case, there is no such certainty because the set of unimodal, but non-concave functions is too diverse. For example, it also contains a function that in the whole interval from 0 to 0.9999 assumes only inadmissible values, and all acceptable values are concentrated only in an extremely narrow interval from from 0.9999 to 1.0. For such cases, in which the maximum cannot be obtained by discrete samples, involve in the arsenal of the methods you mastered also the following method, which is, in a sense, dual to the methods we have already studied and which will be called *direct methods* by us.

Dual methods have in common that it is not the domain of a function that is made discrete, but the set of its values. Let us introduce the basic idea of dual methods. Assume that we have found two values u_{down} and u_{up} with such a feature that there exists a value $t \in T$ with which $f(t) \geq u_{\text{down}}$, and that there is no value $t \in T$ for which $f(t) \geq u_{\text{up}}$ holds. In this case it is necessary to check if a value $t \in T$ exists for which $f(t) \geq u_{\text{mid}} = \frac{1}{2}(u_{\text{down}} + u_{\text{up}})$ holds. If it exists then the pair $(u_{\text{down}}, u_{\text{up}})$ is changed to become the pair $(u_{\text{mid}}, u_{\text{up}})$. If not then the pair $(u_{\text{down}}, u_{\text{up}})$ is changed to become the pair $(u_{\text{down}}, u_{\text{mid}})$. When during the running of the algorithm the difference $u_{\text{up}} - u_{\text{down}}$ is decreased below the predetermined value ε then the algorithm stops with the certainty that the maximum sought was determined with an error which is not greater than ε. The advantage of dual methods compared to the direct methods lies in that the maximum is estimated with a given accuracy. The advantage of direct methods is that it is the position of the maximum that is estimated with a given accuracy. The most important question in applying dual methods certainly is how easily one can find out if the inequality $f(t) \geq u$ has a solution.

The function we would intend to maximise is quite well known for us to solve the inequality $f(t) \geq u$ rather easily. Let us ask if the inequality

$$\min_{j\in J} \frac{\langle \alpha\cdot(1-t)+t\cdot\Delta\alpha, \mu^j\rangle}{\sqrt{\langle \alpha\cdot(1-t)+t\cdot\Delta\alpha, \sigma^j\cdot(\alpha\cdot(1-t)+t\cdot\Delta\alpha)\rangle}} \geq c \tag{5.92}$$

has its solution in the interval from 0 to 1, and if it is so then an interval of the values of the quantity t is to be found for which the inequality (5.92) is satisfied. If we make use of the expression (5.90) then the inequality (5.92) can be rewritten as a system of inequalities

$$\frac{s^j + t\cdot\Delta s^j}{\sqrt{(1-t)^2\cdot\sigma^j_\alpha + 2t\cdot(1-t)\cdot\sigma^j_{\alpha\Delta} + t^2\cdot\sigma^j_\Delta}} \geq c\,, \quad j\in J\,, \quad \theta\leq t\leq 1\,, \tag{5.93}$$

where $s^j \geq 0$, Δs^j, σ^j_α, $\sigma^j_{\alpha\Delta}$, σ^j_Δ are known numbers. We are interested in the system of inequalities (5.93) only for positive c, i.e., for such values t, only at which the inequalities

$$s^j + t\cdot\Delta s^j \geq 0 \tag{5.94}$$

are valid for all $j \in J$. The inequalities from the system (5.94) for which $\Delta s^j \geq 0$ need not be taken into consideration because on the interval $0 \leq t \leq 1$ they are always satisfied. Let us denote by J^0 a set of indices j for which $\Delta s^j < 0$ holds. The system (5.94) is equivalent to the system

$$s^j + t\cdot\Delta s^j \geq 0\,, \quad j\in J^0,$$

or, written in another way,

$$t \leq \frac{-s^j}{\Delta s^j}\,, \quad j\in J^0\,,$$

or, at last,

$$t \leq \min_{j\in J^0} \frac{-s^j}{\Delta s^j}\,.$$

We will denote

$$T = \min\left(1,\ \min_{j\in J^0}\frac{-s^j}{\Delta s^j}\right)\,.$$

We will take into consideration that now on the condition $0 \leq t \leq T$ in the inequalities (5.93) no numerator is negative and rewrite (5.93) into the form

$$\left.\begin{array}{c} \left(s^j + t\cdot\Delta s^j\right)^2 \geq c^2\left((1-t)^2\cdot\sigma^j_\alpha + 2t\cdot(1-t)\cdot\sigma^j_{\alpha\Delta} + t^2\cdot\sigma^j_\Delta\right)\,, \\ j\in J\,,\quad \theta\leq t\leq T\,. \end{array}\right\} \tag{5.95}$$

In this way the question whether the inequality (5.92) has a solution has been reduced to the solvability of the system (5.95), which contains quadratic inequalities of one variable t only on a sufficiently simple condition $0 \le t \le T$. The solvability of (5.95) can be easily verified. Each inequality in the first line of the system (5.95) can be rewritten into the form

$$A^j \cdot t^2 + B^j \cdot t + C^j \ge 0 \,,$$

where the coefficients A^j, B^j, C^j are calculated from numbers s^j, Δs^j, c, σ^j_α, $\sigma^j_{\alpha\Delta}$, σ^j_Δ which are already known. The system (5.95) will assume the form

$$A^j \cdot t^2 + B^j \cdot t + C^j \ge 0 \,, \quad j \in J \,, \quad \theta \le t \le T \,. \tag{5.96}$$

According to whether the quadratic equation $A^j \cdot t^2 + B^j \cdot t + C^j = 0$ has its solution (this can be easily verified), and according to the sign of the coefficient A^j the set of values t satisfying the j-th inequality in the system (5.96) will be either empty or will be expressed by one interval or by a pair of intervals open from one side. The intersection of the set of values t with the interval $0 \le t \le T$ will be either empty or will be expressed by one interval. If this intersection consisted of two intervals then it would be contradictory to the proved fact that the set t satisfying (5.92) is convex for positive c. The set of values t satisfying the system (5.96), and thus also the inequality (5.92) is an intersection of intervals which are calculated for each $j \in J$. The result is an interval again.

Thus, we can see that in our case the maximal value of one variable with an arbitrary beforehand given accuracy can be estimated without making this variable discrete.

And now it is the turn of another question inspired by the lecture. It was said that in modern applied mathematics the methods of optimisation of non-smooth or non-differentiable functions had been thoroughly examined and that the fundamental concept was that of the generalised gradient. I completely agree with you that these methods are not sufficiently known in the pattern recognition sphere. I would like to learn more, at least about the most fundamental concepts, and see what the complete analysis of Anderson task would look like if the results of the non-smooth optimisation were applied.

The main core of the theory of the non-differentiable optimisation can be rather briefly explained, as it is with any significant knowledge. But this brief explanation opens a wide space for thinking.

Let X be a finite-dimensional linear space on which a concave function $f: X \to \mathbb{R}$ is defined. Let x_0 be a selected point and $g(x_0)$ such a vector that the function

$$f(x_0) + \big\langle g(x_0), (x - x_0) \big\rangle$$

dependent on the vector x is not less than $f(x)$ for any arbitrary $x \in X$. Then we can write

$$f(x_0) + \big\langle g(x_0), (x - x_0) \big\rangle \ge f(x) \tag{5.97}$$

or, in another, form

$$\big\langle g(x_0), (x - x_0)\big\rangle \geq f(x) - f(x_0)\,. \tag{5.98}$$

The vector $g(x_0)$ is referred to as the *generalised gradient* of the function f in the point x_0.

Let us show why the generalised gradient is important:

1. If the function f in the point x_0 can be differentiated then the gradient of function f satisfies the conditions (5.97) and (5.98). Therefore the gradient is, in a usual sense, a special case of the generalised gradient.
2. The definition of the generalised gradient is by no means based on the concept of partial derivative, and therefore, it is not dependent on the differentiability of the function.
3. Even if it is not evident at first glance, we will later make sure in an informal way that an arbitrary concave function has the generalised gradient (sometimes not a single one) in all points of its domain of definition.
4. And last but not least, the validity of many gradient optimisation methods is not based on the knowledge that the gradient is a gradient, but on a weaker property stating that the gradient satisfies the conditions of (5.97), (5.98). This means that the properties of many gradient optimisation algorithms can be transferred also to the algorithms in which instead of a gradient the generalised gradient is used. This is stated in the following theorem.

Theorem 5.7 On gradient optimisation with the generalised gradient. *Let $f\colon X \to \mathbb{R}$ be a concave function and γ_i, $i = 1, 2, \ldots, \infty$, be a sequence of positive numbers which converge to zero and the sum of which is infinite; f^* is*

$$\max_{x \in X} f(x)$$

and X_0 is a set of such points $x \in X$, in which the maximum is reached; x_0 is any point in the space X and

$$x_i = x_{i-1} + \gamma_i \cdot \frac{g(x_i)}{|g(x_i)|}\,,$$

where $g(x_i)$ is the generalised gradient of the function f in the point x_i. In this case

$$\lim_{i \to \infty} f(x_i) = f^*\,,$$

$$\lim_{i \to \infty} \min_{x \in X_0} |x_i - x| = 0\,. \qquad \blacktriangle$$

Proof. We will not prove Theorem 5.7 and refer to [Shor, 1979; Shor, 1998]. ■

Let us now try to work out, on an informal level, the generalised gradient and its properties. Imagine that the concave function $y = f(x)$ is defined in an n-dimensional space X and represented by its graph in an $(n+1)$-dimensional space. In this $(n+1)$-dimensional space n coordinates are original coordinates of the space X and coordinate $(n+1)$ corresponds to the value y. The function f is mapped in this space as a surface that contains all the points (x, y), for which $y = f(x)$ holds. Such a surface will be referred to as the graph of the function

f and denoted by D; $D = \{(x, y) \mid y = f(x)\}$. Furthermore, let us take into consideration the set of points (y, x), which lie 'below the graph', call it the body of the function f and denote it by T. To be accurate, $T = \{(x, y) \mid y \leq f(x)\}$. Since the function f is concave, the body T is convex.

Similarly as we represented the function f by the graph D, an arbitrary linear function can be represented. The graph of any linear function is a hyperplane. And vice versa, each hyperplane which is not orthogonal to the space X, is the graph of a linear function. Assume that a hyperplane passes through a point $(x_0, f(x_0))$ in the graph D in such a way that the whole body T is below the hyperplane. For each point on the surface of the body T such a hyperplane can be constructed for the very reason that the body T is convex. An exact formulation of this property and its proof are included in Farkas' excellent lemma. The hyperplane will be referred to as a tangent to the body T in the point x_0, which lies on the graph D. By the symbol L a linear function will be denoted the graph of which is the hyperplane. From the abovementioned definition of the generalised gradient it follows that the generalised gradient is a gradient of a linear function that is represented as a tangent. In some points only one tangent exists. Informally speaking, it is in those points where the graph D is not 'broken', and is 'smooth'. In other points, where the graph D is 'broken', the tangents can be several different hyperplanes. If in the point x_0 two different hyperplanes are tangents to the body T, then any hyperplane in between is also a tangent to the body T in the same point. It is expressed more exactly in the following theorem.

Theorem 5.8 On tangents in between. *The set of vectors that are generalised gradients of a certain function f in a point x_0 is convex.* ▲

Proof. Let g_1 and g_2 be generalised gradients in the point x_0. For all points $x \in X$ there holds

$$\langle g_1, x - x_0 \rangle \geq f(x) - f(x_0) ,$$

$$\langle g_2, x - x_0 \rangle \geq f(x) - f(x_0) .$$

Let us construct a convex combination with the numbers α_1 and $\alpha_2 = 1 - \alpha_1$

$$\langle \alpha_1 \cdot g_1 + \alpha_2 \cdot g_2, \, x - x_0 \rangle \geq f(x) - f(x_0) .$$

This means that the vector $\alpha_1 \cdot g_1 + \alpha_2 \cdot g_2$ is also the generalised gradient of the function f in the point x_0. ■

In an informal understanding of the generalised gradient as a vector parameter of a tangent hyperplane, the following fundamental theorem of non-smooth optimisation, which states the necessary and sufficient condition for the existence of a maximum, seems to be evident.

Theorem 5.9 Necessary and sufficient condition for the existence of a minimum in non-smooth optimisation. *The point x_0 maximises the concave function if and only if the zero vector is its generalised gradient in the point x_0.* ▲

Proof. Instead of the proof, let us give an informal interpretation of Theorem 5.9. It states that the point (y, x_0) is the 'highest' point of the body T

then and only then when there exists a 'horizontal' tangent hyperplane in that point. Note that even the formal proof of the theorem is rather brief. ■

Let us now make use the acquired knowledge for an analysis of our task.

Theorem 5.10 On the generalised gradient in a selected point. *Let $\{f^j(x),\ j \in J\}$ be a set of concave differentiable functions, x_0 be a selected point, and $g^j(x_0)$, $j \in J$, be a gradient of the function $f^j(x)$ in the point x_0.*

Let J_0 be a set of indices j for which there holds

$$f^j(x_0) = \min_{\iota \in J} f^\iota(x_0) \tag{5.99}$$

and γ^j, $j \in J_0$, are non-negative numbers the sum of which is 1. Then the convex combination $g_0 = \sum_{j \in J^0} \gamma^j \cdot g^j(x_0)$ is the generalised gradient of the function

$$f(x) = \min_{j \in J} f^j(x)$$

in the point x_0. And vice versa, if any vector g_0 is a generalised gradient of the function $f(x)$ in the point x_0 then the vector g_0 belongs to the convex hull of the vectors $g^j(x_0)$, $j \in J_0$. ▲

Proof. First, let us prove the first statement of the theorem. The assertion that the vectors $g^j(x_0)$, $j \in J_0$, are gradients of the functions f^j means

$$\big\langle g^j(x_0), x - x_0 \big\rangle \geq f^j(x) - f^j(x_0)\,, \quad j \in J_0\,,$$

or in another form

$$\big\langle g^j(x_0), x - x_0 \big\rangle \geq f^j(x) - \min_{j \in J} f^j(x_0)\,, \quad j \in J_0\,.$$

From that follows

$$\sum_{j \in J_0} \gamma^j \cdot \big\langle g^j(x_0), x - x_0 \big\rangle \geq \sum_{\iota \in J_0} \gamma_\iota \cdot f^\iota(x) - \min_{j \in J} f^j(x_0)\,. \tag{5.100}$$

Since there holds

$$\sum_{j \in J_0} \gamma^j \cdot f^j(x) \geq \min_{j \in J_0} f^j(x) \geq \min_{j \in J} f^j(x)\,,$$

from (5.100) the following relations result

$$\begin{aligned} \left\langle \sum_{j \in J_0} \gamma^j \cdot g^j(x_0),\ x - x_0 \right\rangle &= \sum_{j \in J^0} \gamma^j \cdot \langle g^j(x_0),\ x - x_0 \rangle \\ &\geq \min_{j \in J} f^j(x) - \min_{j \in J} f^j(x_0) = f(x) - f(x_0)\,, \end{aligned}$$

which can be summed up in the inequality

$$\left\langle \sum_{j \in J^0} \gamma^j \cdot g^j(x_0),\; x - x_0 \right\rangle \geq f(x) - f(x_0)\,,$$

which is the proof of the first statement of Theorem 5.10.

The second statement of the theorem will then be proved by a contradiction. Let the vector g_0 satisfy the inequality $g_0 \neq \sum_{j \in J_0} \gamma^j\, g^j(x_0)$ at arbitrary non-negative coefficients satisfying the condition $\sum_{j \in J_0} \gamma^j = 1$. Because the vector g_0 does not belong to a convex polyhedron whose vertices are the vectors $g^j(x_0)$, $j \in J_0$, the existence of a hyperplane follows that separates the vector g_0 from all vectors $g^j(x_0)$, $j \in J_0$. More exactly, there exists a vector x' and a threshold value θ satisfying the inequalities

$$\langle g_0, x' \rangle < \theta\,, \tag{5.101}$$

$$\left\langle g^j(x_0), x' \right\rangle > \theta\,, \quad j \in J_0\,. \tag{5.102}$$

If we subtract the inequality (5.101) from every inequality of the system (5.102) we obtain

$$\left\langle g^j(x_0), x' \right\rangle - \langle g_0,\, x' \rangle > 0\,, \quad j \in J_0\,. \tag{5.103}$$

The left-hand part in each j-th given inequality is a derivative of a continuous differentiable function $f^j(x_0 + tx') - \langle g_0, x_0 + tx' \rangle$ with respect to the variable t in the point $t = 0$. Since the derivatives are, thanks to the inequality (5.103), positive the following inequality will be satisfied, at least at small positive values t for all $j \in J_0$

$$f^j(x_0 + tx') - \langle g_0,\, x_0 + tx' \rangle > f^j(x_0) - \langle g_0, x_0 \rangle\,,$$

from which there follows the inequality

$$\min_{j \in J_0} \left(f^j(x_0) + tx') \;-\; f^j(x_0) \right) > \langle g_0,\, x_0 + tx' \rangle - \langle g_0, x_0 \rangle.$$

From the assumption (5.99) there follows that at $j \in J_0$ the value $f^j(x_0)$ does not depend on j and is equal to $f(x_0)$. Thus

$$\min_{j \in J_0} f^j(x_0 + tx') - f(x_0) > \langle g_0,\, x_0 + tx' \rangle - \langle g_0, x_0 \rangle\,. \tag{5.104}$$

All functions $f^j(x)$, $j \in J_0$, are continuous, therefore for the index $j \in J$, $j \notin J_0$, for which the equality $f^j(x_0) = \min_{j \in J} f^j(x_0)$ is not satisfied, neither the equality $f^j(x_0 + tx') = \min_{j \in J} f^j(x_0 + tx')$ will be satisfied, at least at small values of t. Thus

$$\min_{j \in J_0} f^j(x_0 + tx') = \min_{j \in J} f^j(x_0 + tx')$$

is valid and the inequality (5.104) assumes the form

$$f(x_0 + tx') - f(x_0) > \langle g_0, x_0 + tx' \rangle - \langle g_0, x_0 \rangle$$

and g_0 is not a generalised gradient of $f(x)$. This proves the second statement of Theorem 5.10. ■

If we now shut our eyes to the contradiction that the above theorems express the properties of concave functions, and the maximised function in Anderson task is not concave, even when it is unimodal, we would see that Theorem 5.2 on necessary and sufficient conditions of the maximum in Andeson task, proved in the lecture, quite obviously follows from the theorems on the generalised gradients presented now. And indeed, Theorem 5.10 claims that any generalised gradient of the function

$$\min_{j \in J} f^j(x) \quad \text{has the shape} \quad \sum_{j \in J_0} \gamma_j \cdot g^j(x)\,.$$

Theorem 5.9 claims that it is necessary and sufficient for the maximisation that some of the generalised gradients should be zero. This means that such positive values γ^j are to exist that

$$\sum_{j \in J_0} \gamma^j \cdot g^j = 0\,.$$

In the lecture, the gradients g^j were proved to be collinear with the positional vector of the contact point. Therefore the statement that the convex hull of the gradients g^j, $j \in J_0$, includes the coordinate origin is equivalent to the statement that the convex hull of contact points contains the coordinate origin. The condition on the maximum, which we introduced at the lecture informally at first, and which was also proved there, could be derived as a consequence of Theorem 5.7 known from the theory of non-smooth optimisation.

Further on it can be easily seen that the actual proposal of maximisation presented at the lecture is one of the possible alternatives of the generalised gradient growth, which is stated in Theorem 5.7. We require a direction to be sought in which the function $f(x) = \min_j f^j(x)$ grows, i.e., in which each function $f^j(x)$, $j \in J_0$ grows. This direction will be one of the possible generalised gradients. But not every generalised gradient has the property that the motion in its direction guarantees the growth of all functions $f^j(x)$, $j \in J_0$. And this is why the recommendation resulting from the lecture is stricter than the recommendation to move in the direction of the generalised gradient. The general theory of non-smooth optimisation claims that such a recommendation is extremely strict. The direction in which a point is to move in its next step when seeking the maximum can be given by any point from the convex hull of the gradients g^j, $j \in J$, and not only by that which secures the growth of the function $f(x) = \min_{j \in J} f^j(x)$. Simply, it can be any of the gradients $g^j, j \in J_0$. The algorithm for solving Anderson task could then have even the following very simple form.

The algorithm creates the sequence of vectors $\alpha_1, \alpha_2, \ldots, \alpha_t, \ldots$ If the vector α_t has already been created then any (!!!) j is sought for which there holds

$$f^j(\alpha_t) = \min_{j \in J} f^j(\alpha_t)\,.$$

The contact point x_0^j is calculated as well as the new position α_{t+1} of the vector α as

$$\alpha_{t+1} = \alpha_t + \gamma_t \cdot \frac{x_0^j}{|x_0^j|},$$

where γ_t, $t = 1, 2, \ldots, \infty$, is a predetermined sequence of coefficients that satisfy the conditions

$$\sum_{t=0}^{\infty} \gamma_t = \infty \quad \text{and} \quad \lim_{t \to \infty} \gamma_t = 0\,.$$

As far as general concepts and theorems were discussed, everything seemed to me to be natural and understandable. Up to the moment when it was stated how a simple algorithm for solving Anderson task resulted from the whole theory. This seems to me to be particularly incredible, including Theorem 5.7, which certainly is of fundamental significance.

I will show now why it seems incredible to me. Assume we want to maximise the function

$$f(x) = \min_{j \in J} f^j(x)$$

and we have reached the point where two functions, say f^1 and f^2, assume an equal value which is less than are the values of all other functions. It is to change x so that the function $f(x)$ should be increased and this means increasing both the functions f^1 and f^2. The algorithm following from the general theory, instead of taking into consideration both functions, deals with one of them only trying to make it larger, and simply ignores the other. In such a case, however, the other function can even decrease, which means that the function f can decrease as well. It is not only possible for such a situation to occur, but it certainly will occur since in approaching the maximum the number of functions which the algorithm should take into consideration is growing. But it looks to be about one function only.

You are right, but it is not because Theorem 5.7 may be wrong. You claim that the algorithm of maximisation does not secure a monotonic growth, but Theorem 5.7 does not state so. It only says that the algorithm converges to a set of points in which the maximum is reached. For this convergence a monotonic growth is not necessary. We will examine, though not very strictly, the counterexample you quoted. When it has already happened that in one of the steps together with the growth of one function, say the function f_1, the other function f_2 has decreased (which can happen since the algorithm did not regard the function f_2 at all) then certainly as soon as in the next step the function f_1 will be not taken into consideration, and it may be just the function f_2 which will have the worst value and the algorithm will have to consider it.

A precise proof of Theorem 5.7 must be damned complicated!

Yes, it really is complicated!

Well, when you once have persuaded me that Theorem 5.7 holds, I will dare to ask an impertinent question. Why did not the theorem become the basis of the lecture? Did not you explain the subject matter as if the theorem had not existed?

We would like to remind you that only a while ago you said that you did not believe what Theorem 5.7 states. Now you may already believe it, but it does not make any difference. We would not intend to base our lecture on a theorem which we together did not completely understand. It would be a lecture based on a trust in a theorem the proof of which was not presented by us, and not based on knowledge.

Furthermore, earlier we shut our eyes to the fact that our task deals with the maximisation of a function which is not concave. Now we can open our eyes again and see that Theorem 5.7 by itself cannot be the basis for solving our task. The theorem may be worth generalising so that the sphere of its validity could cover even the functions which had occurred in our task. But that would be quite another lecture.

In pattern recognition publications, in papers from journals and conferences, I frequently find results which are identical with the results known for long in other spheres of applied mathematics. I do not know what standpoint toward such results to take. Is it a repeated invention of the wheel or is it a normal thing for us to use better or less well known mathematical methods in a qualified way? It is rather a philosophical problem. When I am once engaged in pattern recognition I cannot help thinking from time to time of whether pattern recognition is a science or art, or a set of clever tricks, or a body of knowledge taken over from other fields.

Not to go too far in looking for examples, I refer to an example from your lecture. It introduces and proves Theorem 5.3, which says that a hyperplane optimally separating two sets of points lies in the middle of the shortest distance between their convex hulls, and is perpendicular to it. This is, however, a known result in convex analysis and convex programming. In computational geometry similar procedures like this are common when seeking the shortest abscissa. Let us admit that I would know these results well. What new knowledge do I acquire when I see them once more in the pattern recognition context?

You have already asked yourself several questions and some of them are fantastically extensive. Let us start from whether pattern recognition is art, science, or yet something else. This question is very extensive, and however seriously intended the answer might be it will inevitably not be precise. The easiest way might be to do away with the question by saying that pattern recognition is what we are just now, together with you, dealing with and will be dealing in our lectures, and not to go back to that question any longer. It is not a very clever answer, but as far as we know, other answers are not much cleverer.

And now as to your second question. What attitude are we to take to the fact that the fundamentals of pattern recognition involve knowledge that has been generally and for long known in other spheres of applied mathematics. We should be pleased at it. It is only a pity that there are still far fewer concepts and procedures taken over by and adopted in pattern recognition from other fields than we would have wished for.

And now to your last question. Let us again go over the result of the lecture you refer to. Let X_1 and X_2 be two finite sets of points in a linear space. A hyperplane is sought which will separate the sets X_1 and X_2 and its distance from the nearest points of both the sets is will be the greatest. In searching for the hyperplane, two least distant points x_1 and x_2 are to be found, the former belonging to the convex hull of the set X_1 and the latter belonging to the convex hull of the set X_2. The hyperplane found should be perpendicular to a straight line passing points x_1 and x_2 and lie at the halfway distance between those points.

This result is known in convex analysis as the divisibility theorem. It is important that the theorem, after being transferred to pattern recognition, answered questions which had their origin in pattern recognition and in their original reading had nothing in common with the theorem known on convex linear divisibility. Thus there were no convincing answers to these questions at a disposal either.

The questions concern, e.g., recognition by virtue of the *minimal distance from the exemplar*. They will be stated in the way as they were originally brought forth in pattern recognition.

Let X_1 and X_2 be two finite sets of points in a metric space X with Euclidean metric $d\colon X \to \mathbb{R}$. Let the classifier operate in the following manner. It has in memory two exemplars (exemplar points) α_1 and α_2 which are points of the space X. The recognised point x is placed in the first class if the distance between x and α_1 is less than the distance between x and α_2. In the opposite case the point is placed in the second class. The question is what the exemplars α_1 and α_2 are to be like to allow all objects from the set X_1 to be placed in the first class and all objects from the set X_2 to be placed in the second class.

Even now, when pattern recognition has reached a certain degree of advancement, you can find different and not very convincing recommendations how to choose the exemplars. For example, the exemplars α_1 and α_2 should be the 'less damaged', or in a sense the best elements of the sets X_1 and X_2. Another time, even artificial exemplars are designed that suit the designer's conception of ideal representatives of the sets X_1 and X_2. Another inaccurate consideration leads to a requirement for the exemplar to lie in average in position that is the nearest to all elements of the set it represents. Therefore the exemplar is determined as the arithmetic average of the elements which corresponds to the centre of gravity of a set. We have seen several other unconvincing answers.

The answer has become convincing only after a precise formulation of the task and by applying Theorem 5.3 on linear divisibility of sets. You may admit that the result is rather unexpected. For an exemplar neither the best nor an average representative of a set is to be chosen. Just the opposite, for an exemplar of one set such a point in the convex hull of the set is to be chosen that is the nearest to the convex hull of the second set. If, for example, we wanted to recognise two letters A and B then such a specimen of the letter A that best resembles the letter B should become the exemplar. In a sense it is the worst specimen. It also holds vice versa, the worst representative of the letters B that best resembles the letter A is chosen for the exemplar of the letter B.

This knowledge is new, from the point of view of pattern recognition it is nontrivial, and thus it was not revealed without applying the theorem on linear divisibility of sets. In convex analysis, which is the home of the theorem, this knowledge was not revealed because the concepts which express this knowledge are quite alien to convex analysis. And thus only after the meeting of one field with the other did the new knowledge originate.

We would like you to notice that even in pattern recognition this accepted theorem borders on other questions, not only on those of linear divisibility of sets. Through being applied in pattern recognition, the theorem has been enriched and has contributed to another third field. In pattern recognition different non-linear transformations of space observation are continually used, which are known as *straightening of the feature space.* The task of decomposing sets by means of a straight line then stops being different from separating sets by means of circles, parabolas, ellipses, etc. Taking over the theorem on linear separability of sets, the pattern recognition has broaden the sphere of its possible applications. For example, there are the algorithms in computational geometry seeking distances between convex hulls as well as linear separation of points in a plane. At the same time you can relatively often notice that even an experienced programmer using computational geometry does not come at once across the algorithm for separating sets by means of circles. You can rarely see a programmer who would not be at a loss if he was expected to divide two sets of points by means of an ellipse. But you will be not at a loss because, as we hope, you do not see any substantial difference between those tasks. And you do not see the difference just thanks to your having used the theorem on separability in pattern recognition. And so the theorem on separability after being used in pattern recognition was enriched, and so enriched it returned to the scientific environment from where it had once come to pattern recognition.

We still do not dare to seriously define whether pattern recognition is a science, art, or a collection of technical and mathematical tricks. But we dare to claim that hardly a field will be found in which the knowledge from different spheres of applied mathematics meets common application so frequently as it is in pattern recognition. And therefore pattern recognition could become attractive not only for young researchers who are engaged in it, as you yourself are, but for everybody who wants to learn all the 'charming features' necessary for work in applied informatics, quickly and from his or her own experience.

The analysis of Anderson task is laborious. In such a case we should do a bit of thinking about the possibility of generalising the results beyond the frame of the particular case for which it was proved. Notably strict is the assumption that in Anderson task the multi-dimensional random variables are Gaussian variables. The user cannot rely on that assumption beforehand: and I also *doubt that it can be experimentally determined whether the multi-dimensional random variable is Gaussian. I am very interested in such formulations of the task that would not be based on the assumption of Gaussian distribution, but on a substantially weaker assumption that it concerns a random variable with an unknown multi-dimensional distribution (well, how could it be known?). But one would know that the random variable has the mathematical expectation μ and the covariance matrix σ. I will try to state the task more precisely.*

Let X be, similarly as in Anderson task, a multi-dimensional space, $\{\mu^j, j \in J\}$ be a set of vectors, $\{\sigma^j, j \in J\}$ be a set of symmetrical positive-definite matrices. Let $\mathcal{P}^j$ be a set of functions $p\colon X \to \mathbb{R}$, such that for each $p \in \mathcal{P}^j$ there holds

$$\sum_X p(x) = 1\,,$$

$$\sum_X p(x) \cdot x = \mu^j\,,$$

$$\sum_X p(x) \cdot x \cdot x^T = \sigma^j\,,$$

$$p(x) \geq 0\,, \quad x \in X\,.$$

The (row) vector x^T denotes the transposed (column) vector x. It is clear to me that in all the formulæ of this system, as well as further on, I should write integrals instead of infinite sums. But this is not essential now.

Let the vector α and the number θ define the set $X(\alpha,\theta) = \{x \mid \langle\alpha, x\rangle \leq \theta\}$. Furthermore, let us have the function $p\colon X \to \mathbb{R}$ and the number

$$\varepsilon(\alpha,\theta,p) = \sum_{x \in X(\alpha,\theta)} p(x)\,,$$

which means the probability of the event that a random vector x with probability distribution $p(x)$ will satisfy the inequality $\langle\alpha, x\rangle \leq \theta$.

The task according to my wish tries to avoid the assumption of the Gaussian character of the random vector. A vector α and the number θ are to be found which minimise the value

$$\max_{j \in J} \max_{p \in \mathcal{P}^j} \varepsilon(\alpha,\theta,p)\,.$$

The pair of parameters sought is thus

$$(\alpha,\theta) = \operatorname*{argmin}_{\alpha,\theta} \max_{j \in J} \max_{p \in \mathcal{P}^j} \varepsilon(\alpha,\theta,p)\,. \tag{5.105}$$

It occurred to me that this task was very similar to the task that had interested me after Lecture 3. There I tried to find a reasonable strategy that, despite a common procedure, is not based on the assumption of the independence of features. At the same time nothing was known about the form of the dependence. With your help I saw at that time that the recognition strategy with the unknown mutual dependence was different from the strategy which was correct on the condition of feature independence. After this lecture a similar question worries me: is the strategy solving the task (5.105) the same as the strategy of solving the same task on a stricter assumption that the corresponding random vectors are Gaussian? I am going to say it in different words: can I also use the solution of Anderson task in a situation when I am not sure that the random vectors are Gaussian, and, moreover, when I know hardly anything about their distribution besides their mathematical expectation and covariance matrix?

Boy! We respect you for your asking from time to time such deeply thought out and precisely formulated questions. To your question we have found a convincing and unambiguous answer. Your question was so well thought out that it does not seem to us that you had not already had an answer to it. All right, you must examine the function $\max_{p \in \mathcal{P}^j} \varepsilon(\alpha, \theta, p)$. When you see that the function is decreasing in a monotonic way when the ratio

$$\frac{\langle \alpha, \mu^j \rangle - \theta}{\sqrt{\langle \alpha, \sigma^j \cdot \alpha \rangle}}$$

increases then the solution of the task (5.105) is identical with the solution of Anderson task. When the function is not decreasing in a monotonic way then further examining is necessary.

I see that, but I do not know the way and I have not made any progress in my research either. It seems to me to be too complicated.

Do not worry and start analysing the question, for example, in the following formulation. Let the mathematical expectation of an n–dimensional random vector $x = (x_1, x_2, \ldots, x_n)$ be zero, i.e.,

$$\sum_{x \in X} p(x) \cdot x = 0 . \tag{5.106}$$

The previous expression is a shortened notation of the following n equations

$$\sum_{(x_1, \ldots, x_n) \in X} p(x_1, x_2, \ldots, x_n) \cdot x_i = 0 , \quad i = 1, 2, \ldots, n .$$

Let the covariance matrix of the same random variable be σ, i.e.,

$$\sum_{x \in X} p(x) \cdot x \cdot x^T = \sigma , \tag{5.107}$$

which is a brief expression of the following $n \times n$ equations

$$\sum_{x \in X} p(x) \cdot x_i \cdot x_j = \sigma_{ij} , \quad i = 1, 2, \ldots, n , \quad j = 1, 2, \ldots, n .$$

Let α be a vector, $\theta < 0$ be a number and $X^-(\alpha,\theta)$, $X^0(\alpha,\theta)$ and $X^+(\alpha,\theta)$ be the sets

$$\begin{aligned} X^-(\alpha,\theta) &= \{x \in X \mid \langle \alpha, x\rangle \le \theta\}\,, \\ X^0(\alpha,\theta) &= \{x \in X \mid \langle \alpha, x\rangle = \theta\}\,, \\ X^+(\alpha,\theta) &= \{x \in X \mid \langle \alpha, x\rangle > \theta\}\,. \end{aligned}$$

The probability $\varepsilon(\alpha,\theta,p)$ is then

$$\varepsilon(\alpha,\theta,p) = \sum_{x \in X^-(\alpha,\theta)} p(x)\,. \tag{5.108}$$

You would like to know what the numbers $p(x)$, $x \in X$, are to be that satisfy the conditions (5.106), (5.107) and further the conditions

$$\sum_{x \in X} p(x) = 1\,, \quad p(x) \ge 0, \quad x \in X,$$

and maximise the number $\varepsilon(\alpha,\theta,p)$ expressed by the equation (5.108). To get used to this task, look first at a one-dimensional case. What are the numbers $p(x)$ which for a one-dimensional variable x maximise the number

$$\sum_{x \le \theta} p(x) \tag{5.109}$$

and satisfy the conditions

$$\begin{aligned} \sum_{x \in X} p(x) &= 1\,, \\ \sum_{x \in X} p(x)\cdot x &= 0\,, \\ \sum_{x \in X} p(x)\cdot x^2 &= \sigma\,, \\ p(x) \ge 0\,, &\quad x \in X\,. \end{aligned}$$

That is the linear programming task. Although the task has infinitely many variables, it is still solvable. It resembles the well known Chebyshev problem, which differs from the case you have in mind only in that instead of (5.109) the sum

$$\sum_{|x| \ge \theta} p(x) \tag{5.110}$$

is to be maximised on the same conditions as those in your task. The solution of Chebyshev task is so clever that we cannot but quote it, even if it is quite known.

$$\begin{aligned} \sum_{|x|\ge\theta} p(x) &= \theta^2 \cdot \frac{1}{\theta^2} \sum_{|x|\ge\theta} p(x) = \frac{1}{\theta^2} \sum_{|x|\ge\theta} p(x)\cdot\theta^2 \\ &\le \frac{1}{\theta^2} \sum_{|x|\ge\theta} p(x)\cdot x^2 \le \frac{1}{\theta^2} \sum_{x\in X} p(x)\cdot x^2 = \frac{\sigma}{\theta^2}\,. \end{aligned}$$

Hence it follows further

$$\sum_{|x|\geq\theta} p(x) \leq \frac{\sigma}{\theta^2} \,. \tag{5.111}$$

Let $\sigma/\theta^2 \leq 1$ hold. Let us have a look at the following function $p^*(x)$,

$$p^*(x) = \begin{cases} \dfrac{1}{2} \cdot \dfrac{\sigma}{\theta^2}\,, & \text{when } |x| = \theta\,, \\ 1 - \dfrac{\sigma}{\theta^2}\,, & \text{when } x = 0\,, \\ 0\,, & \text{for all other values of } x\,. \end{cases}$$

For the function $p^*(x)$ the sum (5.110) is σ/θ^2, and for all other functions, thanks to (5.111), it is not greater. Therefore the maximal value of the sum (5.110) at constraints (5.111) corresponds to the value σ/θ^2. You can see then that the task of maximising the functions (5.110) need not be difficult, even when it depends on an infinite number of arguments. Do you not think that Chebyshev perfectly mastered the task? Why could not you as well master first the maximisation (5.109) on the conditions (5.111), and then solve a multi-dimensional task?

I mastered it! My objective is to solve the following linear programming problem

$$\left.\begin{array}{ll} & \max \sum\limits_{x\in X(\alpha,\theta)} p(x) \\ \lambda_0 & \sum\limits_{x\in X} p(x) = 1, \\ \lambda_1 & \sum\limits_{x\in X} p(x)\cdot x = 0, \\ \lambda_2 & \sum\limits_{x\in X} p(x)\cdot x\cdot x^T = \sigma, \\ & p(x) \geq 0, \quad x\in X. \end{array}\right\} \tag{5.112}$$

In the first line of (5.112) the function to be maximised is written. The designation $X(\alpha,\theta)$ in this objective function of the above linear task formulation means $X^-(\alpha,\theta)\cup X^0(\alpha,\theta)$. In the second line the constraint is stated to which the dual variable λ_0 corresponds. The third line briefly shows n constraints related to n dual variables which are denoted by the vector λ_1. The fourth line yields $n\times n$ constraints and the corresponding ensemble of dual variable is represented by the matrix λ_2.

The variables in the task (5.112) are the numbers $p(x)$, $x\in X$. To each such variable a constraint corresponds which the dual variable λ_0, the vector λ_1 and the matrix λ_2 must satisfy. To the variable $p(x)$, $x\in X(\alpha,\theta)$, the following constraint corresponds

$$\langle x, \lambda_2\cdot x\rangle + \langle\lambda_1, x\rangle + \lambda_0 \geq 1\,, \quad x\in X(\alpha,\theta), \tag{5.113}$$

and to the variable $p(x)$, $x \in X^+(\alpha,\theta)$, the following constraint corresponds

$$\langle x, \lambda_2 \cdot x\rangle + \langle \lambda_1, x\rangle + \lambda_0 \geq 0\,, \quad x \in X^+(\alpha,\theta)\,. \tag{5.114}$$

The constraints (5.113) and (5.114) are to be understood in such a way that the ensemble of dual variables defines on the space X a quadratic function which will be denoted by F,

$$F(x) = \langle x, \lambda_2 \cdot x\rangle + \langle \lambda_1, x\rangle + \lambda_0\,,$$

which must not be less than 1 on the set $X(\alpha,\theta)$ and must not be negative on the set $X^+(\alpha,\theta)$. As a whole, the function $F(x)$ is positive-semidefinite.

I will analyse the problem (5.112) only for the situations in which the mathematical expectations of random variable x belongs to the set $X^+(\alpha,\theta)$. This means that $\theta < 0$.

Let p^* be the solution of the task. At least in one point $x \in X^+(\alpha,\theta)$ there must be $p^*(x) \neq 0$, since in the opposite case

$$\begin{aligned}\left\langle \alpha, \sum_{x\in X} p^*(x)\cdot x\right\rangle &= \left\langle \alpha, \sum_{x\in X(\alpha,\theta)} p^*(x)\cdot x\right\rangle \\ &= \sum_{x\in X(\alpha,\theta)} p^*(x)\cdot\langle\alpha, x\rangle \\ &\leq \sum_{x\in X(\alpha,\theta)} p^*(x)\cdot\theta = \theta < 0\,,\end{aligned} \tag{5.115}$$

would occur and thus

$$\left\langle \alpha, \sum_{x\in X} p^*(dx)\cdot x\right\rangle < 0$$

would hold, which would contradict the constraint in the third line of (5.112). The point $x \in X^+(\alpha,\theta)$ for which $p^*(x) \neq 0$ holds will be denoted by x_0 and the scalar product $\langle \alpha, x_0\rangle$ will be denoted by the symbol Δ.

Now I will prove that the scalar product $\langle\alpha, x_1\rangle$ in an arbitrary point $x_1 \in X^+(\alpha,\theta)$, for which $p^*(x_1) \neq 0$ holds, is also Δ. In other words, all points $x_1 \in X^+(\alpha,\theta)$ for which $p^*(x) \neq 0$ holds lie in a hyperplane parallel to the hyperplane $\langle\alpha, x\rangle = \theta$ denoted $X^0(\alpha,\theta)$. I assume that it is wrong and consider a pair from the hitherto point x_0 and another point x_1 so that for them there holds

$$\left.\begin{aligned} &p^*(x_0) \neq 0\,,\\ &p^*(x_1) \neq 0\,,\\ &\langle\alpha, x_0\rangle \neq \langle\alpha, x_1\rangle\,. \end{aligned}\right\} \tag{5.116}$$

I will examine how the function $F(x)$ behaves on a straight line which passes through the points x_0 and x_1. This straight line is not parallel with the hyperplane $X^0(\alpha,\theta)$. Therefore there certainly exists a point x_2 which lies on that *straight line and at the same time it lies in the hyperplane $X^0(\alpha, \theta)$. Thanks to the constraint (5.113) in this point there holds*

$$F(x_2) \geq 1 \, .$$

By the second duality theorem the first and the second inequalities in (5.116) imply correspondingly

$$F(x_0) = 0 \, , \quad F(x_1) = 0 \, .$$

Any convex combination of the points x_0 and x_1 belongs to the set $X^+(\alpha, \theta)$, and thus for each point of this abscissa

$$F(x^*) \geq 0$$

holds by (5.114). Thus on the abscissa passing points x_0 and x_1 the quadratic function F must behave in the following way. On the abscissa between points x_0 and x_1 the function F must not be negative, in the extreme points x_0 and x_1 it must be zero, and in a certain point x_2 outside the abscissa it must not be less than 1. Since such a quadratic function does not exist the assumption (5.116) is not satisfied. Thus we have proved by contradiction that in all points x in the set $X^+(\alpha, \theta)$ for which $p^*(x) \neq 0$ holds the scalar product $\langle \alpha, x \rangle$ is the same and equal to the number which I denoted by Δ,

$$\left[\left(x \in X^+(\alpha, \theta)\right) \, \wedge \, \left(p^*(x) \neq 0\right)\right] \Rightarrow \left[\langle \alpha, x \rangle = \Delta\right] \, .$$

I will now prove the statement

$$[(x \in X(\alpha, \theta))] \, \wedge \, \left(p^*(x) \neq 0\right) \Rightarrow [\langle \alpha, x \rangle = \theta] \, , \tag{5.117}$$

which is equivalent to the statement

$$\left(\langle \alpha, x \rangle < \theta\right) \Rightarrow \left(p^*(x) = 0\right) \, . \tag{5.118}$$

Let us assume that the statement (5.118) is wrong. Let for a point x_1 hold

$$p^*(x_1) \neq 0 \, , \quad \langle \alpha, x_1 \rangle < \theta \, . \tag{5.119}$$

I will examine how the function F behaves on the straight line which passes through the points x_1 and x_0. For the point x_0 the existence of which has already been proved by me, there holds

$$p^*(x_0) \neq 0 \, , \quad \langle \alpha, x_0 \rangle > \theta \tag{5.120}$$

This straight line must intersect the hyperplane $X^0(\alpha, \theta)$ in a point which will be denoted x^*. Resulting from (5.113) there holds

$$F(x^*) \geq 1 \, .$$

Thanks to the second theorem on duality (Theorem 2.2) from the first inequality in (5.119)

$$F(x_1) = 1$$

holds and from the first inequality in (5.120) there holds

$$F(x_0) = 0\,.$$

The function F must behave on the straight line examined as follows. In a selected point x_1 the function F assumes the value 1, in another point x_0 it assumes the value 0 and in the interjacent point x^* it assumes a value that is not less than 1. Since no positive-semidefinite quadratic form can behave in this way, the assumption (5.119) is not valid. Thus I have proved (5.117).

A set of points x for which $\langle \alpha, x\rangle = \Delta$ holds is a hyperplane $X^0(\alpha,\Delta)$. Furthermore, I will denote

$$p^*_\theta = \sum_{x\in X^0(\alpha,\theta)} p^*(x)\,,$$

$$p^*_\Delta = \sum_{x\in X^0(\alpha,\Delta)} p^*(x)\,,$$

where p^* represents, as before, the function which solves the task (5.112). The function p^*, therefore, must also satisfy the conditions of this task from which there follows

$$\left.\begin{aligned} p^*_\theta \quad &+ p^*_\Delta &&= 1\,,\\ p^*_\theta\cdot\theta \quad &+ p^*_\Delta\cdot\Delta &&= 0\,,\\ p^*_\theta\cdot\theta^2 &+ p^*_\Delta\cdot\Delta^2 &&= \langle\alpha,\sigma\cdot\alpha\rangle\,. \end{aligned}\right\} \qquad (5.121)$$

I will show how (5.121) follows from the conditions (5.112). For the first condition from (5.121) it is quite evident because all points x with $p(x)\neq 0$ are either in $X^0(\alpha,\Delta)$ or in $X^0(\alpha,\theta)$.

The equation $\sum_{x\in X} p(x)\cdot x = 0$ in (5.112) is only a brief representation of n equations

$$\sum_{x\in X} p^*(x)\cdot x_i = 0\,, \quad i = 1,2,\dots,n\,.$$

From the previous n equations there follows that

$$\alpha_i\cdot\sum_{x\in X} p^*(x)\cdot x_i = 0\,, \quad i = 1,2,\dots,n\,,$$

and further

$$\begin{aligned} 0 &= \sum_i \alpha_i\cdot\sum_{x\in X} p^*(x)\cdot x_i = \sum_{x\in X} p^*(x)\cdot\Big(\sum_i \alpha_i\cdot x_i\Big)\\ &= \sum_{x\in X} p^*(x)\cdot\langle\alpha,x\rangle = \sum_{x\in X^0(\alpha,\theta)} p^*(x)\cdot\theta + \sum_{x\in X^0(\alpha,\Delta)} p^*(x)\cdot\Delta\\ &= p^*_\theta\cdot\theta + p^*_\Delta\cdot\Delta\,. \end{aligned}$$

The condition $\sum_{x \in X} p^*(x) \cdot x \cdot x^T = \sigma$ in (5.112) is also only a brief expression for $n \times n$ equations

$$\sum_{x \in X} p^*(x) \cdot x_i \cdot x_j = \sigma_{ij} \,, \quad i = 1, \ldots, n \,, \quad j = 1, 2, \ldots, n \,.$$

From the previous $n \times n$ equations there follows

$$\alpha_i \cdot \alpha_j \cdot \sum_{x \in X} p^*(x) \cdot x_i \cdot x_j = \alpha_i \cdot \sigma_{ij} \cdot \alpha_j \,, \quad i = 1, \ldots, n \,, \quad j = 1, \ldots, n \,,$$

and further

$$\begin{aligned}
\langle \alpha, \sigma \cdot \alpha \rangle &= \sum_{i,j} (\alpha_i \cdot \sigma_{ij} \cdot \alpha_j) = \sum_{i,j} \alpha_i \cdot \alpha_j \cdot \sum_{x \in X} p^*(x) \cdot x_i \cdot x_j \\
&= \sum_{x \in X} p^*(x) \cdot \sum_{i,j} \alpha_i \cdot \alpha_j \cdot x_i \cdot x_j \\
&= \sum_{x \in X} p^*(x) \cdot \Big(\sum_i \alpha_i \cdot x_i \Big) \cdot \Big(\sum_j \alpha_j \cdot x_j \Big) \\
&= \sum_{x \in X} p^*(x) \cdot \Big(\sum_i \alpha_i \cdot x_i \Big)^2 \\
&= \sum_{x \in X^0(\alpha,\theta)} p^*(x) \cdot \theta^2 + \sum_{x \in X^0(\alpha,\Delta)} p^*(x) \cdot \Delta^2 \\
&= p^*_\theta \cdot \theta^2 + p^*_\Delta \cdot \Delta^2 \,.
\end{aligned}$$

The system (5.121) consists of three scalar equations only in which the variables are the numbers p^*_θ, p^*_Δ and Δ. The system has one single solution for which the following equality holds

$$p^*_\theta = \frac{\langle \alpha, \sigma \cdot \alpha \rangle}{\theta^2 + \langle \alpha, \sigma \cdot \alpha \rangle} \,.$$

The number p^*_θ is just the value of the sum $\sum_{x \in X(\alpha,\theta)} p(x)$, after substituting the function $p^*(x)$ in it which maximises the sum in our task, and therefore is an explicit expression for solving the task (5.112). The result is the number

$$\frac{\langle \alpha, \sigma \cdot \alpha \rangle}{\theta^2 + \langle \alpha, \sigma \cdot \alpha \rangle}$$

which decreases monotonically when the quantity

$$\frac{\theta}{\sqrt{\langle \alpha, \sigma \cdot \alpha \rangle}}$$

grows.

I then ask a half-hearted question. Have I enhanced Chebyshev inequality?

Of course, you have not. The Chebyshev inequality cannot be enhanced because, as we have already seen, it defines the exact upper limit for a certain probability. This means that even such a random variable exists at which Chebyshev inequality becomes an equation. The Chebyshev inequality estimates the probability of a two-sided inequality $|x - \mu| \geq \theta$, whereas your inequality estimates the probability of a one-sided inequality $x - \mu \leq -\theta$. You have managed to prove that your estimate is also exact. You have not enhanced the Chebyshev inequality, but you have avoided its often wrong or inaccurate application. The first application of this kind is based on a correct estimate

$$P(x - \mu \leq -\theta) \leq P(|x - \mu| \geq \theta) \leq \frac{\sigma}{\theta^2}$$

which is not exact and instead of which you can now use an exact estimate on the basis of the inequality of yours. The second common application is based on the following consideration

$$P(x - \mu \leq -\theta) \approx \frac{1}{2} P(|x - \mu| \geq \theta) \leq \frac{\sigma}{2\theta^2} ,$$

which is not correct. The estimate which results from your considerations is

$$p(x - \mu \leq -\theta) \leq \frac{\sigma}{\theta^2 + \sigma} . \tag{5.122}$$

Since you have proved that in the relation (5.122) the equality can be obtained, your estimate, as well as that of Chebyshev, cannot be enhanced.

So we have together proved that all algorithms which were proved for Anderson task on the assumption that the random variables were Gaussian can be used even in a case where the Gaussian assumption is not satisfied. But be careful and use this recommendation only in the sense that has been stated here. Use it particularly when you are not sure that the random variables are Gaussian and when you doubt over other assumptions as well. If you know for certain that a random variable is not Gaussian and if, moreover, you are sure that it belongs to another class of random variables, and you know the class, even more effective algorithms can be created.

I thank you for your important advice, but I must say that other different algorithms do not worry me much at the moment. I am interested in actual recommendations of what to do when I have two finite sets X_1 and X_2 and want to separate them by means of a hyperplane in a reasonable way. I see that here I can proceed in at least two directions.

In the first case I can try to separate two sets by means of the Kozinec algorithm. If the number of points in the sets X_1, X_2 is very large then I can even go in the other direction, i.e., calculate the vectors μ^1, μ^2 and matrices σ^1, σ^2. For searching for the separating hyperplane I can use the algorithms for solving Anderson task. I can afford to do so because I have proved that for the correctness of such a procedure it is not necessary to assume the Gaussian character of random variables. Anderson task, which I use here for replacing *the original task, is quite simple because one operates with two classes only,* $|J| = 2$.

Here, in practice, we do not recommend you anything. We would like to remark only that your being at a loss that you know several approaches of how to solve an application task is of a quite different character now than as if you knew none. But the worst situation is when someone knows one approach only, and therefore he or she uses it without any hesitation or doubt.

But here several approaches could be considered. I need not, for example, represent the whole set X_1 *by means of a single pair* μ^1, σ^1, *but I can divide it, in some reasonable way, into subsets and so express* X_1 *by means of more vectors and matrices. But I do not know how this division of a set into subsets can be done.*

You are already interfering with the subject matter of our next lecture, where these problems will be dealt with. We are sure that after the lecture you will again have interesting comments.

July 1997.

5.7 Link to a toolbox

The public domain Statistical Pattern Recognition Toolbox was written by V. Franc as a diploma thesis in Spring 2000. It can be downloaded from the website `http://cmp.felk.cvut.cz/cmp/cmp_software.html`. The toolbox is built on top of Matlab version 5.3 and higher. The source code of algorithms is available. The development of the toolbox has been continued.

The part of the toolbox which is related to this lecture implements linear discriminant functions, e.g., separation of the finite point sets, perceptron learning rule, Kozinec algorithm, ε-solution by the Kozinec algorithm, Support Vector Machines (linearly separable case), Fisher classifier, modified Perceptron rule, modified Kozinec algorithm, generalized Anderson task, original Anderson–Bahadur solution, ε-solution. The quadratic discriminant functions are implemented, too, through non-linear data mapping.

5.8 Bibliographical notes

The formulation and solution of the original Anderson–Bahadur task is in the paper [Anderson and Bahadur, 1962]. Schlesinger [Schlesinger, 1972a; Schlesinger, 1972b] proposed a solution for the case in which classes are not characterised by one Gaussian distribution but by several Gaussian distributions.

The algorithm of the linear decomposition of finite sets of points comes from Rosenblatt perceptron [Rosenblatt, 1962]. The climax of this research was Novikoff theorem [Novikoff, 1962]. In the lecture a proof of Novikoff theorem was taken over after [Vapnik and Chervonenkis, 1974].

Another, but quite close, view of the decomposition of finite sets is represented by potential functions [Ajzerman et al., 1970]. The approach deals with decomposition of finite or infinite sets and uses nonlinear discriminant functions. Essentially the idea of feature space straightening is generalised. Note that space straightening has been introduced in Lecture 3. The main idea of the potential functions method states that the straightening can be used even if the feature space straightened has infinite dimension, and not only finite dimension. The potential function method discovers that the main question asks if the scalar product can be constructively calculated in the space straightened, and not if the space straightened has finite or infinite dimension. The scalar product in space straightened is a function of two variables defined in original space, and it is just a potential function. In many applications the potential function can be implied directly from the content of the problem solved. The convergence of the potential function method is proved similarly to the proof of Novikoff theorem. Another algorithms separating finite and infinite sets of points were suggested by Kozinec [Kozinec, 1973] and Jakubovich [Jakubovich, 1966; Jakubovich, 1969].

The transformation of the task separating linearly and optimally a set of points to the quadratic programming task was explained in the lecture. The approach has roots in known results of Chervonenkis and Vapnik. Their method of generalised portraits was published in [Vapnik and Chervonenkis, 1974]. The method is well known now in western publication as a Support Vector Machine [Boser et al., 1992], [Vapnik, 1995], [Vapnik, 1998].

Class of Fisher strategies, synthesis of which can be reduced to synthesis of linear discriminant functions, were originally introduced in [Fisher, 1936].

A modification of the perceptron and Kozinec algorithms for dividing infinite sets and their use for ε-solution of tasks is in [Schlesinger et al., 1981].

Let us note that Kozinec algorithm can be extended to non-separable data [Franc and Hlaváč, 2001].

The mathematical basis for non-smooth optimisation used in the discussion has been taken over from [Shor, 1979; Shor, 1998]. Considerations on Fibbonaci numbers can be found in [Renyi, 1972].

Lecture 6

Unsupervised learning

Messieurs, lorsqu'en vain notre sphère
Du bonheur cherche le chemin,
Honneur au fou qui ferait faire
Un rêve heureux au genre humain.

Qui découvrit un nouveau monde?
Un fou qu'on raillait en tout lieu.
Sur la croix, que son sang inonde,
Un fou qui meurt nous lègue un Dieu.
Si demain, oubliant d'éclore,
Le jour manquait, eh bien, demain,
Quelque fou trouverait encore
Un flambeau pour le genre humain.

Part of the song Fools *by* Pierre-Jean de Branger, *1833.*

6.1 Introductory comments on the specific structure of the lecture

This lecture is devoted to unsupervised learning, which in the theory of recognition is sometimes referred to as self-learning, learning without a teacher, or is even given some other name. The nature of the lecture will be somewhat different from the previous explanation. The previous five lectures had the form of more or less convincing considerations which were based on the most possible unambiguous formulation of the task and led to its solution. If the centre of a lecture is the formulation of the task then the long term and extensive research which preceded the clear understanding of the task will remain behind the stage. This life cycle of scientific thought is typically imbued with an extraordinary emotional tension and drama.

The previous explanation was devoted to those chapters of pattern recognition theory which were inherited from other scientific disciplines. In them the results assumed a completely accomplished form in which they assimilated into the pattern recognition theory. Therefore today we can only remain in the dark about why and how T. Bayes, for instance, discovered the excellent formulation of the statistical decision task. We can only imagine how he worked his way through dim ideas and doubts, which there certainly were a lot of, how he stood up to the notes of criticism by his colleagues, etc., etc..

Unlike the fundamental knowledge acquired in pattern recognition, but adopted from other scientific disciplines, the concept of unsupervised learning was created as late as in pattern recognition theory itself. Therefore unsupervised learning cannot pride itself upon such a respectable age as that enjoyed by the Bayesian and non-Bayesian approaches. However, a relatively young age does render to unsupervised learning one advantage. Unlike the case of established methods, the whole period of existence of the scientific field of unsupervised learning can be surveyed at one glance, from its birth till to the present day state of the art, and we can see all the pitfalls and valleys through which it had to pass to manage, at last, to express it in the form of unambiguously formulated task and its solution.

The consistency and inevitability of the certain inconvenience which accompanies the birth of new knowledge is really remarkable. The dramatic background of scientific development is such an attractive topic for a discussion that we avoid only with difficulty the temptation to show a long series of well known cases which illustrate, in spite of the time and spacial remoteness, a much similar destiny for the new pieces of knowledge. Not quite seriously, but not as a joke either, one could say that the example of the discovery of America by Columbus joins together, in its core, the following features which characterise scientific discovering as well.

- The discovery of America by Columbus was made possible only thanks to the powerful financial support of the whole project.
- Columbus managed to gain resources for the project only after he had given a guarantee that he would find a new way from Europe to India.
- Columbus' confidence that he would solve the practical task was based upon false premises. That is why the goal of the project was not reached. Conversly, during the project an unexpected and quite substantial obstacle appeared (a whole continent!) which spoiled reaching the practical goal of the project.
- Columbus managed, despite all the apparently negative outcomes of the project, to convince his sponsors and customers that all that had been expected from the project had been successfully accomplished.
- After the project was accomplished nobody including the author (of the project) noticed what a significant discovery had been actually made. Moreover, if someone had drawn attention to the importance of the discovery, hardly anyone would have realised its actual significance: the investors in the project were interested not in the new (and up to that time unknown territories), but only in new ways to the territories already known.

- When the knowledge which Columbus had already discovered began to be needed nobody remembered it. To re-discover it a completely new project had to be started and a new grant obtained.
- The continent discovered was not eventually named after Columbus but after someone else who just repeated that what Columbus found earlier.
- Columbus himself did not learn until the end of his lifetime what he had actually discovered and remained convinced (or did he only pretend?) that he had discovered a new way to India.
- Now it is generally known that Columbus was not the first European to have sailed to America. Its existence had already been known for centuries before Columbus, not just to individuals, but to entire nations of Scandinavia who made practical use of that knowledge.

The form of this lecture has been influenced not only by the content presented, but also by the effort of the authors to show how stony is the path along which something new comes into existence, even if the new is rather imperceptible.

6.2 Preliminary and informal definition of unsupervised learning in pattern recognition

Let us make a brief survey of the subject matter of the previous lectures and on this basis we will classify pattern recognition algorithms into those which

- function without learning,
- learn,
- are based on unsupervised learning.

The classification of recognition algorithms presented depends on the degree of completeness of available knowledge about the statistical model of the recognised object or on the way in which to compensate for the lack of this knowledge.

A complete description of the statistical model of an object is the function $p_{XK}: X \times K \to \mathbb{R}$, whose value $p_{XK}(x, k)$, $x \in X$, $k \in K$, denotes a joint probability that the object is in the state k and is characterised by the feature x. The probability distribution p_{XK} is the basis for the formulation of some *Bayesian task* (according to the penalty function used). The solution of the Bayesian task is the strategy $q: X \to K$ which a certain (non-learning) pattern recognition device implements. Here the task is formulated as a search for a strategy which is suited for one single statistical model p_{XK}. The formal properties of Bayesian strategies have been described in Lecture 1.

Incomplete knowledge of the statistical model is expressed in such a way that the function p_{XK} is not known and only a certain set of models $\mathcal{P}$ is known which contains the function p_{XK}. The incompleteness of the knowledge can be due to two reasons which fundamentally differ from each other.

In the first case the lack of knowledge can be in no way compensated for since the function p_{XK} is changeable and depends on a certain non-random, but also changeable intervention. The recognition strategy cannot, in this case, take into consideration one particular statistical model p_{XK}, but requires the whole group of models $\mathcal{P}$. These tasks have been known as non-Bayesian recognition tasks and were the content of Lecture 2.

In the second case the reason of the incomplete knowledge for the probability p_{XK} appears not to matter so much. The function p_{XK} can be unchanging, but unknown only because the recognised object has not yet been carefully enough investigated. It is now possible to get on with examining it. Lectures 4 and 5 have demonstrated how to formalise additional investigation. The outcome of the formalisation is a procedure which is referred to as *learning in pattern recognition* and is based on the property that a *training multi-set* in the form (x_1, k_1), (x_2, k_2), ..., (x_n, k_n), which consists of mutually independent members (x_i, k_i), is at our disposal. Each member of the training multi-set is an instance of a random pair (x, k), given by the probability distribution p_{XK} which is unknown. Let us remember that we use a more exact term—the training multi-set—instead of frequently used terms the training set (it is not a set because the same members can be repeated several times), or the training sequence (no fixed order is given here).

The training multi-set is obtained from two different sources. The observations x are acquired in a training phase from the same source which the observations x are aquired in the recognition phase. Therefore the observations x can be obtained more easily than the hidden states k. In theoretical research the question from where the particular state is to be obtained is waved aside with an easy going answer that this type of information is provided by the teacher. In applications complex and frequently expensive provisions are necessary to obtain the states k. The state, as it is known, is a hidden parameter of an object and is inaccessible for immediate measurements. Therefore the effort to find the state k at all costs can lead to an unusual intervention into the object at which its action is interrupted, or the object is completely destroyed.

This difficulty leads to a number of questions. Are both the observations and corresponding states needed during learning to compensate for the deficient knowledge of the statistical model p_{XK}? Is not the multi-set of observations sufficient? Is it possible to do without the information provided by the teacher during learning? Questions asked in this way brought about the situation in which some conjectural abilities of recognition devices had been called unsupervised learning even before it was well understood what should be meant by this term. It took some time to successfully express in the form of a concrete mathematical task the vague wish concerning what were the required features of pattern recognition devices capable of unsupervised learning. The solution of this task should look like a strategy that improves according to how the device analyses more and more objects. The teacher's information is not used.

We will demonstrate how present day ideas on such self-improvement gradually originated owing to the confluence of powerful streams which had existed independently of each other in different fields of science, to eventually meet in pattern recognition. There were three streams: perceptron, empirical Bayesian method, and cluster analysis.

6.3 Unsupervised learning in a perceptron

The American mathematician Halmos [Halmos, 1971] seems to have been right, we are afraid, when saying that the best motivation towards a better performance of a demanding piece of work was a situation where this hard work had previously been done incorrectly by somebody else. There are fare more examples supporting this rule than we would wish for. In pattern recognition too, there are more research projects that act more as a powerful stimulation for further research than as the results which would seem to be faultlessly complete. It can be even predicted that it will be so for some time.

Among the research results stimulating further exploration the foremost position is, beyond dispute, held by the perceptron and various neural networks. Among the positive features of the direction of neural networks in pattern recognition, which undoubtedly justify its existence, one feature is worth being extensively discussed, be it positive or negative, without arriving at any conclusion. The problem is that by the aid of neural networks people often try to solve pattern recognition tasks which are really difficult. The results of their solutions gained with the aid of neural networks are, however, sometimes so bad that they immediately call for an effort to solve them in a better way.

The dissension and mutual lack of understanding between the supporters of neural networks and the followers of other trends became apparent immediately after the first information by F. Rosenblatt about the perceptron in the late fifties [Rosenblatt, 1957]. Since the dissension already lasts more than 40 years it can be assumed not to be a result of a temporary misunderstanding, but based on more substantial ground. The reasons of the mutual misunderstanding were well expressed by the author of the perceptron himself, F. Rosenblatt, in his monograph Principles of Neurodynamics [Rosenblatt, 1962], which summarised many years' research of the perceptron.

According to Rosenblatt there are two main research models in cybernetics which he called the *monotype* model and the *genotype* model. The objective of the research in both models is to find a mutual relation between a certain device, one would call it an algorithm now, and the task which is solved by means of this device. This objective, however, is achieved in the monotype and genotype research in substantially different ways. In the monotype model, the task to be solved is defined first, and only then the device that is to solve the task is created. It is a style we tried to adhere to in our previous lectures and we will continue in this trend even in further explanation.

The genotype model is simply the other way round. At first a certain device is constructed, at least mentally, this is then examined in an effort to comprehend for which task solution it is actually suited. From the point of view of usual engineering practice, the genotype model appears entirely absurd. In most cases an engineer, perhaps due to his/her long term education, simply does not begin to construct a machine before making clear the purpose for which the machine is to be used. Among physiologists and other natural scientists who examine living organisms, the genotype approach is a usual one. At first, certain ideas on information processing mechanisms in an already existing device come into being, such as that in a living organism, and only then the researcher tries to understand the purpose of this processing.

Rosenblatt accounts for the lack of understanding of the perceptron by stating that every supporter of a certain model, be it of the genotype, or monotype kind, acts as if his/her model were the only possible one. After Rosenblatt, the perceptron is a significant example of the genotype research. The main reason for the lack of understanding for it is an effort to evaluate it from the monotype point of view. The perceptron is neither a device for image analysis nor is it a device for speech recognition. It is not even a device that could be defined in a way a monotypist would expect it. It is because the common feature of all perceptrons, we consider a class of neural networks, is not expressed by virtue of formulating the tasks being solved, but by describing its construction. We will give the description in the original form in which the perceptron was defined in early publications, with only such an amount of detail that is necessary for our lecture.

The *perceptron* is a device which consists of a certain set of elements referred to as neurons. Each neuron can be in either of two states: in the excited state or in the inhibited state. The state of a neuron is unambiguously determined by the image which is at the input of the perceptron. Let the set of images which can occur at the input of the perceptron consist of l images. This set can be classified into two classes in 2^l ways. The first class images will be called positive, and the second class images will be called negative. There exists an authority—the *teacher*—who selects from these 2^l classifications a single one which is then regarded as unchanging further on. This classification is called the *teacher's classification*, or the correct classification. In addition to the teacher's classification, there are also other input image classifications which in the general case need not be identical to the correct one. One of these classifications, which will be called *perceptron classification*, will be implemented in the following way. Each neuron is characterised by its numerical parameter which is called the weight of the neuron. The weight can be any real number. The image at the perceptron input is evaluated as negative or positive according to whether the sum of weights of the neurons that have been excited by the image are positive or negative, respectively.

The classification by the perceptron depends on neuron weights. The weights vary by means of *reinforcement* and *inhibition* of a neuron, which represent the respective increase or decrease of a neuron weight by a certain constant quantity Δ. In observing each image a particular neuron is reinforced or inhibited according to

- whether the given neuron was excited by the observed image;
- which class the image was classified by the teacher;
- which class the image was classified by the perceptron itself.

Different combinations of these conditions determines the following three algorithms for the alteration of neuron weights.

Algorithm 6.1 Learning after the teacher's classification

1. If the teacher included the image in the positive class then all excited neurons are reinforced.
2. If the teacher included the image in the negative class then all excited neurons are inhibited.

Algorithm 6.2 Learning after the classification by both the teacher and the perceptron

1. If the perceptron included the input image in the same class as the teacher did then none of the neurons are either reinforced or inhibited.
2. If the perceptron included the input image in the positive class and the teacher included it in the negative class then all excited neurons are inhibited.
3. If the perceptron included the input image in the negative class and the teacher included it in the positive class then all excited neurons are reinforced.

Algorithm 6.3 Unsupervised learning as classification by the perceptron

1. If the perceptron included the input image in the positive class then all excited neurons are reinforced.
2. If the perceptron included the input image in the negative class then all excited neurons are inhibited.

It can be seen that Algorithm 6.3 differs from Algorithm 6.1 only in that the role of the teacher is performed by the perceptron itself. This difference is substantial. To implement Algorithm 6.1 or Algorithm 6.2 the perceptron must have two inputs: one for observing the input image and the other for information from the teacher. Algorithm 6.3 can be implemented without any contact with the teacher. Therefore if the first two Algorithms 6.1 and 6.2 have been called perceptron learning then the third Algorithm 6.3 is named unsupervised learning.

Rosenblatt's main premise was in that a perceptron controlled by any of the three presented algorithms reaches the state in which it classifies all input images in a correct way.

It is hardly possible to describe the universal enthusiasm which was evoked by Rosenblatt's first publications concerning these premises. With the perceptron everything seemed to be marvellous: the charming simplicity of the algorithm, the application of terms unusual in computer technology of those times, such as 'neuron', 'learning', and the word 'perceptron' itself. A romantic atmosphere was created as when one seems to stand in front of an unlocked, but still not open door. Moreover, one is convinced that behind the door something is waiting for him/her what has been long expected, even though one does not yet know, what particularly it will be. Should Rosenblatt's premise appear correct it would mean that the painful and tedious work of constructing pattern recognition devices could be easily avoided. It would only suffice to build a perceptron and then show it some examples of how to operate, and it would proceed automatically. In addition, the assumed ability of the perceptron for unsupervised learning would make a correct performance possible even without showing certain examples to the perceptron: it would suffice for the perceptron to examine images which it should recognise and it would find out by itself, how to classify them.

As nearly always happens, only an insubstantial part of the outlined beautiful fairy tale becomes true. The realisable part is expressed by Novikoff theorem which was quoted and proved in the previous lecture. Novikoff theorem has confirmed Rosenblatt's assumption in Algorithm 6.2, controlled both by the teacher's and perceptron's classification. Let us note that the theorem holds under the condition that such an ensemble of neuron weights exists in which the correct classification is realised.

For an algorithm controlled only according to the teacher's classification we can easily prove that Rosenblatt's assumption is erroneous. Though it is not the main purpose of this lecture, let us look over a simple counterexample which proves Rosenblatt's assumption wrong.

Example 6.1 Rosenblatt Algorithm 6.1 need not converge to correct classification. *Let perceptron consist of three neurons. Each positive image will excite all three neurons, and each negative image will excite either the second or the third neuron only. This situation is favourable for the perceptron since there exists an ensemble of neuron weights* w_1, w_2, w_3, *with which the perceptron faultlessly classifies all the input images. Such weights can be, for example,* $w_1 = 3$, $w_2 = -1$, $w_3 = -1$. *The sum of the weights of neurons excited by a positive image will be* $+1$, *and that by the negative image will be* -1.

Assume that the perceptron learning starts with exactly those weight values with which the perceptron can correctly recognise all input images. Let us see what happens when the weights are changed after learning on n *positive and* n *negative images. During learning each neuron will be* n *times reinforced, whilst, the second neuron will be* n_2 *times, and the third neuron* n_3 *times inhibited, where* $n = n_2 + n_3$. *The neuron weights will be*

$$\begin{aligned} w_1 &= \;\; 3 + n\,\Delta\,, \\ w_2 &= -1 + \Delta\,(n - n_2)\,, \\ w_3 &= -1 + \Delta\,(n - n_3)\,. \end{aligned}$$

The sum of the neuron weights that are excited by a positive image will be $1 + 2n\,\Delta$ *which is a positive number for any* n. *This means that the recognition of positive images does not deteriorate due to the corrections of neuron weights. But the sum of neuron weights excited by a negative image is either* $-1+\Delta\,(n-n_2)$ *or* $-1 + \Delta\,(n - n_3)$. *The sum of these two numbers is* $-2 + n\,\Delta$ *which is a positive number at a sufficiently great* n. *From this it follows that at least one of the summands* $-1 + \Delta\,(n - n_2)$ *or* $-1 + \Delta\,(n - n_3)$ *is positive. We can see that the perceptron after learning passed from the state in which it had been correctly recognising all the images, to a state in which it is not recognising all of the images correctly.* ▲

Rosenblatt's assumption on the convergence of the learning algorithm toward a correct classification appeared correct for perceptrons controlled by both the teacher's and the perceptron's classification, i.e., for Algorithm 6.2 only. The assumption is wrong for a perceptron controlled solely by the teacher's classification. If we try to find out now, if Rosenblatt's assumption is also correct for a perceptron controlled only by the perceptron classification, we will be embarrassed since the assumption is neither correct nor wrong.

We will explain it at an example. The statement $2 \cdot 2 = 4$ is correct, and the statement $2 \cdot 2 = 5$ is wrong. But the statement $2 \cdot 2 = \frac{\sqrt{\cdot}}{2}$ is neither correct nor wrong. It is evident that the previous statement cannot be considered as correct. At the same time it cannot be considered wrong since if we considered the statement as a wrong one then we would immediately admit that the statement $2 \cdot 2 \neq \frac{\sqrt{\cdot}}{2}$ is correct, but that would not be our intention. The statement $2 \cdot 2 = \frac{\sqrt{\cdot}}{2}$ is simply nonsense and he/she who wrote it did not add anything important to it, and therefore it is an improper statement.

Rossenblatt's assumption on unsupervised learning in a perceptron is explained ambiguously because it does not say which of the 2^l classifications of input images is regarded as correct. According to how this classification is defined one can already consider the feasibility or infeasibility of unsupervised learning by means of the perceptron.

It is evident that if the teacher's classification is regarded as correct then the assumption on unsupervised learning is wrong. The classification attained by the perceptron by means of unsupervised learning depends on the set of input images only. But the teacher can classify this set in different ways. What is constant cannot be identical to what is varying.

If the classification attained by the perceptron is regarded as correct then the statement on unsupervised learning becomes valid. It is, however, a statement of the form: The perceptron attains correct classification, and that classification is correct which is attained by the perceptron. This validity has an empty content as does a statement of the kind $2 \cdot 2 = 2 \cdot 2$.

Therefore some understandable properties of the classification attained by the perceptron during unsupervised learning need to be formulated. In other words, a task is to be formulated that is solved in unsupervised learning by means of the perceptron. Even such a retrospective analysis of an algorithm might evoke an understandable aversion by specialists of technological orientation, we are not afraid of it since such a research strategy is usual in disciplines of natural sciences.

Rosenblatt made a number of experiments with the aim of understanding the features of the classification which is reached by the perceptron in unsupervised learning. The outcome of the experiments was unexpected. From any state, the perceptron stubbornly passed to a state in which all the images were included in one class, be it positive or negative. Such a classification is really good for

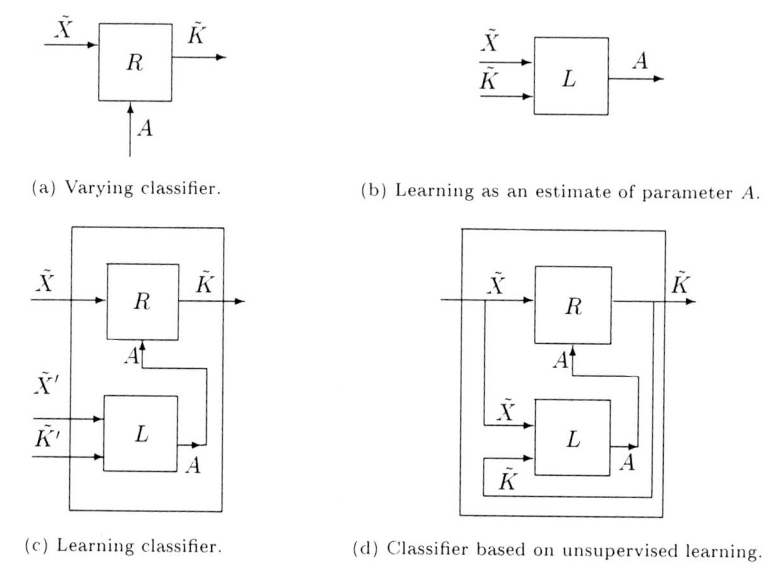

(a) Varying classifier. (b) Learning as an estimate of parameter A.

(c) Learning classifier. (d) Classifier based on unsupervised learning.

Figure 6.1 Different classifier configurations.

nothing. And as it frequently happens, only when the experiments had led to that negative result, it was clear that the result could not have been different. When the perceptron has reached the state in which it includes all the images in one class, to be specific let it be the positive class, then no subsequent unsupervised learning can alter the situation. After processing a new image, the weights of the excited neurons are only rising, and thus the sum of weights of the excited neurons is rising as well. The image so far evaluated as positive can be classified as positive henceforth.

That was the first part of our discourse on how the procedure known as unsupervised learning in pattern recognition had been gradually created and then ended with such disillusionment. In spite this ideas had been present there which have become an important constituent of the present day understanding of unsupervised learning. These ideas, however, were unnecessarily encumbered with neural, physiological, or rather pseudo-physiological considerations. From the present day point of view it may seem to have been purposely done to hopelessly block the way to reasonable outcomes. We will clear away from the idea all that seems to be useless in the context of this lecture and present the definition of a certain class of algorithms which is not very strict. We will call the class *Rosenblatt algorithms* owing to deep respect to the author of the perceptron.

Let X and K be two sets members of which are the observations x and recognition results k. The function $q: X \to K$ is considered as the recognition strategy. Further let Q be a set of strategies. Let a device (see Fig 6.1a) be assigned to implement each strategy from the set Q. The particular strategy, the device is performing at that moment, is given by the value of a parameter A which leads to the input assigned to it. The observations lead to another input.

The sequence of observations $(x_1, x_2, \ldots, x_n)$, $x_i \in X$, which will be denoted $\widetilde{X}$ is transformed by the device to a sequence of recognition results $(k_1, k_2, \ldots, k_n)$, $k_i \in K$, which will be denoted $\widetilde{K}$. The choice of the concrete transformation which will be performed by the pattern recognition device depends on the value of the parameter A. The device, which will be called a *varying classifier* and denoted by the symbol R, implements the function of two variables $\widetilde{X}$, A. Thus the varying classifier performs the decision $\widetilde{K} = R(\widetilde{X}, A)$.

Let another machine transform each pair $\widetilde{X}$, $\widetilde{K}$ to the value of the parameter A (see Fig. 6.1b). The function that is implemented by this machine will be denoted L, and we will call it *learning*, $A = L(\widetilde{X}, \widetilde{K})$. We will connect both the devices as can be seen in Fig. 6.1c, and create a device which we will call a *learning classifier*. The inputs to the learning classifier are three sequences $\widetilde{X}$, $\widetilde{X}'$ and $\widetilde{K}'$, and they are transformed to the output sequence $\widetilde{K}$. The transformation is determined by two algorithms, the *learning algorithm* L and the *algorithm of varying classification* R. First, the value of the parameter A is calculated on the basis of the sequences $\widetilde{X}'$ and $\widetilde{K}'$ by the algorithm L. The value of A obtained is then led to the input of the classifier R. By virtue of the parameter A the concrete decision rule is determined. Then the classifier R transforms the sequence $\widetilde{X}$ to the sequence $\widetilde{K}$.

This rather general description is by no means a strict definition of the learning classifier. It only states of what parts the learning classifier should consist, what the mutual relations between the parts account to, and what the data handled by the classifier means. As the learning classifier is concerned it is necessary for the algorithms R and L to be well set up, in a certain sense. The classification $\widetilde{K}$ of the sequence $\widetilde{X}$ should not be very much different (again in a certain sense) from the teacher's classification $\widetilde{K}'$ of the sequence $\widetilde{X}'$. These rather important requirements were made concrete in Lecture 4.

Now, Rosenblatt's idea on unsupervised learning which was nearly buried in the impenetrable thicket of neural networks can be expressed even without this concretisation. Let us have a learning classifier which is given in Fig. 6.1c and assume that the algorithms R and L are correctly built up. *Unsupervised learning* can be defined as an algorithm which consists of the same algorithms R and L, but which are connected in a different way than in the case of learning, see Fig. 6.1d. The new configuration is to be understood as follows.

At any initial value A_0 of the parameter A the input sequence $\widetilde{X}$ is to be classified, i.e., the sequence $\widetilde{K}_0$ is to be obtained. This sequence is processed together with the sequence $\widetilde{X}$ as if it were information from the teacher. The value $A_1 = L(\widetilde{X}, \widetilde{K}_0)$ is to be calculated. For this new value $A_1 = L(\widetilde{X}, \widetilde{K}_0)$ of the parameter A the input sequence $\widetilde{X}$ is to be classified again, i.e., the sequence $\widetilde{K}_1 = R(\widetilde{X}, A_1)$ is to be calculated. This procedure is repeated so that at the base of the classification $\widetilde{K}_{t-1}$ the classification $\widetilde{K}_t = R\big(\widetilde{X}, L(\widetilde{X}, \widetilde{K}_{t-1})\big)$ is obtained, or similarly at the base of the value A_{t-1} of the parameter A the new value $A_t = L\big(\widetilde{X},\, R(\widetilde{X},\, A_{t-1})\big)$ is calculated.

The algorithm formulated in this way is defined up to recognition algorithm R and learning algorithm L. To each pair R, L a particular unsupervised learning algorithm corresponds. The given definition covers an extent class of algorithms to which the term unsupervised learning applies, since the data which comes from the teacher in the process of learning is created by the classifier itself. Rosenblatt's contribution to the theory of unsupervised learning could be evaluated from the present point of view as a design of an algorithm class which, according to his opinion, solves some intuitively considered tasks. These tasks have not been precisely formulated so far.

In the meantime in the scientific disciplines neighbouring pattern recognition, tasks were formulated that had remained without solution for a long time. Only when these tasks were adopted by the pattern recognition theory, a solution was hatched in its framework in the form of algorithms that are quite close to Rosenblatt's algorithms. But let us postpone this until Section 6.6.

6.4 Empirical Bayesian approach after H. Robbins

Another powerful stream which was infused into pattern recognition and within its framework resulted in modern ideas on unsupervised learning is the so called empirical Bayesian approach by H. Robbins. It originated as an effort to fill in a gap between Bayesian and non-Bayesian methods in the theory of statistical decision making to which we devoted space in Lectures 1 and 2. After Robbins, the statistical methods of decision making should not be divided into two classes, the Bayesian and non-Bayesian ones, but into the following three classes.

Bayesian classes. The domain of Bayesian decision making methods consists of situations in which the state k as well as the observation x are random variables, for which *a priori* probabilities of the state $p_K(k)$ and conditional probabilities $p_{X|K}(x \mid k)$ of the observation x under the condition that the object is in the state k are known.

Non-Bayesian classes. The domain of non-Bayesian decision making methods consists of situations in which only conditional probabilities $p_{X|K}(x \mid k)$ are known. The state k is varying, it is true, but it is not random, and therefore the *a priori* probabilities $p_K(k)$ are not only unknown, but they simply do not exist. In Lecture 2 we showed a variety of non-Bayesian tasks. In spite of this we will further consider a special case of the non-Bayesian tasks in its minimax formulation in the same way as Robbins did.

Robbins' classes. Finally, the domain of the newly proposed methods by Robbins are cases in which both the states k and the observations x are random quantities. Similarly those in a Bayesian task have a certain probability distribution, but only conditional probabilities $p_{X|K}(x \mid k)$ are known. The *a priori* probabilities of the state $p_K(k)$ are unknown even if they exist and have a value. Here Robbins recommends applying special methods which are to fill in the gap between the Bayesian and non-Bayesian approaches.

The main idea of the proposed methods is explained by Robbins in the following simple example.

Let X be a set of real numbers. Let the set K consist of two states, $K = \{1, 2\}$ and the feature x be a one-dimensional random variable with normal

(Gaussian) distribution, with variance $\sigma^2 = 1$, and mathematical expectation which depends on the state k. If the object is in the first state then the mathematical expectation is 1. If it is in the second state then the mathematical expectation is -1. Let $q\colon X \to \{1,2\}$ be any strategy and α be the *a priori* probability of the object being in the first state. Let $R(q,\alpha)$ be the probability of a wrong decision for the state k. This probability depends on the probability of the first state α and on the strategy q used. Thus the probability $R(q,\alpha)$ is

$$R(q,\alpha) = \alpha \int\limits_{\{x|q(x)=2\}} \frac{1}{\sqrt{2\pi}} \mathrm{e}^{-\frac{1}{2}(x-1)^2}\,\mathrm{d}x + (1-\alpha) \int\limits_{\{x|q(x)=1\}} \frac{1}{\sqrt{2\pi}} \mathrm{e}^{-\frac{1}{2}(x+1)^2}\,\mathrm{d}x\,.$$

By the symbol $q(\alpha)$ a Bayesian strategy will be denoted which could be created if the probability α was known and the probability of the wrong decision was minimised. This means that

$$q(\alpha) = \operatorname*{argmin}_{q' \in Q} R(q',\alpha)\,,$$

where Q is the set of all possible strategies. Let q^* denote the strategy which decides that the object is in the first state if $x \geq 0$, and in the opposite case it decides for the second state. Even though it is rather obvious that the strategy q^* is a minimax one, and in this sense an optimal one, we will briefly prove this statement.

The probability of the wrong decision $R(q^*,\alpha)$ is

$$\begin{aligned} R(q^*,\alpha) &= \alpha \int\limits_{-\infty}^{0} \frac{1}{\sqrt{2\pi}} \mathrm{e}^{-\frac{1}{2}(x-1)^2}\,\mathrm{d}x + (1-\alpha) \int\limits_{0}^{\infty} \frac{1}{\sqrt{2\pi}} \mathrm{e}^{-\frac{1}{2}(x+1)^2}\,\mathrm{d}x \\ &= \alpha \int\limits_{-\infty}^{-1} \frac{1}{\sqrt{2\pi}} \mathrm{e}^{-\frac{1}{2}x^2}\,\mathrm{d}x + (1-\alpha) \int\limits_{1}^{\infty} \frac{1}{\sqrt{2\pi}} \mathrm{e}^{-\frac{1}{2}x^2}\,\mathrm{d}x\,. \end{aligned}$$

Since the integrals

$$\int\limits_{-\infty}^{-1} \frac{1}{\sqrt{2\pi}} \mathrm{e}^{-\frac{1}{2}x^2}\,\mathrm{d}x \qquad \text{and} \qquad \int\limits_{1}^{\infty} \frac{1}{\sqrt{2\pi}} \mathrm{e}^{-\frac{1}{2}x^2}\,\mathrm{d}x$$

are the same (they are approximately equal to 0.16), the number $R(q^*,\alpha)$ does not depend on the probability α, and thus it holds

$$R(q^*,\alpha) = \max_{\alpha} R(q^*,\alpha) \cong 0.16\,. \tag{6.1}$$

On the other hand the strategy q^* is identical with the Bayesian strategy $q(0.5)$ which is based on the assumption that *a priori* probabilities of states are the same. This means that for any strategy q which is not identical with the strategy q^* the following inequality holds

$$R(q,\ 0.5) > R(q^*,\ 0.5) \tag{6.2}$$

which is strictly satisfied in the case examined. By joining the equality (6.1) and the inequality (6.2) together

$$\max_\alpha R(q,\alpha) \geq R(q,\ 0.5) > R(q^*,\ 0.5) = \max_\alpha R(q^*,\alpha)$$

we obtain

$$\max_\alpha R(q,\alpha) > \max_\alpha R(q^*,\alpha) \approx 0.16$$

which means that the strategy q^* has the following two features.

1. The strategy q^* makes a wrong decision about the state k with the probability 0.16 independently of what the *a priori* probabilities of the states are.
2. Any other strategy does not have this property any longer, and moreover, for every other strategy there exist such *a priori* probabilities of the states in which the probability of the wrong decision is greater than 0.16.

In the above sense the strategy q^* is optimal in a situation in which the *a priori* probabilities of the object states are arbitrary. However, if the *a priori* probabilities were known, i.e., the probability α was known then the strategy $q(\alpha)$ could be built in the following form. The strategy decides for the first state when

$$\frac{\alpha}{\sqrt{2\pi}}\,\mathrm{e}^{-\frac{1}{2}(x-1)^2} \geq \frac{1-\alpha}{\sqrt{2\pi}}\,\mathrm{e}^{-\frac{1}{2}(x+1)^2}\,,$$

or (which is the same)

$$x \geq \frac{1}{2}\ln\frac{1-\alpha}{\alpha}\,. \tag{6.3}$$

In the opposite case the strategy makes a decision that the object is in the second state.

The probability of a wrong decision $R\big(q(\alpha),\alpha\big)$ provided by this strategy is 0.16 only at $\alpha = 0.5$ and at any other value α it is smaller. Fig. 6.2 shows how the probabilities $R\big(q(\alpha),\alpha\big)$ and $R(q^*,\alpha)$ depend on the *a priori* probability α. The strategy $q(\alpha)$ applies knowledge about the probability α. The strategy q^* is a minimax strategy built up without knowing this probability. Fig. 6.2 once more illustrates that the quality of recognition through the strategy q^* does not depend on α and constantly remains on the level 0.16. It is a level, however, that is reached through the Bayesian strategy only in the worst case.

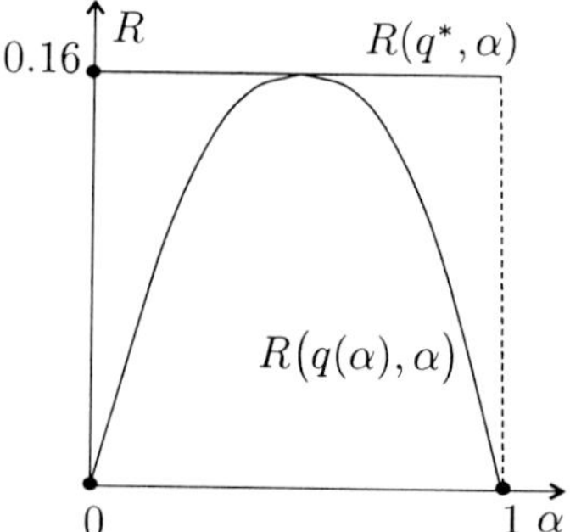

Figure 6.2 The dependence of a probability of a wrong decision R on the *a priori* probability α for the Bayesian and minimax strategy.

The facts described evoke a feeling of sharp dissatisfaction. Why actually, when the facts are not known, the strategy should be oriented particularly to the worst case? Well, the reality need not always be the worst one. The strategy q^* described behaves as if no reality existed other than the worst one all the time. Is it possible to find such a way of classifying the observation

x, that would not use the explicit knowledge about *a priori* probabilities, but yield wrong answers with the probability 0.16 only in the case in which the *a priori* probabilities are the worst ones? Is it possible to attain better quality recognition in the case in which the reality is better than the worst one? Or, asked in a quite sharp form now: is there a possible strategy which would not be worse than the Bayesian strategy, but would not be based on the complete knowledge about the *a priori* probabilities of states?

The answer to this question be it in one or another form is indubitably negative. The Bayesian strategy $q(\alpha)$ substantially depends on *a priori* probabilities (it is evident, e.g., from the expression (6.3)). Therefore a strategy which is independent of *a priori* probabilities cannot be identical with all Bayesian strategies which do depend on *a priori* probabilities.

Notice that the reasoning by virtue of which we arrived at the negative answer is nearly the same as the reasoning used in proving that the perceptron by itself was not able to get at a correct classification. In both cases the negative answer is based on an evident fact that a constant object cannot ever be identical with some other object that is varying. The significance of Robbins' approach which Neyman [Neyman, 1962] ardently valued as a breakthrough in the Bayesian front is in that he did not try to reply with a positive answer to a question for which a negative answer was nearly evident. Instead of that he changed the question, from the practical point of view rather slightly, but in such a way that the reply to the modified question ceased to be so obviously negative.

Imagine that the task of estimating the state of an object is not to be solved only once, but many times in some n moments $i = 1, 2, \ldots, n$. Assume as well that the state of the object is random at these moments. The sequence $k_1, k_2, \ldots, k_n$ consists of mutually independent random elements k_i. The probability of the event $k_i = 1$ is equal to the probability α which we do not know. But the *a priori* probability α is known to be the same for all moments of observation. Let us also imagine that the classifier need not decide about the state k_1 at once at the first moment in which only the first observation x_1 is known. The decision can be delayed, and the decision made only when the entire sequence $x_1, x_2, \ldots, x_n$ is available. In this case the state k_1 is evaluated not only on the basis of one single observation x_1, but on the basis of the entire sequence $x_1, x_2, \ldots, x_n$. In the same way if the entire observation sequence is known then a decision can be made about the state k_2 at the second moment, and then that about the states k_3, k_4, etc.. In this case strategies of the form $X \to K$ need not be referred to, but it concerns strategies of a more general form $X^n \to K^n$.

Let us repeat now the old question, but modified after H. Robbins: Is there a strategy of the form $X^n \to \{1, 2\}^n$ which does not use *a priori* probabilities of states, and is not worse than the Bayesian strategy of the form $X \to \{1, 2\}$ which uses such information? Now it is not quite evident that the answer to the question in such a formulation must be negative.

Robbins proposed a specific strategy which proves that at least with large values of n the answer to this question is positive. The strategy decides for the state 1 if

$$x_i \geq \frac{1}{2} \ln \frac{n - \sum_{i=1}^{n} x_i}{n + \sum_{i=1}^{n} x_i}\,, \tag{6.4}$$

and in the opposite case, it decides for the state 2. As our case is concerned, for $n \to \infty$ the strategy (6.4) becomes arbitrarily close to the strategy (6.3) in the sense that the random quantity in the right-hand side of (6.4) converges to a constant given in the right-hand side of (6.3). Indeed, it holds that

$$\lim_{n\to\infty} \frac{n - \sum_{i=1}^{n} x_i}{n + \sum_{i=1}^{n} x_i} = \lim_{n\to\infty} \frac{1 - \frac{1}{n}\sum_{i=1}^{n} x_i}{1 + \frac{1}{n}\sum_{i=1}^{n} x_i} = \frac{1 - \lim\limits_{n\to\infty} \frac{1}{n}\sum_{i=1}^{n} x_i}{1 + \lim\limits_{n\to\infty} \frac{1}{n}\sum_{i=1}^{n} x_i}\,. \tag{6.5}$$

With respect to $\lim\limits_{n\to\infty} \frac{1}{n}\sum_{i=1}^{n} x_i$ is the mathematical expectation $E(x)$ of the random variable x and since this mathematical expectation in the first state is 1 and in the second state it is -1, it holds that $E(x) = \alpha \cdot 1 + (1-\alpha)(-1) = 2\alpha - 1$. Thus we can resume deriving (6.5) which has been interrupted

$$\lim_{n\to\infty} \frac{n - \sum_{i=1}^{n} x_i}{n + \sum_{i=1}^{n} x_i} = \frac{1 - (2\alpha - 1)}{1 + (2\alpha - 1)} = \frac{2\,(1 - \alpha)}{2\,\alpha} = \frac{1 - \alpha}{\alpha}\,.$$

Robbins explains that the strategy (6.4) is the result of the double processing of the input observation sequence which yields information of two different types. On the one hand each observation x_i is a random variable which depends on the state k_i of the object at the i-th moment, and provides certain information on the state. On the other hand, the sequence $x_1, x_2, \ldots, x_n$ on the whole is a sequence of random samples of the population. The probability distribution p_X on this population depends on the *a priori* probabilities $p_K(1) = \alpha$ and $p_K(2) = 1 - \alpha$, since

$$p_X(x) = \frac{p_K(1)}{\sqrt{2\pi}}\, e^{-\frac{1}{2}(x-1)^2} + \frac{p_K(2)}{\sqrt{2\pi}}\, e^{-\frac{1}{2}(x+1)^2}\,.$$

Therefore the sequence $x_1, x_2, \ldots, x_n$ provides certain information on unknown *a priori* probabilities too. The sequence is to be processed in two passes:

1. *A priori* probabilities are estimated that are not known in advance.
2. The result of this more or less approximate estimate is used for the decision on the states $k_1, k_2, \ldots, k_n$ with the help of the Bayesian strategy as if the estimated values of the *a priori* probabilities are the true ones.

The strategy (6.4) was formed on the basis of indirect evaluation of *a priori* probabilities. It starts from the fact that the mathematical expectation $E(x)$ of the observation x uniquely determines unknown probabilities so that

$$p_K(1) = \frac{1}{2}\Big(1 + E(x)\Big)\,, \qquad p_K(2) = \frac{1}{2}\Big(1 - E(x)\Big)\,,$$

and that the number $\frac{1}{n}\sum_{i=1}^{n} x_i$ at sufficiently large n is an acceptable estimate of the mathematical expectation $E(x)$.

Such an indirect procedure is not the best one. Robbins formulated the task seeking such values of *a priori* probabilities $p_K^*(k)$, $k \in K$, which maximise the probability of the multi-set $x_1, x_2, \ldots, x_n$, i.e., they are

$$\begin{aligned}\left(p_K^*(k),\ k \in K\right) &= \operatorname*{argmax}_{(p_K(k)\,|\,k\in K)} \prod_{i=1}^{n} \sum_{k\in K} p_K(k)\, p_{X|K}(x_i\,|\,k) \\ &= \operatorname*{argmax}_{(p_K(k)\,|\,k\in K)} \sum_{i=1}^{n} \log \sum_{k\in K} p_K(k)\, p_{X|K}(x_i\,|\,k)\,. \end{aligned} \tag{6.6}$$

Robbins task (6.6) remained unsolved for rather a long time. Its exact solution was found only in pattern recognition by means of clever algorithms which rather strongly resemble Rosenblatt's algorithms, and which were proposed in pattern recognition as a model of unsupervised learning.

Before introducing and proving algorithms which solve Robbins task in this lecture we would like to say that Robbins approach should not be considered in a limited manner in any case. Our explanation set forth just a recommendation for a concrete situation with one-dimensional Gaussian random quantities which Robbins described only as an illustration of his approach. The formulation (6.6) itself is far more general. Robbins' approach understood in a far more general way can be used even in those situations which, formally speaking, do not belong to the framework of the formulation (6.6). We will present an example of this situation. For this purpose we will repeat all Robbins' considerations that led to the formulation (6.6), and then formulate the task which formalises Robbins' approach in its full generality.

Let us go back to the example with one-dimensional Gaussian variables. This time let us assume that the *a priori* probability of each state is known and it has the value 0.5 for each state. On the other hand conditional mathematical expectations are not known under the condition that the object is in one or the other state. It is only known:

- If the object is in the first state then the mathematical expectation lies in the interval from 1 to 10;
- If the object is in the second state then the mathematical expectation is a number in the interval from -1 to -10.

In this example as well as in the previous one, the best strategy in the minimax sense is the one which decides $q^*(x) = 1$ if $x \geq 0$, and $q^*(x) = 2$ in the opposite case. The strategy q^* is oriented to the worst case, when the first mathematical expectation is 1 and the second is -1. The actual situation, however, can be better than the worst one. The first mathematical expectation can be, for example, 1 and the second -10. For this case a far better strategy than q^* can be thought of. This possibility can be used if we do not start recognising the observations x immediately, but wait for some time until a sufficient number of observations for recognition has been accumulated. The accumulated observations x are first used for the more exact estimation of the mathematical expectations, and only then are the observations classified all at once.

It is possible to create further more complex examples since the 'breakthrough in the Bayesian front' has already set in. The tasks that occur in all situations of this kind are expressed in the following *generalised form*.

Let X and K be two sets. Their Cartesian product forms a set of values of a random pair (x, k). The probability of the pair (x, k) is determined by the function $p_{XK}: X \times K \to \mathbb{R}$. The function p_{XK} is not known, but a set $\mathcal{P}$ which contains the function p_{XK} is known. In addition, the sequence $x_1, x_2, \ldots, x_n$ of random, mutually independent observations x is known. Its probability distribution is

$$\sum_{k \in K} p_{XK}(x, k) .$$

The objective is to find such a distribution p^*_{XK} which belongs to the set $\mathcal{P}$ and for which the occurrence probability of the abovementioned multi-set $x_1, x_2, \ldots, x_n$ is the greatest. Therefore

$$p^*_{XK} = \operatorname*{argmax}_{p_{XK} \in \mathcal{P}} \sum_{i=1}^{n} \log \sum_{k \in K} p_{XK}(x_i, k) . \tag{6.7}$$

The solution of the formulated task allows construction of pattern recognition devices which are capable of enhancing their strategy of recognition only through the analysis of patterns submitted for recognition, without any additional teacher's information on the states the object was in during the observations. The task formally expresses intuitive ideas on unsupervised learning which were already described in examining the perceptron. We will see in Section 6.6 that even the algorithms solving this task resemble those by Rosenblatt. Prior to the long expected liaison between Rosenblatt's algorithms and Robbins tasks, Rosenblatt's algorithms lived their own life, and they met the long known clustering tasks. In the following section, we will show the fruitful outcomes of this meeting.

6.5 Quadratic clustering and formulation of a general clustering task

At the end of Section 6.3 we presented a class of algorithms which we called Rosenblatt's unsupervised learning algorithms. We did not formulate the algorithms in a sufficiently accurate way, but did it by means of four schematic diagrams in Fig 6.1. The unsupervised learning algorithms consist, as do the supervised learning algorithms, of two parts: learning and classification. But these two parts are in the cases of supervised learning and unsupervised learning connected in a different way. In the case of supervised learning the input of the learning algorithm obtains information from the teacher on a certain correct classification of an input observation. In the case of unsupervised learning the teacher's information is replaced by the results of the classification itself. Due to Rosenblatt's research a hypothesis arose that by feeding the results of intrinsic classification to the input of the learning algorithm some reasonable data processing tasks could be solved.

This hypothesis was so strong that it could hardly be shaken by a disaster caused by the undoubtedly negative results of experiments with perceptron learning. The negative results could also have been accounted for by the assumption that the perceptron algorithms applied for classification and learning had not been well set up. Recall that a perceptron was not able to operate in the case in which the two mentioned algorithms were applied in supervised learning. Classification and learning algorithms of that sort continued to be objects of experiments until an applicable classifier was successfully created. As will be seen the experiments were worth making because they successfully resulted in creating a new algorithm for unsupervised learning which is known under the name of ISODATA.

Let us look at the following statistical model of an object, which is perhaps one of the simplest. Let the object be in one of two possible states with the same probabilities, i.e., $K = \{1, 2\}$, $p_K(1) = p_K(2) = 0.5$. Let x be an n-dimensional vector of features. As long as the object is in the first state, x is a random Gaussian vector with mathematical expectation μ_1. Components of the vector x are mutually independent, the variance of each component being 1. The vector x has the same properties under the condition that the object is in the second state, with one single difference that its mathematical expectation is μ_2 in this case.

The Bayesian strategy which on the basis of the observation x decides on the state k with the least probability of the wrong decision, selects the first state if

$$(x - \mu_1)^2 \le (x - \mu_2)^2 , \tag{6.8}$$

and in the opposite case it selects the second state. This strategy will be denoted by the symbol q. If the mathematical expectations μ_1 and μ_2 are not known then the strategy is considered as a varying strategy $q(\mu_1, \mu_2)$. This is fully determined if the vectors μ_1 and μ_2 are defined.

The learning algorithm has to find the maximum likelihood estimate μ_1^*, μ_2^* of mathematical expectations μ_1, μ_2 on the basis of the multi-set (x_1, k_1), (x_2, k_2), ...,(x_m, k_m) of random pairs (x, k) with the probability density

$$p_{XK}(x, k) = p_K(k)\, \frac{1}{\left(\sqrt{2\pi}\right)^n}\, \mathrm{e}^{-\frac{1}{2}(x-\mu_k)^2} .$$

The most likely values are

$$\begin{aligned}
(\mu_1^*, \mu_2^*) &= \operatorname*{argmax}_{(\mu_1,\mu_2)} \sum_{i=1}^{m} \log\left(p_K(k_i)\, \frac{1}{\left(\sqrt{2\pi}\right)^n}\, \mathrm{e}^{-\frac{1}{2}(x_i-\mu_{k_i})^2} \right) \\
&= \operatorname*{argmax}_{(\mu_1,\mu_2)} \sum_{i=1}^{m} -(x_i - \mu_{k_i})^2 = \operatorname*{argmin}_{(\mu_1,\mu_2)} \left(\sum_{i\in I_1} (x_i - \mu_1)^2 + \sum_{i\in I_2} (x_i - \mu_2)^2 \right) \\
&= \left(\operatorname*{argmin}_{\mu_1} \sum_{i\in I_1} (x_i - \mu_1)^2 ,\ \operatorname*{argmin}_{\mu_2} \sum_{i\in I_2} (x_i - \mu_2)^2 \right) \\
&= \left(\frac{1}{|I_1|} \sum_{i\in I_1} x_i ,\ \frac{1}{|I_2|} \sum_{i\in I_2} x_i \right) ,
\end{aligned} \tag{6.9}$$

where I_k, $k = 1, 2$, denotes the sets of indices i for which $k_i = k$ holds.

Authors of the ISODATA algorithm, Hall and Ball [Ball and Hall, 1967], joined the strategy (6.8) and the learning algorithm (6.9) in such a way that they created the following unsupervised learning algorithm. The algorithm examines the multi-set of input observations $x_1, x_2, \ldots, x_n$ many times. The parameters μ_1 and μ_2 change after each pass through the data in the general case, and thus the results of classification are also changed.

The values μ_1^0 and μ_2^0 can be nearly arbitrary before the first passage (the right-hand superscript denotes the number of iteration). Only the strategy $q(\mu_1^0, \mu_2^0)$ in the analysis of the input sequence is prohibited from placing all observations into one class. The initial values can, for example, be $\mu_1^0 = x_1$ and $\mu_2^0 = x_2$. If after the examination step number $(t-1)$ the vectors μ_1 and μ_2 were assuming the values $\mu_1^{(t-1)}$ and $\mu_2^{(t-1)}$ then in the step t of the analysis two procedures will be performed.

1. *Classification.* The observation x_i, $i = 1, \ldots, n$, is placed into the first class if

$$\left(x_i - \mu_1^{(t-1)}\right)^2 \le \left(x_i - \mu_2^{(t-1)}\right)^2 ,$$

   and into the second class in the opposite case. The result of the procedure is the decomposition of the set of indices $i = 1, 2, \ldots, m$ into two subsets $I_1^{(t)}$ and $I_2^{(t)}$.
2. *Learning.* New values $\mu_1^{(t)}$ and $\mu_2^{(t)}$ are calculated as the average of vectors included in the first and second classes which is

$$\mu_k^{(t)} = \frac{1}{|I_k^{(t)}|} \sum_{i \in I_k^{(t)}} x_i , \quad k = 1, 2 .$$

To create and experimentally examine this algorithm required courage since after unsuccessful experiments in the perceptron with unsupervised learning the problems of unsupervised learning seemed never to recover. Courage was rewarded at last by the following, experimentally attained, positive results.

1. The algorithm converges after a finite number of steps, independently of the initial values μ_1^0 and μ_2^0, to the state in which the values μ_1 and μ_2 do not change. The values μ_1 and μ_2 and the result of the classification I_1 and I_2 are the solution of the system of equations

$$\left.\begin{aligned} \mu_1 &= \frac{1}{|I_1|} \sum_{i \in I_1} x_i , \\ \mu_2 &= \frac{1}{|I_2|} \sum_{i \in I_2} x_i , \\ I_1 &= \{ i \mid (x_i - \mu_1)^2 \le (x_i - \mu_2)^2 \} , \\ I_2 &= \{ i \mid (x_i - \mu_1)^2 > (x_i - \mu_2)^2 \} . \end{aligned}\right\} \tag{6.10}$$

2. The system of equations (6.10) can have more than one solution. With one of them the classification I_1, I_2 of the input sequence $x_1, x_2, \ldots, x_n$ is rather close to the one which would be attained in supervised learning. We

will call it a good solution. Unfortunately the algorithm does not converge necessarily to the good solution. But if the length of the sequence n is large enough as well as the ratio of the distance $|\mu_1 - \mu_2|$ to the variance then the convergence of the algorithm to the good solution becomes quite probable.

The algorithm quoted finds, at least in the abovementioned particular case, the correct classification of the input observations only on the basis of an analysis of the input observations without any additional information from the teacher. Unfortunately, the algorithm could not be successfully enhanced or generalised to such an extent to yield similar desirable results in other cases. For example, when the *a priori* probabilities of the states are not equal, or when the variances are not the same, etc.. But it has been found that even without an enhancement, the system of equations (6.10) is usable. It expresses a certain understandable task which is, however, completely different from the recognition task concerning the unknown state k of the object on the basis of the observations x. The formulated task belongs to a class of tasks of the following form.

Let x be a random object from the set of objects X with the probability distribution $p_X : X \to \mathbb{R}$. Let D be a set elements of which will be referred to as *decisions*, such that for each object $x \in X$ a certain decision $d \in D$ is assigned. Let $W : X \times D \to \mathbb{R}$ be a penalty function value $W(x, d)$ of which represents losses in the case in which for the object x the decision d is chosen. Note that this penalty function is of a quite different form than the penalty function in Bayesian recognition tasks. With a firmly chosen decision d the penalty here depends on the known observation x and not on an unobservable state. The term unobservable state does not occur in this construction at all.

Assume preliminarily that the aim of the task is to construct the strategy $q: X \to D$ which minimises the value

$$\sum_{x \in X} p(x)\, W\big(x, q(x)\big)\,,$$

i.e., the mathematical expectation of losses. It can be seen that the task in this formulation is solved by the strategy

$$q(x) = \operatorname*{argmin}_{d \in D} W(x, d)\,.$$

The strategy $q(x)$ does not depend on the probability distribution p_X and for some forms of the penalty function W it can be easily found. But the task becomes substantially more complicated if the strategy q must satisfy a constraint of the following form. The strategy is to assume only the assigned number of values on the set X which are, for example, only two values d_1 and d_2. But these are, unfortunately, not known beforehand, and are to be chosen in an optimal way. The strategy q should not, therefore, have the form $q: X \to D$, but it is given by the representation $q: X \to \{d_1, d_2\}$. If the values d_1, d_2 were known then the strategy q^* would decide for $q^*(x) = d_1$ if

$W(x, d_1) \le W(x, d_2)$, and for $q^*(x) = d_2$ in the opposite case. The penalty would be evidently equal to

$$W\big(x, q^*(x)\big) = \min\big(W(x, d_1), W(x, d_2)\big) .$$

If the values d_1, d_2 are not known then the aim is to find d_1^* and d_2^* which minimise the value

$$\sum_{x \in X} p(x) \min\big(W(x, d_1), W(x, d_2)\big) . \tag{6.11}$$

The solution of this task does not lie in the found decisions d_1^* and d_2^*, but also in the decomposition of the set X into the subsets X_1 and X_2. The decomposition means that afterwards objects of the same set X_i will be handled in the same way, and thus a certain diversity of objects can be ignored. Let us see two examples of such tasks.

Example 6.2 Locating water tanks. *Let us imagine a village in which the water supply was damaged and must be compensated for by several drinking-water tanks. Let x be the location of a house in the village, X be the set of points in which the houses are placed, and $p_X(x)$ be the number of inhabitants in a house placed at point x. Let $W(x, d)$ evaluate the loss of an inhabitant from the house at point x who fetches water from the tank at point d.*

The strategy $q\colon X \to D$ which should without additional constraints set down the position $q(x)$ of the water tank for the inhabitant from the house at point x would be simple. For each x the position of the tank $q(x)$ would be identical with x. It is quite natural, since it would be best for each inhabitant to have the tank next to his/her house.

If however the constraint had to be taken into account that the whole locality could have only two water tanks at their disposal then a far more difficult task of the sort in (6.11) would be created. First a decision must be made in which points d_1 and d_2, the tanks, are to stand. Then an easier task is to be solved which will classify the locality into two subclasses so that each house could be assigned to either of the two tanks. ▲

Example 6.3 Clustering of satellite images. *Let X be a set of images of the Martian surface scanned by a satellite orbiting around the planet. Let $x \in X$ be a random image and let the probability $p_X(x)$ of the occurrence of the image x in the set X be known for each image x. Imagine that the image x is evaluated by a device on board the satellite which is to pass the information over to a device on Earth. Assume that because of an insufficient transmission capacity within the communication channel it is not possible to transmit each image x to Earth. But if there is some knowledge about the purpose for which the images of Mars' surface are to serve, the demands on the transmission channel can be softer.*

In this situation a set of fewer images $d_1, d_2, \ldots, d_m$ can be created which will be known to both the device on the satellite and the device on the ground. A loss function $W(x, d)$ will be introduced indicating the damage caused by receiving *an image d instead of the actual desired image x. The information transmission would now look like this: in the satellite the observed image x is replaced by one of the images $d_1, d_2, \ldots, d_m$. To the device on the Earth only the number $i = 1, 2, \ldots, m$ defining the image will be passed over. The device on the Earth then acts as if the image was d_i instead of the actual observed image x. The loss $W(x, d_i)$ is brought about. A task was created here with the aim to select a set $\{d_1, d_2, \ldots, d_m\}$ of images which can best replace the set X in the sense of the task (6.11).* ▲

In these examples as well as in the general formulation (6.11) we can see that the task is composed in such a way that its solution can lead to a useful and reasonable classification of the observation set x into subsets. In this respect the task resembles the tasks of statistical pattern recognition, where the final result is the classification of the observation set as well. But the nature of the mentioned tasks is substantially different.

In the task of statistical pattern recognition the existence of a certain factor k is assumed which affects the observed parameters x. The factor is not, however, immediately observable, but by means of a demanding provision one can, sooner or later, obtain some knowledge about it. The classification which is the solution of these kind of tasks can be regarded to be good or bad, according to what extent a correct estimate is obtained of what is not known, but what actually exists.

In the task presented now in the form (6.11) and illustrated by the two examples no actually existing factor is assumed. The aim of the classification is completely different. For the user's convenience a rather large set of objects is classified into subsets so that afterwards the objects from the same subset are managed in the same way. It results in that the differences between some objects are ignored and quality of managing the objects becomes worse. The problem is to classify the objects so that the loss in quality will be as small as possible.

Informally, the tasks of this form have been long known as *clustering* tasks, taxonomy tasks, classification tasks, etc.. The formulation (6.11) can be regarded as a formal expression of such tasks. The ISODATA algorithm is an algorithm that solves a particular case of this task in which $W(x, d) = (x - d)^2$.

Experimentally discovered properties showing that the clustering algorithms approach the Bayesian classification of input observations on some conditions (e.g., in the case of Gaussian shapes of conditional probability distributions), is a mere coincidence. The situation would be the same if someone tried to create an algorithm for calculating the function $\log_2 x$, but by chance had an algorithm implementing the function $\sqrt{x}$ at his/her disposal. The first experiments in which the algorithm had been used for the values not much different from $\sqrt{4} = 2 = \log_2 4$, would encourage him/her. But later it would be found out that this algorithm could be regarded neither as an approximate calculation, nor as the calculation of the function $\log_2 x$, but that it was a precise algorithm for calculating a quite different function, even if the function was also necessary and perhaps beautiful.

We have seen that Rosenblatt's algorithms asserted themselves for solving certain clustering tasks which are something quite different from the unsupervised learning tasks as were formulated in Section 6.2. For solving unsupervised learning tasks Rosenblatt's tasks had to be somewhat modified. As will be seen in the following section this modification is very slight on the one hand, but rather substantial on the other.

6.6 Unsupervised learning algorithms and their analysis

Let X and K be two sets. The probability distribution p_{XK} is defined on Cartesian product of X and K. To avoid certain theoretical complications which are not important here we will assume as before that the set X and the set K are finite.

6.6.1 Formulation of a recognition task

Let function p_{XK} be known and a sequence of observations $(x_1, x_2, \ldots, x_n)$ be submitted to recognition. The recognition of this sequence cannot be considered as creating the sequence $k_1, k_2, \ldots, k_n$ because the observation x does not unambiguously determine the state k. In the general case each value x can occur at different states of the object, but with different probabilities. Therefore, if the penalty function is not determined, and if a Bayesian evaluation of the state k is not sought then the knowledge of the state can be expressed only by calculating the *a posteriori* probabilities $\alpha(i,k)$, $k \in K$, that the observed object was in the state k at the moment at which x_i was being observed,

$$\alpha(i,k) = \frac{p_{XK}(x_i,k)}{\sum\limits_{k\in K} p_{XK}(x_i,k)} = \frac{p_K(k)\,p_{X|K}(x_i\,|\,k)}{\sum\limits_{k\in K} p_K(k)\,p_{X|K}(x_i\,|\,k)}\,. \tag{6.12}$$

The ensemble of quantities $\alpha(i,k)$, $i = 1,2,\ldots,n$, $k \in K$, will be regarded as the result of recognition.

6.6.2 Formulation of a learning task

The values $p_{XK}(x,k)$ of the function $p_{XK}\colon X \times K \to \mathbb{R}$ mean a joint probability of the observation x and the state k. The function p_{XK} unambiguously determines a pair of functions $p_K\colon K \to \mathbb{R}$ and $p_{X|K}\colon X \times K \to \mathbb{R}$. The former means the distribution of *a priori* probabilities of the states k. The latter expresses the distribution of conditional probabilities of the observation x under the condition of the state k. For further explanation we will change the notation for these conditional probabilities in such a way that they will not be denoted by means of a single function $p_{X|K}\colon X \times K \to \mathbb{R}$ of two variables x and k, but by means of $|K|$ functions of the form $p_{X|k}\colon X \to \mathbb{R}$ of one variable x. Functions from this group will be indexed by k which uses values from the set K. The conditional probability $p_{X|k}(x)$ denotes the observation x under the condition the object is in the state k.

Assume that *a priori* probabilities $p_K(k)$ are unknown and the conditional probabilities $p_{X|k}(x)$ for each $k \in K$ are known up to the value of a parameter

$a \in A$, where A is a set of values for the unknown parameter. We will express this incomplete knowledge by means of the function $p\colon X \times A \to \mathbb{R}$ of two variables x and a. This function represents the partial knowledge of a statistical model in the sense that for every $k \in K$ such a value a_k exists (not known beforehand) that the conditional probability $p_{X|k}(x)$ is equal to $p(x, a_k)$. The ensemble of all unknown statistical parameters of an object will be denoted by m and will be referred to as the *statistical model of an object*, or briefly the model; $m = \big((p_K(k), a_k),\ k \in K\big)$. The model m unambiguously determines the probabilities $p_{XK}(x, k)$ which are necessary for recognition.

Assume that a multi-set (x_1, k_1), (x_2, k_2), ..., (x_n, k_n) of random and mutually independent pairs (x, k) is at our disposal. Each pair has occurred in agreement with the probability distribution p_{XK} which is unknown. However, it can be found how the probability of the multi-set (x_1, k_1), (x_2, k_2), ..., (x_n, k_n) depends on the model m, i.e on the *a priori* probabilities $p_K(k)$ and the values a_k, $k \in K$. This probability is

$$l(m) = \prod_{i=1}^{n} p_{XK}(x_i, k_i) = \prod_{i=1}^{n} p_K(k_i) \prod_{i=1}^{n} p_{X|k_i}(x_i) = \prod_{i=1}^{n} p_K(k_i) \prod_{i=1}^{n} p(x_i, a_{k_i}) \,.$$

If we denote by the symbol $L(m)$ the logarithm of the number $l(m)$, we will obtain

$$L(m) = \sum_{i=1}^{n} \log p_K(k_i) + \sum_{i=1}^{n} \log p(x_i, a_{k_i}) \,.$$

The information on the states $k_1, k_2, \ldots, k_n$ received from the teacher can be expressed by an ensemble of numbers $\alpha(i, k)$, i=1,...,n, $k \in K$. Here $\alpha(i, k)$ is equal to 1, if $k = k_i$ and it is equal to 0 in any other case. Applying the denotation introduced here, the function $L(m)$ assumes the form

$$L(m) = \sum_{i=1}^{n} \sum_{k \in K} \alpha(i, k) \log p_K(k) + \sum_{i=1}^{n} \sum_{k \in K} \alpha(i, k) \log p(x_i, a_k) \,. \tag{6.13}$$

Through the *learning task* the search for such a model $m = \big((p_K(k), a_k),\ k \in K\big)$ is understood in which the maximum of the logarithm of probability $L(m)$ is achieved. The values $p_K(k)$, $k \in K$, must of course satisfy the condition that their sum $\sum_{k \in K} p_K(k)$ is 1. It is not difficult to concede (we will return to it, anyhow) that the best estimate of the probability $p_K(k)$ is

$$p_K(k) = \frac{\sum_{i=1}^{n} \alpha(i, k)}{n} \,, \quad k \in K \,, \tag{6.14}$$

and the task concerning the maximisation of $L(m)$ with respect to the ensemble $(a_k,\ k \in K)$ is reduced to $|K|$ independent tasks. The following value is to be maximised in each of them

$$\sum_{i=1}^{n} \alpha(i, k) \log p(x_i, a_k)$$

with respect to the value a_k. Thus, we can write

$$a_k = \operatorname*{argmax}_a \sum_i \alpha(i,k) \log p(x_i, a) . \tag{6.15}$$

Note that the learning task formulated in this way is defined not only for a case of a *fully informed teacher* (supervisor) who at each observation x_i correctly indicates the actual state of the object. The values $\alpha(i,k)$ need not always be ones or zeroes, but they can be any real numbers within the interval $0 \le \alpha(i,k) \le 1$ sum $\sum_{k \in K} \alpha(i,k)$ of which is 1 for each i. In this manner the information from an *incompletely informed teacher* (supervisor) who does not precisely know the actual state of the object, but who knows, from some source, the probabilities of each state, can be expressed. And this is information of the same form as that provided by a recognition device as was defined in the previous paragraph.

6.6.3 Formulation of an unsupervised learning task

Let $p_{XK}(x,k) = p_K(k)\, p_{X|k}(x)$, $x \in X$, $k \in K$, be a joint probability of the observation x and the state k. *A priori* probabilities $p_K(k)$ of states are unknown, and conditional probabilities $p_{X|k}(x)$ are known up to the value of a certain parameter a_k. This means that a certain function $p\colon X \times A \to \mathbb{R}$ is known such that at a certain value a_k of the parameter a, the conditional probability $p_{X|k}(x)$ is equal to the number $p(x, a_k)$. Let $x_1, x_2, \ldots, x_n$ be a multi-set of mutually independent observations whose probability distribution is

$$p_X(x) = \sum_{k \in K} p_{XK}(x,k) = \sum_{k \in K} p_K(k)\, p(x, a_k) .$$

The probability of this multi-set is

$$l(m) = \prod_{i=1}^{n} \sum_{k \in K} p_K(k)\, p(x_i, a_k) . \tag{6.16}$$

The probability $l(m)$ is thus dependent on *a priori* probabilities $p_K(k)$, $k \in K$, and on the ensemble of values a_k, $k \in K$. The logarithm of this probability is

$$L(m) = \sum_{i=1}^{n} \log \sum_{k \in K} p_K(k)\, p(x_i, a_k) . \tag{6.17}$$

The unsupervised learning task is formulated as a search for such an ensemble of values of *a priori* probabilities and unknown parameters

$$m^* = \big(p_K^*(k),\ a_k^* \,|\, k \in K\big)$$

which maximises the expression (6.17). This means that the probability of occurrence of the observed multi-set $x_1, x_2, \ldots, x_n$ is maximised.

In a particular case in which only maximum likelihood estimates of *a priori* probabilities $p_K(k)$ are to be found, the task being formulated becomes Robbins task.

6.6.4 Unsupervised learning algorithm

Let $m^0 = \big(p_K^0(k), a_k^0 \mid k \in K\big)$ be an initial ensemble of unknown parameters. The unsupervised learning algorithm is an algorithm which stepwise builds a sequence of models $m^1, m^2, \dots, m^t, m^{t+1}, \dots$ according to the following rules.

Let $m^t = \big(p_K^t(k), a_k^t \mid k \in K\big)$ be an ensemble which is obtained at step t of the unsupervised learning algorithm. The next model m^{t+1} is calculated in two stages. The following numbers are calculated in the first stage

$$\alpha^t(i,k) = \frac{p_K^t(k)\, p(x_i, a_k^t)}{\sum\limits_{k' \in K} p_K^t(k')\, p(x_i, a_{k'}^t)} \tag{6.18}$$

for each $i = 1, 2, \dots, n$ and each $k \in K$. These are the numbers which should be calculated in recognition stage, provided the values $p_K^t(k)$ and a_k^t were actual values. The numbers $\alpha^t(i,k)$ resemble the *a posteriori* probabilities of the state k under the condition of observing the signal x_i from the presented multi-set. However, they are not the *a posteriori* probabilities because one cannot claim with certainty that the values $p_K^t(k)$ and a_k^t are actual *a priori* probabilities and values of the parameter a. Nevertheless, in the second stage of the $t+1$-th algorithm iteration a new model m^{t+1} is to be calculated by means of the already quoted learning algorithm (6.14) and (6.15) in such a way as if the numbers were actual probabilities $\alpha(i,k)$ provided by the teacher, i.e.,

$$p_K^{t+1}(k) = \frac{\sum_{i=1}^n \alpha^t(i,k)}{\sum\limits_{k' \in K} \sum_{i=1}^n \alpha^t(i,k')} = \frac{\sum_{i=1}^n \alpha^t(i,k)}{n}, \quad k \in K, \tag{6.19}$$

$$a_k^{t+1} = \operatorname*{argmax}_a \sum_{i=1}^n \alpha^t(i,k) \log p(x_i, a), \quad k \in K. \tag{6.20}$$

It can be easily seen that the described algorithm is very similar to Rosenblatt's algorithms, though it also markedly differs from them. Both Rosenblatt's algorithms and the algorithms described here are arranged as a multiple repetition of recognition and learning. For learning, the data obtained from the teacher is not used, but the results of one's own recognition. The difference lies in that in the described algorithm both recognition and learning are considered in a somewhat wider sense than in Rosenblatt's algorithms. Recognition is not strictly considered as a unique inclusion of the observation into just one class. The algorithm behaves so as if it breaks each observation into parts proportional to the numbers $\alpha(i,k)$ and then includes the observation x_i partly into one class and partly into another. In a similar way, the concept of learning, i.e., the maximum likelihood estimation of unknown parameters of a random variable is modified. Unlike learning in Rosenblatt's algorithms, it is not necessary to know in which state exactly the object was during the observation x_i. It is sufficient to know only the probability of this or that state. These differences are substantial for the success of the unsupervised learning algorithm described and will be presented in the following explanation.

We think it necessary to point out the immense generality of both the formulation of a task and the algorithm for its solution. We will see that even the properties of the algorithm are proved either at not very restraining premises, or even without any additional premises. This concerns the *basic relation between three extensive classes of tasks: the recognition itself, the supervised learning and unsupervised learning.* Thanks to the fact that this relation is expressed in an illustrative form, which becomes easily fixed in one's memory, its discovery belongs to the most significant outcomes not only in pattern recognition, but also in the modern analysis of statistical data. The theory of pattern recognition thus shows a certain amount of maturity when it no longer merely absorbs the outcomes of the neighbouring scientific fields, but is able to enrich its neighbouring disciplines with the results of its own.

6.6.5 Analysis of the unsupervised learning algorithm

Let us deal with the most important property of the unsupervised learning algorithm which is valid in the most general case without any additional premises. This property is that in the sequence of models $m^0, m^1, \ldots \ldots, m^t, m^{t+1}, \ldots$, which is created in unsupervised learning, every succeeding model m^t is *better than the preceding one*, if of course the extreme possible situation did not occur, i.e., that the equality $m^t = m^{t-1}$ was satisfied. This means that during unsupervised learning it cannot happen in any case that the once achieved level could deteriorate. To prove this feature we will first introduce a lemma and then a theorem which formulates the desired feature exactly.

Lemma 6.1 Shannon. *Let α_i, $i = 1, \ldots, n$, be positive constants and x_i, $i = 1, \ldots, n$, positive variables for which it holds $\sum_{i=1}^{n} x_i = 1$. In this case the inequality holds*

$$\sum_{i=1}^{n} \alpha_i \log x_i \le \sum_{i=1}^{n} \alpha_i \log \frac{\alpha_i}{\sum_{j=1}^{n} \alpha_j} \tag{6.21}$$

and the equality comes only when $x_i = \alpha_i / \sum_{j=1}^{n} \alpha_j$ for all i. ▲

Proof. We will denote $F(x_1, x_2, \ldots, x_n) = \sum_{i=1}^{n} \alpha_i \log x_i$ and find for which values x_i the function F reaches its maximum under the condition $\sum_i x_i = 1$. Since the function F is a concave one, the point $x_1, x_2, \ldots, x_n$ in which the maximum is reached is the solution of the system of equations

$$\left.\begin{aligned} \frac{\partial \Phi(x_1, x_2, \ldots, x_n)}{\partial x_i} &= 0\,, \quad i = 1, \ldots, n\,, \\ \sum_{i=1}^{n} x_i &= 1 \end{aligned}\right\} \tag{6.22}$$

where Φ is Lagrange function

$$\Phi(x_1, x_2, \ldots, x_n) = \sum_{i=1}^{n} \alpha_i \log x_i + \lambda \sum_{i=1}^{n} x_i\,.$$

The system of equations (6.22) can be expressed in the form

$$\left.\begin{aligned} \frac{\alpha_i}{x_i} + \lambda &= 0\,, \quad i = 1, \dots, n\,, \\ \sum_{i=1}^{n} x_i &= 1 \end{aligned}\right\}$$

which means that

$$x_i = \frac{-\alpha_i}{\lambda}\,, \quad \sum_{j=1}^{n} x_j = \frac{-\sum_{j=1}^{n} \alpha_j}{\lambda}\,, \quad 1 = \frac{-\sum_{j=1}^{n} \alpha_j}{\lambda}\,, \quad \lambda = -\sum_{j=1}^{n} \alpha_j\,.$$

The result is

$$x_i = \frac{\alpha_i}{\sum_{j=1}^{n} \alpha_j}\,. \tag{6.23}$$

It can be seen that the system of equations (6.22) has only one solution. The resulting point x_i given by the equation (6.23) is one single point in the hyperplane $\sum_{i=1}^{n} x_i = 1$ where the maximum is reached. ■

Theorem 6.1 On monotonous nature of unsupervised learning. *Let* $m^t = \big(p_K^t(k),\, a_k^t \mid k \in K\big)$ *and* $m^{t+1} = \big(p_K^{t+1}(k), a_k^{t+1} | k \in K\big)$ *be two models computed after step* t *and step* $(t+1)$ *of the unsupervised learning algorithm. Let the following inequality be satisfied at least for one* i *and one* k

$$\frac{p_K^t(k)\, p(x_i, a_k^t)}{\sum\limits_{k' \in K} p_K^t(k')\, p(x_i, a_{k'}^t)} \neq \frac{p_K^{t+1}(k)\, p(x_i, a_k^{t+1})}{\sum\limits_{k \in K} p_K^{t+1}(k')\, p(x_i, a_{k'}^{t+1})}\,. \tag{6.24}$$

Then also the inequality

$$\sum_{i=1}^{n} \log \sum_{k' \in K} p_K^t(k')\, p(x_i, a_{k'}^t) < \sum_{i=1}^{n} \log \sum_{k' \in K} p_K^{t+1}(k')\, p(x_i, a_{k'}^{t+1}) \tag{6.25}$$

is satisfied. ▲

Proof. Let $\alpha(i,k)$, $i = 1, \dots, n$, $k \in K$, be any non-negative numbers which for each i satisfy the equality

$$\sum_{k \in K} \alpha(i,k) = 1\,. \tag{6.26}$$

In this case for the function $L(m)$ the following holds

$$\begin{aligned} L(m) &= \sum_{i=1}^{n} \log \sum_{k' \in K} p_K(k') p(x_i, a_{k'}) = \sum_{i=1}^{n} \sum_{k \in K} \alpha(i,k) \log \sum_{k' \in K} p_K(k')\, p(x_i, a_{k'}) \\ &= \sum_{i=1}^{n} \sum_{k \in K} \alpha(i,k) \log p_K(k) + \sum_{i=1}^{n} \sum_{k \in K} \alpha(i,k) \log p(x_i, a_k) \\ &\quad - \sum_{i=1}^{n} \sum_{k \in K} \alpha(i,k) \log \frac{p_K(k)\, p(x_i, a_k)}{\sum_{k' \in K} p_K(k)\, p(x_i, a_{k'})}\,. \end{aligned} \tag{6.27}$$

The decomposition (6.27) of the function $L(m)$ into three summands is valid for any numbers $\alpha(i,k)$ which satisfy the constraint (6.26), and thus also for the numbers $\alpha^t(i,k)$. We write this decomposition for the numbers $L(m^t)$ and $L(m^{t+1})$. In both cases the same coefficients $\alpha^t(i,k)$ will be used. We change the order of addition $\sum_{i=1}^n$ and $\sum_{k\in K}$ in the first and second summands and obtain

$$\begin{aligned} L(m^t) = {} & \sum_{k\in K}\sum_{i=1}^{n} \alpha^t(i,k)\,\log p_K^t(k) + \sum_{k\in K}\sum_{i=1}^{n} \alpha^t(i,k)\,\log p(x_i, a_k^t) \\ & - \sum_{i=1}^{n}\sum_{k\in K} \alpha^t(i,k)\,\log \frac{p_K^t(k)\,p(x_i,a_k^t)}{\sum_{k'\in K} p_K^t(k')\,p(x_i,a_{k'}^t)}\;; \end{aligned} \tag{6.28}$$

$$\begin{aligned} L(m^{t+1}) = {} & \sum_{k\in K}\sum_{i=1}^{n} \alpha^t(i,k)\,\log p_K^{t+1}(k) + \sum_{k\in K}\sum_{i=1}^{n} \alpha^t(i,k)\,\log p(x_i, a_k^{t+1}) \\ & - \sum_{i=1}^{n}\sum_{k\in K} \alpha^t(i,k)\,\log \frac{p_K^{t+1}(k)\,p(x_i,a_k^{t+1})}{\sum_{k'\in K} p_K^{t+1}(k')\,p(x_i,a_{k'}^{t+1})}\;. \end{aligned} \tag{6.29}$$

Because the values of *a priori* probabilities $p_K^{t+1}(k)$, $k \in K$, are chosen according to the definition (6.19),

$$p_K^{t+1}(k) = \frac{\sum_{i=1}^n \alpha^t(i,k)}{\sum\limits_{k'\in K}\sum_{i=1}^n \alpha^t(i,k')}\,,$$

and by virtue of Lemma 6.1 we obtain

$$\sum_{k\in K}\sum_{i=1}^{n} \alpha^t(i,k)\,\log p_K^t(k) \le \sum_{k\in K}\sum_{i=1}^{n} \alpha^t(i,k)\,\log p_K^{t+1}(k)\,. \tag{6.30}$$

This means that the first summand on the right-hand side of (6.28) is not greater than the first summand on the right-hand side of (6.29).

According to definition (6.20) there holds that

$$\sum_{i=1}^{n} \alpha^t(i,k)\,\log p(x_i,a_k^t) \le \sum_{i=1}^{n} \alpha^t(i,k)\,\log p(x_i,a_k^{t+1})\,, \quad k\in K\,.$$

If we sum up these inequalities over all values $k \in K$ we obtain

$$\sum_{k\in K}\sum_{i=1}^{n} \alpha^t(i,k)\,\log p(x_i,a_k^t) \le \sum_{k\in K}\sum_{i=1}^{n} \alpha^t(i,k)\,\log p(x_i,a_k^{t+1}) \tag{6.31}$$

which means that the second summand on the right-hand side of (6.28) is not greater than the second summand on the right-hand side of (6.29).

Owing to definition (6.18) we can write

$$\alpha^t(i,k) = \frac{p_K^t(k)\, p(x_i, a_k^t)}{\sum\limits_{k' \in K} p_K^t(k')\, p(x_i, a_{k'}^t)} \,.$$

Because $\sum\limits_{k \in K} \alpha^t(i,k) = 1$ and owing to Lemma 6.1 the following inequality holds

$$\sum_{k \in K} \alpha^t(i,k) \log \frac{p_K^t(k)\, p(x_i, a_k^t)}{\sum\limits_{k' \in K} p_K^t(k')\, p(x_i, a_{k'}^t)} \geq \sum_{k \in K} \alpha^t(i,k) \log \frac{p_K^{t+1}(k)\, p(x_i, a_k^{t+1})}{\sum\limits_{k' \in K} p_K^{t+1}(k')\, p(x_i, a_{k'}^{t+1})} \tag{6.32}$$

which is satisfied for all $i = 1, 2, \ldots, n$. At the same time owing to assumption (6.24) the inequality

$$\frac{p_K^t(k)\, p(x_i, a_k^t)}{\sum\limits_{k' \in K} p_K^t(k')\, p(x_i, a_{k'}^t)} \neq \frac{p_K^{t+1}(k)\, p(x_i, a_k^{t+1})}{\sum\limits_{k' \in K} p_K^{t+1}(k')\, p(x_i, a_{k'}^{t+1})}$$

is satisfied at least at some i and k. So the inequality (6.32) can be rewritten in a strict form,

$$\sum_{i=1}^{n} \sum_{k \in K} \alpha^t(i,k) \log \frac{p_K^t(k)\, p(x_i, a_k^t)}{\sum\limits_{k' \in K} p_K^t(k')\, p(x_i, a_{k'}^t)}$$
$$> \sum_{i=1}^{n} \sum_{k \in K} \alpha^t(i,k) \log \frac{p_K^{t+1}(k)\, p(x_i, a_k^{t+1})}{\sum\limits_{k' \in K} p_K^{t+1}(k')\, p(x_i, a_{k'}^{t+1})} \,. \tag{6.33}$$

This means that the negatively given summand on the right-hand side of (6.28) is greater than the corresponding element on the right-hand side of (6.29).

The inequality $L(m^t) < L(m^{t+1})$ is a quite evident consequence of inequalities (6.30), (6.31) and (6.33). ■

We can see that during unsupervised learning the logarithm of probability $L(m)$ is growing in a monotonic way. The consequence is that repeated recognition accompanied by unsupervised learning on the basis of the same recognition will not deteriorate the knowledge expressed by the initial model m^0. Just the opposite, the knowledge is enhanced in a sense and leads to the enhancement of recognition. It follows from the chain $L(m^0) < L(m^1) < \ldots < L(m^t) < L(m^{t+1})$ while $m^{t+1} \neq m^{t+2}$ that the sequence $L(m^t)$ converges at $t \to \infty$ since $L(m)$ cannot take positive values. But this does not imply that the sequence of models m^t converges as $t \to \infty$. Moreover, at the level of generality of our analysis, the statement itself that the sequence of models m^t converges is not sufficiently understandable.

Until now no metric properties of the parametric set of models have been assumed. Therefore it has not been defined yet what convergence of models means. A parameter of the model can be, for example, a graph, and thus it is not clear what the convergence of a sequence of graphs to a fixed graph means. We call attention to this fact not only because the next analysis will be necessarily supported by some assumptions. It also means that the outcome of this analysis will also be less general. We would simply like to draw attention to the generality of Theorem 6.1 which is valid without any of such additional assumptions.

We will show that the unsupervised learning algorithm has certain asymptotic properties which can be considered similarly to convergence. In spite of it, we will avoid setting the form of the parameters according to which the asymptotic behaviour of the algorithm is to be evaluated. We will attain it by examining the behaviour of the numbers $\alpha^t(i,k)$, not the model m^t. For such examination some assumptions will be needed, which on one side will narrow the scope of the results proved further on, but not so strongly as it would happen if we attempted to prove the convergence of the sequence m^t. From a certain point of view the numbers $\alpha(i,k)$ are more important than the model m, since particularly the numbers $\alpha(i,k)$ indicate how the observations $x_1, x_2, \ldots, x_n$ will be eventually recognised. The model m plays only an auxiliary role in achieving this goal.

For further explanation we will need new denotations. The ensemble of numbers $\alpha(i,k)$, $i = 1,2,\ldots,n$, $k \in K$, will be denoted α without brackets. The symbol α^t denotes the ensemble of numbers $\alpha^t(i,k)$ obtained in the step t in the unsupervised learning algorithm. The assigned ensemble α can be fed to the input of the learning algorithm which is defined by expressions (6.19) and (6.20). Based on the ensemble α the algorithm will calculate a new model m. We will denote this transformation by the symbol U (update) so that $m^{t+1} = U(\alpha^t)$.

On the basis of model m a new ensemble α can be computed by means of the expression (6.18). We will denote this transformation by R (recognition) so that $\alpha^t = R(m^t)$. The symbol S will mean the application of transformation R to the result of transformation U, so that $\alpha^{t+i} = S(\alpha^t) = R\big(U(\alpha^t)\big)$. The transformation S is performed during one iteration of an unsupervised learning algorithm, the later being now formulated in a somewhat different way than that in the formulations (6.18), (6.19) and (6.20). Each iteration now begins and ends with certain newly obtained ensembles α. Before, we thought that the result of each iteration was a new model m. It is evident that this tiny change does not change the unsupervised algorithm in any way.

The ensemble α will be considered as a point in a normed linear space with a naturally assigned norm $|\alpha|$ as

$$|\alpha| = \sqrt{\sum_i \sum_k \big(\alpha(i,k)\big)^2}\,.$$

The following analysis is supported by the main premise on the *continuous character* of the function S, i.e., for small changes of α, $S(\alpha)$ should also undergo a small change. Let us express this assumption more exactly. Let α^i, $i = 1,2,\ldots\infty$, be any convergent sequence, not only the sequence obtained with the unsupervised learning. Let α^* be a limit point of this sequence. Then the sequence $S(\alpha^i)$, $i = 1, 2, \ldots \infty$, also converges, namely towards $S(\alpha^*)$.

The ensemble α which satisfies the condition $\alpha = S(\alpha)$ will be referred to as the *fixed point of unsupervised learning*. The validity of further analysis is constrained by another important premise on the finite number of the fixed points of unsupervised learning. These *two premises suffice* for the sequence of $\alpha^1, \alpha^2, \ldots, \alpha^t, \ldots$ to *converge towards a fixed point during unsupervised learning*. To prove this statement we will need a number of auxiliary statements.

Lemma 6.2 Kullback. *Let α_i and x_i, $i = 1, 2, \ldots, n$, be positive numbers for which there hold $\sum_{i=1}^n \alpha_i = \sum_{i=1}^n x_i = 1$. In this case*

$$\sum_{i=1}^n \alpha_i \ln \frac{\alpha_i}{x_i} \geq \frac{1}{2} \sum_{i=1}^n (\alpha_i - x_i)^2 . \tag{6.34}$$

▲

Proof. Let $\delta_i = x_i - \alpha_i$, $i = 1, \ldots, n$. Let us define two functions φ and ψ dependent on the variable γ which assumes the values $0 \leq \gamma \leq 1$,

$$\varphi(\gamma) = \sum_{i=1}^n \alpha_i \ln \frac{\alpha_i}{\alpha_i + \gamma \delta_i} ,$$

$$\psi(\gamma) = \frac{1}{2} \sum_{i=1}^n (\gamma\, \delta_i)^2 .$$

It is obvious that $\varphi(0) = \psi(0) = 0$. For all i and γ the inequalities $0 < \alpha_i + \gamma\, \delta_i < 1$ hold, except for the case $n = 1$, which is, however, trivial. Let us state derivatives of the functions φ and ψ with respect to the variable γ,

$$\frac{d\varphi(\gamma)}{d\gamma} = -\sum_{i=1}^n \frac{\alpha_i\, \delta_i}{\alpha_i + \gamma\, \delta_i} ,$$

$$\frac{d\psi(\gamma)}{d\gamma} = \sum_{i=1}^n \gamma\, \delta_i^2 .$$

Since $\sum_{i=1}^n \delta_i = 0$, the following relations are valid for the derivative $d\varphi/d\gamma$,

$$\frac{d\varphi(\gamma)}{d\gamma} = -\sum_{i=1}^n \frac{\alpha_i\, \delta_i}{\alpha_i + \gamma\, \delta_i} + \sum_{i=1}^n \delta_i = \sum_{i=1}^n \frac{\gamma\, \delta_i^2}{\alpha_i + \gamma\, \delta_i} \geq \sum_{i=1}^n \gamma\, \delta_i^2 = \frac{d\psi(\gamma)}{d\gamma} .$$

The correctness of the inequality in this derivation results, as we stated before, from the property that the number $\alpha_i + \gamma\, \delta_i$ is less than 1 for any values i and γ. The number $\varphi(1)$ is $\sum_{i=1}^n \alpha_i \log(\alpha_i / x_i)$. The number $\psi(1)$ is $\frac{1}{2} \sum_{i=1}^n (\alpha_i - x_i)^2$. The relationship between these two numbers is determined by the relations

$$\varphi(1) = \varphi(0) + \int_0^1 \frac{d\varphi}{dt} dt = \psi(0) + \int_0^1 \frac{d\varphi}{dt} dt \geq \psi(0) + \int_0^1 \frac{d\psi}{dt} dt = \psi(1) ,$$

from which the inequality (6.34) follows. ■

In Kullback's Lemma 6.3 a logarithm with natural base was used because the proof is more concise. Further on, we will use logarithms of other bases which will not bring about difficulties, since at an arbitrary logarithm base there hold

$$\sum_{i=1}^{n} \alpha_i \log \frac{\alpha_i}{x_i} \geq c \sum_{i=1}^{n} (\alpha_i - x_i)^2 .$$

The chosen base of the logarithm affects only the constant c. In this sense we will refer to Kullback's lemma in proving the following lemma.

Lemma 6.3 *Let α^t and α^{t+1} be two ensembles obtained in the steps t and $(t+1)$ of the unsupervised learning algorithm, respectively. In this case the number*

$$\sum_i \sum_k \left(\alpha^t(i,k) - \alpha^{t+1}(i,k)\right)^2 \tag{6.35}$$

converges towards zero at $t \to \infty$. ▲

Proof. The sequence $L(m^t)$ monotonically rises and does not assume positive values. Therefore it is sure to converge, which means that the difference $L(m^{t+1}) - L(m^t)$ converges towards zero,

$$\lim_{t\to\infty} \left(L(m^{t+1}) - L(m^t)\right) = 0 . \tag{6.36}$$

Let us write the expression for $L(m^{t+1}) - L(m^t)$ in more detail using somewhat modified decompositions (6.28) and (6.29) to which we have substituted from (6.18).

$$\begin{aligned}
&L(m^{t+1}) - L(m^t) \\
&= \sum_k \sum_i \alpha^t(i,k) \log p_K^{t+1}(k)\, p(x_i, a_k^{t+1}) - \sum_i \sum_k \alpha^t(i,k) \log \alpha^{t+1}(i,k) \\
&\quad - \sum_k \sum_i \alpha^t(i,k) \log p_K^t(k)\, p(x_i, a_k^t) + \sum_i \sum_k \alpha^t(i,k) \log \alpha^t(i,k) \\
&= \left(\sum_k \sum_i \alpha^t(i,k) \log p_K^{t+1}(k)\, p(x_i, a_k^{t+1}) - \sum_k \sum_i \alpha^t(i,k) \log p_K^t(k)\, p(x_i, a_k^t) \right) \\
&\quad + \left(\sum_i \sum_k \alpha^t(i,k) \log \frac{\alpha^t(i,k)}{\alpha^{t+1}(i,k)} \right) .
\end{aligned}$$

We can see that the difference $L(m^{t+1}) - L(m^t)$ consists of two summands closed by large brackets in the last step of the preceding derivation. Both summands are non-negative. The non-negativeness of the former follows immediately from the definitions (6.19) and (6.20) and the non-negativeness of the latter is proved by Lemma 6.2. Since the sum of two non-negative summands converges towards zero, there also holds that either of these two summands converges towards zero as well. It is important from our point of view that the second summand converges towards zero,

$$\lim_{t\to\infty} \sum_i \sum_k \alpha^t(i,k) \log \frac{\alpha^t(i,k)}{\alpha^{t+1}(i,k)} = 0 .$$

Thanks to Kullback's Lemma 6.2

$$\lim_{t\to\infty} \sum_i \sum_k \left(\alpha^t(i,k) - \alpha^{t+1}(i,k)\right)^2 = 0\,.$$

■

Lemma 6.4 *Let function $S(\alpha)$ be a continuous function of the ensemble*

$$\alpha = \big(\alpha(i,k)\,,\; i = 1,2,\ldots,n\,,\; k \in K\big)\,;$$

$\alpha^1, \alpha^2, \ldots, \alpha^t, \ldots$ *be an infinite sequence of ensembles α which are the result of unsupervised learning, i.e., $\alpha^t = S(\alpha^{t-1})$.*

Then the limit of each convergent infinite subsequence

$$\alpha^{t(1)}, \alpha^{t(2)}, \ldots, \alpha^{t(j)}, \ldots\,,\quad \textit{with } t(j) > t(j-1),$$

is a fixed point *in unsupervised learning.* ▲

Proof. The sequence $\alpha^1, \alpha^2, \ldots, \alpha^t, \ldots$ is an infinite sequence of points in a limited and closed subset of a linear space. Therefore, as it is known from the mathematical analysis, this sequence is sure to contain the convergent subsequence $\alpha^{t(1)}, \alpha^{t(2)}, \ldots, \alpha^{t(j)}, \ldots$. We will denote its limit as α^*

$$\lim_{j\to\infty} \alpha^{t(j)} = \alpha^*\,.$$

Let $S(\alpha^{t(1)}), S(\alpha^{t(2)}), \ldots, S(\alpha^{t(j)}), \ldots$ be set up from those points that in the sequence α^1, α^2,...,α^t, ... immediately follow after the elements $\alpha^{t(1)}$, $\alpha^{t(2)}, \ldots$ $\ldots, \alpha^{t(j)}$, ... The limit of the selected sequence $S(\alpha^{t(j)})$, $j = 1, 2, \ldots$, is also α^*, i.e.,

$$\lim_{j\to\infty} S(\alpha^{t(j)}) = \alpha^*\,, \tag{6.37}$$

since Lemma 6.3 claims that $\lim_{t\to\infty} |\alpha^t - S(\alpha^t)| = 0$. The premise of the lemma being proved is the continuity of the mapping S, and thus it follows from (6.37) that $S(\alpha^*) = \alpha^*$. ■

Lemma 6.5 *Let Ω be a set of fixed points for the algorithm;* $\min_{\alpha^*\in\Omega} |\alpha - \alpha^*|^2$ *be the distance of the point α to the set Ω. If the function $S(\alpha)$ is continuous then*

$$\lim_{t\to\infty} \min_{\alpha^*\in\Omega} |\alpha^t - \alpha^*|^2 = 0\,. \tag{6.38}$$

▲

Proof. Assume that the relation (6.38) is not correct. Let us write the formal meaning of the relation (6.38) and then the meaning of its negation and what results from it. The relation (6.38) is a concisely written statement

$$\forall \varepsilon > 0\,,\; \exists T\,,\; \forall t > T\;:\; \min_{\alpha^*\in\Omega} |\alpha^t - \alpha^*|^2 < \varepsilon$$

and its negation corresponds to the statement

$$\exists \varepsilon > 0\,,\; \forall T\,,\; \exists t > T\;:\; \min_{\alpha^*\in\Omega} |\alpha^t - \alpha^*|^2 \geq \varepsilon\,. \tag{6.39}$$

The statement (6.39) means that there exists such $\varepsilon > 0$ and such an infinite subsequence

$$\alpha^{t(1)}, \alpha^{t(2)}, \ldots, \alpha^{t(j)}, \ldots, \quad \text{where} \quad t(j) > t(j-1)$$

for which an inequality

$$\min_{\alpha^* \in \Omega} |\alpha^{t(j)} - \alpha^*|^2 \geq \varepsilon \tag{6.40}$$

holds for each element $t(j)$. Since this subsequence is an infinite sequence on a closed and limited set, it also contains a convergent subsequence. The limit of this new sequence owing to (6.40) will not belong to Ω, and thus it will not be a fixed point of unsupervised learning. We have arrived at a result which is in contradiction with Lemma 6.4. Thus the assumption (6.39) is wrong. ■

Theorem 6.2 On convergence of unsupervised learning. *If the function S which denotes one iteration of the unsupervised learning algorithm is continuous, and the set of fixed points is finite then the sequence*

$$\alpha^1, \alpha^2, \ldots, \alpha^t, \ldots, \quad \textit{with} \quad \alpha^t = S(\alpha^{t-1}),$$

converges and its limit is a fixed point of the algorithm. ▲

Proof. We will denote by the symbol Δ as the distance between two nearest points. The number Δ is not zero since a finite number of fixed points is assumed. We will prove that in the sequence $\alpha^1, \alpha^2, \ldots, \alpha^t, \ldots$ it can occur only a finite number of times that

$$\operatorname*{argmin}_{\alpha^* \in \Omega} |\alpha^t - \alpha^*| \neq \operatorname*{argmin}_{\alpha^* \in \Omega} |\alpha^{t+1} - \alpha^*| \,.$$

Let us admit that it would happen an infinite number of times that the distance α^t from the nearest fixed point α^* would be less than a certain δ, and the distance α^{t+1} from the nearest, but now from another fixed point, would also be less than δ. Thanks to Lemma 6.5 this situation would occur at any positive value δ, thus even at a rather small one. As a result this would mean it would occur an infinite number of times that the distance between α^t and α^{t+1} would be greater than $\Delta - 2\delta$. But this is not possible because Lemma 6.3 states that the distance between α^t and α^{t+1} converges towards zero.

We have proved that after some finite t the fixed point

$$\operatorname*{argmin}_{\alpha^* \in \Omega} |\alpha^t - \alpha^*|$$

which is the closest to the point α^t ceases to change. Such a fixed point will be denoted α^{**} and the proved relation (6.38) assumes the form $\lim_{t\to\infty} |\alpha^t - \alpha^{**}|^2 = 0$, or similarly $\lim_{t\to\infty} \alpha^t = \alpha^{**}$. ■

With some additional assumptions it could be shown that the fixed points of unsupervised learning have certain properties from the standpoint of the logarithm likelihood

$$\sum_i \log \sum_k p_K(k) p(x_i \,|\, a_k) \,.$$

In some cases it could be proved that the values $p_K(k)$, a_k, $k \in K$, through which a fixed point is characterised, are in a sense the best values in their neighbourhood. A general account of these properties for a rather extensive class of models is not difficult, but is not very interesting either. Therefore we recommend analysing, in each particular case, the properties of the fixed points using all specific features of a particular case. As an example of such an analysis of a particular case of an unsupervised learning task a situation will be discussed in which the conditional probabilities $p_{X|K}(x \mid k)$ are completely known, and only the *a priori* probabilities $p_K(k)$ of states are unknown. This example is valuable in itself. It corresponds, indeed, to Robbins task in its complete generality as well as in its original formulation. We will prove that for quite self-evident assumptions the unsupervised learning algorithm converges towards the globally most likely estimates of *a priori* probabilities $p_K(k)$.

6.6.6 Algorithm solving Robbins task and its analysis

The algorithm for solving Robbins task has the following form. Let $p_K^0(k)$, $k \in K$, be initial estimates of *a priori* probabilities and $p_K^t(k)$ be the values after the iteration t of the algorithm. In agreement with the general unsupervised learning algorithm, the numbers $\alpha^t(i,k)$, $i = 1,2,\ldots,n$, $k \in K$, are to be calculated first,

$$\alpha^t(i,k) = \frac{p_K^t(k)\, p_{X|K}(x_i \mid k)}{\sum\limits_{k' \in K} p_K^t(k')\, p_{X|K}(x_i \mid k')}\,, \tag{6.41}$$

and then new estimates $p_K^{t+1}(k)$, $k \in K$,

$$p_K^{t+1}(k) = \frac{\sum_i \alpha^t(i,k)}{n}\,. \tag{6.42}$$

We can see that the algorithm for solving Robbins task is expressed quite explicitly. This is the difference compared with the unsupervised learning algorithm in the general case. There the learning task in its optimising formulation (6.20) has to be solved in every particular case of its construction. We can also see that the algorithm itself, described by the relations (6.41), (6.42), is incredibly simple. The calculation of the values $p_{X|K}(x_i \mid k)$ is anyway expected to be algorithmically supported because it is necessary for recognition even if learning or unsupervised learning is not used.

It seems to be plausible that the algorithm given by the relations (6.41), (6.42) converges towards the point in which the global maximum of likelihood function

$$\sum_i \log \sum_k p_K(k)\, p_{X|K}(x_i \mid k)$$

has been attained since this function is a concave one. The algorithm that converges towards the local maximum of a concave function provides its maximisation also in the global sense, since a concave function has (roughly speaking) only one local maximum. These considerations are, of course, only preliminary and cannot replace the following exact formulation and the proof with which we will end this lecture.

Theorem 6.3 On solving Robbins task for the general case. *Let values $p_K^*(k)$, $k \in K$, be parameters of the fixed point of the algorithm (6.41), (6.42). Furthermore let none of these values be 0, i.e., $p_K^*(k) \neq 0$ for all $k \in K$. Then the following inequality is satisfied,*

$$\sum_{i=1}^{n} \log \sum_{k \in K} p_K^*(k)\, p_{X|K}(x_i \mid k) \geq \sum_{i=1}^{n} \log \sum_{k \in K} p_K(k)\, p_{X|K}(x_i \mid k)$$

for any a priori probabilities $p_K(k)$, $k \in K$. ▲

Proof. We will use the relations (6.41), (6.42) and exclude the auxiliary variables $\alpha^t(i,k)$. So we express how the probabilities $p_K^{t+1}(k)$, $k \in K$, depend on the probabilities $p_K^t(k)$, $k \in K$,

$$p_K^{t+1}(k) = \frac{1}{n} \sum_i \frac{p_K^t(k)\, p_{X|K}(x_i \mid k)}{\sum\limits_{k' \in K} p_K^t(k')\, p_{X|K}(x_i \mid k')}\,, \qquad k \in K\,.$$

Let the ensemble $\big(p_K^*(k),\ k \in K\big)$ represent the fixed point of the algorithm and so $p_K^*(k) = p_K^{t+1}(k) = p_K^t(k)$, $k \in K$. With respect to the property that no probability $p_K^*(k)$ is zero, we obtain

$$n = \sum_i \frac{p_{X|K}(x_i \mid k)}{\sum\limits_{k \in K} p_K^*(k)\, p_{X|K}(x_i \mid k)}\,, \qquad k \in K\,. \tag{6.43}$$

Let $p_K(k)$, $k \in K$, be any positive numbers the sum of which is 1. We will multiply each equality from the system (6.43) by the number $p_K(k) - p_K^*(k)$, sum up all the equations and obtain the relation

$$n \sum_{k \in K} \big(p_K(k) - p_K^*(k)\big) = \sum_{i=1}^{n} \frac{\sum\limits_{k \in K} p_K(k)\, p_{X|K}(x_i \mid k) - \sum\limits_{k \in K} p_K^*(k)\, p_{X|K}(x_i \mid k)}{\sum\limits_{k \in K} p_K^*(k)\, p_{X|K}(x_i \mid k)}$$

which is equivalent to the relation

$$\sum_{i=1}^{n} \frac{\sum\limits_{k \in K} p_K(k)\, p_{X|K}(x_i \mid k) - \sum\limits_{k \in K} p_K^*(k)\, p_{X|K}(x_i \mid k)}{\sum\limits_{k \in K} p_K^*(k)\, p_{X|K}(x_i \mid k)} = 0\,, \tag{6.44}$$

since both the sum $\sum_{k \in K} p_K(k)$ and the sum $\sum_{k \in K} p_K^*(k)$ are 1.

To be brief we will denote the ensemble $\big(p_K(k),\ k \in K\big)$ by the symbol p_K and introduce the denotation $f_i(p_K)$,

$$f_i(p_K) = \sum_{k \in K} p_K(k)\, p_{X|K}(x_i \mid k)\,. \tag{6.45}$$

The relation (6.44) can be expressed in a simpler way

$$\sum_{i=1}^{n} \frac{f_i(p_K) - f_i(p_K^*)}{f_i(p_K^*)} = 0 \,. \tag{6.46}$$

We will create the following function dependent on the scalar variable γ

$$Q(\gamma) = \sum_{i=1}^{n} \frac{f_i(p_K) - f_i(p_K^*)}{f_i(p_K^*) + \gamma\big(f_i(p_K) - f_i(p_K^*)\big)} \,. \tag{6.47}$$

It is clear that $Q(0)$ is the left-hand part of the expression (6.46), and thus

$$Q(0) = 0 \,. \tag{6.48}$$

It is also evident that at any value γ the derivative $\mathrm{d}Q(\gamma)/\mathrm{d}\gamma$ is not positive since

$$\frac{\mathrm{d}Q(\gamma)}{\mathrm{d}\gamma} = \sum_{i=1}^{n} -\left(\frac{(f_i(p_K) - f_i(p_K^*))^2}{\Big(f_i(p_K^*) + \gamma\big(f_i(p_K) - f_i(p_K^*)\big)\Big)^2} \right) \leq 0$$

which with respect to (6.48) means that

$$Q(\gamma) \leq 0$$

at any non-negative value $\gamma \geq 0$. From that it follows further that the integral $\int_0^1 Q(\gamma)\mathrm{d}\gamma$ is not positive. Let us write it in greater detail, see (6.47),

$$\begin{aligned} \int_0^1 Q(\gamma)\,\mathrm{d}\gamma &= \sum_{i=1}^{n} \log\Big(f_i(p_K^*) + \gamma\,\big(f_i(p_K) - f_i(p_K^*)\big)\Big)\Bigg|_{\gamma=0}^{1} \\ &= \sum_{i=1}^{n} \log f_i(p_K) \;-\; \sum_{i=1}^{n} \log f_i(p_K^*) \;\leq\; 0 \,. \end{aligned}$$

We will write the inequality in even greater detail using the definition (6.45)

$$\sum_{i=1}^{n} \log \sum_{k \in K} p_K(k)\, p_{X|K}(x_i \,|\, k) \;\leq \sum_{i=1}^{n} \log \sum_{k \in K} p_K^*(k)\, p_{X|K}(x_i \,|\, k) \,.$$

Theorem 6.3 has been proved. ■

6.7 Discussion

It seems to me that something is missing in this lecture, something that would sound like the final chord in a composition; what would evoke the impression of completed work, and a clear feeling what the 'net weight' is. I belong to those who are interested more in the outcomes than in the historical pathways along which one had to go to attain the outcomes. It seems to me that the important subject matter that I will need in future is constrained by the formulation of the unsupervised learning task in Subsection 6.6.3, by the unsupervised learning algorithm in Subsection 6.6.4 and by Theorem 6.1 in Subsection 6.6.5. It is a relatively small part of the lecture, and that is why I dare to ask directly and plainly what more from the lecture should be, according to your opinion, necessary for me, and whether such an extended introduction to these essential results is not simply valueless.

We are answering your question directly and plainly. In this course we will still use the unsupervised learning algorithm in the form it was presented in Subsection 6.6.4. It will be of benefit to you when you understand it well enough and not forget about its existence, at least until when we develop on the basis of it new algorithms for solving certain specific tasks. Regardless of whether you will need these results, you should know them. They are the kind of results which, because of their generality, belong to the gamut of fundamental knowledge in random data processing. Therefore everyone who claims to be professionally active in this field should know them. It is, simply, a part of one's education. Naturally, from the demand that everyone should know these results it does not follow that you in particular should know them. There are quite enough people who do not know the most necessary things.

As to the rest of the lecture, we agree with you that it is a rather long introduction to the main results. But we are not so resolute and we would not like to completely agree that a detailed introduction is useless. Everything depends on from which side you intend to view the results that both you and we regard as necessary. You know that any product can be evaluated from two sides: from the standpoint of him who will use the product, and from the standpoint of him who makes the product. Even a simple product, such as beef steak, looks from the eater's standpoint completely different than it does from the standpoint of a cook. You must decide yourself where you stand in the kitchen called pattern recognition, whether amongst the eaters or the cooks. In the former case it is quite needless for you to know the entire pathway that unsupervised learning had gone through before it was formed into its present day shape. In the latter case you will realise sooner or later, how very low the efficiency of scientific research can be. You will also realise a small ratio of results which manage to get established in science for some time to the erroneous or little significant results, which a researcher must rummage through before he/she gets across a result worth anything. Furthermore, you will see that it needs the patience of Job to rear an idea, powerless at its birth, liable to being hurt or destroyed by anybody, from its swaddling clothes, and to lead it to maturity, when it starts living a life of its own. The sooner you also realise the diversity of negative sides of scientific research, the better for you, even though the process of realising it is no pleasant thing in itself. We have used this lecture to warn you what is definitely in store for you.

Try to imagine clearly that you will be attracted by a vague supposition, of which you do not have the least idea where to start the process of proving it. And moreover, you do not know any means of how to express it. There will be a lot of colleagues around you, but you have none to approach with your problem, since every good colleague of yours quickly makes your vague suppositions concrete so that they result in complete nonsense. An so you will remain alone, facing the problem till your perseverance is rewarded and you come across a precise formulation of what you actually want. But the most probable situation to occur will be that your problem will appear incapable of being solved. Your pains will end with the poor consolation that a negative result is also a kind of result. But you are lucky, as ever is your case, and so quite quickly after the formulation of your problem you will find out its solution.

But now the second stage of your calvary comes. Before you found the formulation and solution of your task, you had scrutinised your problem in a criss-cross way and had found its fantastic complexity. Therefore you naturally consider its solution as quite simple. But your colleagues did not go along this path and thus the entire construction built up seems to them clumsy, shapeless, requiring efficient computers your laboratory does not own yet, or demanding high-tech hardware. And unless everything is substantially simplified it will be difficult to obtain an order from a customer and it will not be industrially realisable. Until you, by chance, come across a reasonable partner who himself has been dealing with similar problems but was not engaged in inventing any new methods. Say, the entire task has been known since about the beginning of the 20th century, when it was solved up by a certain Dutchman or Moravian, nobody now knows the author. Since that time these methods have been applied in crystallography, in medical care, etc.. Though you are not so much interested in it, you will plunge into back publications and find out that the actually new ideas you have invented during your research are hardly 5 per cent of what you have done, and not more than 0.5 per cent of the whole problem.

Several years will yet elapse, computer efficiency continues rising, the scientific level in pattern recognition is rising too. Slowly at first, but then increasingly often, particularly your solution of the task will be used, and not the procedure designed in the early years of the century, since just the 0.5 per cent you brought to the research have influenced the viability of the whole construct. It will be your victory in a sense. But let us now bet that you will not be delighted at all, and rather urgently need some support. Only then read this lecture again from that passage where we not very seriously, but not quite by way of a joke either, wrote about Columbus, and his discovery of America, up to these lines.

I did not wait so long; I read the passage about Columbus once more straightaway. Now I am absolutely sure that the whole historical introduction is useless for me, because I will somehow avoid a development like this. And moreover, for some persons this topic is not only useless but simply detrimental. A picture drawn by you, which I begin to understand only now, has an unnecessarily dramatic character. When I was reading your comments on the discovering of America for the first time, they seemed to me rather as a joke, which is attained by arranging the known events deliberately in such an order so that they sound jocular. But when I should enter into the outlined scenario as a dramatis persona of those events, I am really seized by horror. Are you not afraid that someone will understand the lecture exactly in the way you would like him or her to, and will lose all the zest to be engaged in scientific research? I can vividly imagine a gourmand who lost appetite for beefsteak when he learnt all what had preceded before the steak appeared on his plate.

We are not afraid of that. First, if it happened so, we would be pleased for his/her sake that he/she had quickly learned that he/she liked something else more than science. Second, many people do not take very seriously that the career of a scientist is so harsh and are convinced that they will be able to avoid all unpleasant circumstances in a way. Third, and this is the saddest of all, that an actual dramatic situation occurs only in the cases when really significant scientific discoveries are at stake, and this happens rather rarely. Thus, the majority of us is rather safely protected from the worst unpleasant cases of this kind.

I would like to look at the positive outcomes of the lecture from the eater's view. The algorithm for solving Robbins task is expressed quite unambiguously. It is, therefore, a product that is prepared for practical application. But I would not say so about the general unsupervised learning algorithm. I would rather say that the general unsupervised learning algorithm is more a semi-finished product than a product ready for use. It is an algorithm that is expressed up to some other algorithm, and this represents immense ambiguity. An algorithm that is to be constructed for a certain optimisation task will be unambiguously expressed only when there is another algorithm for a further optimisation task at our disposal. This auxiliary algorithm is to be inserted into the algorithm that is being constructed. I cannot clearly see what I will practically gain from such a recommendation when there is nothing to account for the statement that the auxiliary task is simpler than the original one.

The best thing for you will be to thoroughly analyse several quite simple examples.

Assume that k is either 1 or 2, $p_K(k)$ are *a priori* probabilities, x is a one-dimensional Gaussian random variable the conditional probability distribution $p_{X|k}(x)$, $k = 1, 2$, of which is

$$p_{X|k}(x) = \frac{1}{\sqrt{2\pi}} e^{-\frac{1}{2}(x-\mu_k)^2} .$$

Assume that the values $p_K(k)$ and μ_k, $k = 1, 2$, are unknown and it is necessary to estimate these values on the basis of a sequence $x_1, \ldots, x_n$, where each x_i is an instance of a random variable x, having a probability distribution

$p_K(1)\, p_{X|1}(x) + p_K(2)\, p_{X|2}(x)$. This means that numbers $p_K(1), p_K(2), \mu_1$ and μ_2 are to be found for which the value

$$\sum_{i=1}^{n} \log \left(p_K(1) \frac{1}{\sqrt{2\pi}} e^{-\frac{1}{2}(x_i-\mu_1)^2} + p_K(2) \frac{1}{\sqrt{2\pi}} e^{-\frac{1}{2}(x_i-\mu_2)^2} \right) \tag{6.49}$$

is maximal.

In order to maximise the function (6.49) I must solve rather simple auxiliary maximisation task

$$\begin{aligned} \mu_k^* &= \operatorname*{argmax}_{\mu} \sum_{i=1}^{n} \alpha(i,k) \log \frac{1}{\sqrt{2\pi}} e^{-\frac{1}{2}(x_i-\mu)^2} \\ &= \operatorname*{argmax}_{\mu} \left(-\sum_{i=1}^{n} \alpha(i,k)\,(x_i-\mu)^2 \right) = \operatorname*{argmin}_{\mu} \sum_{i=1}^{n} \alpha(i,k)\,(x_i-\mu)^2 . \end{aligned} \tag{6.50}$$

Since the function $\sum_{i=1}^{n} \alpha(i,k)\,(x_i-\mu)^2$ is convex with respect to μ, the minimising position μ_k^* is obtained by solving the equation in three steps:

$$\left. \frac{\mathrm{d}\left(\sum_{i=1}^{n} \alpha(i,k)\,(x_i-\mu_k)^2\right)}{\mathrm{d}\mu_k} \right|_{\mu=\mu^*} = 0 ,$$

$$-2 \sum_{i=1}^{n} \alpha(i,k)\,(x_i-\mu_k^*) = 0 ,$$

$$\mu_k^* = \frac{\sum_{i=1}^{n} \alpha(i,k)\, x_i}{\sum_{i=1}^{n} \alpha(i,k)} .$$

Thus the algorithm for maximising the function (6.49) is to have the following form: Let the initial values be, for example, $p_K^0(1) = p_K^0(2) = 0.5$, $\mu_1^0 = x_1$, $\mu_2^0 = x_2$. The algorithm is to iteratively enhance the above four numbers. Assume that after the iteration t the numbers $p_K^t(1)$, $p_K^t(2)$, μ_1^t, μ_2^t have been attained. The new values $p_K^{t+1}(1)$, $p_K^{t+1}(2)$, μ_1^{t+1}, μ_2^{t+1} are to be calculated on the basis of the following explicit formulæ:

$$\left.\begin{aligned} \alpha(i,1) &= \frac{p_K^t(1)\, e^{-\frac{1}{2}(x_i-\mu_1^t)^2}}{p_K^t(1)\, e^{-\frac{1}{2}(x_i-\mu_1^t)^2} + p_K^t(2)\, e^{-\frac{1}{2}(x_i-\mu_2^t)^2}}, \quad i = 1,2,\ldots,n ; \\ \alpha(i,2) &= \frac{p_K^t(2)\, e^{-\frac{1}{2}(x_i-\mu_2^t)^2}}{p_K^t(1)\, e^{-\frac{1}{2}(x_i-\mu_1^t)^2} + p_K^t(2)\, e^{-\frac{1}{2}(x_i-\mu_2^t)^2}}, \quad i = 1,2,\ldots,n ; \\ p_K^{t+1}(1) &= \frac{\sum_{i=1}^{n} \alpha(i,1)}{n} ; \qquad p_K^{t+1}(2) = \frac{\sum_{i=1}^{n} \alpha(i,2)}{n} ; \\ \mu_1^{t+1} &= \frac{\sum_{i=1}^{n} \alpha(i,1)\, x_i}{\sum_{i=1}^{n} \alpha(i,1)} ; \qquad \mu_2^{t+1} = \frac{\sum_{i=1}^{n} \alpha(i,2)\, x_i}{\sum_{i=1}^{n} \alpha(i,2)} . \end{aligned}\right\} \tag{6.51}$$

I have not written the superscripts t with the variables $\alpha(i,k)$. I believe that it is obvious that they vary at every iteration of the algorithm.

So you can see that you have mastered the optimisation task (6.50) quite quickly. You needed just several lines of formulæ, whereas the original task (6.49) may have scared you.

The optimisation function (6.49) is really not a very pleasant function, but I believe that with a certain effort I would also manage its optimisation.

We do not doubt it. But with the aid of general recommendations you have written the algorithm (6.51) without any effort, and this is one of the outcomes of the lecture, if we wanted to view it from the eater's standpoint. We think that in a similar manner you would master even algorithms for more general cases. For example, for cases in which also the conditional variances of random variables were unknown or for multi-dimensional cases, and the like. But in all these cases, as well as in many others, you will clearly notice that the auxiliary task is substantially simpler than the original one.

Go once more through another rather easy example which is profuse because of its consequences. Let the state k be a random variable again which assumes two values: $k = 1$ with the probability $p_K(1)$ and $k = 2$ with the probability $p_K(2)$. Let x be a random variable which assumes values from the set X with the probabilities $p_{X|1}(x)$, provided the object is in the first state; and with the probabilities $p_{X|2}(x)$, provided the object is in the second state. Let y be another random variable which assumes values from the set Y with the probabilities $p_{Y|1}(y)$, provided the object is in the first state; and with the probabilities $p_{Y|2}(y)$ in the opposite case. The numbers $p_K(1)$, $p_K(2)$, $p_{X|1}(x)$, $p_{X|2}(x)$, $x \in X$, and the numbers $p_{Y|1}(y)$, $p_{Y|2}(y)$, $y \in Y$, are unknown. This means that there is no knowledge about the dependence of any of these features on the state k. But it is known that under the condition the object is in the first state, as well as under the condition it is in the second state, the features x and y do not depend on one another, i.e., the equality

$$p_{XY|k}(x,y) = p_{X|k}(x)\, p_{Y|k}(x)$$

is valid for any triplet x, y, k, $x \in X$, $y \in Y$, $k = 1, 2$. The denotation $p_{XY|k}(x,y)$ expresses the joint conditional probability of the features x and y under the condition the object is in the state k.

Assume that due to the observations of the object a sequence of features $(x_1, y_1), (x_2, y_2), \ldots, (x_n, y_n)$ has been obtained. On its basis a statistical model of the object is to be evaluated, i.e., the numbers $p_K(k)$, $p_{X|k}(x)$, $p_{Y|k}(y)$, are to be found that maximise the probability

$$\sum_{i=1}^{n} \log\Bigl(\sum_{k=1}^{2} p_K(k)\, p_{X|k}(x_i)\, p_{Y|k}(y_i)\Bigr) .$$

The auxiliary task consists in that the numbers $p^*_{X|k}(x)$, $x \in X$, $p^*_{Y|k}(y)$, $y \in Y$, $k = 1, 2$, *are to be found which maximise the function*

$$\sum_{i=1}^{n} \alpha(i,k) \log \left(p_{X|k}(x_i)\, p_{Y|k}(y_i) \right) \tag{6.52}$$

at the known numbers $\alpha(i,k)$, $i = 1, \ldots, n$, $k = 1, 2$, *and the assigned sequence* (x_1, y_1), $(x_2, y_2), \ldots, (x_n, y_n)$. *This auxiliary task can be solved quite simply*

$$\begin{aligned}
\left(p^*_{X|k}, p^*_{Y|k}\right) &= \operatorname*{argmax}_{(p_{X|k}, p_{Y|k})} \sum_{i=1}^{n} \alpha(i,k) \log p_{X|k}(x_i)\, p_{Y|k}(y_i) \\
&= \operatorname*{argmax}_{(p_{X|k}, p_{Y|k})} \left(\sum_{i=1}^{n} \alpha(i,k) \log p_{X|k}(x_i) + \sum_{i=1}^{n} \alpha(i,k) \log p_{Y|k}(y_i) \right) \\
&= \left(\operatorname*{argmax}_{p_{X|k}} \sum_{i=1}^{n} \alpha(i,k) \log p_{X|k}(x_i),\ \operatorname*{argmax}_{p_{Y|k}} \sum_{i=1}^{n} \alpha(i,k) \log p_{Y|k}(y_i) \right) \\
&= \left(\operatorname*{argmax}_{p_{X|k}} \sum_{x \in X} \sum_{i \in I_X(x)} \alpha(i,k) \log p_{X|k}(x)\,,\ \operatorname*{argmax}_{p_{Y|k}} \sum_{y \in Y} \sum_{i \in I_Y(y)} \alpha(i,k) \log p_{Y|k}(y) \right) \\
&= \left(\operatorname*{argmax}_{p_{X|k}} \sum_{x \in X} \left(\sum_{i \in I_X(x)} \alpha(i,k) \right) \log p_{X|k}(x)\,, \right. \\
&\qquad \left. \operatorname*{argmax}_{p_{Y|k}} \sum_{y \in Y} \left(\sum_{i \in I_Y(y)} \alpha(i,k) \right) \log p_{Y|k}(y) \right) .
\end{aligned}$$

The first equality in the preceding derivation merely repeats the formulation (6.52) of the auxiliary task. The second equality takes advantage of the rule that the logarithm of a product is the sum of the logarithms. The third equation is valid because a sum of two summands is to be maximised, where each summand is dependent on the group of variables of its own, and therefore the sum can be maximised as the independent maximisations of each particular summand separately. The fourth equation uses the denotation $I_X(x)$ *for the set of those indices* i, *for which it holds that* x_i *has assumed the values* x. *A similar denotation is used for the set* $I_Y(y)$. *The summands* $\alpha(i,k) \log p_{X|k}(x_i)$ *can thus be grouped in such a way that the addition is first done over the indices* i, *at which the observed feature* x_i *assumed a certain value, and then it is done over all values* x. *The sum* $\sum_{i=1}^{n}$ *can be changed to*

$$\sum_{x \in X} \sum_{i \in I_X(x)} \quad \text{or to} \quad \sum_{y \in Y} \sum_{i \in I_Y(y)} .$$

And finally, in the last equality advantage was taken of the property that in the sums

$$\sum_{i \in I_X(x)} \alpha(i,k) \log p_{X|k}(x) \quad \text{and} \quad \sum_{i \in I_Y(y)} \alpha(i,k) \log p_{Y|k}(y)$$

the values $p_{X|k}(x)$ and $p_{Y|k}(y)$ do not depend on the index i according to which the addition is done, and thus they can be factored out behind the summation symbol $\sum_i$.

We will find that as a consequence of Lemma 6.1 the numbers $p^*_{X|k}(x)$ and $p^*_{Y|k}(y)$, which maximise the sums

$$\sum_{x\in X}\left(\sum_{i\in I_X(x)}\alpha(i,k)\right)\log p_{X|k}(x) \quad\text{and}\quad \sum_{y\in Y}\left(\sum_{i\in I_Y(y)}\alpha(i,k)\right)\log p_{Y|k}(y)\,,$$

and thus also the sum (6.52), are the probabilities

$$p^*_{X|k}(x) = \frac{\sum_{i\in I_X(x)}\alpha(i,k)}{\sum_{x\in X}\sum_{i\in I_X(x)}\alpha(i,k)}\,,$$

$$p^*_{Y|k}(x) = \frac{\sum_{i\in I_Y(y)}\alpha(i,k)}{\sum_{y\in Y}\sum_{i\in I_Y(y)}\alpha(i,k)}\,.$$

The algorithm for solving the original maximisation task has the following explicit expression. Let $p^t_K(k), p^t_{X|k}(x)$, $p^t_{Y|k}(y)$, $k = 1,2$, $x \in X$, $y \in Y$, be the values of unknown probabilities after the iteration t of unsupervised learning. The new values of these probabilities are to be calculated according to the formulæ

$$\alpha(i,1) = \frac{p^t_K(1)\,p^t_{X|1}(x_i)\,p^t_{Y|1}(y_i)}{p^t_K(1)\,p^t_{X|1}(x_i)\,p^t_{Y|1}(y_i) + p^t_K(2)\,p^t_{X|2}(x_i)\,p^t_{Y|2}(y_i)}\,;$$

$$\alpha(i,2) = \frac{p^t_K(2)\,p^t_{X|2}(x_i)\,p^t_{Y|2}(y_i)}{p^t_K(1)\,p^t_{X|1}(x_i)\,p^t_{Y|1}(y_i) + p^t_K(2)\,p^t_{X|2}(x_i)\,p^t_{Y|2}(y_i)}\,;$$

$$p^{t+1}_K(1) = \frac{\sum_{i=1}^n \alpha(i,1)}{n}\,; \qquad p^{t+1}_K(2) = \frac{\sum_{i=1}^n \alpha(i,2)}{n}\,;$$

$$p^{t+1}_{X|1}(x) = \frac{\sum_{i\in I_X(x)}\alpha(i,1)}{\sum_{i=1}^n \alpha(i,1)}\,; \qquad p^{t+1}_{X|2}(x) = \frac{\sum_{i\in I_X(x)}\alpha(i,2)}{\sum_{i=1}^n \alpha(i,2)}\,;$$

$$p^{t+1}_{Y|1}(x) = \frac{\sum_{i\in I_Y(y)}\alpha(i,1)}{\sum_{i=1}^n \alpha(i,1)}\,; \qquad p^{t+1}_{Y|2}(x) = \frac{\sum_{i\in I_Y(y)}\alpha(i,2)}{\sum_{i=1}^n \alpha(i,2)}\,.$$

In a similar way as in the previous case (6.51) I have omitted superscripts t with the variables $\alpha(i,k)$, even when in the iterations they are changed.

I do not seem to have made a mistake anywhere, the more so that I used only the simplest mathematical tools for the derivation. Except for Lemma 6.1, which does not exhibit anything much complicated either, no astonishing mathematical tricks were used. I was surprised that the most difficult part of the algorithm was connected with the calculation of the 'a posteriori' probabilities $\alpha(i,k)$, i.e., in the part which cannot be avoided and which must be present even in the recognition itself. The hyperstructure, which adds the capability *of unsupervised learning to plain recognition, is so simple that it is not worth mentioning. This simplicity makes me think that I may not understand everything in a proper manner. I do not know well how to express what worries me, but I have a feeling that something important has slipped away and the very algorithm is not unsupervised learning, but a self-delusion. I cannot manage to set up a question you could answer to me because I do not know what has remained hidden from me. Is it not an incredible trick? Well now, on the basis of the general recommendations I have quite formally created an algorithm, which appeared to be absurdly simple. In spite of being simple, the algorithm claims to solve up very ambitious tasks.*

Primarily, the matter is to analyse the behaviour of a parameter, i.e., the behaviour of the state k, to find which state occurs more often and which less often, namely in a situation in which this state has never been openly observed. Certainly, it could be found if some other parameter was observed the dependence of which on the unobservable parameter is known. There are special methods of indirect measurement for this purpose. But here we have a fundamentally different situation. We have two features, x and y, which are known to depend in a way on the unobservable state, but nothing is known about kind of this dependence. And now you say that on the basis of observing only these two features one can find how the features depend on something what has never been observed. Moreover, the behaviour of that unobserved entity can be revealed. Is it not, by chance, the very nonsense about which Robbins said with self-criticising humour (I read it in [Robbins, 1951]) that it was an effort to pull oneself by one's own hair. In any case, I think that this problem deserves being more thoroughly discussed by us.

You see, it actually seems to be nonsense, but only at first and second glance. But at third and fourth glance, after scrutinising the problem with great attention, which it justly deserves, even here certain intrinsic regularities can be revealed. You are not quite right when saying that nothing is *a priori* known about the dependence of the features x and y on the state k. Although nothing is known about how either of the features x and y depends on the state k, we still know something very substantial about their joint dependence on the state. It is namely known that the features x and y depend on the state k independently of each other. In other words, if the object is in one fixed state k then the features themselves cease to depend on each other. If you are interested in it then we can discuss it later in greater detail.

Now we will only roughly answer your question. You and we altogether should not see such a great nonsense in that one can learn about something which has never been observed. The entire intellectual activity of individuals, as well as that of large human communities, has for long been turned to those parameters which are inaccessible to safe observation. We will not be speaking about such grandiose parameters as good and evil. We will choose something much simpler at first glance, for example the temperature of a body which is regarded as an average rate of motion of the body's molecules. Even though the average rate of motion of molecules has not ever been observed, it is now quite precisely known how the volume of a body, its state of aggregation, radiation depend on the average rate of molecule motion. It is also known how the temperature of the body itself depends on the temperature of surrounding bodies, which is, by the way, also unobservable. Many other properties of the body temperature are well known though they never have been directly observed.

The path leading to knowledge about directly unobservable phenomena is nothing else than an analysis of parameters which can be observed, and a search for a mechanism (model) explaining the relations between the parameters. This means an effort of exploring the relations between the observed parameters and the impossibility to explain them in another way (or more simply) than as an existence of a certain unobservable factor that affects all the visible parameters and thus is the cause of their mutual dependence. Recall astronomers who have been predicting a still unobservable planet by encountering discrepancies in observations from assumed elliptical orbits of observable planets since Kepler laws have been known. Such an approach is a normal procedure for analysing unknown phenomena. The capability of doing such exploring has since long ago been considered to be a measure of intelligence.

Could it be understood as a certain decorrelation of features?

The word decorrelation could be well matched to the purpose if it had not been already used for tasks of quite a different sort, i.e., in which the eigenvectors of covariance matrices had served as a new orthogonal base of a linear space. The method is also referred to as Karhunen–Loeve expansion. By this method, random quantities can be transformed to a form where their correlation is equal to zero.

If the decorrelation is meant as searching for an invisible influence of a phenomenon, the presence of which causes a dependence of visible parameters which would be independent if the invisible parameter did not change, then the case we can see in our example would be that very decorrelation.

Your reference to all that a human can manage cannot be any argument in judging whether the formulated task is, or is not a nonsense. I am afraid I have not yet got the answer to my question. Now, at least, I am able to formulate my question more precisely.

Let x, y, $k = 1, 2$, be three random variables the probability distribution $p_{XYK}(x, y, k)$ of which has the form of the product

$$p_K(k)\ p_{X|K}(x \mid k)\ p_{Y|K}(y \mid k)\,.$$

Let the sequence of observations of a random pair (x, y) be of such a length that for each pair of values (x, y) the probability $p_{XY}(x, y)$ can be estimated precisely enough. These data obtained empirically are in a certain mutual relation

$$p_{XY}(x, y) = \sum_{k=1}^{2} p_K(k)\ p_{X|K}(x \mid k)\ p_{Y|K}(y \mid k) \tag{6.53}$$

with unknown probabilities $p_K(k)$, $p_{X|K}(x)$, $p_{Y|K}(y)$ which are of interest for us, but cannot be directly stated because the parameter k is not observable. Assume we have chosen appropriate values $p'_K(k)$, $p'_{X|K}(x \mid k)$, $p'_{Y|K}(y \mid k)$, which satisfy the condition

$$p_{XY}(x,y) = \sum_{k=1}^{2} p'_K(k)\; p'_{X|K}(x \mid k)\; p'_{Y|K}(y \mid k)\,, \tag{6.54}$$

and thus they explain the empirically obtained data that are expressed by means of the numbers $p_{XY}(x,y)$. And now it is the turn of my question. Can I be sure that a wilful explanation $p'_K, p'_{X|K}, p'_{Y|K}$, which satisfies the relation (6.54), will be identical with the reality $p_K, p_{X|K}, p_{Y|K}$? Or, if I use a milder question: In what relation will be the explanation and the reality?

Your fears are sufficiently justified. They appear in literature in a general form referred to as the 'problem of compound mixture identifiability'. In our case the equation (6.53) is, in fact, not always sufficient for the numbers $p_{XY}(x,y)$ to unambiguously define the functions $p_{X|K}$ and $p_{Y|K}$ and the numbers $p_K(1)$ and $p_K(2)$. Everything depends on what these functions and *a priori* probabilities are actually like. In some cases hardly anything can be said about them based only on the knowledge of the statistics $p_{XY}(x,y)$. These cases, as can be seen later, are so exotic that we need not take them into consideration. Even if we somehow managed to obtain the functions necessary for us, we could see that nothing can be recognised on the basis of them. In other situations, which occur more frequently, the necessary statistical dependencies can be found, except for some ambiguity, which does not make any difference in the practical solution of some problems. And finally, on certain, but not so much restricting conditions either, the statistical model of an object can be uniquely determined.

Let us describe one method for determining the function $p'_K(k)$, $p'_{X|K}, p'_{Y|K}$ on the assumption that the probabilities $p_{XY}(x,y)$ are known. You must not think in any case that it is a method to be applied in practice. The purpose of this method is only to understand to what extent the relation (6.53) defines the functions sought. For practical application, the most appropriate algorithm is the one you have already developed.

On the basis of numbers $p_{XY}(x,y)$, $x \in X$, $y \in Y$, the following numbers can be calculated

$$\frac{p_{XY}(x,y)}{\sum\limits_{x \in X} p_{XY}(x,y)}\,, \qquad x \in X\,, \quad y \in Y\,,$$

which are nothing else than conditional probabilities $p_{X|Y}(x \mid y)$ that the value x of the first feature occurred in the experiment under the condition that the second feature assumed the value y. As before, we will express the function $p_{X|Y}$ of two variables x and y as an ensemble of several functions of one variable $p_{X|y}, y \in Y$. If we regard each function $p_{X|y}$ of this ensemble as a point in an $|X|$-dimensional linear space then we can immediately notice that all the functions $p_{X|y}$, $y \in Y$, lie on one straight line passing through the points corresponding to unknown functions $p_{X|1}$, $p_{X|2}$. It is so because

$$p_{X|y}(x) = p_{K|y}(1)\, p_{X|1}(x) + p_{K|y}(2)\, p_{X|2}(x)\,, \quad y \in Y\,, \quad x \in X\,, \qquad (6.55)$$

where $p_{K|y}(k)$ is the *a posteriori* probability of the state k at the observation y. Let us denote this straight line by the symbol Γ. The straight line Γ represents the shape of the dependence between the visible parameters x and y, which is affected by an invisible parameter k. Think its meaning over well, and then we will proceed further.

In certain cases the straight line Γ can be uniquely determined without the functions $p_{X|1}$ and $p_{X|2}$ being known, namely on the basis of empirical data $p_{XY}(x,y)$. If the set $\{p_{X|y} \mid y \in Y\}$ contains more than one function then any pair of non-equal functions uniquely determines the straight line Γ. But it can happen that the set $\{p_{X|y} \mid y \in Y\}$ contains only one single function. It happens when all functions $p_{X|y}$, $y \in Y$, are the same. In this case, the straight line Γ is not determined in a unique way, and that is the insoluble case mentioned above. It concerns the first of the abovementioned situations in which a reconstruction of the statistical model of an object based on empirical data $p_{XY}(x,y)$ is not feasible. Let us look at this situation in greater detail.

The function $p_{X|y}$ is the same for all values y in three cases (cf. (6.55)):

1. The functions $p_{X|1}$ and $p_{X|2}$ are the same; in this case the function $p_{X|y}$ does not depend on probabilities $p_{K|y}(2)$ and consequently, the set $\{p_{X|y} \mid y \in Y\}$ consists of only one single function.
2. The functions $p_{Y|1}$ and $p_{Y|2}$ are the same; in this case *a posteriori* probabilities $p_{K|y}(k)$ do not depend on y and the set $\{p_{X|y} \mid y \in Y\}$ contain again only one single function.
3. One of the *a priori* probabilities $p_K(1)$ or $p_K(2)$ is zero; in this case the probabilities $p_{K|y}(k)$ do not depend on y and moreover, one of them is always zero.

All three cases are degenerate. From an observation no information on the state can be extracted, nor in a case if the statistical model of the object was known. It is clear that no great harm is done when the function $p_K, p_{X|K}, p_{Y|K}$ cannot be reconstructed in such case. Even if they could be reconstructed, they would not be helpful for recognition.

Let us now discuss a normal situation, in which the function $p_{X|y}$ depends on the value y, which means that the set $\{p_{X|y} \mid y \in Y\}$ includes more than one function. The straight line Γ can be thus uniquely determined and the unknown functions $p_{X|1}$ and $p_{X|2}$ can no longer be of an arbitrary character. Functions must correspond to the points lying on the straight line Γ. Assume for a while that the position of these points on the straight line Γ is known. We will introduce a coordinate system on the straight line (one single coordinate) so that the unit coordinate is represented by the coordinate of the point in which the function $p_{X|1}$ is located, and zero coordinate is represented by the coordinate of the point which corresponds to the function $p_{X|2}$. If the coordinate of the point corresponding to the function $p_{X|y}$ is denoted $e(y)$ then on the basis of the relation (6.55) we can claim that the coordinate $e(y)$ is the *a posteriori* probability $p_{K|y}(1)$ of the first state on the condition of the observation y.

In this way we have made sure that the set Y of the observations y can be naturally ordered in agreement with the position of the function $p_{X|y}$ on the straight line Γ. At the same time this order is identical with the order according to the the *a posteriori* probability $p_{K|y}(1)$ of the first state. From this it then follows that any Bayesian strategy (according to the penalty function) will have just one single parameter, which will be the coordinate of a point Θ on the straight line Γ. All points on one side with respect to the point Θ are to be included in one class, and all points on the other side are to be included in the other class. And the most important of all is that for this ordering the functions $p_{X|1}$ and $p_{X|2}$ need not be known. The order which is made only on the basis of how the functions $p_{X|y}$ are placed on the straight line, i.e., on the knowledge of empirical data $p_{XY}(x, y)$, is sure to be identical either with the order according to the *a posteriori* probability $p_{K|y}(1)$ of the first state, or with the order according to the *a posteriori* probability of the second state.

Now we are able to quantitatively express the information on the classification of the set Y which can be extracted from mere empirical data. Let n be the number of values of the variable y. The set Y can be separated into two classes in 2^n ways. To express the correct classification it is necessary to have n bits. These n bits can be considered as n binary replies of a certain teacher to the question in which of the two classes each of n observations is to be included.

After appropriate examination of the empirical data the overwhelming amount of these 2^n classifications can be rejected, since the correct classification is one of the $2n$ classifications which are already known. To obtain a correct classification only $1 + \log_2 n$ bits are needed. This additional piece of information can be considered as a reply of a certain teacher to the question in which class not all but only properly selected observations are to be included.

Note that even in the case in which the functions $p_{Y|1}$, $p_{Y|2}$ and the numbers $p_K(1)$, $p_K(2)$ are completely known, the classification of the set Y into two classes will not be uniquely determined. Only a group of $2n$ classifications would be determined, where each of them, according to the penalty function, can claim that it is just the very function to be correct. Even though we can see that on the basis of empirical data the statistical model of an object is not always capable of being uniquely determined, the empirical data contain the same information about the required classification as the complete knowledge of a statistical model.

Now, let us assume that the statistical model of an object has to be determined not because of the succeeding classification but for other purposes when it is necessary to determine just the actual model. When is such a unique determination possible? We will find out on what additional conditions the system of equations

$$p_{XY}(x, y) = \sum_{k=1}^{2} p_K(k)\, p_{X|k}(x)\, p_{Y|k}(y)\,, \quad x \in X\,, \quad y \in Y\,,$$

has only one solution with respect to the functions p_K, $p_{X|k}$, $p_{Y|k}$. It is quite natural that two models which differ only with the name permutation of the states k, will be considered identical. More precisely speaking, the two models p_K, $p_{X|k}$, $p_{Y|k}$ and p'_K, $p'_{X|k}$, $p'_{Y|k}$ will be considered identical even in the case in which

$$\begin{aligned} p_K(1) &= p'_K(2)\,, & p_K(2) &= p'_K(1)\,,\\ p_{X|1} &= p'_{X|2}\,, & p_{X|2} &= p'_{X|1}\,,\\ p_{Y|1} &= p'_{Y|2}\,, & p_{Y|2} &= p'_{Y|1}\,. \end{aligned}$$

Assume that under the conditions $p_K(1) \neq 0$, $p_K(2) \neq 0$, $p_{X|1} \neq p_{X|2}$, which were assumed in the preceding analysis, another additional condition is satisfied. Let it be called the condition of ideal representatives' existence. Such a value of y_1 of the feature y is assumed to exist that can occur only when the object is in the first state. Further on, such a value y_2 exists which has a non-zero probability only when the object is in the second state. This means that

$$p_{Y|1}(y_1) \neq 0\,, \quad p_{Y|2}(y_1) = 0\,, \quad p_{Y|1}(y_2) = 0\,, \quad p_{Y|2}(y_2) \neq 0\,. \tag{6.56}$$

From the assumption (6.56) it follows that

$$p_{K|y_1}(1) = 1\,, \quad p_{K|y_1}(2) = 0\,, \quad p_{K|y_2}(1) = 0\,, \quad p_{K|y_2}(2) = 1\,,$$

and thus on the basis of (6.55) there holds

$$p_{X|y_1} = p_{X|1}\,, \quad p_{X|y_2} = p_{X|2}\,. \tag{6.57}$$

The assumption about ideal representatives applies to their existence only, and not to the knowledge of what values are being the representatives at that time.

If the assumption about ideal representatives satisfies the functions $p_{X|1}$ and $p_{X|2}$ then the representatives can be reconstructed in the following quite simple way. They can be only the first and the last elements in the set $\{p_{X|y} \mid y \in Y\}$ ordered according to the selected orientation of the straight line Γ. The functions $p_{X|1}$ and $p_{X|2}$ are, therefore, uniquely determined (except for the name permutation of the states k, which was already mentioned above). Similarly, the functions $p_{Y|K}$ and p_K can be determined.

I hope I have understood the main core of your considerations. It is that the set of functions $\{p_{X|y} \mid y \in Y\}$ cannot be of any kind, but only such a set that lies on a one-dimensional straight line. I could simply generalise this property even for the case in which the number of states does not equal two but it can be any integer number. If the number of states is n then the set of functions $\{p_{X|y} \mid y \in Y\}$ fits completely into an $(n-1)$-dimensional hyperplane where also all sought functions $p_{X|k}$, $k \in K$, which are not known beforehand, are contained.

You are right, there is the very rub!

One can still think a great deal about it. But I would not bother you with that. I hope I will be able to ferret out all possible consequences of this result myself. Now I would rather make use of the time I have at my disposal for a discussion with you to make clear for myself a more important question.

I understand that any considerations about how to create the functions p_K, $p_{X|K}$, and $p_{Y|K}$ by no means suit practical application, but serve only for explaining how the information on these functions is hidden within the empirical data $p_{XY}(x,y)$, $x \in X$, $y \in Y$. In spite of that, I would still like to pass from these idealised thoughts to real situations. Therefore I am asking, and answering myself, a key question: Why cannot these considerations be a foundation for solving a task in a real situation? It is certainly because of that in an ideal case the straight line Γ can be sought on the basis of an arbitrary pair of different points which lie on the straight line being sought, since at the end all the points lie on this straight line. But if the sequence of observations is finite, even though quite large then the probabilities $p_{X|y}(x)$ cannot be considered as known. Only some other numbers $p'_{X|y}(x)$ are known which state how many times the observation x occurred under the condition that observation y occurred. The set $\{p'_{X|y} \mid y \in Y\}$ of functions formed in this way naturally need not lie on one straight line. Thus even the straight lines which pass across different pairs of functions need not be the same. The search for the straight line Γ and the very definition of this straight line in this case is already not very easy. It is necessary to find a straight line which would appropriately approach the set $\{p'_{X|y} \mid y \in Y\}$ and appropriately approximate it. Therefore the practical solution of the task should start from formulating the criterion which quantitatively determines the way how well the straight line Γ replaces the empirically observed set $\{p'_{X|y} \mid y \in Y\}$. Afterwards that best straight line should be sought.

The result you would like to attain has already actually been presented in the lecture when the unsupervised learning task was formulated as seeking a model which is in a certain sense the best approximation of empirical data, i.e., of the finite sequence of observations. The unsupervised learning algorithm is just the procedure for the best approximation of empirical data in a situation in which actual probabilities are not available, for whose calculation indefinitely many observations of the object would be needed. We have only a finite sequence at our disposal, at the basis of which these probabilities can be calculated with a certain inevitable error. We are not certain this time what else would you like to know because we seem to have already done what you desire for.

I will try to describe my idea once more. On the one hand, we have a task to find such numbers $p_K(k)$, $p_{X|K}(x \mid k)$, $p_{Y|K}(y \mid k)$, where $k = 1, 2$, $x \in X$, $y \in Y$, which maximise the number

$$\sum_{i=1}^{n} \log \sum_{k=1}^{2} p_K(k)\, p_{X|K}(x_i|k)\, p_{Y|K}(y_i|k)$$

on the known sequence (x_i, y_i), $i = 1, 2, \ldots, n$. The value

$$\underset{(p_K, p_{X|K}, p_{Y|K})}{\operatorname{argmax}} \sum_{i=1}^{n} \log \sum_{k=1}^{2} p_K(k)\, p_{X|K}(x_i|k)\, p_{Y|K}(y_i \mid k) \tag{6.58}$$

has to be found. I wrote the algorithm for this calculation quite formally as a particular case of a more general algorithm which was presented and proved (formally as well) in the lecture. In the labyrinth of formalism I have completely lost clear understanding of how it can happen that one finds statistical parameters of a variable which has never been observed. Thanks to you, things have cleared up for me, but only for the ideal case in which the observation sequence is indefinitely long. This elucidation is supported by the property that a system of equations

$$p_{XY}(x, y) = \sum_{k=1}^{2} p_K(k)\, p_{X|K}(x \mid k)\, p_{Y|K}(y \mid k)\,, \quad x \in X\,, \quad y \in Y\,, \tag{6.59}$$

cannot have too diverse solutions with respect to the functions $p_K, p_{X|K}, p_{Y|K}$ at the known numbers $p_{XY}(x, y)$. The main factor of this elucidation is the straight line Γ which is built up in a certain manner. But a straight line cannot be seen at all in the formulation (6.58). Therefore I am not able to transfer my way of thinking, attained with your help for the ideal case (6.59), to the real case expressed by the requirement (6.58). And now I would like to beat a path from the ideal requirement (6.59), which I well understand, to the real task which I understand only formally. In working my way from the task (6.59) to the task (6.58), I would not like to lose the straight line Γ out of my view. It is for me the single clue in the problem.

I seem to see the first step in this working out of the way. In a real case the system of equations (6.59) has no solution. The straight line Γ, which is uniquely expressed in the ideal case, simply does not exist here. The task (6.59) is to be re-formulated in such a way that the straight line Γ should be defined even in the case in which the ensemble of functions $p_{X|y}$, $y \in Y$, does not lie on one straight line. Could you, please, help me make this step, but in such a way that I should not lose the straight line Γ from my considerations?

We think we could. The preliminary formulation of the task can be, for example, as follows. Let $(p'_{X|y} \mid y \in Y)$ be an ensemble of points which do not lie on one straight line. Another ensemble of points $(p_{X|y} \mid y \in Y)$ which lies on one straight line and rather strongly resembles the ensemble $(p'_{X|y} \mid y \in Y)$ is to be found. It would be natural to define the resemblance of the ensemble as a sum of a somehow defined resemblance of its elements, i.e., by means of a function which has the following form

$$\sum_{y \in Y} p'_Y(y)\, L(p'_{X|y}, p_{X|y})\,, \tag{6.60}$$

where $L(p'_{X|y}, p_{X\,|\,y})$ is the 'similarity' of the functions $p'_{X|y}$ and $p_{X|y}$, and the number $p'_Y(y)$ states how often the value y occurred in the finite sequence (x_i, y_i), $i = 1, \ldots, n$, i.e.,

$$p'_Y(y) = \sum_{x \in X} p'_{XY}(x, y) \, .$$

The number $p'_{XY}(x, y)$ states how often the pair (x, y) has occurred in the sequence of observations. Let us still recall that recall holds

$$p'_{X|y}(x) = \frac{p'_{XY}(x, y)}{p'_Y(y)} \, .$$

Now, let us consider what could be regarded as the similarity L of the functions $p'_{X|y}$ and $p_{X|y}$. The function $p'_{X|y}$ is the result of a finite observation of the object and $p_{X|y}$ is a function assumed to be the result of an infinite observation. It seems to be natural that the resemblance measure of the result of a finite experiment should be the logarithm of probability of this result, therefore

$$L(p'_{X|y}, p_{X|y}) = \sum_{x \in X} p'_{X|y}(x) \log p_{X|y}(x) \, . \tag{6.61}$$

The straight line Γ, which you would not like to lose from your considerations, can be expressed through the following formulation of the task.

Functions $p_{X|y}$, $y \in Y$, are to be found to lying on one (not known beforehand) straight line and at the same time to maximise the number (6.60), which is, with respect to the definition (6.61), the value

$$\sum_{y \in Y} p'_Y(y) \sum_{x \in X} p'_{X|y}(x) \log p_{X|y}(x) \, , \tag{6.62}$$

where p'_Y and $p'_{X|y}$ are empirical data obtained from the finite observation sequence. The straight line Γ, on which the best functions $p_{X|y}$ in the sense of (6.62), are expected to lie, is exactly that straight line you like so much.

You can see that the number (6.62) resembles in a way the number (6.58) to which we intend to work our way. We will make another step in this direction. The straight line Γ will be determined by means of two functions $p_{X|1}$ and $p_{X|2}$, which are assumed to lie on this straight line. The position of the function $p_{X|y}$, $y \in Y$, which is expected also to lie on the straight line Γ, will be denoted by means of two numbers $p_{K|y}(1)$ and $p_{K|y}(2)$. Using these two numbers we can replace the expression $p_{X|y}(x)$ in (6.62) by equivalent expression

$$p_{X|y} = p_{K|y}(1)\, p_{X|1} + p_{K|y}(2)\, p_{X|2} = \sum_{k=1}^{2} p_{K|y}(k)\, p_{X|k}(x) \, .$$

The number (6.62) which is to be maximised is thus

$$\sum_{y \in Y} p'_Y(y) \sum_{x \in X} p'_{X|y} \log \sum_{k=1}^{2} p_{K|y}(k)\, p_{X|k}(x) \, . \tag{6.63}$$

The selection of the best ensemble $(p_{X|y},\ y \in Y)$ is to be understood in such a way that the position of the straight line Γ (determined by the numbers $p_{X|K}(x \mid k)$, $x \in X$, $k = 1, 2$) and the position of the point $p_{X|y}$ for each $y \in Y$ on the straight line (which is given by the pair of numbers $p_{K|y}(1)$ and $p_{K|y}(2)$) is sought. You can see, therefore, that the maximisation (6.63) according to the numbers $p_{K|y}(k), p_{X|K}(x \mid k)$, where $x \in X$, $y \in Y$, $k = 1, 2$, is nothing else than searching for the straight line Γ which in a sense well approximates empirical data $(p'_{X|y},\ y \in Y)$. Note also that in the task (6.63) the straight line Γ has not got lost. It is expressed by the pair of functions $p_{X|1}$ and $p_{X|2}$.

We will demonstrate now that the original task (6.58) incorporates also the task (6.63). The number to be maximised according to the requirement (6.58), can be transformed in the following manner.

$$
\begin{aligned}
&\sum_{i=1}^{n} \log \sum_{k=1}^{2} p_K(k)\, p_{X|K}(x_i \mid k)\, p_{Y|K}(y_i \mid k) \\
&= n \sum_{x \in X} \sum_{y \in Y} p'_{XY}(x, y) \log \sum_{k=1}^{2} p_K(k)\, p_{X|K}(x \mid k)\, p_{Y|K}(y \mid k) \\
&= n \sum_{x \in X} \sum_{y \in Y} p'_{XY}(x, y)\, \log \left(p_Y(y) \sum_{k=1}^{2} \frac{p_K(k)\, p_{Y|K}(y \mid k)}{p_Y(y)}\, p_{X|K}(x \mid k) \right) \\
&= n \sum_{x \in X} \sum_{y \in Y} p'_Y(y)\, p'_{X|y}(x)\, \log \left(p_Y(y) \sum_{k=1}^{2} p_{K|y}(k)\, p_{X|K}(x \mid k) \right) \\
&= n \sum_{y \in Y} p'_Y(y) \log p_Y(y) + n \sum_{y \in Y} p'_Y(y) \sum_{x \in X} p'_{X|y}(x) \log \sum_{k=1}^{2} p_{K|y}(k)\, p_{X|K}(x \mid k)\,.
\end{aligned}
$$

We can see that maximisation (6.58) with respect to the functions p_K, $p_{X|K}$, $p_{Y|K}$ is equivalent to the maximisation of the number

$$
\sum_{y \in Y} p'_Y(y) \log p_Y(y) + \sum_{y \in Y} p'_Y(y) \sum_{x \in X} p'_{X|y}(x) \log \sum_{k=1}^{2} p_{K|y}(k) p_{X|K}(x \mid k)\,, \quad (6.64)
$$

according to the functions p_Y, $p_{K|y}$, $p_{X|K}$. Since either of the two summands in the expression (6.64) depends on its group of variables, the maximisation of their sum can be satisfied by maximisation of either of these two summands. The first summand is maximised according to p_Y and the second according to $p_{K|y}$ and $p_{X|K}$. We can also see that the second summand is identical with the number which is to be maximised in seeking the optimal straight line Γ. So we have beaten the path from the task (6.59) to the task (6.58). Are you now happy?

Quite happy. Perhaps, except that when we had beaten the path to the task (6.58) transformed to the task (6.64), we saw that it incorporated, besides *seeking the straight line Γ in the task (6.58), even something more. What could that be?*

That is evident. Naturally, besides seeking the straight line that approximates the ensemble of functions $\{p_{X|y},\ y \in Y\}$, another straight line must be sought which properly approximates the ensemble of functions $\{p_{Y|x},\ x \in X\}$. Without supplying this the whole procedure would use asymmetrically the information which either of the features x and y bears.

Now at last I feel that I understand the tasks and algorithms presented as if I had found them out myself. I see that these tasks deserve that I think them over well and find the algorithm for their solutions. I am not sure that the algorithm I have designed is really the right one. It is only an adaptation of a general algorithm for a particular case. The general algorithm at the lecture was proved only to converge monotonically to some fixed point. Can I be sure that in our particular case the global maximum of the number (6.58) is reached in the fixed point, similarly as it was in Robbins task?

You can be certain of that, but this certainty is based only on frequent experimental checking of this algorithm. In the theoretical way, the certainty has not yet been achieved. It might be a nice task for you.

I would definitely not want to do that. I would rather formulate the task in such a way that it should be solvable with certainty, even if it did not appear so well reputed as the tasks based on the maximum likelihood estimate. What would you say to the following procedure?

It transforms empirical data $p'_{XY}(x,y)$, $x \in X$, $y \in Y$, to the form

$$\sum_{k=1}^{2} p_K(k)\, p_{X|k}(x)\, p_{Y|k}(y)$$

provided the system of equations

$$p'_{XY}(x,y) = \sum_{k=1}^{2} p_K(k)\, p_{X|k}(x)\, p_{Y|k}(y)\,, \quad x \in X\,, \quad y \in Y\,,$$

has no solution. What can be more natural in this case than seeking such numbers $p_K(k)$, $p_{X|k}(x)$ and $p_{Y|k}(y)$, $x \in X$, $y \in Y$, $k = 1,2$, which minimise the sum

$$\sum_{x\in X}\sum_{y\in Y}\left(p'_{XY}(x,y) - \sum_{k=1}^{2} p_K(k)\, p_{X|k}(x)\, p_{Y|k}(y)\right)^2 . \tag{6.65}$$

The task formulated like this is not based on any statistical considerations but its advantage is that it has been thoroughly examined and its solution is known. It is again a task about the Karhunen–Loeve decomposition. In my context, however, this sounds a bit unusual.

This idea is new, and therefore we do not intend to restrain it at its birth. Such a formulation does not seem to us to be very natural, since it is difficult to explain the meaning of the second power of differences of probabilities. Furthermore, in further formal manipulation with the expression (6.65) quantities appear such as
'length' $\sum_x \left(p_{X|k}(x)\right)^2$,
'scalar product' $\sum_x p_{X|1}(x)\, p_{X|2}(x)$,
'matrix' $|X| \times |Y|$ the elements of which are the probabilities $p'_{XY}(x,y)$,
'covariance matrix' of the dimension $|X| \times |X|$ having elements
$\sum_y p'_{XY}(x',y)\, p'_{XY}(x'',y)$,
and other different mathematical objects that can be hard to interpret in terms of our original task. You will need a certain amount of patience to rear this idea and put it adrift to the world.

I have exploited nearly all I could from this lecture. Now I would like to place the subject matter of this lecture in the total framework of statistical pattern recognition, to which the previous lectures were devoted. You pointed out that Robbins' methods the generalisation of which is the unsupervised learning presented, had originated as an effort to fill the gap between Bayesian and non-Bayesian methods. It seems to me that this effort has succeeded only to a small extent. Already from the formulations of the tasks, we can see the great difference between the Bayesian and non-Bayesian methods on one side, and the empirical Bayesian approach and unsupervised learning on the other side. In spite of all their distinctness, the Bayesian and non-Bayesian methods have a common property. The purpose of either of them is to seek a certain recognition strategy. Unsupervised learning tasks (and, in fact, even the supervised learning) in the formulation as was presented in the lecture do not lead to any recognition strategy, but require only the most likely evaluation of a priori *unknown statistical parameters of an object. The mutual relation between supervised learning and unsupervised learning tasks, formulated in this way, and particularly the mutual relation between algorithms for their solution is undoubtedly elegant, so that I may never forget it.*

But (this unpleasant 'but' must ever occur) a quite visible gap remains between the maximum likelihood estimate of unknown parameters and the searching for an optimal strategy. It does not follow from anywhere that in the case of an incompletely known statistical model of an object the recognition strategy is to be built exactly as a Bayesian strategy, to which instead of the actual values of unknown parameters their most likely values are substituted. Such a procedure has, therefore, the form of a postulate which is accepted without giving any reasons. But a postulate should have a far simpler formulation so that a question of the type 'why exactly in this way' might not arise. Here this question is justified. The reply to it is usually based only on rather general and imprecise considerations, for example, that at a quite extent training multi-set the most likely values differ only slightly from the actual ones. Then also the strategy which uses these values differs only slightly from the best one.

Let me make an observation based on intuition. If the training multi-set is rather small, say of the length 1 for clarity, then the Bayesian strategy, in which the most likely values of unknown parameters are included, can be worse than the non-Bayesian strategy, which is based on the statement that the parameter values are simply unknown. This means that in certain cases the use of supervised and unsupervised learning, in the form formulated in the lecture, can do greater harm than if they had not been used at all. It happens so because these methods start to be used not in the case for which they have been designed (i.e., for infinitely long sequences), but they are applied where a gap appears for the time being. In practice a user must decide, only on the basis of intuition, if his/her experimental data are so rich that they can be used as if they were infinite, or if they are so short that they can lead the way to the gap. If we are enough strict in evaluating then we will come to the conclusion that empirical Bayesian methods do not fill the gap between the Bayesian and non-Bayesian methods at all. They only occupy a point in the gap, when the training multi-set is infinite.

I can see that the present day statistical pattern recognition theory is built up of three groups of methods: (1) Bayesian, (2) non-Bayesian, and (3) supervised and unsupervised learning methods. Between these three groups clearly visible gaps exist. In my view the classification of a theory to spheres having gaps is a sign of a certain incompleteness in the structure of the theory. The known methods have not yet become parts of a well elaborated hierarchical structure. I would like to build up such a theory of statistical pattern recognition that would not consist of isolated spheres at all, but which would closely cover, by methods elaborated, the whole spectrum of tasks from Bayesian to non-Bayesian ones so that they should approach the Bayesian methods in cases of the increasing length of learning; and the non-Bayesian methods in cases of shortening the length of learning. It should be something similar to the theme you presented at the end of Lecture 3.

There are now three of us who have these ambitious desires. We may sometime manage to build up such a theory. We will now follow together the well known and wise advice that one has to seek for truth, even though sometimes one has to search for the truth nearly in the dark—but one should run away fast from those who have already found that truth.

January 1998.

6.8 Link to a toolbox

The public domain Statistical Pattern Recognition Toolbox was written by V. Franc as a diploma thesis in Spring 2000. It can be downloaded from the website `http://cmp.felk.cvut.cz/cmp/cmp_software.html`. The toolbox is built on top of Matlab version 5.3 and higher. The source code of algorithms is available. The development of the toolbox has been continued.

The part of the toolbox which is related to this lecture implements the unsupervised learning algorithm for normally distributed statistical models (the

Expectation Maximisation algorithm) and the minimax learning algorithm for normally distributed statistical models.

6.9 Bibliographical notes

The research into unsupervised learning was begun by Rosenblatt with his publications [Rosenblatt, 1957; Rosenblatt, 1959] and a general interest in this problem was encouraged. A view, sober in tone, of the unsupervised learning was presented in Rosenblatt's summarising monograph [Rosenblatt, 1962]. The interest in perceptron unsupervised learning was gradually becoming weaker. A sharp criticism of perceptron learning was furnished by Glushkov [Glushkov, 1962b; Glushkov, 1962a]. In spite of that, the concept of perceptrons was further developing and was analytically examined [Minsky and Papert, 1969].

The first impulse for unsupervised learning came to pattern recognition from statistics by applying methods of clustering [Schlesinger, 1965; Ball and Hall, 1967]. In the sphere of clustering many publications have come into existence since that time.

The second impulse was the concept that unsupervised learning could be formalised similarly as in statistics, where the statistical parameters of a mixture of probability distributions were estimated. The task of criterion maximisation (6.17) is also known as the problem of a mixture parameter estimate. It was first formulated, perhaps by Pearson in 1894, for two normal densities and solved by applying the momentum method. The problems of mixture estimate was brought to pattern recognition by the publication [Cooper and Cooper, 1964].

From another view the task was seen in the fundamental statistical publications known as Robbins' empirical Bayesian approach [Robbins, 1951; Robbins, 1956; Neyman, 1962].

For many years the task withstood the efforts of mathematical statisticians to find the most likely estimate of mixture parameters. Only in the sixties of the 20th century did different authors independently of one another propose an iteration scheme generally applicable in the multi-dimensional case of a normal mixture. Iteration relations were originally intuitively derived by modifying likelihood equations which will result from the necessary condition of criterion maximisation (6.17). An interesting and practically important feature of the resulting iteration scheme was a monotonous convergence to a local or global maximum, which was first proved by Schlesinger in the year 1968 in the publication [Schlesinger, 1968] and later by others [Demster et al., 1977]. Today the procedure is known as EM (Expectation and Maximisation) algorithms. From further publications let us quote [Wu, 1983; Grim, 1986].

Lecture 7

Mutual relationship of statistical and structural recognition

7.1 Statistical recognition and its application areas

The generality of the results explained in the previous lectures has its positive and negative aspects. We have pointed to it several times. Thanks to their generality, the results have the air of laws. This must be taken into account in solving any application tasks, be it a diagnosis of the human heart according to an electrocardiogram, an evaluation of a turner's tool according to the sound it produces, processing of microscopic images in analysing blood, or the study of natural resources from satellite images. We would like to stress that diversity is not only in the interpretation of the observed and hidden parameters in applications, but also in their abstract properties. The sets X and K, from which the observed and hidden parameters assume their values, can vary greatly even in the formal mathematical sense. We will call it, not very precisely, for the time being a varied structure of these sets.

When we say, for example, that such a parameter as the weight of an object assumes the values from a well structured set we mean that the members of this set, i.e., real numbers, can be added, multiplied, inverted and many other operations of this kind can be performed with them. A set from which the mark of a pupil's school report assumes its value is to be understood in a completely different way. It is the set $\{1, 2, 3, 4, 5\}$ (in some countries, at least). The members of this set, however, are not numbers because two pupils, who were evaluated by 2 and 3 are in no sense equivalent to the one whose usual mark is 5. The set of possible values of school marks has a different structure from that of a set of natural numbers. For school marks the relations $=$, $<$, and $>$ are defined, other operations which are meaningful with the set of numbers are not defined. In other words, school marks form a completely ordered set, and nothing more. An even weaker structure is that of tram route numbers. Unlike real numbers, a tram Nr 12 cannot be replaced by two trams numbered 6. Unlike a completely ordered set, the tram Nr 2 is no better than the tram Nr 3, but, at the same time, not a shred worse than the tram Nr 1. When now speaking about the diversity of the statistical recognition theory, we mean mainly the diversity of formal properties of sets that play a part in this theory.

In pattern recognition tasks there often occurs that an observation x does not consist of one, but several measurements $x_1, x_2, \ldots, x_n$. We can speak not only about the structure of sets of values pertaining to individual features, but also about the structure of relations between the features, which is different with different applications. We will examine two cases which illustrate the different character of the structure of features (just of the set of features and not of the set of their values).

In the first example, the features $x_1, x_2, \ldots, x_n$ are answers in a medical questionnaire which is filled in at a patient's first visit at a doctor's surgery. It concerns data about the patient's body temperature, blood pressure, pulse, sex, say, n answers altogether. The second example is a case in which after the medical treatment one particular data about a patient, say the body temperature, is measured n-times at regular time intervals. The outcome of such an observation is again an ensemble of indexed values x_i, where the index i assumes values from the set $\{1, 2, \ldots, n\}$ as it was in the first case. However, the dissimilarity between the structure of this ensemble and the structure in the previous case must be evident. It was not essential in the first case for, e.g., the age to be represented exactly by the third feature, because no essential change would occur in the task if the features were numbered in another way. This means that in the first case the set of features is simply void of any structure and its members $1, 2, \ldots, n$ are considered not as numbers, but only as symbols in an abstract alphabet.

In the second case the matter is quite different. Here the feature index has just the meaning of a number. The ensemble of the measured values of the feature forms a sequence and the numbering of sequence elements cannot be arbitrary. The set of indices has now a clearly visible structure which the set of indices in the first case was empty of.

We will go back to these problems more than once and speak about them in a more concrete way. At the moment, we only want a cursory taste from the immense diversity expressed in the words 'let the sets of observations X and states K be two finite sets' which used to be quoted refrain-like at the beginning of formal reasoning in previous lectures. The results obtained are not supported by any concretisation of the form of the sets X and K. A very positive consequence is that the results are valid even in the case in which, owing to a concrete context of an application, the mathematical form of the sets X and K must be expressed more precisely. The negative consequence follows from the mentioned generality too, because the sets X and K have to be expressed as specifically as possible when the statistical methods have to be used in a useful way. This means that from the vast set of potentials a case must be chosen that corresponds to the original application task. This is by no means easy to do.

Fortunately, some applications can be expressed through formalism which has already been thoroughly examined in applied mathematical statistics. Its most developed part is the statistics of random numbers. The overwhelming majority of applied statistics recommendations is based on concepts such as mathematical expectation, variance, correlation and covariance matrices, which are meaningful only when the random object is represented by a number. There are, however, lots of applications in which the result of a feature measurement cannot be expressed by a number. General recommendations of numerical statistics cannot be applied to practical problems of such a type. If somebody wants to squeeze at any price such applied problems into the framework of a random numbers statistics then he or she deforms their original properties and eventually solves quite a different problem, not the one which had to be solved.

Perhaps, the most unlucky field of application in this respect was recognition of images (this means the recognition of two-dimensional brightness functions obtained, e.g., by a TV camera). An image is a rather unusual object for a formal analysis. Sets of images, important from one or another application aspect, do not belong to the class of sets which have been mathematically thoroughly examined outside of pattern recognition. They are not convex sets, subspaces, or something else which is well known.

The specificity of images as objects of a formal analysis is quite considerable. Substantial results in image recognition cannot be successfully achieved using merely general statistical recommendations without carefully taking into account peculiarities of images as an object of formal analysis.

7.2 Why is structural recognition necessary for image recognition?

7.2.1 Set of observations

General pattern recognition theory does not rest upon the assumption of a concrete form of the observation set X. Despite this, many users assume as self-evident that the set X is a linear space and that such understanding of the set X suits every application. An illusory logic on the background of such consideration is quite simple. Recognition is to be carried out based on a number of measurements of the object. The outcome of every measurement is a number. Thus the input information for recognition is an ensemble of n numbers. It can be regarded as a point in an n-dimensional linear space the i-th coordinate of which is the number obtained in the i-th measurement.

As far as recognition is understood as an estimation of a hidden parameter which assumes only two values, it could be described as a decomposition of the space X into two subsets X_1 and X_2. The boundary between these subsets could be interpreted as a surface in the space X defined by the equation $f(x) = 0$, where f is a function of n variables $x_1, x_2, \ldots, x_n$ and x_i is the result of the i-th measurement. It seems to be natural to regard the value $f(x)$ as positive or negative according to whether the point x belongs to the set X_1 or X_2, respectively. In this way the fundamental concepts of pattern recognition have been expressed in quite an illustrative manner from which a number of fruitful results follow. A part of them were presented in Lecture 5.

Nothing wrong can be seen in such a formalisation of the set of observations X, since every formalisation has the right to exist, as far as it does not pretend to be the only possible and universally usable one for any application. Some researchers, and there were not few of them, could not avoid this temptation.

For example, it was assumed that an image could also be represented by a point in a linear space in a natural way. An image was regarded as a function $f(x, y)$, where x, y are coordinates of a point in the domain of definition of the function f which is a square with the dimensions $D \times D$. The value of the function $f(x, y)$ was the brightness (intensity) in the corresponding point. The square becomes covered by N^2 smaller squares of the dimensions $\Delta \times \Delta$, $\Delta = D/N$. The observation of the image f corresponds to the measured value of the average brightness in each smaller square. The outcome of the observation is an ensemble of N^2 numbers considered as a point in N^2-dimensional linear space.

A long time and a lot of effort were needed to understand that such formalisation of images is highly deceptive. This representation actually deceived the correct trend of analysis. The identification of an image with a point in a multi-dimensional linear space involuntarily invited one to use such sets, transformations and functions that are well examined and verified for linear spaces, i.e., convex sets, hyperplanes, half-spaces, linear transformations, linear functions, etc.. And it is these mathematical means that are least suitable for images. That is also why the concept of linear space in the processing and recognizing of images has not started such an avalanche of fruitful results as it was the case, for example, in linear programming. Pure geometrical relations, such as the possibility of passing along the edges of a polyhedron from an arbitrary vertex of the polyhedron to any other vertex of it, greatly supported the researcher's intuition, i.e., they made it possible to be certain that this or that statement is right before it was formally proved. In the case of image recognition it was just the opposite. An avalanche of results following from understanding the image as a point in a linear space appeared destructive rather than becoming a contribution. This situation was evaluated quite sharply by M. Minsky and S. Papert [Minsky and Papert, 1969], when saying that nothing has brought about more damage to the machine analysis of images than the multi-dimensional geometrical analogies.

Let us try to understand what the fundamental difference between an image and a point in a multi-dimensional space consists of. Therefore let us first formulate both concepts so as to see what they have in common. Both the vector and the image, can be considered as a function of the form $T \to V$ the domain of definition of which is a finite set T. The function itself assumes its values from the set V. If this function is an n-dimensional vector, the coordinates of which are the numbers $x_i, i = 1, 2, \ldots, n$, then T is a set of indices, and V is a set of real numbers. If the function $T \to V$ is an image then the set T is a rectangle in a two-dimensional integer lattice, i.e., $T = \{i, j \mid 1 \leq i \leq n;\ 1 \leq j \leq n\}$, and V is a kind of a set of observations, as a rule a finite one.

If we consider the function $T \to V$ as a multi-dimensional vector then we can see that it assumes its values from a well structured set for which addition and multiplication make sense, where special members 0 and 1 exist, and many other things which provide a certain mathematical structure to the set. The domain of definition of this function, however, can be understood as an abstract alphabet only, without any structure or ordering.

With the function $T \to V$ representing an image, the case is just the opposite. There are, of course, also applications in which the domain of the values V must be considered just as a set of real numbers. An example is the tasks where the brightness in a point of the image represents the result of a direct measurement of a physical quantity, such as that in examining the temperature of a product on the rolling train, or the overall density of a body in a particular direction in computer tomography. But these are tasks of such a kind that we would not regard as typical image analysis by humans. There exist many tasks in which an observation in a certain point need not be considered so strictly. For example, for the image $x: T \to V$ and a certain monotonically increasing function $f: V \to V$ it commonly occurs that a change in brightness $x(t)$ in the point $t \in T$ to the brightness $f(x(t))$ in the same point does not alter anything from the point of view of information which one is observing within an image, on the assumption that the transformation f is the same for all points $t \in T$. This means that the domain of values V need not be considered as a set of real numbers, but it can be interpreted as a set with a far weaker structure, e.g., as a completely ordered set. Situations for which the domain of values V has an even weaker structure or it is void of any structure are not rare. Consider, for example, the analysis of the color graphical documents.

At a cursory glance it might seem that applying a richer formalism than is needed for the examined reality does not do any harm. The application of a richer formalism, however, adds to the original task properties it does not actually have. Then involuntarily algorithms are created based on additional properties such as minimisation of mean square deviation of brightness, or linear transformations of an image. Thus operations are applied which make sense within the accepted formalism, but which do not provide a reasonable explanation in terms of the initial application task.

Now let us have a look at what differentiates the domains of definition of the function $T \to V$ in the case in which the function is considered as a vector, and when it represents an image. The transition from an image to a vector is connected with a loss of immensely important properties of the image, particularly those, which account for the specificity of the image as an information medium. For the vector $T \to V$ the domain of definition is an abstract alphabet of indices without any structure. For the image $T \to V$ the domain of definition T is a rectangle in a two-dimensional integer lattice, which is a set with a clear structure. Outside this structure one can hardly imagine properties such as the connectivity of objects in an image, the symmetry of an image, and other concepts important for image analysis. We can only admire the optimism with which pattern recognition in its young days hoped that the machine analysis of images would be successfully solved without taking into account the structure of the domain of definition of the functions characteristic for the images.

In the constructive application of the general theory of pattern recognition for image analysis, in the first place a set of observations X must be concretised as a set of functions the domain of definition of which has a completely clear structure.

7.2.2 Set of hidden parameter values for an image

The statistical theory of pattern recognition has been explained in the previous lectures using a permanent assumption that the hidden parameter k of an object assumes its values from a finite set K. Earlier we have mentioned how greatly varied possibilities are contained in this brief formulation. The results of the previous lectures are valid for very heterogeneous hidden parameters. In spite of this, by many authors the *set K was implicitly understood in the following narrower sense*, in which applied domain of statistical pattern recognition theory also gets rather narrow.

1. It is assumed that the set K is a list of labels which are assigned to the object. Thus the set K is an abstract alphabet void of any structure.
2. The number of members in the set K is assumed to be so small that an exhaustive search in this set does not bring about any serious computational problems.
3. An assumption that $|K|$ is small leads to the concept of decomposition of an observation set into the classes X_k, $k \in K$. Even when any strategy $q\colon X \to K$ defines a set of subsets $X_k = \{x \in X \mid q(x) = k\}$, $k \in K$, on the set X, the application itself of the concept of 'decomposition', or 'classification' is for some functions unnatural, and thus these strategies are implicitly taken out of consideration.

The abovementioned assumptions substantially narrow the extent of what is practically done nowadays in image recognition, and so lead to an impression that there is a visible gap between pattern recognition theory and image recognition practice.

A rather rare outcome of image recognition is the image labeling by a label out of a finite and small alphabet. In usual applications the aim of recognition can be to find the position of an object (a number), to calculate its geometric parameters (an ensemble of numbers), to create a sequence of names in recognizing a text document (a sentence in a certain alphabet), to create a list of components pertaining to an electrical device and the links between them, as it is in machine interpretation of an electrical circuit diagram or of another technical drawing (a graph), or last, to create a map according to an aerial photograph (an image again). It is unnatural to consider all the above forms of the outcome of recognition as labels of a recognized image; it is unnatural, as well, to consider as a result the assignment of the input image to one of the classes.

The outcome of recognition in the case of an image is not only a symbol from a finite alphabet, but is a rather complicated mathematical object from a set with a well ordered structure. Such a concept of recognition would not go into the narrow ideas mentioned identifying pattern recognition with classification, which leads to a wrong conclusion that the statistical theory of pattern recognition has nothing in common with image recognition tasks. It would be right if someone said, 'That quite small part which I understand well in the statistical pattern recognition theory has nothing in common with the applications that I am engaged in'.

Let us summarise; for a practical application of statistical pattern recognition theory in analysing images it is necessary to thoroughly concretise the set of K values of hidden parameters which are to be found in the analysis.

7.2.3 The role of learning and unsupervised learning in image recognition

In cases in which a multi-dimensional linear space X is assumed to be a suitable mathematical model for an observation set, it is, at the same time, assumed that the strategy $q\colon X \to K$ is simple in the sense that the computation of the value $q(x)$ for each known observation x does not bring about insurmountable computational difficulties. With these optimistic assumptions the solution of the task itself would not be a complicated matter. If an algorithm for computing the values of a known linear, quadratic, or cubic function q for each previously defined point $x \in X$ is to be created then it is a not very difficult task for a programmer. No problem arises that could be an object of mathematical research. Therefore within such completely erroneous ideas, as will be seen later, the conclusion is arrived at that the purpose of scientific research in pattern recognition is limited to problems of learning and unsupervised learning. It is a serious underestimate of the problem of recognition without learning which is expressed in bombastic statements such as 'if the strategy $q\colon X \to K$ is known then the pattern recognition problem has been already solved', or 'if the statistical model $p_{XK}\colon X \times K \to \mathbb{R}$ is known, then the recognition problem becomes trivial'. We will be dealing with a simple concrete example which demonstrates how deceptive such prejudices are.

Example 7.1 Vertical and horizontal lines.
Let the set $T = \{(i,j) \,|\, 1 \le i \le m;\ 1 \le j \le n\}$ be a rectangle of size $m \times n$ in a two-dimensional integer lattice and $V = \{0,1\}$ (i.e., white, black) be a set of observation results. A function of the form $T \to V$ will be called an image. We will denote by h_i an image which we will call the i-th horizontal line, $i = 1,\dots,m$. Within this image the observation $h_i(i',j)$ has the value 1 if and only if $i' = i$. In a similar way we will denote by v_j, $j = 1,\dots,n$, an image which we will call the j-th vertical line, in which $v_j(i,j') = 1$ if and only if $j' = j$. We will denote by the symbol h the set of all horizontal lines and by the symbol v the set of all vertical lines, $h = \{h_i \mid i = 1,\dots,m\}$, $v = \{v_j \mid j = 1,\dots,n\}$.

Let k be a subset of horizontal and vertical lines which does not contain all horizontal and all vertical lines, i.e., $k \subset v \cup h$, $v \not\subset k$, $h \not\subset k$. The set of all possible subsets created in this way will be denoted K, $|K| = (2^m - 1)(2^n - 1)$. For each group $k \in K$ of horizontal and vertical lines an image x will be created which depends on the group of images k in the following manner: $x(i,j) = 0$, if and only if within each image of the group k the observation in the point

(i, j) is zero. Simply speaking, the image x is created in such a way that several horizontal and vertical lines are coloured in black, but not all possible horizontal and not all possible vertical lines.

Let us formulate the pattern recognition task so that, based on the knowledge of the image x, it is necessary to find k, i.e., to tell what lines are drawn in the image.

The task is actually a trivial one. Since the ensemble k does not include all horizontal lines there exists a row i^* in which there is not a horizontal line. In addition, there exists a column j^* in which there is not a vertical line. Thus in the image x there is a point i^*, j^* which is white. Therefore the solution of the task is the following: the ensemble k contains a horizontal line h_i if and only if $x(i, j^*) = 1$, and it contains a vertical line v_j if and only if $x(i^*, j) = 1$.

We will now take into consideration the inevitable fact that the image x cannot be observed without noise. We will see that even with the simplest model of noise the task not only ceases to be trivial, but it is not solvable in polynomial time.

Let the observed image be changed as a result of the noise to become the image $x' \colon T \to \{0, 1\}$ so that in each point $(i, j) \in T$ the equation $x(i, j) = x'(i, j)$ is satisfied with the probability $1 - \varepsilon$, and the inequality $x(i, j) \neq x'(i, j)$ is satisfied with the probability ε. Not to diverge from the very simplest model of the noise we will assume that the noise affects different points of the image independently. When we assume that all ensembles k in the set K are equally probable the statistical model $p(x', k)$, i.e., the joint probability of the ensemble k and observation x', is then uniquely determined, and for each k and x' can be easily calculated according to the formulæ

$$p(x', k) = \frac{(1-\varepsilon)^{m\,n}}{(2^m - 1)(2^n - 1)} \prod_{i=1}^{m} \prod_{j=1}^{n} \left(\frac{\varepsilon}{1-\varepsilon} \right)^{|x'(i,j) - \alpha(i)\,\beta(j)|},$$

where $\alpha(i) = 1$, if $v_i \in k$, $\alpha(i) = 0$, if $v_i \notin k$, $\beta(j) = 1$, if $s_j \in k$, $\beta(j) = 0$, if $s_j \notin k$. The calculation according to this formula for either pair x' and k is not even worth mentioning. The function $p(x', k)$ is thus not only uniquely determined, but moreover it has a very simple form. Nevertheless, the pattern recognition task remains unsolvable. The task could be, in this case, defined as seeking the most probable ensemble k which consists of horizontal and vertical lines under the conditions of the known observation x'. It is not difficult to understand that this task is reduced to a minimisation task

$$\Big(\alpha^*(1), \ldots, \alpha^*(m)\,; \beta^*(1), \ldots, \beta^*(n)\Big) = \operatorname*{argmin}_{\substack{(\alpha(1),\ldots,\alpha(m),\\ \beta(1),\ldots,\beta(n))}} \sum_{i=1}^{m} \sum_{j=1}^{n} |x'(i,j) - \alpha(i)\,\beta(j)|$$

which is to be solved under the condition that the variables $\alpha(i)$ and $\beta(j)$ assume only the values 0 or 1. It is not very difficult to write an algorithm for this minimisation problem. However, this algorithm will require a computing time which is proportional to $2^{\min(m,n)}$, and thus cannot be practically applied. No one in the world today seems to know substantially better algorithms. The *minimisation task quoted is of such a fantastic complexity that if anybody were to solve it successfully, without needing an enormously long time for the minimisation, it would be a worldwide sensation.* ▲

In the previous example it can be seen how a seemingly easy task appears to be practically unsolvable. We are speaking about an exact solution of an exactly formulated task, and not about so called practically acceptable suggestions which, though they do not guarantee a solution, but in the majority of practical cases they ... etc.. Neither do we mention here the so called practically acceptable algorithms, since they can be discussed only in the case in which the algorithm is intended for solving a task, which is of real practical significance. Our example is mere child's play in which it is to be found if there are ideal lines in the image. Practical tasks are substantially more difficult, because in the image far more intricate objects on a more complicated background are to be sought. The situation is further complicated by not fully known errors which affect the observed image. When coming across such tasks the incomplete theoretical understanding of the applied algorithm can be excused, since the requirements for theoretical perfection give way to far harder requirements of a practical character. In contrast to practical tasks, a theoretical clarity should be achieved in the extremely simplified case quoted, be it positive or negative.

It is typical for the task mentioned that the set K of hidden parameter values is so immensely large that one cannot apply algorithms in which finding and examining of each member of the set K would be expected, even though the examination of each single member is extremely simple. The immense extent of the set K is nothing extraordinary in image recognition. In image recognition, the set K is, as a rule, so extensive that its exhaustive enumeration is impossible in practice. The complexity of image recognition tasks, however, does not only consist in this extent. In the example mentioned, we can formulate the aim of the recognition, e.g., in such a way that it is only to find whether a horizontal line passes through the fifth row of the image. The task in this case would just be the classification of the image set into two classes: into images in which the particular line occurs, and into the rest of the images. Unfortunately, the property that the hidden parameter assumes only two values does not bring about any simplification of the problem. The choice of the most probable value of the hidden parameter would not give rise to any difficulties in this case, since the parameter assumes two values only. However, insurmountable obstacles would appear in computing the *a posteriori* probabilities of these two values.

The example mentioned demonstrates the usual difficulties in image recognition which do not consist in the lack of knowledge about necessary relations, sets, or probabilities. This knowledge might be available, of course, but it does not belong to the well respected classes of mathematical objects, which owing to centuries' old study and research, have assumed a perfect and elegant form. They are, simply, something new.

7.3 Main concepts necessary for structural analysis

The peculiarity of images as objects of machine processing requires also specific mathematical methods for their analysis which will be explained in the following lectures. In the same way as in the previous lectures we will do our best so that the explanation may have a form usual in mathematical research, i.e., it should be based on unambiguously formulated definitions, premises, and results. Such an explanation necessarily anticipates a certain degree of abstraction. Abstract concepts are inclined to live their own lives after being born, and to get into contact not merely with those application problems which gave birth to them. Therefore the ideas described in the following lectures are known not as the theory of image recognition, but as the theory of structural pattern recognition, and it is right that it is so. With the greatest possible generality of constructions which we will be speaking about, the structural methods cannot cover all the richness of problems in image analysis, and, on the other hand, some outcomes are a contribution not only for images.

We will now present the main concepts which will be used in the following explanation.

Let an object be characterised by a certain set of parameters. Even though it need not always be quite natural, let us imagine that the parameter values are written in memory which belongs to the object itself. Let us denote the memory by the symbol T and call it the *object field* (the field of the object). We would like to point out that T means only the memory, and not what is written into the memory. The memory as a set of parameters T consists of *memory cells* t which correspond to individual parameters. For the time being we will assume that all parameters take their values from one single set S which is the same for all parameters.

An object is *completely described* when for each parameter $t \in T$ its value is known, which will be denoted $s(t)$, $s(t) \in S$. Formally speaking, the description of an object is the function $s \colon T \to S$ the domain of which is the set of parameters T (which is the same as the memory T), and its value domain is the set S.

The *recognition task* here and in other parts of the lecture will be understood in this way: based on the knowledge of the memory's contents pertaining to its part $T' \subset T$ something meaningful is to be said about the contents of the rest of the memory. Or in another formulation: based on the knowledge of values pertaining to some parameters, the values of others are to be found. Even from this informal definition we can feel its affinity with statistical pattern recognition, but here from the very beginning respect has been taken that both the observed and the hidden parameters are considered as sets of parameters.

For a more precise formulation of the concepts needed we will accept the following notation. Let X and Y be sets. The set of all possible functions of the form $X \to Y$ is denoted by Y^X. It is clear that $|Y^X| = |Y|^{|X|}$. Let $f \in Y^X$ be a function of the form $X \to Y$ and $X' \subset X$ be a subset in X. The *restriction of the function* $f \colon X \to Y$ to the subset $X' \subset X$ is defined as the function $f' \colon X' \to Y$, where for all $x \in X'$ the relation $f'(x) = f(x)$ holds.

For the restriction of a function we introduce a slightly unusual notation. The restriction of the function $f: X \to Y$ to the subset $X' \subset X$ will be denoted by $f(X')$. Let us have one point $\sigma \in X$. The symbol $f(\sigma)$ denotes the value of the function in the point σ which is the restriction of the function $f: X \to Y$ to the subset $\{\sigma\}$, where $\{\sigma\} \subset X$. We will make one further step. We will use the notation $f(X')$ even in the case in which X' is a subset, not a point. The notation $f(X')$ is no longer one value, but a function defined on the subset X', see Fig. 7.1. A misunderstanding cannot occur since according to the argument of the function $f(X')$ in parentheses it can be easily seen if the result is one single value (when the argument is a single point), or a function (when the argument is a subset).

Let us consider a decomposition of the set T (i.e., the field of objects) into two subsets: the set Tx of observable parameters (observable field) and the remaining set Tk of hidden parameters (hidden field). The symbols x, k in the introduced notation are understood as a part of the symbol and not as an index.

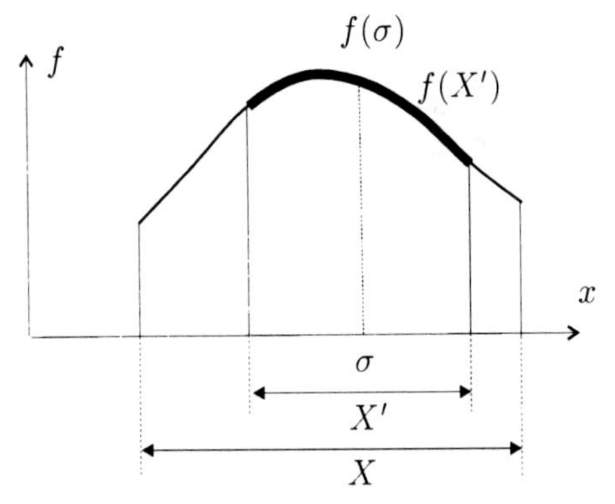

Figure 7.1 Restriction of the function $f(X')$.

The *recognition task*, still understood informally, assumes the following form: There is a function $s: T \to S$ which is defined on the known set T and assumes values from the known set S. The function $s: T \to S$ is not known, but the restriction $x: Tx \to S$ of the function s to the known observed field $Tx \subset T$ is known. The task is to determine the restriction $k: Tk \to S$ of the function s to the hidden field $Tk \subset T$. The function $x: Tx \to S$ represents the same notion that was understood as an observation in the previous lectures. The function $k: Tk \to S$ corresponds to the previous hidden state of an object. The function $s: T \to S$, which is nothing else but a pair created from functions $x: Tx \to S$ and $k: Tk \to S$, will be called *complete description of the object.*

It is obvious that the formulation mentioned is not the task definition. The hidden function $k: Tk \to S$ could be found on the basis of the observed function $x: Tx \to S$ only in the case in which the relation between functions $k: Tk \to S$ and $x: Tx \to S$ were to be known *a priori*, i.e., if a constraint to the complete description $s: T \to S$ of the object would be known. We will define this relation by means of two different but still similar ways. In the first case a subset $L \subset S^T$ of parameters' values ensembles will be determined that are admissible for the object, i.e., the subset of functions $s: T \to S$ that may occur. In the second case the function $p_T: S^T \to \mathbb{R}$ will be determined. This function decides for any function $s: T \to S$, i.e., for each ensemble of parameters, what is the probability $p_T(s)$ of the function (as well as the ensemble) occurrence.

The foreshadowed *formulation of the relation between observable and hidden parameters* is exactly the *link connecting our forthcoming explanation of structural pattern recognition with foregoing elucidation on statistical pattern recognition.* Structural pattern recognition is not torn away from statistical pattern recognition, actually, the former is part of the latter. In structural pattern recognition the set of admissible pairs (x, k) and their probabilities are defined using its own specific means. Let us bring them up.

The set $\mathcal{T} = \{T_1, T_2, \ldots, T_m\}$ of subsets of the object field T will be called a *field structure* and each element of the field structure will be called a *field fragment.* The number of elements in the largest fragment of the structure $\mathcal{T}$, i.e., the number $\max_{T' \in \mathcal{T}} |T'|$, will be called the *order of the structure.* For instance, when the structure $\mathcal{T}$ contains only a few pairs of elements from T, the structure has order two, i.e., the structure is simply the (unoriented) graph. We shall show which means will be needed to define the subset $L \subset S^T$ of admissible description of the object and then how the probability distribution on the set of descriptions can be defined.

Let $\mathcal{T}$ be a structure and for each fragment $T' \in \mathcal{T}$ let a subset $L_{T'} \subset S^{T'}$ be defined. The function $s: T \to S$ is defined as admissible when its restriction to each fragment T' of the structure $\mathcal{T}$ belongs to $L_{T'}$. So the relation between observable and hidden parameters of the object, i.e., the subset $L \subset S^T$, is determined by means of the ensemble of subsets $L_{T'} \subset S^{T'}$, $T' \in \mathcal{T}$. It is obvious that not every subset $L \subset S^T$ can be expressed in this way. *The domain of structural recognition contains only those cases in which the complex set can be decomposed into several simpler ones.* We will see that even this constraint leaves the domain of structural recognition quite broad.

The probability distribution $p_T: S^T \to \mathbb{R}$ is defined in another, but still similar manner. For each fragment T' from the structure $\mathcal{T}$ the probability distribution $p_{T'}: S^{T'} \to \mathbb{R}$ is determined, i.e., the function that for each function of the form $T' \to S$ assigns its probability. What is the relation between probability distribution p_T and probability distributions on the fragments T'? Let us assume that the random function $s: T \to S$ is generated according to the original probability distribution p_T. The restriction of this random function into the fragment T' is $s(T')$. This restriction is random too, and its probability is $p_{T'}\big(s(T')\big)$.

Example 7.2 Probabilities on fragments and on the entity. *Let the object field be $T = \{1, 2, 3\}$ and the structure $\mathcal{T}$ contain the pairs $\{1,2\}$, $\{1,3\}$, $\{2,3\}$. From the assignment it follows that we consider three random variables s_1, s_2, s_3. The distribution of their joint probabilities is described by the function $p_T(s_1, s_2, s_3)$. This function is not known, but we know three distributions of joint probabilities $p_{\{12\}}(s_1, s_2)$, $p_{\{13\}}(s_1, s_3)$ and $p_{\{23\}}(s_2, s_3)$. These three functions constrain the possible function $p_T(s_1, s_2, s_3)$, since it must satisfy three equations,*

$$p_{\{12\}}(s_1, s_2) = \sum_{s_3} p_T(s_1, s_2, s_3)\,, \quad p_{\{13\}}(s_1, s_3) = \sum_{s_3} p_T(s_1, s_2, s_3)\,,$$

$$p_{\{23\}}(s_2, s_3) = \sum_{s_3} p_T(s_1, s_2, s_3)\,.$$

▲

The assumption postulated in the paragraph before the preceding Example 7.2 does not define the probability distribution p_T uniquely. The unambiguity is achieved using additional assumptions which are of the Markovian form. A precise formulation of these assumptions will be left for future lectures. Here we point out just once more the characteristic feature of structural pattern recognition: a complex function with a wide domain s^T is defined by means of a number of simpler functions with a restricted domain of definition. And as before, the reduction of a complex concept into a number of simpler ones is not always possible. And it is the possibility of such a reduction that defines the action radius of structural methods which is very extensive even despite this limitation.

We will illustrate the concepts mentioned and make use of the task outlined in Example 7.1 of horizontal and vertical lines.

Example 7.3 Horizontal and vertical lines, illustration of concepts. *To store the full description of an object it is necessary to have* $(mn + m + n)$ *memory cells in which information on the image* $(mn$ *cells), on the set of horizontal lines* $(m$ *cells), and the set of vertical lines* $(n$ *cells) is stored. The set of cells* T, *i.e., the object field, is a set* $\{(i,j) \mid 0 \le i \le m,\ 0 \le j \le n,\ i+j \ne 0\}$. *Only a part of the cells in this set is observable, namely, the part containing the information on the image. This part is the set* $\{(i,j) \bigm| 1 \le i \le m,\ 1 \le j \le n\}$ *and it represents the observed field* Tx. *The other hidden part of the field* T *in which the information on horizontal and vertical lines is stored, contains the cells* $(i,0)$, $i = 1,\dots,m$, *and the cells* $(0,j)$, $j = 1,\dots,n$. *Thus the hidden field* Tk *is the set* $\{(i,0) \bigm| 1 \le i \le m\} \cup \{(0,j) \bigm| 1 \le j \le n\}$. *The set of values which can be stored in the cells is evidently* $\{0,1\}$, *where the numbers* 0 *and* 1, *written in the hidden cells, carry information on whether a certain line occurs in the image. The same numbers, written in the observed cells, inform on the values of brightness (here black and white only) in certain positions of the image.*

The relationship between all parameters of the object, be the noise taken into account or not, is expressed by virtue of a structure of the third order, and thus through the structure which contains triplets of cells of the form $\big((i,0),(0,j),(i,j)\big), 1 \le i \le m,\ 1 \le j \le n$. *So the structure of the object is the set* $\mathcal{T} = \big\{\big((i,0),(0,j),(i,j)\big) \mid 1 \le i \le m,\ 1 \le j \le n\big\}$.

In the case in which the noise is not taken into consideration the restriction of the description $s\colon T \to S$ *is defined by the set of permitted triplets* $\big(s(i,0), s(0,j), s(i,j)\big)$ *of values that can occur in the fragment* $\big((i,0),\ (0,j),\ (i,j)\big)$. *This set is the same for all fragments, which is* $\{(1,1,1),\ (1,0,1),\ (0,1,1),\ (0,0,0)\}$. *This set formally expresses the relationship between hidden and observable parameters which was informally stated before. It is indicated that the cell* (i,j) *can be black only when the* i*-th horizontal line or the* j*-th vertical line is represented in the image.*

When the noise is taken into account its probability then must be given for each triplet $\big(s(i,0),\ s(0,j),\ s(i,j)\big)$. *In our case we assume that the probabilities do not depend on the coordinates* (i,j). *It is therefore necessary to have* $|S|^3 = 8$ *numbers which are presented in Table 7.1.* ▲

		$s(i,j)=1$	$s(i,j)=0$
$s(i,0)=0$	$s(0,j)=0$	$\frac{1}{4}\varepsilon$	$\frac{1}{4}(1-\varepsilon)$
$s(i,0)=0$	$s(0,j)=1$	$\frac{1}{4}(1-\varepsilon)$	$\frac{1}{4}\varepsilon$
$s(i,0)=1$	$s(0,j)=0$	$\frac{1}{4}(1-\varepsilon)$	$\frac{1}{4}\varepsilon$
$s(i,0)=1$	$s(0,j)=1$	$\frac{1}{4}(1-\varepsilon)$	$\frac{1}{4}\varepsilon$

Table 7.1 Eight probabilities describing noise in Example 7.2.

The concepts illustrated in the previous example do not yet define any pattern recognition task, but make us acquainted with the 'characters' which will be acting in the task. These are the concepts the formulations of which will precede the formulation of structural recognition problems, similarly as the sentence 'let X and K be two finite sets, the function $p_{XK}\colon X \times K \to \mathbb{R}$ being defined on their Cartesian product' preceded the formulation of a task in the general statistical theory of pattern recognition.

In further explanation, the previous brief sentence will be replaced by the following more detailed introductory sentence. Let T and S be two finite sets, where T is the set of parameters and S is the set of values which each parameter assumes. Let $Tx \subset T$ be a subset of observed parameters, and the parameters from the set $Tk = T \setminus Tx$ be hidden. Let $\mathcal{T}$ be a finite set of subsets from T. Let for each set $T' \in \mathcal{T}$ the following be determined:

either the subset $L_{T'} \subset S^{T'}$;

or the probability distribution $p_{T'}: S^{T'} \to \mathbb{R}$.

Various tasks will be formulated after these introductory sentences. Generally the tasks will require that according to the known function $x\colon Tx \to S$ defined on an observed field, the function $k\colon Tk \to S$ defined on the hidden field, should be found somehow. The following lectures will be devoted to the analysis of tasks of this kind for different classes of structures, starting from the simplest cases and switching over to the analysis of the problem in its complete generality.

7.4 Discussion

Lecture 6 is a kind of dividing line in that we completed a section of pattern recognition theory. I would now like to ask you a favour. Upon my supervisor's recommendation, I acted as a tutor for a seminar of the optional subject Pattern Recognition for students of Applied Informatics. As a basis I used your lectures. Some questions I have asked you in our discussions were questions asked by my students in the seminars.

It has occurred to me that it would be helpful if I organised a seminar to summarise and verify what the students had learned. Could you possibly make a list of questions for me to use in the seminar?

The previous lectures were to ensure that she/he who studied them would be well oriented in the following concepts and tasks.

1. Observable and hidden parameters of an object, strategy, penalty functions, risk, Bayesian tasks of statistical decision making.
2. Probability of the wrong recognition (decision) as a special case of Bayesian risk; the strategy which minimises the probability of the wrong decision.
3. Bayesian strategy with allowed non-decision.
4. The deterministic character of Bayesian strategies.
5. The Neyman–Pearson task and its generalisation to a case in which the number of states is greater than two.
6. Minimax task.
7. Wald task of the form presented in Lecture 2.
8. Testing of complex hypotheses, statistical decision making after Linnik. (Let us note that the form of strategy for solving the tasks under items 5-8 is to be derived by using two theorems of duality).
9. Risk, empirical risk, Chervonenkis–Vapnik theorem on the necessary and sufficient condition of convergence of the empirical risk to the risk itself.
10. The growth function and the capacity of a set of strategies, sufficient conditions for convergence of the empirical risk to the risk itself.
11. Anderson task in generalised form, necessary and sufficient conditions for optimality of strategies in Anderson task.
12. Linear strategy, algorithm of perceptron and Novikoff theorem.
13. Linear strategy, Kozinec algorithm and its validation.
14. Fisher strategy by means of linear separation algorithm.
15. ε-solution of Anderson task by means of linear separation algorithm.
16. Formulation of clustering tasks, taking the ISODATA algorithm as an example.
17. Empirical Bayesian approach by Robbins, Robbins task and its generalisation.
18. Mutual relationship of learning and unsupervised learning in pattern recognition, unsupervised learning algorithm, a theorem about its monotonous convergence.

That may be all. We wonder how successful your students will be in seminars and at examinations.

I expect that future lectures will be substantially different from the previous ones. It may be a question of an entirely different subject matter which will be based on new concepts. That is why I would like to know if the next subject matter can be understood without knowing the previous one. Some students who did not attend my seminars would like to join them. Is not now the right moment for them to join the course?

No, it is not. The following subject matter cannot be understood without a thorough knowledge of the previous one. Your question reminded us once more of the rooted, but erroneous concept of statistical and structural pattern recognition as two mutually independent areas of pattern recognition. With such a conception only a small part of application tasks can be mastered. The image recognition problems, in which complex structural relations between image fragments need not be taken into account, are quite rare. Similarly, rare are tasks where a random character of images need not be considered. The randomness may be, at least, inevitably affected by noise in the observed image. And thus, in solving practical tasks both of the images' properties must be considered, their complex structure as well as their random character.

I am afraid you do not seem to have understood my question in the right way. I did not ask whether for a practical activity in the pattern recognition field both, the previous and the future subject matter, are necessary. I have some doubts about it as well, but I am going to ask my question later. I am now above all interested in whether the subject matter of future lectures can be understood without the knowledge of what was explained in the preceding ones.

We have understood your question in the right way and we answer once more: The following lectures cannot be delivered to those who did not properly master the previous subject matter. In solving practical tasks, two parts of the images' nature must be considered in a more ingenious way than by merely applying purely statistical methods in a certain stage of processing, and then applying purely structural methods in the later stage. The applied algorithms are to make use of both, statistical and structural features of the image. Therefore one cannot say if the probability of the wrong decision is being minimised at some step of the algorithm, or the structural relations between fragments of the image are being analysed. Both activities are performed together at each moment. In our future lectures we will direct our attention towards designing such algorithms and therefore a competence in statistical methods is a prerequisite.

You are saying that for the future explanation everything that was previously dealt with is needed. Could you, perhaps, select the key concepts? I would like to recommend them to the students who are going to join us now so they can study the topics by themselves.

Boy, you are rather strongly insistent today. You want, at all costs, to find something that was unnecessary in previous lectures. Well, as the saying goes, discretion is the better part of valour.

To understand the subsequent subject matter it would be required to know the Bayesian formulation of the pattern recognition task, the formulation of learning and unsupervised learning tasks as the maximum likelihood estimation of unknown statistical parameters. The designing of unsupervised learning algorithms should be understood almost automatically. Moreover, it is necessary to master the procedure seeking the linear discriminant function by means of Kozinec or perceptron algorithms which is nearly the entire contents of Lecture 5.

I would rather be less insistent. But the subject matter is far too extensive. I am not afraid of it since I have carefully studied and understood all of the subject matter. At the same time I wonder if we will still need algorithms linearly separating subsets in a linear space. Methods based on formulations by a set of observations in a linear space were subject to such a sweeping criticism from your side that I believed that I should forget them as something that allures the student astray. When you are saying that we will still need these methods in structural analysis, I cannot come to terms with it.

It might sound strange even to us, if we did not get used to peculiar destinies of some scientific ideas that develop in a quite unpredictable and unexpected way and start to live lives of their own. Let us again recall Minsky and Papert [Minsky and Papert, 1969], who compared pattern recognition to a mathematical novel in which the characters can disappear for some time and reappear at the right moment. Only in a lapse of time, could it be seen that the contribution of those ideas is far richer than it seemed at first glance. Even when the criticism against the linear approach is justified we will, despite this, actively apply methods seeking separating hyperplanes in linear spaces as soon as we deal with problems of learning in structural pattern recognition. Do not try to cope with this contradiction now. When we arrive at this subject matter, the knowledge will settle down in a desirable way.

I already wonder! It is an unexpected turning point in the events. Rosenblatt invents algorithms seeking automatically a proper strategy in a certain class of strategies. After a time, this entire class is found to be not very well adapted for image analysis, and as an alternative, the structural methods appear. Due to it, even Rosenblatt algorithms disappear from the stage and the learning problem retires from the scene. After a time, when the learning methods start to be applied again, Rosenblatt algorithms will appear to assert themselves in quite another class of strategies. Do I understand it correctly?

Yes, you do.

I understood the mathematical contents of the previous lectures. Could you tell me, please, what other mathematical concepts will be used in further explanation so that I may prepare?

Besides probabilities, which we have actively used, we will need a number of additional concepts. The fundamental and best known of them are: Markovian chain, graph, the shortest path in a graph, dynamic programming, automaton, regular expression, NP-completeness, Chomsky's formalisms for languages and grammars, regular and context free language. The development of some algorithms of structural analysis will be based on several concepts of abstract algebra, such as semi-rings with an idempotent addition operation, modules on these semi-rings and isomorphisms on them.

Do not worry about the new concepts for the time being. They are the most elementary concepts, and the meaning of each of them will be explained in the lecture itself. Roughly speaking, to understand the next subject matter, no special mathematical knowledge is required other than that which was dealt with in previous lectures. But what will be helpful is the capability and zest for acquiring knowledge.

It seems to me that I have understood the main concepts of the lecture which will become the basis for future formal constructions. The objects of research will be functions of the form $T \to S$, where T is a finite set for which a structure $\mathcal{T} \subset 2^T$ is defined, where 2^T denotes a system of all subsets of T. Why was nothing said about the structure of the set S?

The structure of the set S will not be taken into consideration anywhere in our lectures. The course was built up, in a sense, in contrast with the theory which formalises the observed object by means of points in a linear space, i.e., by means of a function which assumes numerical values, but is defined on a set void of any structure. We will examine the opposite of such an assumption and will see what results follow from a case, where the domain of definition of the function is a well structured set, and despite this, the structure of the value set is not taken into consideration.

It is natural that this diametrically opposed view has its drawbacks. But you must not forget that the anticipated view of the tasks is rather general. From the fact that the structure of the value set of a function is not considered, it does not follow that the results of such an examination hold only for sets void of any structure. The other way round, there will be results that are valid for sets with whatever structure they may have. The other thing is that on certain assumptions about this or that set of values which the function under examination assumes, other additional results could still be obtained. These open questions, however, will not be dealt with in our lectures.

I have revealed a certain contradiction connected with the case of horizontal and vertical lines which was quoted in the lecture. On the one hand, it has been said that in the image only such a group of horizontal and vertical lines is regarded as admissible which does not include all the horizontal and vertical lines. This restriction was essential in deriving a pattern recognition algorithm of a noise-free image.

On the other hand, if a set of admissible images was defined by means of structural methods then this restriction was not taken into consideration. The structure $\mathcal{T}$ and the constraint on the fragments of this structure were defined in such a way that any subset of lines was admissible, including those which had been regarded as inadmissible. I would not bother you with this contradiction if I thought that it was just a mere oversight from your side. When I tried, however, to do away with it, I was unable to, and it seemed to me that it might be something more important than merely a slight mistake. The matter is, in fact, that the ensemble $(s(i,0),\ i = 1,2,\ldots,m)$ cannot be an arbitrary one, but one in which there exists at least one number $s(i,0)$, different from 1. This requirement, which restricts the ensemble of m variables, cannot be reduced to *constraints of smaller groups of variables. I may be able to prove it because I hardly made an error in my considerations. I assume you know it even without my comments. If I am right then I can be afraid that the case of horizontal and vertical lines is just the one which cannot be coped with even by the theory which is to be explained later.*

We will reply in the style of the Clever highlander girl (a character from the popular Czech fairy tale). Yes as well as no. Do not ask for a more precise answer from us now. We would like to draw your attention to the fact that the question, whether the restriction to a certain set of variables can be expressed by means of relations in their subsets, is not so simple. We will examine this question in detail later.

We will now only demonstrate how, by the means mentioned in the lecture, a requirement is to be expressed that the ensemble of numbers $\big(s(1,0)$, $s(2,0)$, ..., $s(m,0)\big)$ must contain one zero at least. You are right that this constraint cannot be reduced to partial constraints. But auxiliary variables $\big(z(0), z(1), \ldots, z(m)\big)$ can be introduced such as to make the restriction possible. A local constraint will be introduced for the variables $z(0)$, $z(m)$ and for the triplets of variables of the form $\big(z(i-1), s(i,0), z(i)\big)$, $i = 1, 2, \ldots, m$. The constraints will result in the following meaning of variables $z(i)$: the quantity $z(i)$, $i = 1, 2, \ldots, m$, is zero if and only if at least one of the quantities $\big(s(1,0), s(2,0), \ldots, s(i,0)\big)$ is zero. Local constraints, therefore, state in which $z(0)$ must be 1, $z(m)$ must be 0, and the triplet of values $\big(z(i-1)$, $s(i,0)$, $z(i)\big)$ must be one of the following four triplets: $(1,1,1)$, $(1,0,0)$, $(0,1,0)$, and $(0,0,0)$.

I see that I have opened a box with rather complicated questions. I realised that some complex 'multi-dimensional' relations can be expressed in a simplified form. This simplification consists of reducing 'multi-dimensional' relations to a larger number of 'less-dimensional' ones. It is also clear to me that there exist such multi-dimensional relations for which the before mentioned reduction is not possible. For the present, can I assume in advance that some multi-dimensional relations can be simplified only when to the original variables additional variables are added?

When we connive at a not very satisfying precision of your statements, you are right.

From what has been said, a lot of new questions arise. But I am afraid that I am in too much of a hurry when I wish to know just now what will be spoken on at the lectures. Still before I start with the questions, I would like to make clear for myself, roughly at least, the relation between the previous and future theories. It may be untimely to make an attempt to do it just before the explanation gets started. But the formulation of concepts itself, such as the set of observations S*, the object field* T*, the observed field* $Tx \subset T$*, the hidden field* $Tk \subset T$*,* $T = Tx \cup Tk$*, the structure of the field* $\mathcal{T} \subset 2^T$ *and the system of subsets* $L_{T'} \subset S^{T'}$*,* $T' \in \mathcal{T}$*, or of functions* $p_{T'}\colon S^{T'} \to \mathbb{R}$*,* $T' \in \mathcal{T}$*, already* *creates the framework of the future theory, and gives an idea of what will be dealt with at the lectures. I have a certain idea at least. For the time being, my idea does not suffice for me to answer the question whether the structural theory is a concretisation of the statistical pattern recognition theory, or its generalisation.*

On the one hand it was said in the lecture that the structural theory is a particular case of the general statistical theory, which concretises the form of the set $X \times K$*, whereas the general theory is not based on the assumption concerning the form of the sets* X *and* K*.*

On the other hand, the general theory as a whole can be regarded as a trivial case of the structural theory in which the observed field Tx *consists of one single cell* tx*, the hidden field consists of a single cell* tk*, the set* S *is* $X \cup K$*. The structure* $\mathcal{T}$ *is* $\{\{tx, tk\}\}$*, i.e., it consists of one single subset. The system of functions* $p_{T'}: S^{T'} \to \mathbb{R}$*,* $T' \in \mathcal{T}$*, also consist of a single function* $p_{XK}: X \times K \to \mathbb{R}$*. In this case the structural pattern recognition can be regarded as an immense generalisation of the previous statistical theory of pattern recognition which in itself is impractically general. Is it really so?*

Certainly, it is. Do not take it too seriously. The statements of the type 'X is a special case of Y' or 'X is a generalisation of Y' express something unambiguously understandable only when the objects X and Y are formally defined in an unambiguous way. In our case these are two rather general theories. An exact expression of the relationship between them would require the construction of a kind of metatheory beforehand, i.e., a theory about theories. But we are not inclined to do it, and so forget the last question, please.

Even though I did not make my previous question with much seriousness, I still cannot dismiss it completely from my mind because I am interested in it for quite earthly, and therefore serious reasons. Now may be the most opportune moment for me to remember why, in fact, I started to study your lectures. If you still remember, my interest in pattern recognition was stimulated by the fact that I had written a program for character recognition. I made it on the basis of considerations that seemed to me quite reasonable, but in spite of that, it did not work satisfactorily.

Now I can tell you that it was a program for recognizing standard printed characters. There are a lot of similar programs commercially available. None work perfectly, but one can make a good profit from such programs. The program I wrote was not based on the results of pattern recognition theory for the very reason that most of them were not known to me. Now, thanks to your lectures, I know the fundamental results of the theory, but I do not have even a vague notion of how to go back, by virtue of these outcomes, to my earthly objectives. The outcomes of the pattern recognition theory are still abstract for me, even when I do not cover up the reason for my interest. I hoped that in the second part we would pass, at last, from the abstract level to a more earthly one. But now I see that instead of returning from the sky to our Earth, we are flying still higher, and I do not know when my quite concrete tasks get their turn. I am almost afraid that I am heading further and further away from my original objective for the sake of which I had set off, together with you, on a laborious journey. Could I, possibly, ask you the question what out of the previous subject matter, and perhaps also out of the future one, may be directed at my not very noble interests?

Why could you not ask? But it may be clear to you that your question is far more difficult than all the previous ones. First, it requires from us the acquisition of your application task, and second, to know what you have actually done. In any case the answer to your question can be neither short, nor complete. If you did not find by yourself what to use from the previous subject matter then no short answer will persuade you. Let us try to reconstruct what troubles you have already undergone and what is still ahead of you. You might feel disappointed seeing that we did not assess or explain your steps adequately. Therefore, you had better imagine that we are not speaking about you, but about someone else.

You certainly tried, at first, to cope with the task on a simplifying assumption that only one single character is present in the image. Thought further simplification, the character can have only one of two labels, say, A or B. The character can be located in any place on the image, it has a certain rotation and size. You assume that when there are more characters arranged in several lines certain complications will arise, but you will leave their solution to a later time. At first, you will cope with primary problems, which lie on the surface and do not seem to be difficult for you.

You will scan several hundred images, each of them consisting of $150 \cdot 100$ pixels. The character in the image covers a rectangle which consists of $30 \cdot 20$ pixels. You made the size of the image five times larger than the size of the character because you anticipate that you will soon intend to recognise simple texts which consist of 25 characters written in five lines by five characters each. For the time being there is only one character in each of the scanned images which can be placed anywhere. A character can assume any out of $(100 - 20) \cdot (150 - 30) = 9600$ positions. Thanks to binary scanning, the image will be represented in the memory of your computer by a two-dimensional array x the element $x(i, j)$ of which is 1 or 0 according to whether the pixel with coordinates (i, j) is white or black. When you display the ensemble on the screen, and you view your characters, you will see how perfectly you can recognise them for your part, and in optimistic spirits you will write a program, which, as you are sure, will also cope with character recognition.

Design the program on the basis of reasonable considerations. You will prepare two small images of dimensions $20 \cdot 30$ by which you represent two recognised characters, A and B, which are, in your opinion, ideal. You call them exemplars. Then you will copy either of the two exemplars 9600 times so that you will obtain 9600 images of dimensions $100 \cdot 150$, where in each image there is a character A in all possible positions. In a similar way, you will obtain 9600 images for a character B. The images will be denoted so that

v_{tk}, $t = 1, 2, \ldots, 9600$, $k = A, B$, will mean the character k in the position t. The image x will be recognised so that you will calculate the dissimilarity of the image x from each of the $2 \cdot 9600$ images. You will find the image $v_{t^*k^*}$ which is least dissimilar from the image x, and you will find that the image x represents the character k^* in the position t^*. You do not think long about how to measure the dissimilarity of the image v_{tk} from the image x. It is clear to you that it should be Hamming distance

$$\sum_{i=1}^{150} \sum_{j=1}^{100} | x(i,j) - v_{tk}(i,j) | , \tag{7.1}$$

and so you define the result k^* of recognising the image x as

$$k^* = \operatorname*{argmin}_k \left(\min_t \sum_{i=1}^{150} \sum_{j=1}^{100} \Big| x(i,j) - v_{tk}(i,j) \Big| \right) . \tag{7.2}$$

You will use the written program for practical experiments and their results will astonish you. They are not only bad. They are unexplainable.

On the one hand, the program makes frequent errors, on the other hand, the errors appear in cases other than those you would expect. You would be able, for example, to explain if an error occurred for the character A, which would be so much distorted by noise that it could be hard to find, in an objective way, if it concerned the character A or B. But such errors are hardly noticed by you. The algorithm quite frequently and with certainty decides for the character A, even in the cases in which you can clearly see that a rather correct character B was presented to it. At first, you assume that an error has crept into the program. After carefully checking the program, you will arrive at a conclusion that you will have to explain the wrong behaviour of the program on the basis of additional experiments. There you will come across difficulties since your program is too slow to be apt for extensive experiments. And so you will arrive at the knowledge that you will have to deal with questions such as the recognition speed which you intended to put off until the time when the algorithm worked properly.

In the effort to make the program faster you will notice, at first, that according to the algorithm, i.e., the relation (7.2), the dissimilarity (7.1) is to be calculated for every possible position t of the character in the image. You assume that it could be possible to find the actual position t_0 of the character by a less expensive means. If the actual position t_0 would be known then the answer k^* sought can be determined by a simpler algorithm

$$k^* = \operatorname*{argmin}_k \sum_{i=1}^{150} \sum_{j=1}^{100} \Big| x(i,j) - v_{t_0 k}(i,j) \Big| . \tag{7.3}$$

instead of the calculation (7.2). We will not deal with the question which of the known algorithms has to be used. The reason is that all of them are rather bad. Actually you used one of them.

It is important for us that the program which implements (7.3) works at about a thousand times higher speed than the program based on (7.2). Now you can examine the algorithm not only for several but for many examples. None of the results will please you. Wrong results are many, but now you start to understand what causes them. You see that Hamming's distance (7.1) rather grossly evaluates what you would like to regard as the dissimilarity between images, because it starts from the assumption that all the pixels in the image are equivalent. But the experiments showed you that different pixels had had different natures. Some pixels in the image A are nearly always black, others are of a rather stable white colour, the third ones are more white than black, the fourth ones are both, white or black, with rather equal frequency. You will sort the pixels of the images into classes, separately for the character A, and separately for the character B. In this way, you will amend the formula for stating the dissimilarity between characters. But neither is this of much help to you, and you are slowly coming to a conclusion, that the diversity of pixels cannot be expressed only by sorting them into classes, but that each pixel must be characterised by a weight coefficient with which the particular pixel is present in the formula (7.1). When you see that in selecting the coefficients based on sound reasoning you still cannot attain correct results, you begin to feel that you are in a tight spot. We may have met in just this moment of your worries and started to discuss our lectures.

After the third lecture when we discussed the two simplest statistical models of observation, you could understand by yourself how to select the weight coefficients. You will easily see that the dissimilarity of the image x from the exemplar v_A of the character A, or from the exemplar v_B of the character B is to be calculated according to the

$$\sum_i \sum_j \alpha_{ij}^k \left(x(i,j) - v_k(i,j)\right)^2, \quad k = A, B \tag{7.4}$$

which can be interpreted in a double sense. If both the observed image x and the exemplars v_A and v_B are binary then the coefficient α_{ij}^k is to be adjusted to the value

$$\log\left(\frac{p_{ij}^k}{1 - p_{ij}^k}\right),$$

where p_{ij}^k is the probability that in the image x, which actually displays the character k, the observation $x(i,j)$ will not have the value $v_k(i,j)$. That means that there will be no observation in the pixel (i,j) which would be expected in an ideal image, i.e., undistorted by noise.

But it can happen that you will find at this stage of analysis that many erroneous decisions were due to wrong binarisation of the original image which was performed by the scanner. You can choose a more exact way of data acquisition by scanning more brightness levels. The brightness is then represented not by a binary, but by an integer number in an interval, say, of 0 through 255. In this case you can express the numbers $v_k(i,j)$ in the formula (7.4) as mathematical expectation of the numbers $x(i, j)$ in the actual image which represents the number k, and the coefficient α_{ij}^k as $(\sigma_{ij}^k)^{-2}$, where σ_{ij}^k is the variance of that number.

It is an awfully primitive method of recognition, but it is still more advanced than the calculation of Hamming distance which seemed to you the only possible one at the beginning. These concrete recommendations are not the most important. More significant is that you have learned in the third lecture from which assumptions about the probability distribution $p_{X|K}$ these recommendations follow. If these assumptions are satisfied then the algorithm recognising characters must (!!!) operate not only in a correct, but also in an optimal way.

Now, when your algorithm has continued yielding unsatisfactory results (and the results will be really bad all the time), you need not examine that algorithm of yours, but you can view the images themselves without their recognition, and try to find what assumptions about the probabilities $p_{X|K}$ from which the application of the formula (7.4) follows are not actually satisfied.

Stop it, please. I apologise for being impolite, but I feel I must interrupt you. Up to now you have managed to give a true picture of my past troubles. It was clear to me that we still keep, in our field of view, our aim which is to create an algorithm for recognising images containing texts. But now you actually recommend me to quit refining the algorithm and to begin a kind of new research. Just now, I started having a feeling that we are beginning to go away from my original task and get stuck in new research from which there would be no way out. I am afraid that in your further explanation my feeling of going away from the original task will be reinforced. When you are at the end of your explanation I will be flummoxed by it. I would not know how to adapt to my task the interesting information which I will certainly learn from you.

You are right to have interrupted us. We are not going away from the solution of your original task, but only reveal that we are still some distance away from the solution. Please realise that when two people are at the same distance from the target, but only one of them knows the distance, it is he that has an advantage over the other. Moreover the second person starts from the wrong assumption that he is quite near to his target.

The path to solving the pattern recognition task inevitably leads through examination of the the model of the object that is to be recognised. Sometimes the path is long, sometimes short, it is not always pleasant, but it is a passage. He who would prefer to avoid it completely can arrive at a marshland. Let us examine the formula (7.4), and try to answer why it does not work properly even if it works a bit better than the relation (7.3). The reply to this question will be sought in examining what premises which validate the procedure (7.4), are not actually satisfied.

If you examine even a small number of images then you will rather quickly discover that at least one premise is not satisfied on the basis of which the formula (7.4) was derived. It is the premise concerning the independence of brightness parameters $x(i, j)$ in different pixels of the observed image. You will see that there are groups of pixels in which the same, or very similar brightness is observed, but which changes from one image to the other. In addition you will notice that in the image such pairs of regions exist within which the brightness is the same, or nearly the same, but the brightness parameters in pixels of different regions are usually different. A dependence can even be observed if brightness parameters in one region rise then in another region fall down, etc..

Primarily you have the idea that dependencies of this kind can be expressed by the general Gaussian model of multi-dimensional random variables. The particular case of this idea is the independence of the components of the multi-dimensional variable which was the basis for deriving the formula (7.4). You are quite entitled to be afraid of expressing the revealed dependence by covariance matrices which would assume huge dimensions in this case, i.e., the number of pixels in the image raised to the power of 2. Already for tiny images of the dimension $30 \cdot 30$ pixels we can speak about a covariance matrix of the dimension $900 \cdot 900$. Of course, if you did not have any knowledge of the nature of brightness dependence in pixels then you would have no other simpler choice than to express the dependence by means of covariance matrices.

But you have quite thoroughly examined and understood your characters. You have arrived at a conclusion that if a character really occurred in the same position in the image then the dependence of brightness in pixels would be far less. The dependence between pixels mostly results from the property that your method of finding the position of a pixel in an image, be it of any kind, is rather inaccurate. The actual position of a character may differ from the position found by your algorithm, say, by one, or two pixels to the left, to the right, downwards, or upwards. That is just why the brightness parameters in the pixels of an actual image do not change independently of each other but all at once for the whole pixel group. It happens because a character as a whole has moved, say, by two pixels to the right. If you know not only the mere fact that brightness parameters in pixels are mutually dependent, but, moreover, you know the mechanism of this dependence then you can state the dependence not by covariance matrices, where your knowledge might dwindle away, but by another model in which the revealed mechanism of the dependence will be explicitly expressed. For this case it would be best to consider the probability distribution $p_{X|K}$ as a mixture of twenty-five partial distributions. The value 25 is the number of all possible positions in which a character can actually occur when your algorithm has defined a certain position for it, i.e.,

$$p_{X|k} = \sum_{t=1}^{25} p^k(t)\, p^t_{X|k}\,, \tag{7.5}$$

and this means that the strategy for recognition should be as follows: it has to decide that the image represents the character A if

$$\sum_{t=1}^{25} p^{\mathrm{A}}(t)\, p^t_{X|A}(x) \geq \sum_{t=1}^{25} p^{\mathrm{B}}(t)\, p^t_{X|B}(x)\,, \tag{7.6}$$

and the character B in the opposite case. In the formulæ (7.5) and (7.6), the number $p^{\mathrm{A}}(t)$ states how often the displacement, numbered by t, of the actual position of the character A from the position found by your algorithm has occurred. A similar meaning is that of the numbers $p^{\mathrm{B}}(t)$ which, of course, are not the same as $p^{\mathrm{A}}(t)$ because the accuracy with which your algorithm finds the position of the character A need not be the same as the accuracy of finding the position of the number B. The function $p^t_{X|k}$ is the probability distribution of the image which displays the character k in the position t. It is a distribution about which one can now assume with greater certainty that the brightness parameters in individual pixels are mutually independent. The influence due to the change of the position of the character, which was the main cause of this dependence, is out of the question.

In applying the formulæ (7.5) and (7.6), you must not assume that the probabilities of each displacement t are the same because your way of finding the position of a character is not altogether bad. Quite often, the position of the character is found in a correct way, and the error in displacement by two pixels does not occur so often as that by one pixel. The recognition algorithm has to take into account these properties of your algorithm which searches for the position of a character, and therefore 50 numbers $p^k(t), t = 1, 2, \ldots, 25$, $k = A, B$ must be known to apply the formulæ (7.6). This is, however, a far more specific task than the task of searching for the recognition algorithm which does not always provide an easy survey. Your attention is drawn to the fact that we deal only with the estimate of the algorithm used to recognise the true position of the character, not with its improvement. The algorithm for locating a position need not be exact but you must know precisely the measure of its inaccuracy.

Your explanation seemed to me so comprehensible that I immediately started writing a program which implements the method (7.6), intending to check the algorithm. But all at once, it appeared that I am not capable of determining those unlucky 50 constants $p^k(t)$ which must be included into the program. To be able to determine them, I am expected to make a number of experiments with images of one class, say, with the characters A, and in each experiment to determine two quantities: the position t found by my algorithm, and the actual position t_0 of the particular character. From these two quantities, only the quantity t is known to me. I cannot find the actual position. When I observe a character on the screen then I can easily estimate if it is the character A or B. But it is completely impossible to determine, if the actual position of a character is the one which the algorithm has found, or if it differs from the actual position by one or two pixels. The actual position is simply unknown to me. Neither have I any algorithm for finding this actual position. If I had it at my disposal then no problems would arise and I would be in much higher spirits.

But this is not the end of my troubles. To apply the algorithm (7.6) I have to know not only the 50 numbers $p^k(t)$, but the ensemble of functions $p^t_{X|k}$ as well. To find the ensemble, I would have to get a set of images in which one character in one position is displayed. I do not have such a set at my disposal, in turn, I do not have the algorithm which would faultlessly indicate the actual *position. I feel like I am in a vicious circle. I cannot do without an algorithm for accurately determining the position of a character, but on the other hand, I do not know how to scan and print a set of characters about which I could claim that they are characters all in the same position. That makes me think that the function* $p^t_{X|k}$ *describes a multi-dimensional random variable which cannot be determined experimentally.*

Eventually I can see that, after all, I become astray in the marshland, even though it was not in the immediate design of the algorithm, but in examining the statistical model of the recognised object, in my case, that of a character image.

You must not be angry with us, but we misled you in the moor land on purpose. We wanted you to notice by yourself that the task waiting to be solved is an unsupervised learning task. Moreover, it is exactly that concrete case you so thoroughly dealt with in the discussion after Lecture 6. We did not bring it to your attention that it was the unsupervised learning task which was waiting for you. It was just for you to see by yourself that unsupervised learning was not a mere intellectual game solved with one's head in the clouds. It is a procedure for solving quite earthly statistical tasks which occur at the first steps of the statistical examination of a complex object, such as an image, if it is examined quite seriously, of course. Well, do not be afraid of getting into the marshland because you know how to get out of it.

For the estimate of parameters in the statistical model (7.5) apply the algorithm which you so brilliantly found in the discussion after Lecture 6. You even wondered at its simplicity. We are already eager to know what will it result in.

I declare victory. Though, I am afraid of speaking too soon. The results of the recognition algorithm into which I included the parameters calculated by means of unsupervised learning appeared to be very satisfactory. I could already regard them as practically applicable. During the relatively long experiments I noticed not a single (!!!) error.

We do not intend to lower your results. But do not forget that the capacity of the set of strategies in the form of (7.6) is quite large in your case. You have to take into consideration all outcomes of the statistical theory of learning that we discussed with you in Lecture 4. They claim that the results of recognition, obtained in learning, can in some cases differ rather substantially from the results you will observe when applying a non-variable algorithm.

Do you not, possibly, think that I have forgotten about the results in Lecture 4? It is clear that I checked my algorithm on data other than that I used for stating the parameters of the model in (7.5).

In advance, and with some caution for the time being, we can already congratulate you. Note that now you have not only the algorithm for recognising characters, but also their statistical model. It is by far more than a mere algorithm since you can continue examining the statistical model for further purposes, and not only for constructing the recognition algorithm.

I have already started doing so. At first I was interested in what the partial probability distributions $p^t_{X|k}$ *looked like. I expected that they would seem as if the algorithm for unsupervised learning itself came to the conclusion that the image is affected by such a parameter as the position of a character. This would be so, if the functions* $p^t_{X|k}$ *at different parameters* t *mutually differed by the displacement. But I did not observe anything like that. I was not able to interpret in a comprehensible manner the group of functions* $p^t_{X|k}$, $k = A, B$, $t = 1, 2, \ldots, 25$ *by which the algorithm approximated the set of observed images of one class.*

Well, as you can see, it was the position of characters that was the influence which first occurred to us when we wanted to explain why the brightness parameters in different pixels were mutually dependent. Actually there are even more parameters of this kind. For example, we can consider yet another parameter which is the thickness of character-lines. If we examined the characters even further, we would come across other parameters as well which are hidden at first glance. The unsupervised learning algorithm devoted more time to this examining than we did, and therefore, within the scope of potentialities presented to it (which were limited by the number 25), it found an appropriate approximation for all parameters which cause the dependence between pixels.

I dare assume that even the statistical model on which recognition is based corresponds to reality at the first approximation. I tried to examine this model even further. I recognised images that were not scanned as such, but which were generated by means of that particular statistical model. For some purposes, these experiments are preferred to the experiments with real images. If I generate an artificial image then I know everything about it. Not only do I know what character is displayed, but I also know at what parameter value t *the character was created. I cannot obtain such information in any way by real experiments because I even do not know the meaning of that parameter. Naturally I even cannot determine its real value.*

When I had completed the experiments with artificial images I found something which arouses some fear in me. I will try to explain this. I found that the probability of the wrong decision substantially depends on the parameter t*. At some values of* t*, and those are the more probable, the recognition is of a far better quality than at those which are less probable. It does not surprise me much because the strategy created is aimed at minimising the average probability of the wrong decision. Primarily, the strategy tries to achieve a correct recognition of images that were generated for values of the parameter* t *with greater probability.*

Though all these considerations are reasonable, I would not like to follow them hastily. After all, I would not like to regard the parameter t *as a random* *one because I do not understand its meaning completely. I would like to create another strategy which would secure, instead of the minimal average probability of the wrong decision, a good magnitude of this error at any value of the parameter t. This means that I must formulate a non-Bayesian task for the already created statistical model and solve it from the very beginning. Shall I make such a decision? If I choose this path then it will mean that I will have to throw the whole of my program away. I regret it. I have already put much effort into it, and finally it is not so bad.*

We start to like you again, because you have found what we wanted to call your attention to. But you are not right when you fear that you have to throw all your work into a waste-paper basket. If you go through Lecture 2 once more then you will see that the task in the new non-Bayesian formulation is solved by a strategy of the same form as the strategy (7.6) which you have already programmed. You just need to include in it other values of the quantities $p^k(t)$ than those you obtained by means of the unsupervised learning algorithm. You must calculate them in a different way. We believe that you will come across it without our help.

But this means I have programmed the unsupervised learning algorithm in vain when I am expected to throw its outcome away.

By no means! The unsupervised learning has provided you not only with the *a priori* probabilities of values of the parameter t, but in addition the conditional probabilities $p^t_{X|k}$ which you will make use of now.

We will answer one more question even though you have not yet asked it. We will not preclude that you will manage to design an algorithm which will secure quite good results of recognition at any value of the parameter t. You can try to simplify the recognition strategy itself, and instead of the strategy of the form (7.6), you can use, say, a linear discriminant function. This can be created as a solution of a generalised Anderson problem. You have the mathematical model of your images at your disposal, and you can apply it for different purposes.

So, you can see that even in your rather simplified task nearly all tasks and their solutions were used that were referred to at a theoretical level in our lectures.

O.K. But in spite of that, the identification of abstract concepts with a concrete application task seems to me to be rather painful and not straightforward.

That is not so. When one has learnt one's application thoroughly and is quite at home in the theory then the connection is clear and it should immediately catch one's eye. We are, therefore, greatly surprised that you needed our explanation concerning the connection of the subject matter from our lecture with your task. We noticed that you are quite at home in the theory. The only explanation may be that you did not carefully examine the images which you were trying to recognise. You may have believed that knowledge of the theory could make up for the lack of knowledge about the application task. Well, you were wrong. Once more we will remind you that the theory of pattern recognition is no magical means for a lazybones who relies on the 'Magic Table' method!

But if one knows one's application rather thoroughly, he or she may solve it even without the theory. Am I not right?

That is true. The only question is how much time it would take him or her. Assume that in an application an area below a graph of a polynomial function of one variable in the interval from 0 to 1 is to be calculated. Your question could be transposed to this example in the following way. Can anybody solve this task without knowing the integral calculus? It is, of course, possible because nothing stands in the way of creating the concept of the derivative, finding formulæ for differentiating the polynomial function, creating the concept of the integral, and proving Newton and Leibnitz theorems. If he or she had known these concepts and relations between them prior to it, he would have solved the task faster.

We analysed my application task in detail. Should I regard it as good luck that in solving it I managed with only the subject matter from the lectures on statistical pattern recognition? Do I not also need the knowledge of structural pattern recognition turn of which will follow?

The task we have just analysed can be regarded as an application task only with a large amount of politeness. We have only analysed a case in which the recognised image consists of one single character. You certainly admit that it is child's play when compared to a real application task.

We have solved the most difficult part of it, I think. To solve an interesting practical task, I have nothing more to do than divide a large image into smaller rectangles circumscribing every character. Then, it could be possible to solve a task of real application significance, could it not?

No, not as a whole, but quite a large part of it. Do not think, however, that you will manage to break up a text into individual characters by means of some simple aids. We again remind you that you should look carefully at the texts you would intend to recognise. Even a cursory glance at real images is an effective means for sobering up. In a real image you can notice that neighbouring characters in one line touch one another now and then and so the space between them is lost; due to noise, individual characters come apart so that the set of black pixels stops being connected; rectangles circumscribing the characters differ in heights and widths; the neighbouring position of two characters in a line forms a configuration, which consists of two halves of two different characters, and this configuration cannot be distinguished from some other character which actually does not occur in that place (such as the pairs oo $\rightarrow$ x, xx $\rightarrow$ o, cl $\rightarrow$ d, ic $\rightarrow$ k, lc $\rightarrow$ k); rectangles that circumscribe individual characters are not disjunctive, etc., etc..

When you get a notion of all these tiny and not so tiny treacheries of an actual image, you will come to the conclusion that recognition of character lines cannot be so easily reduced to recognition of individual characters of which the line consists. Then you will deduce the correct conclusion which is that you have to formulate your task as recognition of a whole line at once. This may drive you mad since the number of possible sequences in such a line is astronomically large. That will be the right moment for you to read the following two lectures and see that there already exist methods which are quite well elaborated for recognising a whole sequence at once. On the basis of these methods, you would be ready to create an algorithm for recognising images with texts that could be respected even in the world of applications, but only on the assumption that there are reasons for believing that you are capable of dividing the image into rectangles circumscribing one and just one line. Only in this case you will manage to reduce the recognition of a whole page to the recognition of individual lines.

In cases in which the lines are closely adjacent to one another, you will probably not manage to break up the text by means of simple tricks into individual lines. You will have to formulate the task as the recognition of a whole page at once. In this case you will have to read further lectures of ours where algorithms for recognising such two-dimensional structures are designed.

I am already looking forward to it. But perhaps one little remark. I do not know how I should express my fear. I would much regret if the entire integral calculus was applicable only for the calculation of the area below a graph of polynomial functions. It might not be worth creating it only for the sake of such a narrowly oriented task.

We understand your fear. The operating radius of structural methods is extensive and covers much more than mere recognition of images with text. But do not ask us to analyse another application area with you. It is now your work, and we would not wish to do it for you.

February 1998.

7.5 Bibliographical notes

This lecture provided an overview on the relation between statistical and structural methods rather than a solid scientific result. A part of the given thoughts was published in [Schlesinger and Gimmel'farb, 1987]. We have not observed a similar approach by other authors. Our approach constitutes a general starting point to the lectures to come. Here, the statistical methods will be used for structural recognition.

The ill-considered representation of images in the multi-dimensional space or equivalently as vectors is criticised in [Minsky and Papert, 1969]. In [Beymer and Poggio, 1996] it is demonstrated that representing images as vectors makes sense only if the correspondence problem is solved. This means that in two or more images the pixels are detected that correspond to the same location in the scene. This location often corresponds to a salient landmark in the scene.

Lecture 8

Recognition of Markovian sequences

8.1 Introductory notes on sequences

A sequence being, perhaps, the simplest structure which for a number of reasons is best suited for us to use to start our talk on structural recognition. Even in such a simple special case it can be shown how to recognise a complex object which consists of many parts and how the knowledge of relations between the parts contributes to better recognition of both the object as a whole and its components as well. We will see that although the algorithms for recognising complex objects are not always trivial, in their realisation no insuperable computational complications will occur. We will realise that the recognition problems, supervised and unsupervised learning which are formulated in the statistical pattern recognition, can be, in a studied particular case, solved exactly and without additional simplifying assumptions.

Formal models described in this lecture are of interest because they express important features of an extensive class of real recognition tasks. We are going to give two examples of such practical tasks. It is not our intention to create the impression that the intrinsic content of these practical tasks is completely covered by expressing them using Markovian sequences. We only wish to show through examples the meaning of abstract concepts which will be used in later explanation.

Example 8.1 Recognition of a line in images. *Let $x_1, x_2, \ldots, x_n$ be a line of text, i.e., a sequence of images representing letters. Each image corresponds to a letter from alphabet K. The aim is to decide what sequence of letters $k_1, k_2, \ldots, k_n$ corresponds to the observed sequence of images $x_1, x_2, \ldots, x_n$. If the letters labeled k_i, $i = 1, \ldots, n$, were mutually independent then the task to estimate the sequence k_1, $k_2, \ldots,$ k_n would be reduced to n mutually independent estimation tasks each yielding the label of the individual letter. The decision concerning which letter stands in the position i would be made only with respect to the image x_i in the sequence $x_1, x_2, \ldots, x_n$. Such a decision would not be the best one in a real situation in which letters in the line of text* *cannot be regarded as mutually independent. The dependence (context) provides such significant information that often, with respect to the knowledge of labels* $k_1, k_2, \dots, k_{i-1}, k_{i+1}, \dots, k_n$ *and the image* x_i*, the label* k_i *can be found even in the case the image* x_i *is severely damaged by noise. The problem is that not a single letter* k_i *is known at the beginning and therefore the task consists in recognising the complete sequence* $k_1, k_2, \dots, k_n$ *at once, i.e., with respect to the complete sequence of images* $x_1, x_2, \dots, x_n$*, but considering the previously known mutual dependence of individual letters.* ▲

Example 8.2 Medical diagnostics. *Let* $x_1, x_2, \dots, x_n$ *be the result of measuring certain parameters of a patient in some particular instants and* $k_1, k_2, \dots, k_n$ *be the sequence of his or her states of health which is not directly observable. The objective is to find the different states of health as exactly as possible with respect to the sequence of observed parameters. Health states of the patient in different moments are not independent. The knowledge of this dependence contributes to a more accurate estimate of individual states and their sequence.* ▲

Methods of sequence recognition for certain formulations of recognition tasks will now be thoroughly examined. The characteristic feature of these methods is the application of dynamic programming to find the optimal sequence of hidden states $k_1, k_2, \dots, k_n$. The original source of these methods are papers by Kovalevski [Kovalevski, 1967] who designed the methods for recognition of lines of text. Vincjuk [Vincjuk,] was the first to apply dynamic programming to speech recognition. The methods spread worldwide in a short time. Nowadays, the respect for these methods is so great that it starts to be detrimental to them. Slowly the original formulation of tasks, for the solution of which dynamic programming is well suited, is being forgotten. Dynamic programming algorithms are applied even in situations in which other algorithms are more suitable.

In this lecture we define a certain class of statistical models of the recognised object and within this class we formulate various Bayesian recognition tasks. Some of these tasks are solved by the well known methods of dynamic programming. Other less known methods are suitable for solving other tasks.

8.2 Markovian statistical model of a recognised object

Let an object be characterised by $2n+1$ parameters that are expressed by two sequences $\bar{x} = (x_1, x_2, \dots, x_n)$ and $\bar{k} = (k_0, k_1, \dots, k_n)$. Sequences will be denoted by the bar over the symbol representing the sequence in this lecture. *Parameters* $k_0, k_1, \dots, k_n$ are *hidden* and parameters $x_1, x_2, \dots, x_n$ are *observable.* The sequences $\bar{x}$ and $\bar{k}$ are random and assume values from the sets X^n and K^{n+1}, where X is a set of all possible values of each observable parameter x_i and K is a set of values of each hidden parameter k_i. The connected subsequence $(x_{i_1}, x_{i_1+1}, \dots, x_{i_2})$ will be denoted $x_{i_1}^{i_2}$ and the subsequence $(k_{i_1}, k_{i_1+1}, \dots, k_{i_2})$ will be denoted $k_{i_1}^{i_2}$. The symbol x_1^n then means $\bar{x}$, the symbol x_i^i represents x_i, the symbol k_0^n means $\bar{k}$, and k_i^i represents k_i.

We will speak of joint and conditional probabilities of the given parameters and different groups of these parameters. All the probabilities will be denoted by a single symbol p. For example, the notation $p(x_i, k_i, k_{i-1})$ will be used for a joint probability that the i-th observable parameter has assumed the value x_i, the $(i-1)$-th hidden parameter has assumed the value k_{i-1}, and the i-th hidden parameter has assumed the value k_i. Along with it, the same symbol p will also be used for the conditional probability $p(x_i, k_i \mid k_{i-1})$ in the event that the i-th observable parameter has assumed the value x_i and the i-th hidden parameter has assumed the value k_i under the condition that the $(i-1)$-th hidden parameter assumed the value k_{i-1}. So, the same symbol p denotes two different functions in two expressions: $p(x_i, k_i, k_{i-1})$ and $p(x_i, k_i \mid k_{i-1})$ and it is not quite correct. Nevertheless, we will use this incorrectness for simplifying the expression. This inaccuracy should not cause misunderstanding since in this lecture we will not use the identifier p without subsequently writing parentheses containing the parameters. The parameters unambiguously determine which function is referred to. In cases in which the incorrectness could lead to ambiguous understanding we will diversify the notation of the probability p by means of indices which will inevitably result in a certain clumsiness of the expressions.

The *statistical model* is determined by the function $X^n \times K^{n+1} \to \mathbb{R}$ which for each sequence $\bar{x}$ and each sequence $\bar{k}$ expresses the probability $p(\bar{x}, \bar{k})$. With this probability we will assume that for each $i = 1, 2, \ldots, n-1$, for each sequence $\bar{k} = (k_0^{i-1}, k_i, k_{i+1}^n)$, and for each sequence $\bar{x} = (x_1^i, x_{i+1}^n)$ the following holds

$$p(\bar{x}, \bar{k}) = p(k_i)\, p(x_1^i, k_0^{i-1} \mid k_i)\, p(x_{i+1}^n, k_{i+1}^n \mid k_i)\,. \tag{8.1}$$

This follows from the assumption that the probability $p(\bar{k})$ can be expressed in the form

$$p(\bar{k}) = p(k_i)\, p(k_0^{i-1} \mid k_i)\, p(k_{i+1}^n \mid k_i)\,. \tag{8.2}$$

The expression is valid for each $i = 1, 2, \ldots, n-1$ and each sequence $\bar{k} \in K^{n+1}$, where $\bar{k} = (k_0^{i-1}, k_i, k_{i+1}^n)$. Equation (8.2) was formed by summing Equation (8.1) over all sequences $\bar{x}$.

A random sequence $\bar{k}$ the probability distribution $p(\bar{k})$ of which satisfies the condition (8.2) is referred to as a *Markovian sequence*, or a *Markovian chain.* In this lecture we will be exclusively concerned with cases in which the random sequence is of the Markovian type.

If the summation in Equation (8.1) will be performed over all sequences k_{i+2}^n and then over all sequences x_{i+2}^n, we will obtain

$$\sum_{x_{i+2}^n} \sum_{k_{i+2}^n} p(\bar{x}, \bar{k}) = p(x_1^i, k_0^i) \sum_{x_{i+2}^n} \sum_{k_{i+2}^n} p(x_{i+1}^n, k_{i+1}^n \mid k_i)\,.$$

This implies that the following holds

$$p(x_1^{i+1}, k_0^{i+1}) = p(x_1^i, k_0^i)\, p(x_{i+1}, k_{i+1} \mid k_i)\,.$$

If we summarise this recursive relation we obtain the following working formula for calculating the joint probability for each sequence $\bar{k}$ and each sequence $\bar{x}$,

$$p(\bar{x}, \bar{k}) = p(x_1, x_2, \ldots, x_n, k_0, k_1, \ldots, k_n) = p(k_0) \prod_{i=1}^{n} p(x_i, k_i \mid k_{i-1})\,. \tag{8.3}$$

We can see that owing to the Markovian assumption (8.1) the definition and calculation of a complex function depending on $2n+1$ variables k_i, $i = 0, \ldots, n$, and x_i, $i = 1, \ldots, n$, is simplified to a definition of n functions $p(x_i, k_i \mid k_{i-1})$ of three variables, and one function $p(k_0)$ of a single variable. The assumption (8.1) therefore specifies a very narrow but important class of statistical models we are going to examine. We consider it useful to understand this specification in an informal manner, as well. We will present some considerations supporting the informal understanding of further ideas.

The property (8.1) can be, for example, understood in the following way. Let, in the universal population of pairs $(\bar{x}, \bar{k}) = (x_1, \ldots, x_n, k_0, k_1, \ldots, k_n)$, the probability distribution $p(\bar{x}, \bar{k})$ have Markovian properties (8.1). We will specify an arbitrary number i, $0 < i < n$, and an arbitrary value σ for the hidden parameter k_i. Let us fixate the selected values i, σ and take from the universal population all pairs (x, k) in which $k_i = \sigma$. Being Markovian then means that the group of parameters $(x_1, x_2, \ldots, x_i)$, $(k_0, k_1, \ldots, k_{i-1})$ in the chosen ensemble is statistically independent of the group of parameters $(x_{i+1}, x_{i+2}, \ldots, x_n)$, $(k_{i+1}, k_{i+2}, \ldots, k_n)$.

This correct interpretation is often expressed in a vulgarised form, i.e., a Markovian sequence is such a sequence in which the future does not depend on the past, but only on the present. The vulgarised form is treacherous, since while being incorrect it is very similar to the correct one.

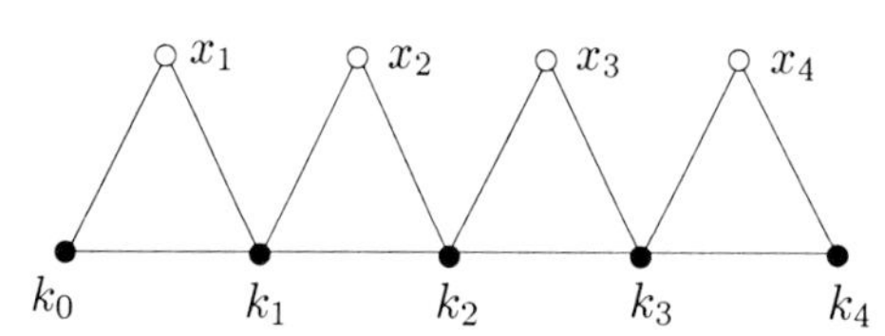

Figure 8.1 A mechanical model of a Markovian sequence.

The following *mechanical model of a Markovian sequence* provides a good intuitive idea. Let the sequences $(k_0, k_1, k_2, k_3, k_4)$ and (x_1, x_2, x_3, x_4) be represented by the positions of points in a plane, Fig. 8.1. Assume some pairs of points are connected by a spring which is denoted by abscissas between the points. Assume that one of the points, say the point x_3, starts for some random reasons to oscillate. By virtue of mechanical links, all (!) the other points of the mechanical system start to oscillate as well, not only the points k_2 and k_3 which are connected to the point x_3 by the spring. In this system each point is dependent on every other point, and the system as a whole does not break up into independent components. Furthermore, if the positions of points x_1, x_2, x_3, x_4, have been fixed then the positions of points k_0, k_1, k_2, k_3, k_4 are determined as well. The position of each point k_i will be affected not only by the positions of the points x_i and x_{i+1} to which the point k_i is immediately connected, but also by the positions of all points x_1, x_2, x_3, x_4.

But if we now imagine that a point, say the point k_3, is fixed immobile on the plane then the whole mechanical system breaks up into two independent parts. One consists of the points k_0, k_1, k_2, x_1, x_2, x_3, and the other of the points x_4 and k_4. Now the oscillation of a point, say the point x_4, does not in any way affect the positions of points k_1, x_2, and of the points to the left of the point k_3.

Further explanation will be based on representing Markovian processes by finite automata. A *finite automaton* is defined as a six-tuplet $(K, V, X, \delta, k_0, F)$, where

K is a finite set referred to as a set of automaton states;

V is a finite alphabet of input symbols;

X is a finite alphabet of output symbols;

k_0 is the initial state, $k_0 \in K$;

$F \subset K$ is a subset of states which are regarded as target states of the automaton;

$\delta\colon K \times V \to K \times X$ is a transfer function.

The above six-tuplet formally defining the finite automaton is interpreted in the following way.

If the automaton is in the state $k \in K$ and the symbol $v \in V$ is brought to its input then the automaton reacts by changing into the state $k' \in K$ and generates the output symbol $x \in X$ where the pair (k', x) is determined through the transition function $\delta\colon K \times V \to K \times X$ so that $(k', x) = \delta(k, v)$. The initial state of the automaton is k_0.

For each sequence $\bar{v}$ of the input symbols $v_1, v_2, \ldots, v_n$, the finite automaton determines one single sequence of the states $\bar{k} = (k_1, k_2, \ldots \ldots, k_n)$ and one single sequence of the output symbols $\bar{x} = (x_1, x_2, \ldots, x_n)$. This is the sequence of states through which the automaton passes and the sequence of symbols which occur at its output if the sequence $\bar{v}$ is brought to the input of the automaton which is in the initial state k_0. At the same time, the state, which the automaton reaches when the sequence $\bar{v}$ is the input, is also uniquely determined. In this way each automaton constructively expresses three mappings of the set of input symbol sequences: (1) into a set of output symbol sequences, (2) into a set of state sequences, and (3) into a set of states. These three representations are given without applying the concept F, i.e., set of final states which is also a characteristic feature of an automaton. The set of final states F determines the mapping of the set of input symbol sequences onto the set $\{0, 1\}$, i.e., it specifies a certain subset of input sequences. These are the sequences that transfer the automaton from the initial state k_0 to one of the target states of F.

The generalisation of the finite automaton has resulted in the *stochastic finite automaton*. In generalising the transition function $\delta\colon K \times V \to K \times X$ is replaced by a more complex function $\delta_s\colon X \times K \times K \times V \to \mathbb{R}$ and the specification of the initial state k_0 is replaced by the function $p\colon K \to \mathbb{R}$. The functions mentioned above have the following meaning. The initial state is random and each state $k \in K$ can become the initial one with the probability $p(k)$. Further behaviour of the automaton is random too. If the automaton is in the state $k \in K$ and the symbol $v \in V$ is its input then the automaton generates a random pair $(x, k') \in X \times K$ with the probability $\delta_s(x, k' \mid k, v)$, transits into the state k', and produces the symbol x at the output. Thus, for each sequence of input symbols the stochastic finite automaton expresses constructively the probability distribution on the set of output symbol sequences, on the set of state sequences, and on the set of states.

Assuming that the alphabet of input symbols consists of only a single symbol, we obtain a construction which is referred to as the *autonomous stochastic automaton*. In this case input symbols need not be taken into account since for each integer number n there is only one input sequence of the length n. The autonomous stochastic automaton is a precise model of processes expressed by Equation (8.1).

Let the autonomous stochastic automaton have a set of states K and a set of input symbols X. The automaton behaves in accordance with the probability distribution $p(k_0)$, $p(x_i, k_i \,|\, k_{i-1})$, $k_0 \in K$, $k_i \in K$, $x_i \in X$, $i = 1, 2, \ldots, n$. The automaton generates a random output sequence of symbols $x_1, x_2, \ldots, x_n$ of length n in the following way. At the beginning the automaton generates a random state k_0 with the probability distribution $p(k_0)$ and transits into it. In the i-th moment, $i = 1, 2, \ldots, n-1$, the automaton generates a random pair (x_i, k_i) according to the probability distribution $p(x_i, k_i \,|\, k_{i-1})$, transits into the state k_i and produces the symbol x_i at its output. The joint probability of the transition of the automaton across the sequence of states $k_0, k_1, \ldots, k_n$ and the generation of the sequence of output symbols $x_1, x_2, \ldots, x_n$ is just given by Equation (8.3).

This model of the autonomous stochastic automaton will be used in our lecture. However, this does not mean that the recognised object must necessarily be an automaton and the sequences $k_0, k_1, \ldots, k_n$ and $x_1, x_2, \ldots, x_n$ must express the development of a state in time.

The formulations of tasks and their solutions which are presented further on are exclusively based on the assumption of their Markovian nature (8.1) and hold for any object that satisfies the assumption of Markovian property.

8.3 Recognition of the stochastic automaton

8.3.1 Recognition of the stochastic automaton; problem formulation

Let a and b be two *autonomous stochastic automata*. Both of them have the same set of states K and the same set of output symbols X, but the statistical properties of the two automata are different. It is obvious that the assumption of the same set of states and output symbols for both the automata is not at the expense of generality. If the sets of states K and the sets of output symbols X were different then we would make them unified and the differences between the automata would concern different statistical parameters.

The first automaton is characterised by the probabilities $p_a(k_0)$ and $p_a(x_i, k_i \mid k_{i-1})$, $k_0 \in K$, $k_i \in K$, $x_i \in X$, $i = 1, 2, \ldots, n$, and the probabilities in the second automaton are $p_b(k_0)$ and $p_b(x_i, k_i \mid k_{i-1})$. Because of this definition we have implicitly accepted that the above probabilities do not depend on the index i. We have done it just to simplify further formulæ. All other considerations can be easily transferred to the general case in which the statistical properties of the automaton depend on i. In a *recognition task* the objective is, based on the knowledge of statistical characteristics of both the automata, to find which of them generated the given sequence $x_1, x_2, \ldots, x_n$.

The recognition task can be expressed as a Bayesian task of minimising the risk of decision making and, in a particular case, it can be a probability of the wrong recognition. The task can be expressed as a Neyman–Pearson task, as a minimax task, or as many others. It has been known from the first two lectures that in any concretisation of a task, the algorithms for solving it have a common part. This is the calculation of probability $p_a(\bar{x})$ of the sequence $\bar{x} = (x_1, x_2, \ldots, x_n)$ under the condition that it was generated by the automaton a and the calculation of the corresponding probability $p_b(\bar{x})$ for the automaton b. The decision benefits the first or second automaton according to the likelihood ratio $p_a(\bar{x})/p_b(\bar{x})$. The calculation of the probabilities $p_a(\bar{x})$ and $p_b(\bar{x})$ is for the given observation $\bar{x}$ the most extensive part of the recognition algorithm and does not depend on the choice of the particular recognition task. Let us now see what the algorithm for this calculation looks like.

8.3.2 Algorithm for a stochastic automaton recognition

The algorithm for calculating the probability $p_a(\bar{x})$ is equal to the algorithm for calculating $p_b(\bar{x})$. Therefore, we will present only one of them and in future explanations we will not give the indices a, b within the symbol p.

According to the definition, the number $p(\bar{x})$ is equal to $\sum_{\bar{k}} p(\bar{k}, \bar{x})$ and by applying (8.3) it can be expressed as a multi-dimensional sum

$$p(\bar{x}) = \sum_{\bar{k}} p(\bar{k}, \bar{x}) = \sum_{k_0} \sum_{k_1} \cdots \sum_{k_{n-1}} \sum_{k_n} p(k_0) \prod_{i=1}^{n} p(x_i, k_i \mid k_{i-1}) \,. \tag{8.4}$$

A direct application of Equation (8.4) for calculating the probability $p(\bar{x})$ is not possible since this number is expressed as a sum of $|K|^{n+1}$ summands. The expression (8.4) can be slightly changed through an equivalent transformation and the calculation becomes constructively realisable. Behind the summation sign according to the variable k_i those factors which do not depend on the variable k_i are factored out, and we obtain

$$\begin{aligned} p(\bar{x}) &= \sum_{k_0} p(k_0) \sum_{k_1} p(x_1, k_1 \mid k_0) \cdots \sum_{k_i} p(x_i, k_i \mid k_{i-1}) \\ &\cdots \sum_{k_{n-1}} p(x_{n-1}, k_{n-1} \mid k_{n-2}) \sum_{k_n} p(x_n, k_n \mid k_{n-1}) \,. \end{aligned} \tag{8.5}$$

If we denote for $i = 1, 2, \ldots, n$,

$$\begin{aligned} f_i(k_{i-1}) &= \sum_{k_i} p(x_i, k_i \mid k_{i-1}) \sum_{k_{i+1}} p(x_{i+1}, k_{i+1} \mid k_i) \cdots \\ &\cdots \sum_{k_{n-1}} p(x_{n-1}, k_{n-1} \mid k_{n-2}) \sum_{k_n} p(x_n, k_n \mid k_{n-1}) \end{aligned}$$

then we obtain the calculation procedure

$$\left.\begin{aligned} f_n(k_{n-1}) &= \sum_{k_n} p(x_n, k_n \mid k_{n-1}); \\ f_i(k_{i-1}) &= \sum_{k_i} p(x_i, k_i \mid k_{i-1})\, f_{i+1}(k_i)\,, \quad i = 1, 2, \ldots, n-1\,; \\ p(\bar{x}) &= \sum_{k_0} p(k_0)\, f_1(k_0)\,. \end{aligned}\right\} \tag{8.6}$$

It can be seen that the number of operations for calculating $p(\bar{x})$ is of the order $|K|^2 n$. First, the numbers $f_n(k_{n-1})$ are to be calculated according to the first row in (8.6), and then gradually the numbers $f_{n-1}(k_{n-2})$, ..., $f_i(k_{i-1})$, ..., $f_1(k_0)$ according to the second row in (8.6), and finally the number $p(\bar{x})$ according to the third row. In this way the task of the stochastic automaton recognition has been solved. According to the procedure (8.6) the probabilities $p_a(\bar{x})$ and $p_b(\bar{x})$ for the automata a and b are to be calculated and then with respect to the ratio $p_a(\bar{x})/p_b(\bar{x})$ a decision is made for the benefit of one of the automata, a or b. As a rule the decision is made by comparing the likelihood ratio to a certain threshold value, though in some tasks the decision making strategy may be more sophisticated.

8.3.3 Matrix representation of the calculation procedure

Even if the procedure (8.6) unambiguously describes the algorithm of automaton recognition, we will express it in a briefer form which is more suitable for further formal analysis.

The procedure (8.6) does not result in an explicit formula for calculating $p(\bar{x})$ because it comprises a calculation of a series of auxiliary quantities $f_i(k_{i-1})$. If we excluded these quantities from the system (8.6) then we would arrive back at the starting formula (8.5). We will express the procedure (8.6) in another way to obtain, after excluding the auxiliary variables, the formula in a form different from that of (8.5).

Probabilities $p(x_i, k_i \mid k_{i-1})$, $k_i \in K$, $k_{i-1} \in K$, can be regarded as the function $K \times K \to \mathbb{R}$ of two variables k_i, k_{i-1}. The quantity x_i is not a variable in our task since it is the result of measuring the i-th observable parameter. In each task x_i is a fixed constant, but changes for different tasks. The function expressing probability can be thought as a square matrix of the dimensions $|K| \times |K|$ in which the (k_i)-th column and the (k_{i-1})-th row contain the number $p(x_i, k_i \mid k_{i-1})$. This matrix will be denoted P_i. The matrix P_i depends on the index i at least because the value x_i is dependent on the index i. Representing a set of probabilities $\big(p(x_i, k_i \mid k_{i-1}),\, k_i \in K,\, k_{i-1} \in K\big)$ by means of a matrix is justifiable since this set will be used later as a factor in matrix multiplications. This will make further analysis of the algorithm more clear and will eventually lead to more efficient algorithms.

The numbers $f_i(k_{i-1})$, $k_{i-1} \in K$, $i = 1, 2, \ldots, n$, which are calculated one after another according to the procedure (8.6) can be regarded as a sequence of $|K|$-dimensional column vectors f_i, $i = 1, \ldots, n$ in which the k-th coordinate is

$f_i(k)$. Let f be a $|K|$-dimensional column vector all the coordinates of which are 1. The probabilities $p(k_0)$, $k_0 \in K$, will be regarded as a $|K|$-dimensional row vector the (k_0)-th coordinate of which is $p(k_0)$. This vector will be denoted φ.

Owing to the designation introduced we can express the calculation procedure (8.6) in a linear algebraic form

$$\begin{aligned} f_n &= P_n f\,, \\ f_i &= P_i\, f_{i+1}\,, \quad i = n-1, n-2, \dots, 2, 1\,, \\ p(\bar{x}) &= \varphi\, f_1\,, \end{aligned}$$

or, after excluding the auxiliary vectors $f_1, f_2, \dots, f_n$, in the form

$$p(\bar{x}) = \varphi\, P_1\, P_2\, \cdots\, P_{n-1}\, P_n\, f\,. \tag{8.7}$$

The notation (8.7) can be made even more concise

$$p(\bar{x}) = \varphi \left(\prod_{i=1}^{n} P_i \right) f\,. \tag{8.8}$$

Strictly speaking the previous equation should not be considered equivalent to the equation (8.7) since the multiplication of matrices is not commutative, which is expressed by the given order of factors in the matrix product (8.7), but which is already hidden in the formula (8.8).

Even if in the matrix representation (8.7) the statistical character of the original task is almost lost, such an expression is appropriate from the computational point of view. Owing to the associativity of the matrix product the formula (8.7) reveals a variety of calculation procedures for calculating the probability $p(\bar{x})$ which was not so evident in the procedure (8.6), and in the formulæ (8.5) and (8.4). From these calculation procedures one can be chosen which is most suitable in one or the other application from the point of view of implementation. For the time being we will show only two alternatives for calculating the product (8.7), and later on we will present others.

For practical illustration let $n = 5$. The formula (8.7) is equivalent to the following two formulæ which differ from the calculation point of view

$$p(\bar{x}) = \varphi\Big(P_1\,(P_2\,(P_3\,(P_4\,(P_5 f))))\Big)\,, \tag{8.9}$$

$$p(\bar{x}) = \Big(((((\varphi P_1)\,P_2)\,P_3)\,P_4)\,P_5\Big)\,f\,. \tag{8.10}$$

The calculation according to the formula (8.9) corresponds to the procedure (8.6) and the calculation according to the formula (8.10) differs from it. Both calculation procedures are correct, but in their formal argument the statistical character of the original task has nearly disappeared from view. The user has already lost a clear idea of what is actually done in each step of the procedure. We will show how the procedures (8.9) and (8.10) can be derived directly from the assumption of the Markovian character of the model (8.1). In addition to understanding formal matrix multiplication the statistical interpretation of the matrix products will be uncovered.

8.3.4 Statistical interpretation of matrix multiplication

We will show how to calculate the probability $p(x_i^n \mid k_{i-1})$ for any i, i.e., the probability that the automaton will generate a sequence $x_i, x_{i+1}, \ldots, x_n$ under the condition that the generation has started in the automaton state k_{i-1}. The calculation of $p(x_n \mid k_{n-1})$ is trivial since

$$p(x_n \mid k_{n-1}) = \sum_{k_n \in K} p(x_n, k_n \mid k_{n-1}) , \tag{8.11}$$

and the numbers $p(x_n, k_n \mid k_{n-1})$ are the known probabilities that represent the stochastic automaton. For the probability $p(x_{n-1}, x_n \mid k_{n-2})$ in the general case the following equation holds

$$\begin{aligned} p(x_{n-1}, x_n \mid k_{n-2}) &= \sum_{k_{n-1} \in K} p(x_{n-1}, x_n, k_{n-1} \mid k_{n-2}) \\ &= \sum_{k_{n-1} \in K} p(x_{n-1}, k_{n-1} \mid k_{n-2})\, p(x_n \mid x_{n-1}, k_{n-1}, k_{n-2}) . \end{aligned} \tag{8.12}$$

Based on Markovian property (8.1) (with the intuitive support of the mechanical model of being Markovian in Fig. 8.1), we have

$$p(x_n \mid x_{n-1}, k_{n-1}, k_{n-2}) = p(x_n \mid k_{n-1}) ,$$

and the expression (8.12) will assume the form

$$p(x_{n-1}, x_n \mid k_{n-2}) = \sum_{k_{n-1} \in K} p(x_{n-1}, k_{n-1} \mid k_{n-2})\, p(x_n \mid k_{n-1}) . \tag{8.13}$$

So, the probabilities $p(x_i^n \mid k_{i-1})$, which we should like to calculate for any i, can be calculated, at least, for $i = n$ and $i = n - 1$ by means of the sums (8.11) and (8.13). Now, we will show how to calculate these probabilities for $i - 1$, assuming that the probabilities $p(x_i^n \mid k_{i-1})$ are already calculated for the value i.

For the probability $p(x_{i-1}^n \mid k_{i-2})$ in the general case there holds that

$$\begin{aligned} p(x_{i-1}^n \mid k_{i-2}) &= p(x_{i-1}, x_i^n \mid k_{i-2}) \\ &= \sum_{k_{i-1} \in K} p(x_{i-1}, x_i^n, k_{i-1} \mid k_{i-2}) \\ &= \sum_{k_{i-1} \in K} p(x_{i-1}, k_{i-1} \mid k_{i-2})\, p(x_i^n \mid x_{i-1}, k_{i-1}, k_{i-2}) . \end{aligned} \tag{8.14}$$

With respect to the property (8.1) (and to an intuitive understanding of the mechanical model of the Markovian property), the sequence x_i^n with the fixed state k_{i-1} does not depend on the previous state k_{i-2} and on the previous observation x_{i-1}, and thus the factor $p(x_i^n \mid x_{i-1},\, k_{i-1},\, k_{i-2})$ in the formula

(8.14) can be changed into $p(x_i^n \mid k_{i-1})$ and the expression (8.14) will then be changed to

$$p(x_{i-1}^n \mid k_{i-2}) = \sum_{k_{i-1} \in K} p(x_{i-1}, k_{i-1} \mid k_{i-2})\, p(x_i^n \mid k_{i-1})\,, \quad i = 2, 3, \dots, n\,. \tag{8.15}$$

By means of the formula (8.11) and the multiply applied formula (8.15) we can calculate the probability $p(x_1^n \mid k_0)$ for each state k_0 and then calculate the probability $p(\bar{x})$ sought according to the relation

$$p(\bar{x}) = \sum_{k_0 \in K} p(k_0)\, p(x_1^n \mid k_0)\,. \tag{8.16}$$

The calculation according to the formulæ (8.11), (8.15) and (8.16) is actually the same as that according to the procedure (8.6). The calculation procedure (8.6), as well as its representation by a matrix product (8.9), is therefore not only formally derived, but can be interpreted from the statistical point of view. The statistical interpretation of a column vector

$$\left(\prod_{j=i}^{n} P_j \right) f$$

means that its coordinates are $|K|$ numbers $p(x_i, x_{i+1}, \dots, x_n \mid k_{i-1})$, $k_{i-1} \in K$, which are the probabilities that the automaton will generate a sequence of symbols x_i, x_{i+1}, $\dots, x_n$ under the condition that the generation started in the state k_{i-1}.

Let as now look at the statistical considerations which will lead to the calculation of the probability $p(\bar{x})$ according to the procedure (8.10). They are nearly the same as the ideas mentioned above. We will state now how the joint probability $p(x_1^i, k_i)$ would be calculated for the event that the automaton generates the sequence $x_1, x_2, \dots, x_i$, and after the end of the generation the automaton will transit into the state k_i. For $i = 1$ the probability is obviously

$$p(x_1, k_1) = \sum_{k_0 \in K} p(k_0)\, p(x_1, k_1 \mid k_0)\,. \tag{8.17}$$

Assume we have already calculated the probabilities $p(x_1^{i-1}, k_{i-1})$ for some i and with respect to them we would like to calculate the probabilities $p(x_1^i, k_i)$. For the probability $p(x_1^i, k_i)$ in the general case holds

$$\begin{aligned} p(x_1^i, k_i) &= p(x_1^{i-1}, x_i, k_i) \\ &= \sum_{k_{i-1} \in K} p(x_1^{i-1}, x_i, k_{i-1}, k_i) \\ &= \sum_{k_{i-1} \in K} p(x_1^{i-1}, k_{i-1})\, p\left(x_i, k_i \mid x_1^{i-1}, k_{i-1}\right)\,. \end{aligned} \tag{8.18}$$

Owing to the Markovian property (8.1) (expressed intuitively by means of a mechanical model), we claim that at the fixed state k_{i-1} the pair (x_i, k_i) does not depend on the observation x_1^{i-1} and thus

$$p(x_i, k_i \mid x_1^{i-1}, k_{i-1}) = p(x_i, k_i \mid k_{i-1}) .$$

If we include the previous expression into the sum (8.18) then we obtain the following recursive expression for the calculation of $p(x_1^i, k_i)$,

$$p(x_1^i, k_i) = \sum_{k_{i-1} \in K} p(x_1^{i-1}, k_{i-1}) \, p(x_i, k_i \mid k_{i-1}) . \tag{8.19}$$

If we have calculated according to the formulæ (8.17) and (8.19) the probabilities $p(x_1^n, k_n)$ then we can calculate the probability $p(\bar{x})$ being sought according to the formula

$$p(\bar{x}) = \sum_{k_n \in K} p(x_1^n, k_n) , \tag{8.20}$$

since x_1^n is simply $\bar{x} = (x_1, x_2, \ldots, x_n)$.

The calculation of the probability $p(\bar{x})$ according to the formulæ (8.17), (8.19) and (8.20) is therefore the same as that in the matrix representation (8.10). On the one hand, this form is the formal consequence of the associativity of matrix multiplication, but on the other hand this can also be statistically conceived. The matrix product

$$\varphi \left(\prod_{j=1}^{i} P_j \right)$$

represents the $|K|$-dimensional row vector whose k_i-th component, $k_i \in K$, is the joint probability $p(x_1, x_2, \ldots, x_i, k_i)$ that the automaton will generate the given observation sequence $x_1, x_2, \ldots, x_i$ and finally traverse to the state k_i.

8.3.5 Recognition of the Markovian object from incomplete data

In previous considerations we identified the recognised object by a stochastic automaton. It meant that we had considered the sequence $x_1, x_2, \ldots, x_n$ as well as the sequence $k_0, k_1, k_2, \ldots, k_n$ to be processes which developed in time, and the index i representing time. Having such a concept of the recognised object facilitated the explanation of the task and the algorithm for its solution. The derived algorithms, however, are not confined only to processes developing in time. The Markovian model described by the relation (8.10) does not require the index i to represent just time. It is important that both the observed and the hidden parameters are sequences, but not necessarily sequences in time.

When the features x_i are measured one after another, and not in accordance with the index i the following situation can occur which requires a different calculation ordering than the two already mentioned procedures.

Assume that the object is described by twenty features $x_1, x_2, \ldots, x_{20}$ and twenty-one hidden parameters $k_0, k_1, \ldots, k_{20}$. Let us also assume that the features $(x_5, x_6, \ldots, x_{10})$ and $(x_{12}, x_{13}, \ldots, x_{17})$ were known at some moment. Waiting for the results of the measurement of the rest of the features takes considerable time. However, when the remaining features become known then the object must be recognised as fast as possible. In such situations a purely technical question arises: How should the features already known be processed before the rest is measured and so the computation time not wasted in waiting? Matrix representation (8.7) of the probability $p(\bar{x})$ provides a clear answer to this question. The expression (8.7) is equivalent to

$$p(\bar{x}) = \varphi\, P_1\, P_2\, P_3\, P_4\, P^*\, P_{11}\, P^{**}\, P_{18}\, P_{19}\, P_{20}\, f\,, \tag{8.21}$$

where

$$P^* = P_5\, P_6\, P_7\, P_8\, P_9\, P_{10}\,, \tag{8.22}$$

$$P^{**} = P_{12}\, P_{13}\, P_{14}\, P_{15}\, P_{16}\, P_{17}\,. \tag{8.23}$$

From the previous relations it can be seen that with the known sequences x_5^{10} and x_{12}^{17} matrix products P^* and P^{**} can be calculated by means of formulæ (8.22) and (8.23). By the time the information for all the other features is available, the probability $p(\bar{x})$ of the formula (8.21) will be calculated. The total number of operations in this case will be greater if compared with the calculation according to the formula (8.9) or (8.10), but on the other hand the number of operations needed for calculating (8.23) with the matrices P^* and P^{**} already known will decrease.

Now let us imagine that in the given example no information about the other features were provided and it was necessary to recognise the object only with respect to the already known features. In this case the probability $p(x_5^{10}, x_{12}^{17})$ should be calculated. We will briefly show how the probability has to be calculated. Let I be a set $\{1, 2, \ldots, n\}$, through which the index i ranges in notation x_i, I' be a subset of indices and for each $i \in I'$ the value x_i is known. The ensemble of known values will be denoted as $(x_i,\, i \in I')$ and the ensemble of not yet known values will be denoted as $(x_i,\, i \notin I')$. Let k be a sequence $k_0, k_1, \ldots, k_n$. The joint probability of the ensemble $(x_i\,,\, i \in I)$ and the sequence $\bar{k}$ is

$$p\big((x_i\,,\, i \in I), \bar{k}\big) = p(k_0) \prod_{i \in I} p(x_i, k_i \mid k_{i-1})\,,$$

and sought probability $p\big((x_i\,,\, i \in I')\big)$ is

$$\begin{aligned}
p\big((x_i\,,\, i \in I')\big) &= \sum_k \sum_{(x_i\,, i \notin I')} p(k_0) \prod_{i \in I} p(x_i, k_i \mid k_{i-1}) \\
&= \sum_k p(k_0) \prod_{i \in I'} p(x_i, k_i \mid k_{i-1}) \sum_{(x_i,\, i \notin I')} \prod_{i \notin I'} p(x_i, k_i \mid k_{i-1}) \\
&= \sum_k p(k_0) \prod_{i \in I'} p(x_i, k_i \mid k_{i-1}) \prod_{i \notin I'} \sum_{x_i} p(x_i, k_i \mid k_{i-1}) \\
&= \sum_k p(k_0) \prod_{i \in I'} p(x_i, k_i \mid k_{i-1}) \prod_{i \notin I'} p(k_i \mid k_{i-1}) \\
&= \sum_k p(k_0) \prod_{i \in I} P_i(k_i \mid k_{i-1})\,,
\end{aligned} \tag{8.24}$$

where

$$P_i(k_i \mid k_{i-1}) = \begin{cases} p(x_i, k_i \mid k_{i-1})\,, & \text{if the value } x_i \text{ is known,} \\ p(k_i \mid k_{i-1}) = \sum\limits_{x_i} p(x_i, k_i \mid k_{i-1}), & \text{if the value } x_i \text{ is not known.} \end{cases}$$

If we regard the function $P_i \colon K \times K \to \mathbb{R}$ as a square matrix of the dimension $|K| \times |K|$ then we can again represent (8.24) for the probability $p\big((x_i\,,\, i \in I')\big)$ as a matrix product

$$p\big((x_i\,,\, i \in I)\big) = \varphi \left(\prod_{i=1}^{n} P_i\right) f\,.$$

When compared with the previous matrix products, the matrix P_i differs in that it depends on whether the value of the feature x_i is known or not known.

Let us now notice the great diversity of object recognition tasks which occur within the framework of the Markovian model. Usually, this class of recognition problems is closely connected to optimisation methods based on dynamic programming. However, in quite meaningful recognition problems considered so far, the dynamic programming has not yet occurred. In this respect we would like to cast doubt upon the naive, but well rooted view which regards dynamic programming to be a universal key opening every door.

Later we would like to draw attention to the importance of representing recognition tasks concerning the Markovian-describable objects by matrix products. This representation keeps all modifications of the task together and does not allow them to be broken into isolated and mutually separated problems. It is not surprising since the stochastic matrix is one of the basic concepts in the general theory of Markovian processes. It is rather strange that the representation through the matrix product has not become thoroughly settled in pattern recognition tasks. Later we will see that matrix products appear even in some well known tasks where hardly anybody would expect them.

8.4 The most probable sequence of hidden parameters

8.4.1 Difference between recognition of an object as a whole and recognition of parts that form the object

The analysis of the task of recognising a Markovian object which was presented in the previous Subsection 8.3.5 provides instructive results. The recognition task was formulated for the object as a whole and not for the recognition of the parts it consists of. It could be subconsciously expected that the formal solution of the task would include the recognition of individual parts of the object too. Then, based on parts, one could decide about the object as a whole. We have thoroughly studied recognition of Markovian object from all aspects, in different modifications of the task, in both the formal and informal way. In spite of that we have not revealed any hierarchy that could be regarded as the recognition of individual parts of the object from which the decision on the object as a whole would be synthesised. The formal solution of the exactly formulated task has a substantially different form than would be intuitively expected. As long as the task was formulated as recognition of the automaton as a whole, the algorithm solving it did not seek the sequence of states the automaton passed through. If we want the algorithm to reveal the sequence of the traversed states then we have to embody this desire in a task formulation (which we will construct in the coming subsection). Solving one task does not mean that another task will be solved at the same time.

8.4.2 Formulation of a task seeking the most probable sequence of states

Let $\bar{x} = x_1^n$ be a sequence of observations $x_1, x_2, \ldots, x_n$ and $\bar{k} = k_0^n$ be a sequence of automaton states. Their joint probability $p(\bar{x}, \bar{k})$ has the form

$$p(\bar{x}, \bar{k}) = p(k_0) \prod_{i=1}^{n} p(x_i, k_i \mid k_{i-1}) ,$$

where $p(k_0)$,$k_0 \in K$, and $p(x_i, k_i \mid k_{i-1})$, $x_i \in X$, $k_i \in K$, $i = 1, 2, \ldots, n$, are known probabilities.

The task is formulated as seeking the sequence $\bar{k}^*$ the *a posteriori* probability of which is greatest under the condition of the sequence $\bar{x}$, i.e.,

$$\bar{k}^* = \operatorname*{argmax}_{\bar{k} \in K^{n+1}} \frac{p(\bar{x}, \bar{k})}{\sum_{\bar{k} \in K^{n+1}} p(\bar{x}, \bar{k})} = \operatorname*{argmax}_{\bar{k} \in K^{n+1}} p(\bar{x}, \bar{k}) . \tag{8.25}$$

8.4.3 Representation of a task as seeking the shortest path in a graph

We will show how an optimisation task (8.25) can be expressed as the known task seeking the shortest path between two given vertices in a graph of special form. The solution of the task is known and uses dynamic programming.

By q_i, $i = 1, 2, \ldots, n$, the function of the form $K \times K \to \mathbb{R}$ will be denoted. Its individual values $q_i(k_{i-1}, k_i)$, $k_{i-1} \in K$, $k_i \in K$, are $-\log p(x_i, k_i \mid k_{i-1})$. By

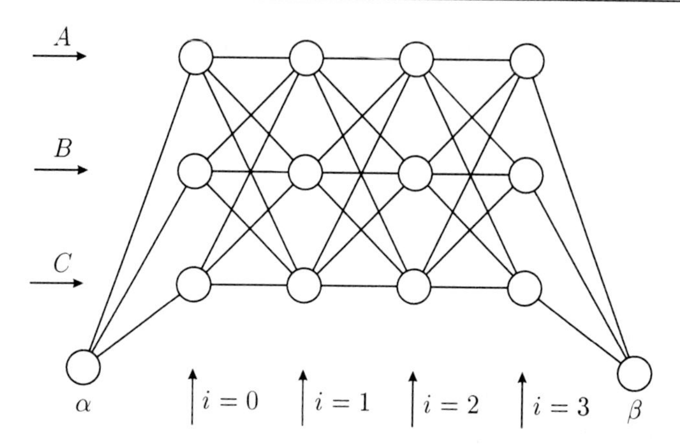

Figure 8.2 The optimisation task concerning the transition through the states is represented as seeking a path in an oriented graph. Edges of the graph are oriented from left to right.

φ_0 the function of the form $K \to \mathbb{R}$ will be denoted the values $\varphi_0(k_0)$, $k_0 \in K$, of which are numbers $-\log p(k_0)$. Thus, the optimisation task (8.25) can be written in the form

$$\bar{k}^* = \underset{k_0,k_1,\ldots,k_n}{\operatorname{argmin}} \left(\varphi(k_0) + \sum_{i=1}^{n} q_i(k_{i-1}, k_i) \right) , \tag{8.26}$$

and represented by means of an oriented graph of a special form.

The *oriented graph* consists of vertices V and oriented edges between them. The set of graph vertices V contains the initial vertex α, goal vertex β, and further $|K|(n+1)$ intermediate vertices of the form (σ, i), $\sigma \in K$, $i = 0, 1, \ldots, n$.

Example 8.3 Representation of a Markovian automaton by a graph. *For simplicity a set of vertices V for the case $n = 3$ is shown in Fig. 8.2. The set of states K of the automaton consists of the states A, B and C.* The vertex *(σ, i) can be considered to be a point in a plane, the coordinates of the point being σ and i. The coordinate σ of the vertex (σ, i) will be referred to as the* label of the vertex. *The label of the vertex corresponds to a state from the set* $K = \{A, B, C\}$. ▲

The oriented edges of the graph are arranged in the following way.

- $|K|$ edges lead from the vertex α to the vertices of the form $(\sigma, 0)$, $\sigma \in K$.
- $|K|$ edges lead to the vertex β. They originate in vertices of the form (σ, n), $\sigma \in K$.
- $|K|$ edges originate in each vertex of the form (σ, i), $\sigma \in K$, $i = 0, 1, \ldots, n-1$. They lead into vertices of the form $(\sigma', i + 1)$, $\sigma' \in K$.
- K edges lead to each vertex of the form (σ, i), $i = 1, 2, \ldots, n$, $\sigma \in K$. They originate in vertices of the form $(\sigma', i - 1)$, $\sigma' \in K$.
- The edge $\big(\alpha, (\sigma, 0)\big)$, $\sigma \in K$, is of the length $\varphi(\sigma)$.
- The edge $\big((\sigma, n), \beta\big)$, $\sigma \in K$, is of the length 0.
- For each $\sigma \in K$, $\sigma' \in K$, $i = 1, 2, \ldots, n$, the edge $\big((\sigma, i-1), (\sigma', i)\big)$ is of the length $q_i(\sigma, \sigma')$.

The graph created in this way defines a set of paths from the vertex α to the vertex β and each path acquires its length given by the sum of lengths of edges which the path consists of. To each path from α to β in the graph a sequence of states corresponds, which is given by the labels of the vertices through which the path goes. Conversely to each sequence $k_0, k_1, \ldots, k_n$ a path in the graph corresponds which passes through the vertices

$$\alpha, (k_0, 0), (k_1, 1), (k_2, 2), \ldots, (k_n, n), \beta .$$

The value $\varphi(k_0) + \sum_{i=1}^{n} q_i(k_{i-1}, k_i)$ is the length of the path corresponding to the sequence $k_0, k_1, \ldots, k_n$. Thus, the optimisation task (8.25) is reduced to seeking the shortest path between a pair of vertices in the graph. This task has been solved successfully by the algorithm based on the Bellman's dynamic programming. In the coming subsection we will provide, for completeness, the algorithm finding the shortest path by means of dynamic programming.

8.4.4 Seeking the shortest path in a graph describing the task

The following informal imagination of the shortest path problem might be useful. Let us imagine that some messengers are to deliver a message from the vertex α to the vertex β. At the beginning the messengers are located at the vertices of the graph in such a way that their number at each vertex of the graph is the same as the number of edges leading out from that particular vertex. When a messenger brings a message to a vertex then the message is immediately handed over to the waiting messengers, and they run out along the edges that have been assigned to them. The speed of all the messengers is assumed to be the same so that the time of transferring the message from one vertex to the other corresponds to the length of the edge of the graph. The target time for a message to go from the initial vertex α to the target vertex β is proportional to the length of the shortest path in the graph between these vertices.

The quoted informal model will be used for a more instructive explanation of the algorithm seeking the shortest path from the vertex α to the vertex β. We will denote by $f_i(\sigma)$ the length of the shortest path from the vertex α to the vertex (σ, i) which corresponds to the shortest time in which the messengers in the vertex (σ, i) receive the message. The numbers $f_0(\sigma)$, $\sigma \in K$, are evidently $\varphi(\sigma)$ since each vertex $(\sigma, 0)$ is connected with the vertex α with only one edge, i.e.,

$$f_0(\sigma) = \varphi(\sigma) . \tag{8.27}$$

The values $f_i(\sigma)$ for the other vertices (σ, i), $\sigma \in K$, $i > 0$, can be calculated by the following informal, but still accurate considerations. The message is brought to the vertex (σ, i) only from the vertex of the form $(\sigma', i-1)$, $\sigma' \in K$. Information from a specific vertex $(\sigma', i-1)$ will be delivered at the moment $f_{i-1}(\sigma') + q_i(\sigma', \sigma)$, where $f_{i-1}(\sigma')$ is the moment when the message was at disposal at the vertex $(\sigma', i-1)$, and the quantity $q_i(\sigma', \sigma)$ specifies the time necessary for the transfer of the message from the vertex $(\sigma', i-1)$ to the vertex (σ, i). The shortest time $f_i(\sigma)$ in which the message will be delivered to the vertex (σ, i) is

$$f_i(\sigma) = \min_{\sigma' \in K} \left(f_{i-1}(\sigma') + q_i(\sigma', \sigma) \right) . \tag{8.28}$$

At the same time it will be indicated from which vertex $(\sigma', i-1)$ the message was delivered to the vertex (σ, i) in the fastest possible way. This preceding vertex will be denoted by the symbol $\text{ind}_i(\sigma)$,

$$\text{ind}_i(\sigma) = \underset{\sigma' \in K}{\text{argmin}} \left(f_{i-1}(\sigma') + q_i(\sigma', \sigma) \right) . \tag{8.29}$$

The variable $\text{ind}_i(\sigma)$ states that the shortest path from the vertex α to the vertex (σ, i) passes through the vertex $(\text{ind}_i(\sigma), i-1)$. Notice that there can be more than one such possibility. One of them can be selected randomly for simplicity.

The moment at which the message is delivered to the target vertex β, i.e., the length of the shortest path from α to β is given by the value

$$\min_{\sigma \in K} f_n(\sigma) . \tag{8.30}$$

The vertex from the group (σ, n) from which the message was first delivered to the end vertex is

$$k_n = \underset{\sigma \in K}{\text{argmin}}\, f_n(\sigma) .$$

The formulæ (8.27), (8.28), (8.29), and (8.30) are the core of the algorithm seeking the shortest path from the vertex α to the vertex β. Let us quote them together

$$f_0(\sigma) = \varphi(\sigma) , \quad \sigma \in K ; \tag{8.31}$$

$$f_i(\sigma) = \min_{\sigma' \in K} \left(f_{i-1}(\sigma') + q_i(\sigma', \sigma) \right) , \quad i = 1, 2, \ldots, n , \quad \sigma \in K ; \tag{8.32}$$

$$\text{ind}_i(\sigma) = \underset{\sigma' \in K}{\text{argmin}} \left(f_{i-1}(\sigma') + q_i(\sigma', \sigma) \right) , \quad i = 1, 2, \ldots, n , \quad \sigma \in K ; \tag{8.33}$$

$$k_n = \underset{\sigma' \in K}{\text{argmin}}\, f_n(\sigma) . \tag{8.34}$$

At first, according to the formula (8.31) the distances to the vertices of the group $(\sigma, 0)$ from the vertex α will be calculated. It is not a matter of calculating, but of transcribing the values of the function $f(\sigma)$ from one memory cell to another. Then gradually by means of the formulæ (8.32) the distances from the vertex α to the vertices of the ensemble $(\sigma, 1), \sigma \in K$, are calculated, then those of the ensemble $(\sigma, 2)$, $\sigma \in K$, and so on, until the distances for the vertices of the group (σ, n), $\sigma \in K$, are calculated. Along with determining the distance $f_i(\sigma)$ the value $\text{ind}_i(\sigma)$ for each vertex will be calculated according to the formula (8.33). This value determines the label of the vertex $(\sigma', i-1)$ which immediately precedes the vertex (σ, i) along the shortest path from the vertex α to the vertex (σ, i).

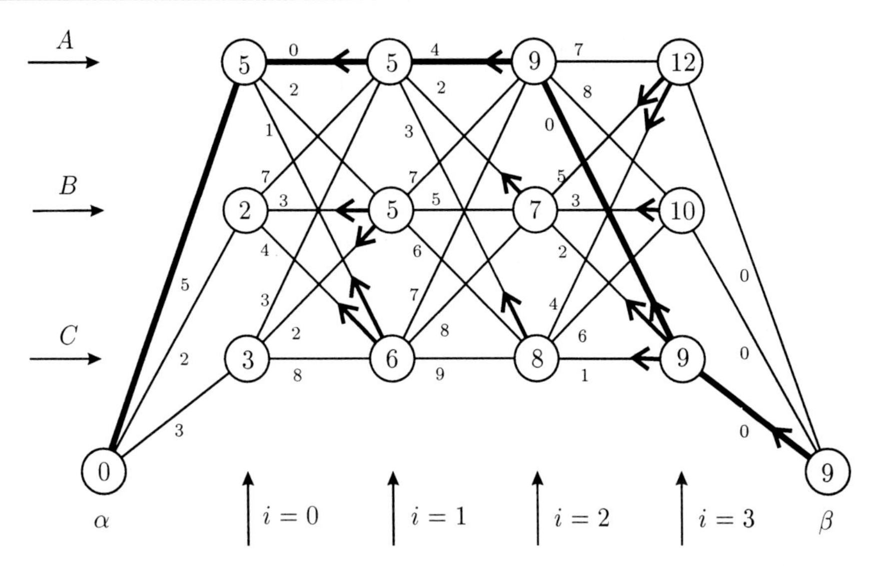

Figure 8.3 Seeking the shortest path in the Markovian sequence.

After completing the calculation according to the formulæ (8.31), (8.32) and (8.33), the length of the shortest path from α to β is determined by the number $\min_\sigma f_n(\sigma)$. The formula (8.34) indicates the label of the last vertex along the shortest path. The value $\text{ind}_n(k_n)$ indicates the label k_{n-1} of the last but one vertex, the value $\text{ind}_{n-1}(k_{n-1})$ indicates the label of k_{n-2}. Expressed in the general way, $\text{ind}_i(k_i)$ indicates the $(i-1)$-th member in the sequence $k_0, k_1, \ldots, k_n$, being sought which minimises (8.26) and maximises (8.25).

The formal notation of the algorithm by virtue of the formulæ (8.31), (8.32), (8.33) and (8.34) does not reveal, at first glance, its immense simplicity and cleverness. If the simplicity is not yet quite evident to the reader then we recommend him or her to study the following example and apply the quoted algorithm for the calculation of all the data given in Fig. 8.3.

Example 8.4 Seeking the shortest path in a graph with three possible states. *The Fig. 8.3 depicts the same situation as in Example 8.3 in which the set of states consists of three states A, B, C. The index i assumes the value $0, 1, 2, 3$. The numbers labelling the edges of the graph provide information about their lengths, i.e., quantities $q_i(\sigma', \sigma)$. Vertices are denoted by circles. The numbers labelling the vertices represent the quantities $f_i(\sigma)$. Values $\text{ind}_i(\sigma)$ are depicted as arrows from the vertex (σ, i) to the vertex $\big(\text{ind}_i(\sigma), i-1\big)$. It is possible to traverse from the target vertex β following the arrows and the shortest path is laid back in the inverse direction, from the target vertex towards the start vertex. Notice that in some cases there can be more than one arrow from a vertex. In such a case just one of the arrows can be selected randomly. Let the selected path traverse through vertices $(C,3)$, $(A,2)$, $(A,1)$, $(C,0)$ and is shown in bold in Fig. 8.3. The sequence that minimises (8.26) and maximises (8.25) is $AAAC$. In this particular case the optimal sequences can be $AABC$ and $AACC$ too.* ▲

8.4.5 On the necessity of formal task analysis

The algorithm mentioned above seeking the shortest path in a graph (and thus also seeking the most probable sequence of states) is usually deduced by informal, but convincing considerations just as we have done in the preceding Subsections 8.4.3 and 8.4.4. The algorithm became favoured far and wide owing to its clearness, and not only in pattern recognition. We do not intend to cast doubt on the positive aspect of the illustrative way of explanation, but we are also going to show why this way is not suited for some tasks.

The way mentioned above refers too often to the illustrative obviousness of considerations which replace their formal reasoning. Algorithms are deduced using sentences of natural language, and not by means of mathematical expressions which are transformed into different, but equivalent forms by applying certain rules. A procedure by means of which an algorithm is deduced is markedly different from a procedure used, say, in solving algebraic equations. The equation solved is given by an algebraic expression. The procedure which solves the equation consists of equivalent transformations. The algebraic expression of the original task is modified to the form in which the solution is obvious.

The rules for equivalent transformations do not make the solved task trivial. But a formal deduction of the solution of a task with the help of a finite number of afore given rules still has some advantages. When all respective equivalent transformations for solving the task are found then it can be easily proved to someone else that the solution is correct. Neither intuition nor informal understanding of the task is expected. It is sufficient only to understand that the applied transformations of the expressions are really equivalent.

The tasks analysed so far have been so transparent that it has been possible to describe them in natural language. However, it become possible only because we have dealt with the easiest problem of structural analysis so far. Now we are going to proceed to the more complex problems. Verbal speculations inevitably would be clumsy and vast and, consequently, less and less convincing. Therefore, in the case in which really difficult problems are considered, their formulations, analysis and solution must be supported by formal deduction. There can be no doubt about the correctness, the analysis of the task can be briefer, and a clarity of an approach is achieved which would be lost in the case of the verbal approach.

In ignoring the actual complexity of a task, there naturally always exists a possibility to avoid the formal analysis and treat the task as if it were quite simple. Then, for solving a certain task an algorithm is willfully used which is assigned for another task, or in an even worse case, it is by no means known what task is solved by that particular algorithm. We will present a rather widespread example of such a solution.

Example 8.5 Unsuitable application of the algorithm seeking the shortest path. *Let $\bar{x} = (x_1, x_2, \ldots, x_n)$ be a sequence of observed symbols, e.g., for $n = 100$. This sequence depends on the sequence $\bar{k} = (k_0, \ldots, k_n)$ of the states the automaton passed through when generating the observed sequence. Thanks* *to this dependence the sequence $\bar{k}$ can be reasonably estimated on the basis of the sequence $\bar{x}$. Now let us assume that we are not interested in the sequence $\bar{k}$ as a whole but only in its last ten elements, k_{91}, k_{92}, ..., k_{100} with the highest a priori probability. This task is not a task seeking the shortest path in a graph, just because the paths correspond to sequences of length 100, but we are only interested in sequences of length 10. The task should be examined from the very beginning, i.e., from its formulation to the proof of its solution.*

With exorbitant faith in knowing how to seek the whole sequence k_0, k_1, ..., k_{100} by means of dynamic programming and having faith in the fact that this knowledge is sufficient for solving any task, one could arrive at different incorrect solutions. For example, it can be an algorithm which seeks the most probable sequence $k_0^*, k_1^*, \dots, k_{100}^*$, and then uses only the last ten elements of it. Such a recommendation is not correct because it is not stated what are the properties of the sequence of the last ten states which were sought in such a way. If the task were formulated as a task of seeking the sequence $k_{91}, k_{92}, \dots, k_{100}$ a posteriori probability of which is greatest under the condition of the observation $x_1, x_2, \dots, x_n$ then the procedure mentioned above would be wrong. The most probable subsequence $k_{91}, k_{92}, \dots, k_{100}$ of the last ten states need not be equal to the last ten states of the most probable sequence $k_0^*, k_1^*, \dots, k_{100}^*$. ▲

We will explain the mathematical apparatus which is suitable for expressing tasks performing structural recognition of sequences. Again we will use the task seeking the most probable sequence even if we have already succeeded in solving it without a new mathematical formalism. Thus, we will better understand the concepts which will be used later on.

8.4.6 Generalised matrix multiplications

We have in mind the optimisation task

$$d = \min_{k_0} \min_{k_1} \min_{k_2} \cdots \min_{k_n} \left(\varphi(k_0) + \sum_{i=1}^{n} q_i(k_{i-1}, k_i) \right). \tag{8.35}$$

To avoid unnecessary complications which would obscure the main idea, we will only concern ourselves with calculating the value of the minimum and we will not seek the sequence $(k_0^n)^*$ in which the minimum is achieved.

The optimisation task (8.35) has a similar form to the expression

$$p(\bar{x}) = \sum_{k_0} \sum_{k_1} \sum_{k_2} \cdots \sum_{k_n} p(k_0) \prod_{i=1}^{n} p(x_i, k_i \mid k_{i-1}) \tag{8.36}$$

which we studied when solving the task recognising the automaton. In both expressions, (8.35) as well as (8.36), a number is calculated, namely the number d in (8.35) and the number $p(\bar{x})$ in (8.36). The number is calculated with respect to the function of the form $K \to \mathbb{R}$ (in (8.35) it is $\varphi(k_0)$ and in (8.36) it is $p(k_0)$, $k_0 \in K$) and n functions of the form $K \times K \to \mathbb{R}$ (in (8.35) they are the functions q_i, $i = 1, 2, \dots, n$, and in (8.36) they are $p(x_i, k_i \mid k_{i-1})$, $k_i \in K$,

$k_{i-1} \in K$, $i = 1, \ldots, n$). The relations (8.35) and (8.36) are calculated by means of different, but yet similar programs. The difference is only in that the program for calculating (8.35) is obtained from the program (8.36) in such a way that wherever a sum of two numbers occurs in the first program, a smaller one of both numbers has to be found in the second program. Moreover, the multiplication of two numbers in the first program is replaced by the sum of the same numbers.

In examining the procedure of calculating the numbers $p(\bar{x})$, see (8.36), we have arrived at a conclusion that when the starting numbers $p(k_0)$ and $p(x_i, k_i \mid k_{i-1})$ are understood as components of a row vector φ and matrices P_i, $i = 1, \ldots, n$, then the calculation according to the formula (8.36) is equivalent to the calculation of the matrix product

$$p(\bar{x}) = \varphi\, P_1\, P_2\, \cdots\, P_n\, f\,. \tag{8.37}$$

This matrix product represents the number being computed, and in this sense converts the problem into creation of an algorithm that must compute the number. This problem is obviously equivalent to the problem (8.36), but is given in a different form. The number to be computed according to the formulated problem is explicitly stated by the expression (8.37). In this sense the matrix expression (8.37) immediately performs the algorithm for its calculation. So, the transformation of the problem of the form (8.36) to the form (8.37) is virtually the solution of the problem because expressing the task in the form (8.37) makes the task trivial.

The equivalence of the expression (8.36) which is the original formulation of the task, and of the matrix product (8.37) is based on properties of adding and multiplying real numbers. These properties are so obvious that usually they go without saying. They are the associativity and distributiveness of multiplication with respect to addition. For any three real numbers x, y and z there hold

$$\left.\begin{aligned} x + (y + z) &= (x + y) + z\,, \\ x\,(y\,z) &= (x\,y)\,z\,, \\ x\,(y + z) &= x\,y + x\,z\,. \end{aligned}\right\} \tag{8.38}$$

In other words a set of real numbers with the operations of addition and multiplication forms an algebraic structure known as a *semi-ring*. This structure satisfies other requirements, but at the moment they are not important for us. The essential observation is that addition and multiplication are not the only pair of operations that satisfy requirements (8.38). It is of key importance for the operation with sequences that a set of non-negative real numbers with operations min and + also forms a semi-ring. There hold

$$\begin{aligned} \min\bigl(x, \min(y, z)\bigr) &= \min\bigl(\min(x, y), z\bigr)\,, \\ x + (y + z) &= (x + y) + z\,, \\ x + \min(y, z) &= \min(x + y, x + z)\,. \end{aligned}$$

When applying matrices and the matrix product we can rely on the fact that a set of non-negative real numbers with a pair of operations $(\min, +)$ constitutes an algebraic structure of a semi-ring. Let $q_i\colon K \times K \to \mathbb{R}$, $i = 1, 2, \ldots, n$, correspond to n functions of two variables which assume their values on the finite set K. Each function q_i can be understood as a matrix of the dimension $|K| \times |K|$ the element of which in the k-th row and k'-th column is $q_i(k, k')$. Let us denote two functions of one variable $\varphi\colon K \to \mathbb{R}$ and $f\colon K \to \mathbb{R}$. The function φ will be understood as a $|K|$-dimensional row vector, and the other a $|K|$-dimensional column vector. Let q' and q'' be two matrices. Their product $q' \odot q''$ will be called a matrix q of the dimension $|K| \times |K|$ the element $q(k, k')$ of which is defined by the expression

$$q(k, k') = \min_{l \in K} \big(q'(k, l) + q''(l, k')\big) . \tag{8.39}$$

The product $\varphi \odot q$ of the row vector φ of the dimension $|K|$ and the matrix q of the dimension $|K| \times |K|$ will be the row vector φ' the k'-th coordinate of which is

$$\varphi'(k') = \min_{k \in K} \big(\varphi(k) + q(k, k')\big) . \tag{8.40}$$

And finally, the product $q \odot f$ of a matrix q having the dimension $|K| \times |K|$ and $|K|$-dimensional column vector f is understood as the column vector f' the k-th coordinate of which is

$$f'(k) = \min_{k' \in K} \big(q(k, k') + f(k')\big) . \tag{8.41}$$

If we denote by $x \oplus y$ the operation $\min(x, y)$ and by $x \odot y$ the operation $x + y$ then we can write the definitions (8.39), (8.40) and (8.41) in the form

$$\left.\begin{aligned} q(k, k') &= \bigoplus_{l \in K} \big(q'(k, l) \odot q''(l, k')\big) , \\ \varphi'(k') &= \bigoplus_{k \in K} \big(\varphi(k) \odot q(k, k')\big) , \\ f'(k) &= \bigoplus_{k' \in K} \big(q(k, k') \odot f(k')\big) \end{aligned}\right\} \tag{8.42}$$

which altogether formally agrees with the conventional definition of the matrix product. So far $(\oplus, \odot)$ has been considered as a pair $(+, \text{product})$ built up using addition and multiplication where the formulæ (8.42) define the matrix products in the usual sense, i.e., the matrix products in the semi-ring $(+, \text{product})$. However, if the $(\oplus, \odot)$ is understood as a pair $(\min, +)$, the same formulæ correspond to a matrix product in the sense we have introduced, i.e., to matrix products in the semi-ring $(\min, +)$.

The original expression (8.35) defining the original optimisation task can be expressed using notation $\oplus$ and $\odot$ as

$$d = \bigoplus_{k_0} \bigoplus_{k_1} \cdots \bigoplus_{k_n} \left(\varphi(k_0) \odot \left(\bigodot_{i=1}^{n} q_i(k_{i-1}, k_i) \right) \right) . \tag{8.43}$$

Based on the same considerations by which the equivalence of the multi-dimensional sum (8.36) and the matrix product (8.37) was proved, we can claim that the expression (8.43), and thus the (8.35) as well is the matrix product

$$d = \varphi \odot q_1 \odot q_2 \odot \cdots \odot q_n \odot f \tag{8.44}$$

in the semi-ring $(\min, +)$, where f is a column vector all the coordinates of which have zero values.

The matrix product (8.44) is just the original optimisation task (8.35) written in an algebraic form. With such a notation, the construction of the actual algorithm is quite trivial since the expression (8.44) directly demonstrates the algorithms of the calculation. Thanks to the associativity of the matrix product we can calculate according to the formula

$$\Big(\big((\ldots((\varphi \odot q_1) \odot q_2) \odot \cdots \odot q_{n-2}) \odot q_{n-1}\big) \odot q_n\Big) \odot f\,,$$

which is just a different notation of the calculation according to the algorithms (8.31)–(8.34), we formulated before by virtue of informal considerations. The expression (8.44) reveals still more possible procedures for calculating d, for example, that according to the formula

$$d = \Big(\varphi \odot \big(q_1 \odot (\,\cdots\, (q_{n-1} \odot (q_n \odot f))\,\cdots)\big)\Big)$$

or

$$\varphi \odot \bigodot_{i=1}^{l} q_i \odot \left(\bigodot_{i=l+1}^{k} q_i\right) \odot \bigodot_{i=k+1}^{n} q_i \odot f\,,$$

and we can choose either of them according to purely technical conditions we know from Section 8.3.

8.4.7 Seeking the most probable subsequence of states

In the task presented in Example 8.5 we have said that seemingly reasonable, but inaccurate considerations can fail. Let us show now how the problem of such type must be dealt with correctly.

Let X and K be two finite sets, $\bar{x} = (x_1, \ldots, x_n) \in X^n$ and $\bar{k} = (k_0, \ldots, k_n) \in K^{n+1}$ be two random sequences the joint probabilities of which are given by

$$p(\bar{x}, \bar{k}) = p(k_0) \prod_{i=1}^{n} p(x_i, k_i \,|\, k_{i-1})\,, \tag{8.45}$$

where $p(k_0)$, $p(x_i, k_i \,|\, k_{i-1})$, $k_i \in K$, $x_i \in X$, $i = 1, 2, \ldots, n$, are known numbers. The quantities x_i, $i = 1, 2, \ldots, n$ are observable, and k_i, $i = 0, 1, \ldots, n$ are hidden parameters of an object.

Let us assume that even when all features x_i are observable, the values of some features were not measured in the experiment. We will denote by the symbol Ix a set of indices of those features the values of which are measured. So the outcome of the experiment is a set of features $(x_i\,,\, i \in Ix)$. On the basis of experimental data, the evaluation should concern the hidden parameters $k_0, k_1, \ldots, k_n$. However, the experimentalist is not interested in values of all the hidden parameters, but only in the values of some of them which are designated by the ensemble of indices Ik. In the task based on the experimental data $(x_i\,,\, i \in Ix)$ an ensemble $(k_i^*\,,\, i \in Ik)$ is to be found the *a posteriori* probability of which is the greatest,

$$(k_i^*\,,\, i \in Ik) = \operatorname*{argmax}_{(k_i\,,\, i \in Ik)} p\big((x_i\,,\, i \in Ix)\,,\, (k_i\,,\, i \in Ik)\big)\,.$$

With respect to general probability properties we have

$$p\big((x_i\,,\, i \in Ix),\, (k_i\,,\, i \in Ik)\big) = \sum_{(x_i\,,\, i \notin Ix)} \sum_{(k_i\,,\, i \notin Ik)} p(x,k)\,.$$

With respect to Markovian property (8.45) of the observed object we write

$$p\big((x_i\,,\, i \in Ix), (k_i\,,\, i \in Ik)\big) = \sum_{(k_i\,,\, i \notin Ik)} \sum_{(x_i\,,\, i \notin Ix)} p(k_0) \prod_{i=1}^{n} p(x_i, k_i \mid k_{i-1})\,. \quad (8.46)$$

This quantity depends only on the values $(k_i\,,\, i \in Ik)$ which are to be determined. It does not depend on the quantities $(x_i\,,\, i \notin Ix)$ and $(k_i\,,\, i \notin Ik)$ since according to them addition is performed. It does not depend on the values $(x_i\,,\, i \in Ix)$ because they are fixed results of the experiment and so they are constants within one task. The number (8.46) which depends on $(k_i\,,\, i \in Ik)$ will be denoted $d((k_i\,,\, i \in Ik))$. The summation with respect to the values $(x_i\,,\, i \notin Ix)$ will be performed in the following manner,

$$\begin{aligned} d\big((k_i\,,\, i \in Ik)\big) &= \sum_{(k_i\,,\, i \notin Ik)} \sum_{(x_i\,,\, i \notin Ix)} p(k_0) \prod_{i=1}^{n} p(x_i, k_i \mid k_{i-1}) \\ &= \sum_{(k_i\,,\, i \notin Ik)} p(k_0) \prod_{i \in Ix} p(x_i, k_i \mid k_{i-1}) \prod_{i \notin Ix} \sum_{x_i \in X} p(x_i, k_i \mid k_{i-1}) \\ &= \sum_{k_i\,,\, i \notin Ik} \varphi(k_0) \prod_{i=1}^{n} q_i(k_{i-1}, k_i) \end{aligned}$$

where

$$q_i(k_{i-1}, k_i) = \begin{cases} p(x_i, k_i \mid k_{i-1}), & \text{if} \quad i \in Ix\,, \\ \sum\limits_{x_i \in X} p(x_i, k_i \mid k_{i-1}), & \text{if} \quad i \notin Ix\,, \end{cases} \quad (8.47)$$

and the number $\varphi(k_0)$ is the probability $p(k_0)$. The objective is to find the maximum value for $d\big((k_i\,,\, i \in Ik)\big)$, i.e.,

$$d = \max_{(k_i\,,\, i \in Ik)} \sum_{(k_i\,,\, i \notin Ik)} \varphi(k_0) \prod_{i=1}^{n} q_i(k_{i-1}, k_i)\,, \quad (8.48)$$

and the ensemble $(k_i^*, i^* \in I\!k)$ by which the maximum value is attained, i.e.,

$$(k_i^*, i \in I\!k) = \operatorname*{argmax}_{(k_i, i \in I\!k)} \sum_{(k_i, i \notin I\!k)} \varphi(k_0) \prod_{i=1}^{n} q_i(k_{i-1}, k_i)\,. \tag{8.49}$$

We will be concerned with the task (8.48) only. According to its solution the solution of the task (8.49) will become clear. As before, the symbol q_i will denote a matrix of the dimension $|K| \times |K|$ in which in the (k_{i-1})-th row and (k_i)-th column we find the number $q_i(k_{i-1}, k_i)$ calculated according to (8.47). We will denote by φ the row vector composed from the coordinates $\varphi(k_0) = p(k_0)$, $k_0 \in K$. We will denote by $\odot_i$ the matrix multiplication in the semi-ring $(+, \text{product})$ if $i \notin I\!k$, and in the semi-ring $(\max, \text{product})$ if $i \in I\!k$. With this notation the number (8.48) is a matrix product

$$d = \varphi \odot_0 q_1 \odot_1 q_2 \odot_2 q_3 \odot_3 \cdots \odot_{n-1} q_n \odot_n f\,, \tag{8.50}$$

where f is a $|K|$-dimensional column vector all coordinates of which are 1.

The expression (8.50) presents the two tasks studied in a unified way. It concerns the recognition of a Markovian object as a whole, as well as the recognition of values of its hidden parameters, including different modifications of the task. An important advantage of expressing tasks in this way is not only that their affinity becomes revealed, but also that the tasks themselves are becoming easy since they are formulated as matrix product which has just to be calculated.

In calculating matrix products of the form (8.50), it must be taken into consideration that in the expression (8.50) the matrix products occur in different semi-rings. This makes them different from the previous two tasks in which the multiplications within the product (8.7) were understood as being in the semi-ring $(+, \text{product})$, and multiplications within (8.44) were considered as being in the semi-ring $(\min, +)$. In both cases thanks to the associativity of matrix multiplication, the calculations according to the formulæ (8.7) or (8.44) could be performed in any arbitrary order, from left to right, from right to left, from the centre, etc.. It is a different matter in the expression (8.50). There the diversity concerning the potential order of calculations is smaller since the products in the semi-ring $(+, \text{product})$ possess priority over the multiplication in the semi-ring $(\max, \text{product})$, and therefore they have to be processed first. It is due to the fact that the product $A \odot (B\,C)$ is not the same as the product $(A \odot B)\,C$, even if $A \odot (B \odot C) = (A \odot B) \odot C$ and $A\,(B\,C) = (A\,B)\,C$. The product $A \odot (B\,C)$ means

$$\max_y \left(a(x,y) \left(\sum_z b(y,z)\, c(z,u) \right) \right),$$

and the product $(A \odot B)\,C$ means

$$\sum_z \left(\max_y \left(a(x,y)\, b(y,z) \right) \right) c(z,u)$$

which are different functions.

Owing to the required order in calculating the product (8.50), the complexity can increase compared with the complexity in calculating the expressions (8.7) and (8.44) which is $\mathcal{O}(|K|^2\, n)$. In the case in which the sequence of parameters to be determined consists of a large number of mutually non-interconnected segments, i.e., if the set $I\!k$ is strongly mixed up with the set $\{0,1,2,\ldots,n\} \setminus I\!k$ then the calculation of the product (8.50) has the complexity $\mathcal{O}(|K|^3\, n)$. In the case, however, if a certain connected subsequence of hidden parameters, i.e., k_l^m is to be determined then the complexity of calculation according to (8.50) will remain $\mathcal{O}(|K|^2\, n)$, i.e., it will not increase with respect to the complexity of calculation according to the formulæ (8.7) and (8.44). The product (8.50) assumes the form

$$d = \varphi \left(\prod_{i=1}^{l-1} q_i \right) \odot \left(\bigodot_{i=l}^{m} q_i \right) \odot \left(\prod_{i=m+1}^{n} q_i \right) f$$

in this case and this form is to be understood as a brief notation to the following calculations.

1. The calculation of a row vector φ' according to the formula

$$\varphi' = \varphi\, q_1\, q_2\, \cdots\, q_{l-1}$$

   which has the complexity $\mathcal{O}(|K|^2\, l)$.
2. The calculation of a column vector f' according to the formula

$$f' = q_{m+1}\, q_{m+2}\, \cdots\, q_n\, f$$

   which has the complexity $\mathcal{O}\big(|K|^2\, (n-m)\big)$.
3. The calculation of the number d, that is looked for, according to the formula

$$d = \varphi' \odot q_l \odot q_{l+1} \odot \cdots \odot q_m \odot f'$$

   which has the complexity $\mathcal{O}\big(|K|^2\, (m-l)\big)$.

The total complexity of the calculation is $\mathcal{O}(|K|^2\, n)$.

We can see that if a matrix operation is considered in a broader sense then a certain group of tasks of structural sequence recognition can be uniformly and concisely expressed by means of matrix multiplications. Certainly, each of the so far quoted tasks could be solved even without applying the formalism presented above. But in a separate analysis of the tasks, one would hardly succeed in finding that all the tasks could be mastered in one lot with one single program which with small alterations can be tuned for solving any of the tasks studied so far.

8.5 Seeking sequences composed of the most probable hidden parameters

The task of estimating a hidden parameter sequence (or subsequence) of an object was formulated as seeking a sequence (or subsequence) which for certain observations outcomes of an object possesses the highest *a posteriori* probability. Such a task formulation is quite natural and was apparently worth the attention. The same attention is to be paid to the aspect that a task in such a formulation is only a special case of a more general task of minimising the Bayesian risk with a concrete penalty function. Let us now examine this special case from a more general point of view, and let us ask the question to what extent the penalty function is natural. This results in the necessity of seeking the most probable sequence.

Seeking for the most probable sequence results from the penalties of the form $W(\bar{k}, \bar{k}') = 0$, $\bar{k} \in K^{n+1}$, $\bar{k}' \in K^{n+1}$, if $\bar{k} = \bar{k}'$, and $W(\bar{k}, \bar{k}') = 1$ in the opposite case. This means that all cases of wrong recognition of the sequence, i.e., the situation when $\bar{k}' \neq \bar{k}$ are given an equal penalty, the sequence $\bar{k}'$ may differ from $\bar{k}$ in one element only, or in all of them. Naturally, such a requirement does not seem to be so self-evident. At least, when considering any application context then it does not seem to be the only possible and universally acceptable requirement. For example, it could be assumed equally natural that the penalty $W(\bar{k}, \bar{k}')$ was equal to the number of elements i with which $k_i \neq k_i'$. It would mean that the penalty function has the form

$$W(\bar{k}, \bar{k}') = \sum_{i=0}^{n} w(k_i, k_i') , \quad \text{where} \tag{8.51}$$

$$w(k_i, k_i') = \begin{cases} 1, \text{ if } & k_i \neq k_i' , \\ 0, \text{ if } & k_i = k_i' , \end{cases}$$

and not the form

$$W(\bar{k}, \bar{k}') = \begin{cases} 1, \text{ if } & \bar{k} \neq \bar{k}' , \\ 0, \text{ if } & k = k' . \end{cases} \tag{8.52}$$

The optimal strategies minimising the risk are different for (8.51) and (8.52). This will be evident in the following example.

Example 8.6 Dissimilar Bayesian solutions for two penalty functions. *Let us have sequences $\bar{k}$ of length $n = 2$. The problem is to estimate the pairs (k_1, k_2). The set of possible values for k_1 as well as for k_2 is $\{A, B, C\}$. There are nine such possible sequences. Assume that by virtue of observation, it has been found that the a posteriori probabilities of these nine possible sequences are those given in the table.*

$\bar{k}_2 \backslash \bar{k}_1$	A	B	C
A	0.30	0	0
B	0.20	0	0
C	0	0.25	0.25

If a sequence $(\bar{k}_1', \bar{k}_2')$ has to be found such that the probability of the event $(\bar{k}_1, \bar{k}_2) \neq (\bar{k}_1', \bar{k}_2')$ should be minimal which corresponds to the penalty function (8.52) then a decision must be made that the sequence (k_1', k_2') is $(A,\ A)$. With *such a decision the probability of the inequality* $(k_1, k_2) \neq (k'_1, k'_2)$ *will be* 0.7, *and with any other decision this probability will be greater. Let us see what risk is present with such a decision with respect to the penalty function (8.51), i.e., in other words, what mathematical expectation of the number of incorrectly recognised sequence elements amounts to. The actual sequence* (k_1, k_2) *can be one of four possibilities* $(A,\ A)$, $(A,\ B)$, $(B,\ C)$ *and* $(C,\ C)$ *the probabilities of which are* $0.3, 0.2, 0.25, 0.25$ *correspondingly. The number of incorrectly recognised elements will be* $0, 1, 2, 2$, *correspondingly and the mathematical expectation of this number will be equal to* 1.2.

Now let us see what the mathematical expectation would be like if the decision were made that $(k'_1, k'_2) = (A,\ C)$. *Let us note that this sequence has a zero a posteriori probability, but in spite of that, at the decision* $(k'_1, k'_2) = (A,\ C)$ *the mathematical expectation of incorrectly recognised elements will have the value* 1. *The actual sequence can rightly be only* $(A,\ A)$, $(A,\ B)$, $(B,\ C)$, $(C,\ C)$, *and with each of these sequences the number of incorrectly recognised elements will be* 1.

We can see that the solution of a Bayesian task at the penalty function (8.51) is not even approximately identical with the solution the penalty function of which is (8.52). Therefore, if the application requires a penalty function of the form (8.51) then the algorithm which is seeking the most probable set of hidden parameters cannot be used. It is suited for other penalty function of the form (8.52). ▲

If the penalty function of the form (8.51) occurs then the Bayesian task has to be solved from the very beginning, i.e., starting from the Bayesian formulation. Let X and K be two finite sets, and $\bar{x}$ and $\bar{k}$ be two sequences of the lengths n and $n+1$, respectively, which are composed from elements of X and K, $\bar{x} = (x_1, x_2, \ldots, x_n)$, $\bar{k} = (k_0, k_1, \ldots, k_n)$. The pair $(\bar{x}, \bar{k})$ is random and assumes the value from the set $X^n \times K^{n+1}$ so that the probability of the pair $(\bar{x}, \bar{k})$ is given by the expression

$$p(\bar{x}, \bar{k}) = p(k_0) \prod_{i=1}^{n} p(x_i, k_i \mid k_{i-1}) \tag{8.53}$$

in which $p(k_0)$, $p(x_i, k_i \mid k_{i-1})$ are known numbers.

Let $W \colon K^{n+1} \times K^{n+1} \to \mathbb{R}$ be a penalty function of the form

$$W(\bar{k}, \bar{k}') = \sum_{i=0}^{n} w(k_i, k'_i)\,, \quad \text{where} \tag{8.54}$$

$$w(k_i, k'_i) = 1\,, \quad \text{if} \quad k_i \neq k'_i\,, \tag{8.55}$$

$$w(k_i, k'_i) = 0\,, \quad \text{if} \quad k_i = k'_i\,. \tag{8.56}$$

For this known data a strategy $q \colon X^n \to K^{n+1}$ is to be created, i.e., an algorithm which for each sequence $x_1, x_2, \ldots, x_n$ determines the sequence $\bar{k}' =$

$(k'_0, k'_1, \ldots \ldots, k'_n)$, minimising the risk

$$\sum_{\bar{k} \in K^{n+1}} p(\bar{k} \mid \bar{x}) \, W(\bar{k}, \bar{k}') \,, \quad \text{i.e., the sequence}$$

$$\begin{aligned} \bar{k}' = (k'_0, k'_1, \ldots, k'_n) &= \operatorname*{argmin}_{\bar{k}'} \sum_{\bar{k}} p(\bar{k} \mid \bar{x}) \, W(\bar{k}, \bar{k}') \\ &= \operatorname*{argmin}_{\bar{k}'} \sum_{\bar{k}} p(\bar{x}, \bar{k}) \, W(\bar{k}, \bar{k}') \\ &= \operatorname*{argmin}_{\bar{k}'} \sum_{\bar{k}} p(\bar{x}, \bar{k}) \sum_{i=0}^{n} w(k_i, k'_i) \,. \end{aligned} \tag{8.57}$$

By using the expression (8.53) we get

$$\begin{aligned} \bar{k}' &= (k'_0, k'_1, \ldots, k'_n) \\ &= \operatorname*{argmin}_{\bar{k}'} \sum_{k_0} \sum_{k_1} \cdots \sum_{k_n} \left(p(k_0) \prod_{i=1}^{n} p(x_i, k_i \mid k_{i-1}) \right) \sum_{i=0}^{n} w(k_i, k'_i) \,. \end{aligned} \tag{8.58}$$

An important feature of this task already results from the mere assumption that the penalty function has the form (8.54), without taking into account the Markovian property (8.53), and its concretisation (8.55) and (8.56). Let us demonstrate this feature.

The risk $\sum_k p(\bar{x}, \bar{k}) \sum_{i=0}^{n} w(k_i, k'_i)$, which the sequence $k'_o, k'_1, \ldots, k'_n$ sought has to minimise, will be denoted by R. We can write

$$\begin{aligned} R &= \sum_{\bar{k} \in K^{n+1}} p(\bar{x}, \bar{k}) \sum_{i=0}^{n} w(k_i, k'_i) = \sum_{i=0}^{n} \sum_{\bar{k} \in K^{n+1}} p(\bar{x}, \bar{k}) \, w(k_i, k'_i) \\ &= \sum_{i=0}^{n} \sum_{k_i \in K} w(k_i, k'_i) \sum_{(k_{i^*},\, i^* \neq i)} p(\bar{x}, k^*_0, k^*_1, \ldots, k^*_i, \ldots, k^*_n) \\ &= \sum_{i=0}^{n} \sum_{k_i \in K} w(k_i, k'_i) \, p(\bar{x}, k_i) \,. \end{aligned}$$

We can see that the function of $n+1$ variables $(k'_0, k'_1, \ldots, k'_n)$, which is to be minimised, is created as a sum of $n+1$ functions, each of them depending on only one variable. The optimisation task (8.57) is thus broken into $n+1$ independent optimisation tasks along a single variable,

$$k'_i = \operatorname*{argmin}_{k'_i} \sum_{k_i} p(\bar{x}, k_i) w(k_i, k'_i) \,. \tag{8.59}$$

When considering the specific forms (8.55), (8.56) of the partial function w then we arrive at the conclusion that

$$k'_i = \underset{k_i \in K}{\operatorname{argmax}}\ p(\bar{x}, k_i)\,. \tag{8.60}$$

The conclusion claims that even when the task was originally formulated as an optimisation one, (see (8.57) and (8.58)), its complexity is not caused by optimisation at all since it is reduced to trivial tasks (8.59) and (8.60). The core of its complexity is in the calculation of the $(n+1)\,|K|$ values $p(\bar{x}, k_i)$, $i = 0, 1, \ldots, n$, $k_i \in K$, according to the general formula

$$p(\bar{x}, k_i) = \sum_{(k_{i^*}\,,\; i^* \neq i)} p(\bar{x}, \bar{k})\ ,$$

or in making use of the Markovian property of the model according to the formula

$$p(\bar{x}, k_i) = \sum_{k_0} \sum_{k_1} \cdots \sum_{k_{i-1}} \sum_{k_{i+1}} \cdots \sum_{k_n} p(k_0) \prod_{i=1}^{n} p(x_i, k_i \mid k_{i-1})\,. \tag{8.61}$$

Computational complexities in counting the previous multi-dimensional sum can be coped with because of the results of the analysis of the automaton recognition problem. In the analysis in Section 8.3 we found that the matrix product $\varphi\big(\prod_{j=1}^{i-1} P_j\big)$ was an ensemble consisting of $|K|$ probabilities $p(x_1, x_2, \ldots, x_i, k_i)$, and the matrix product $\big(\prod_{j=i}^{n} P_j\big) f$ expressed the probabilities $p(x_{i+1}, \ldots, x_n \mid k_i)$. The numbers $p(\bar{x}, k_i)$ defined by the expression (8.61) can be calculated using the formula

$$p(\bar{x}, k_i) = p(x_1^i, x_{i+1}^n, k_i) = p(x_1^i, k_i)\, p(x_{i+1}^n \mid k_i)\ ,$$

correctness of which results both from the expression in relation (8.61) and directly from the Markovian property of the model.

The complexity of calculating the values $p(\bar{x}, k_i)$ for one particular i and for all $k_i \in K$ is identical with the complexity of calculating the matrix products $\varphi(\prod_{j=1}^{i-1} P_j)$ and $(\prod_{j=i}^{n} P_j) f$ and is $\mathcal{O}(|K|^2\, n)$. The complexity of computing the numbers $p(\bar{x}, k_i)$ for all $i = 0, 1, \ldots, n$, and for all $k_i \in K$ will by no means be $\mathcal{O}(|K|^2\, n^2)$, but will remain $\mathcal{O}(|K|^2\, n)$. This will be the same situation as in the previous task of automaton recognition and in the task of the most likely estimation of the sequence of hidden parameters.

We have analysed three recognition tasks which can be formulated within the Markovian model of the recognised object. These are the task of recognising the object as a whole, seeking the most probable sequence of hidden states of the object, and seeking a sequence of the most probable hidden states of the object. Even if we have analysed diverse varieties of these tasks, we do not intend to create an impression that the tasks analysed cover a vast variety of possible applications. Rather the opposite, one of the aims of this lecture is to rouse a feeling that we know only a small part of the relevant tasks. In this way we wish to impair the widespread and pleasantly self-delusive ideas that it is sufficient to know only one method of Markovian sequence recognition, and this is seeking the shortest path in a graph by the methods of dynamic programming.

A significant breakthrough in structural recognition appeared when the solution of problems, insurmountable before, proved successful with the aid of dynamic programming. This deserves credit even after several decades. In this context, however, we wished to point out that significant as the knowledge may be, it need not be an actual contribution, when, without forethought, it begins to be considered as generally valid.

8.6 Markovian objects with acyclic structure

8.6.1 Statistical model of an object

We have seen that if a recognised object can be successfully expressed as a Markovian sequence of its observable and hidden parameters then for such a model the classic Bayesian recognition tasks can be solved. The most important property is that the solution of these tasks does not require calculations of any fantastic complexity. The rooted apprehension that the calculation complexity increases exponentially with the increase in the number of observations (features) has not materialised. The increase of calculation complexity has been linear for Markovian models.

All these pleasant features are owed to one simple assumption about the form of the joint probability $p(\bar{x}, \bar{k})$ which for sequences was expressed by the relation (8.1), equivalent to the relation (8.3), and informally represented by a mechanical model (Fig. 8.1). Now we will demonstrate that constructive recognition of a complex object is possible even at weaker assumptions than those in (8.1), or in an equivalent manner, in (8.3). This means that complex objects can be constructively recognised even in cases in which their parts cannot be one-dimensionally ordered into a sequence, and have a more complex structure.

The sequence $\bar{k} = (k_0, k_1, \ldots, k_n)$ can be regarded as a function of the form $I \to K$ defined on a set of indices $\{0, 1, \ldots, n\}$, which consists of integer numbers. We will now generalise the concept of the sequence in such a way that the set I will not be regarded as a set of integers but as a set of vertices of a connected unoriented acyclic graph G, i.e., of a tree. The ensemble k will be regarded, as before, to be the function $I \to K$ defined on the set of the graph vertices. The sequence is a particular case of such a function if the graph G is a chain, i.e., a connected graph in which from each vertex one edge at least and two edges at most go out.

We will generalise the concept of the sequence $\bar{x} = (x_1, x_2, \ldots, x_n)$ in such a way that x will be regarded as a function $H \to X$ defined on a set of edges H of the graph G. So the observation x is the ensemble $(x_h,\ h \in H)$ built up from indexed quantities x_h, where h represents an edge of the graph G, or which is the same as a certain pair (i, i') of vertices. The observation $\bar{x}$ is a sequence provided the graph G is a chain, and thus also the set of its edges forms a chain.

The main precondition of the joint probability $p(\bar{x}, \bar{k})$ of the observation $\bar{x}\colon H \to X$ and of the hidden parameters $\bar{k}\colon I \to K$ which generalises the

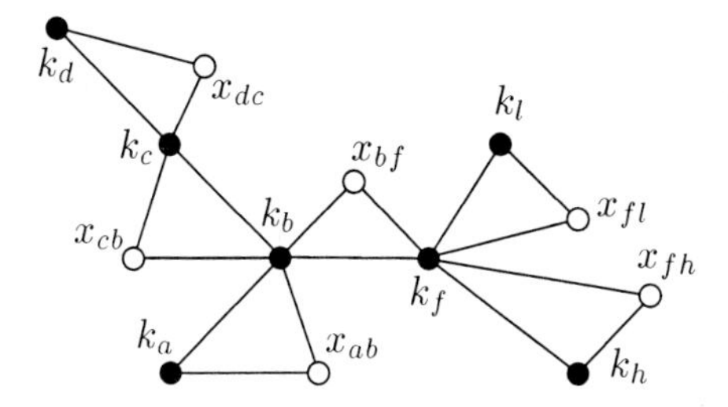

Figure 8.4 Mechanical model of a Markovian acyclic graph.

Markovian model (8.1), consists of the assumption that for any vertex $0 \in I$, let it be called a 0-th one, the probability $p(\bar{x}, \bar{k})$ has the form

$$p(\bar{x}, \bar{k}) = p(k_0) \prod_{i \in I \setminus \{0\}} p\left(x_{\{i,g(i)\}}, k_i \,|\, k_{g(i)}\right) , \tag{8.62}$$

where $g(i)$ is the vertex connected with the vertex i by an edge which pertains to the path from the 0-th vertex to the i-th. The property (8.3) formulated before is a particular case of the property (8.62), for $I = \{0, 1, 2, \ldots, n\}$ and $g(i) = i - 1$.

The property of the model (8.62) can be informally represented by the model in Fig. 8.4 which generalises the mechanical model in Fig. 8.1 used for informal representation of the Markovian sequence properties.

In Fig. 8.4 the values x_h, $h \in H$, and k_i, $i \in I$, are represented by means of points in a plane which are connected by line segments. If we visualise each straight line as a flexible rod then it can be seen that the position of each point affects the positions of all other points. If we fixate a point that corresponds to some quantity k_i, say the point k_f, then the mechanical model in that given case breaks into three independent parts: one consists of the points $k_a, k_b, k_c, k_d, x_{ab}, x_{cb}, x_{dc}$ and x_{bf}, the second consists of the points k_l, x_{fl}, and the third of the points k_h and x_{fh}. It is this property of conditional independence of individual parts of a complex object that is formally expressed by the assumption (8.62). This assumption has become a basis for the formulation and exact solution of Bayesian recognition tasks, similar to those which we analysed for the case of sequences. We will briefly examine only two tasks: the task of calculating the probability $p(\bar{x})$ for the given observation $\bar{x}$ and the task of calculating the number $\max_{\bar{k} \in K^I} p(\bar{x}, \bar{k})$ for the given observation $\bar{x}$. After completing this analysis, the solution will become quite clear for other tasks and their modifications which we have dealt with in detail in the case of sequences.

8.6.2 Calculating the probability of an observation

The calculation of the probability $p(\bar{x})$ means the calculation of a multi-dimensional sum

$$p(\bar{x}) = \sum_{(k_i, i \in I)} p(k_0) \prod_{i \in I \setminus \{0\}} p(x_{\{i,g(i)\}}, k_i \,|\, k_{g(i)}) . \tag{8.63}$$

Since the quantities x_h, $h \in H$, are fixated the expression (8.63) will be written so that these quantities should not be present in it. The probability $p(x_{\{i,g(i)\}}, k_i \mid k_{g(i)})$ will be denoted $f_i(k_i, k_{g(i)})$. To achieve symmetry of the expression (8.63) with respect to the indices i, and also for further reasons which will become clear later, we will introduce the notation $\varphi_i(k_i)$. This means $p(k_0)$, if $i = 0$, and $\varphi_i(k_i) = 1$ for all $i \neq 0$ and $k_i \in K$. The expression (8.63) thus assumes the form

$$d = \sum_{(k_i\,,\, i \in I)} \prod_{i \in I} \varphi_i(k_i) \prod_{i \in I \setminus \{0\}} f_i(k_i, k_{g(i)})\,. \tag{8.64}$$

In this expression the variable k_i for each i is present in a single factor $\varphi_i(k_i)$. As to the factors $f_i(k_i, k_{g(i)})$, the variable k_{i^*} can be present depending on the index i^* in one, two, or more factors. The variable k_{i^*} is naturally present in the factor $f_{i^*}(k_{i^*}, k_{g(i^*)})$, but also in those factors $f_i(k_i, k_{g(i)})$ for which $g(i) = i^*$. There certainly exists such an index i^*, that the variable k_{i^*} is present only in one factor of the form $f_i(k_i, k_{g(i)})$. It is such an index i^* for which $i^* = g(i)$ is valid for no index i. The existence of such an index results from the property that a vertex exists in an acyclic graph from which only one edge goes out. For the index i^* defined in this way the formula (8.64) will be rewritten in the form

$$d = \sum_{(k_i \mid i \in I, i \neq i^*)} \prod_{i \in I \setminus \{i^*\}} \varphi_i(k_i) \prod_{i \in I \setminus \{0, i^*\}} f_i(k_i, k_{g(i)}) \sum_{k_{i^*}} \varphi_{i^*}(k_{i^*})\, f_{i^*}(k_{i^*}, k_{g(i^*)})\,. \tag{8.65}$$

We will denote $g(i^*)$ as i' and calculate new values of the numbers $\varphi_{i'}(k_{i'})$ according to the assignment

$$\varphi_{i'}(k_{i'}) := \varphi_{i'}(k_{i'}) \sum_{k_{i^*}} \varphi_{i^*}(k_{i^*})\, f_{i^*}(k_{i^*}, k_{i'})\,. \tag{8.66}$$

The denotation $\varphi_{i'}(k_{i'})$ on the right-hand side of the expression (8.66) is considered to be the value of the number $\varphi_{i'}(k_{i'})$ before the operator has been satisfied; the same denotation on the left-hand side is considered to be the new value obtained through this operator. The calculation complexity of the operator (8.66) is $\mathcal{O}(|K|^2)$.

The obtained numbers $\varphi_{i'}(k_{i'})$ can be substituted into the expression (8.65) and it can be written in the form

$$\sum_{(k_i \mid i \in I_1)} \prod_{i \in I_1} \varphi_i(k_i) \prod_{i \in I_1 \setminus \{0\}} f_i(k_i, k_{g(i)})\,, \tag{8.67}$$

where $I_1 = I \setminus \{i^*\}$. The expression (8.67) has the same form as the original expression. Only the number of variables k_i, according to which the addition is performed decreased by one. Among the reduced number of variables there is at least one that it is present only in one factor of the form $f_i(k_i, k_{g(i)})$, and can be eliminated in the way already described, i.e., by the operator (8.66). After the $(|I| - 1)$-th elimination of the variables the expression (8.65) will assume

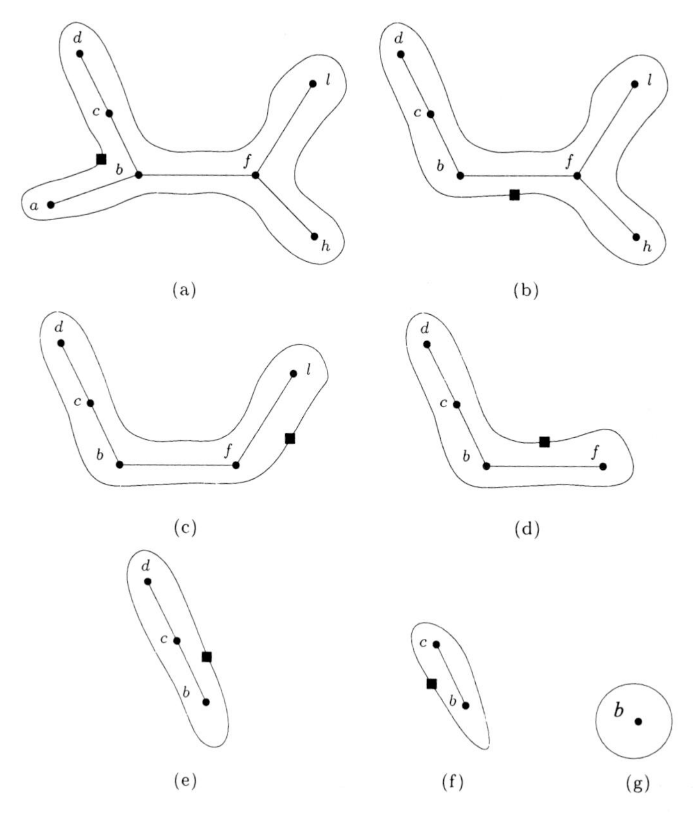

Figure 8.5 Reduction of the graph in calculating $p(x)$.

the form

$$d = \sum_{k_0} \varphi(k_0)$$

which can be easily calculated since the addition is done only along the values of one variable. Therefore, we have proved that the calculation complexity of the number $p(\bar{x})$ is $\mathcal{O}(|K|^2\,|I|)$, i.e., the same as the complexity in the case in which the recognised object had the structure of a sequence.

Example 8.7 Calculating the probability of observation for an acyclic graph. *We will demonstrate the procedure presented above using an example of a graph from Fig. 8.4. The structure of the input data for calculating the number $p(x)$ corresponds to the initial configuration of the graph, see Fig. 8.5(a). For each vertex i the memory for $|K|$ numbers $\varphi_i(k_i)$, $k_i \in K$, is reserved, and for each edge $h = (i, i')$ the memory for $|K|^2$ numbers $f_i(k_i, k_{i'})$, $k_i \in K$, $k_{i'} \in K$ is reserved.*

A calculation diagram for processing this data can be expressed in the following form. In Fig. 8.5(a) the graph was enclosed by a curve (an outline). We will choose a starting point on the outline which is represented by a filled square, and passes along the outline anticlockwise. During the passage we will create a sequence of graph vertices around which the path is led. The sequence of vertices will be $(b, a, b, f, h, f, l, f, b, c, d, c, b)$. *Furthermore, this sequence passes through one vertex after another, and the filled square in Fig. 8.5 indicates the position at that moment. Some vertices are passed by without any change in the data, i.e., nothing is calculated. The data is changed in other vertices including the graph. The changes are brought about by those vertices from which only one edge goes out in the particular momentary graph, but they are not the starting vertex. In our case the vertex* b *is the starting one. The sequence of vertices which change the data is* (a, h, l, f, d, c). *We will show what changes will occur in the data at each of these vertices. After each change the graph is modified in a corresponding way.*

Vertex a. *New values of the number* $\varphi_b(k_b)$ *are calculated according to the formula*

$$\varphi_b(k_b) := \varphi_b(k_b) \sum_{k_a} \varphi_a(k_a)\, f_a(k_a, k_b)\,.$$

The vertex a *is eliminated from the graph and the algorithm continues using the graph in Fig. 8.5(b).*

Vertex h. *New values of the number* $\varphi_f(k_f)$ *are calculated according to the formula*

$$\varphi_f(k_f) := \varphi_f(k_f) \sum_{k_h} \varphi_h(k_h)\, f_h(k_h, k_f)\,.$$

The vertex h *is eliminated from the graph and the algorithm continues using the graph as in Fig. 8.5(c).*

Vertex l. *New values of the number* $\varphi_f(k_f)$ *are calculated according to the formula*

$$\varphi_f(k_f) := \varphi_f(k_f) \sum_{k_l} \varphi_l(k_l)\, f_l(k_l, k_f)\,.$$

The vertex l *is eliminated from the graph and the algorithm continues using the graph as in Fig. 8.5(d).*

Vertex f. *New values of the numbers* $\varphi_b(k_b)$ *are calculated according to the formula*

$$\varphi_b(k_b) := \varphi_b(k_b) \sum_{k_f} \varphi_f(k_f)\, f_f(k_f, k_b)\,.$$

The vertex f *is eliminated from the graph and the algorithm continues using the graph as in Fig. 8.5(e).*

Vertex d. *New values of the number* $\varphi_c(k_c)$ *are calculated according to the formula*

$$\varphi_c(k_c) := \varphi_c(k_c) \sum_{k_d} \varphi_d(k_d)\, f_d(k_d, k_c)\,.$$

The vertex d *is eliminated from the graph and the algorithm continues using the graph as in Fig. 8.5(f).*

Vertex c. *New values of the number* $\varphi_b(k_b)$ *are calculated according to the formula*

$$\varphi_b(k_b) := \varphi_b(k_b) \sum_{k_c} \varphi_c(k_c)\, f_c(k_c, k_b)\,.$$

The vertex c is eliminated from the graph and the algorithm arrived at the elementary graph in Fig. 8.5(g) which contains a single point b.

When the graph was simplified then the probability of the observation x is calculated as the sum $\sum_{k_b} \varphi_b(k_b)$. ▲

8.6.3 The most probable ensemble of hidden parameters

This subsection need not have been written as it would be sufficient to substitute the word 'addition' in the previous subsection by the word 'maximisation'. The new subsection would be correct in the same way as the previous subsection. The correctness of previous subsection is owing to the distributivity of multiplication with respect to addition. The correctness of present section results from distributivity of multiplication even with respect to maximisation on the set of positive numbers. It really holds that $x \max(y, z) = \max(x\,y\,,\ x\,y)$. Thus, the algorithm for the calculation

$$d = \max_{(k_i\,,\, i \in I)} \prod_{i \in I} \varphi_i(k_i) \prod_{i \in I \setminus \{0\}} f_i(k_i, k_{g(i)}) \tag{8.68}$$

has the same structure and is based on the same considerations as the algorithm for calculating

$$d = \sum_{(k_i\,,\, i \in I)} \prod_{i \in I} \varphi_i(k_i) \prod_{i \in I \setminus \{0\}} f_i(k_i, k_{g(i)}) \tag{8.69}$$

which we have just analysed. Let us briefly recapitulate these considerations, but this time with respect to the calculation (8.68), and not to that of (8.69).

Let i^* be such an index that for no $i \in I$ the condition $g(i) = i^*$ is satisfied. This means that the variable k_{i^*} occurs only in one factor of the form $f_i(k_i, k_{g(i)})$, namely in $f_{i^*}(k_{i^*}, k_{g(i^*)})$. Therefore, the formula (8.68) can be rewritten in the form

$$d = \max_{(k_i\,,\, i \in I, i \neq i^*)} \prod_{i \in I \setminus \{i^*\}} \varphi_i(k_i) \prod_{i \in I \setminus \{0, i^*\}} f_i(k_i, k_{g(i)}) \max_{k_{i^*}} \Big(\varphi_{i^*}(k_{i^*})\, f_{i^*}\big(k_{i^*}, g(i^*)\big)\Big) .$$

We will denote $g(i^*)$ as i' and calculate new values of the numbers $\varphi_{i'}(k_{i'})$ by the operator

$$\varphi_{i'}(k_{i'}) \ := \varphi_{i'}(k_{i'}) \max_{k_{i^*}} \big(\varphi_{i^*}(k_{i^*})\, f_{i^*}(k_{i^*}, i'))\big) . \tag{8.70}$$

We will make use of the calculated value and write the number d in the form

$$d = \max_{(k_i \mid i \in I_1)} \prod_{i \in I_1} \varphi_i(k_i) \prod_{i \in I_1 \setminus \{0\}} f_i(k_i, k_{g(i)}) \tag{8.71}$$

in which $I_1 = I \setminus \{i^*\}$. In expression (8.71) the maximisation is performed along the variables the number of which is 1 less than that in the expression (8.68). The operator (8.70) is used $(|I| - 1)$ times and every time one of the variables is eliminated after which the minimisation is done. So the original task is reduced to the calculation

$$d = \max_{k_0} \varphi_0(k_0)$$

which is trivial since the maximisation is done after one variable. The total number of operations needed for solving the task (8.68) has the complexity $\mathcal{O}(|K|^2\, n)$ which is the same as that for a sequence.

8.7 Formulation of supervised and unsupervised learning tasks

In Lecture 4, three learning tasks have been formulated in the general form, through which a reasonable estimation of a statistical model of the object under examination can be found. In Lecture 6, we formulated the unsupervised learning task and solved it in the general form. We will express analogous tasks for the class of Markovian models which we are now dealing with. We will discuss a case in which an ensemble of parameters has the structure of a sequence, and not the general structure of an acyclic graph because here the most essential properties of the analysed tasks can be revealed without being overshadowed by unnecessary details. The results we obtain for sequences can be easily generalised for the general case of acyclic structures.

In the same way as before we assume that a complete description of a recognised object is formed by two sequences: the sequence of observable features $\bar{x} = (x_1, x_2, \ldots, x_n)$ of the length n and the sequence of hidden parameters $\bar{k} = (k_0, k_1, \ldots, k_n)$ of the length $n+1$. The pair $(\bar{x}, \bar{k})$ is random and is described by the probability distribution $p(\bar{x}, \bar{k})$ which is not arbitrary but has the form

$$p(\bar{x}, \bar{k}) = p(k_0) \prod_{i=1}^{n} p_i(k_i, x_i \mid k_{i-1}) \,.$$

This means that the function $p\colon X^n \times K^{n+1} \to \mathbb{R}$ is uniquely determined by n functions p_i, $i = 1, 2, \ldots, n$, of the form $K \times X \times K \to \mathbb{R}$ the value $p_i(k', x, k)$ of which means the joint probability of the $(i-1)$-th hidden parameter having the value k', the i-th hidden parameter having the value k, and the i-th observed feature having the value x. The function p_i uniquely determines the probability that the i-th hidden parameter has the value $k \in K$. This probability is $\sum_{k' \in K} \sum_{x \in X} p_i(k', x, k)$. The same probability is determined by the function p_{i+1} as $\sum_{x \in X} \sum_{k' \in K} p_{i+1}(k, x, k')$. It is quite natural that these two determinations must not be in contradiction which means that the functions p_i, $i = 1, \ldots, n$, have to satisfy the conditions

$$\sum_{k' \in K} \sum_{x \in X} p_i(k', x, k) = \sum_{x \in X} \sum_{k' \in K} p_{i+1}(k, x, k') \,, \quad i = 1, 2, \ldots, n-1 \,, \quad k \in K \,.$$

The ensemble of functions $(p_i\,,\ i = 1, \ldots, n)$ which satisfy the equation above will be denoted as P, and will be referred to as a statistical model of the recognised object. Since the ensemble P uniquely determines also the function $p(\bar{x}, \bar{k})$, i.e., the probability distribution of the pairs $\bar{x} = (x_1, \ldots, x_n)$ and $\bar{k} = (k_0, \ldots, k_n)$, the function p will also be referred to as the *statistical model of the object.*

If the statistical model of an object is known then various pattern recognition tasks can be formulated and solved the examples of which were analysed in the previous parts of the lecture. If the statistical model of an object is not known then it must be found either by experimentally examining the object, or on the basis of the user's information in which he states his or her ideas either about the recognised object, or about the desired behaviour of the recognition algorithm.

The creation of the statistical model is mostly called learning, supervised or unsupervised. The information on the basis of which the model is created is termed training information. A precise formulation of the learning tasks depends on the properties of the training information. These formulations will be presented now.

8.7.1 The maximum likelihood estimation of a model during learning

Let us assume special experimental conditions for examining an object in which all parameters of the object are known, both the observable and the hidden ones. Let l experiments be made with the examined object. The outcome of the j-th experiment, $j = 1, 2, \ldots, l$, was the sequence $\bar{x}^j = (x_1^j, x_2^j, \ldots, x_n^j)$ and the sequence $\bar{k}^j = (k_0^j, k_1^j, \ldots, k_n^j)$. From this point up until the end of the lecture the denotation x_i^j, $i = 1, 2, \ldots, n$, $j = 1, 2, \ldots, l$, means the i-th element in the j-th sequence. The denotation k_i^j has a similar meaning.

Provided we have good reasons for assuming that the outcomes of the l experiments were mutually independent, and each of them was a random instance from the general ensemble of pairs $(\bar{x}, \bar{k})$ with the probability distribution p then the model $P = (p_1, p_2, \ldots, p_n)$ can be estimated as that which maximises the probability of the outcome of the experiment as a whole, i.e., the probability $\prod_{j=1}^{l} p(\bar{x}^j, \bar{k}^j)$. Learning is conceived as finding the ensemble

$$\begin{aligned} P^* &= (p_1^*, p_2^*, \ldots, p_n^*) \\ &= \underset{p_1}{\operatorname{argmax}} \cdots \max_{p_n} \prod_{j=1}^{l} p_1(k_0^j, x_1^j, k_1^j) \prod_{i=2}^{n} \frac{p_i(k_{i-1}^j, x_i^j, k_i^j)}{\sum\limits_{k \in K} \sum\limits_{x \in X} p_i(k_{i-1}^j, x, k)} . \end{aligned} \tag{8.72}$$

Let us note that the previous expression for the joint probability $p(\bar{x}, \bar{k})$ has a slightly different form than that used before. The previous expression involves the joint probabilities $p_i(k_{i-1}^j, x_i^j, k_i^j)$, not the conditional probabilities $p_i(k_i^j, x_i^j \mid k_{i-1}^j)$.

8.7.2 Minimax estimate of the model

Often it is quite difficult to secure or check the conditions under which the maximum likelihood estimate of the model is an appropriate one. These conditions require that the outcome of the experiment is a multi-set of random and mutually independent instances. If these conditions are not satisfied then the task should be formulated in a following way which does not rely on the assumptions mentioned above.

It is assumed that the training set consisting of pairs $(\bar{x}^1, \bar{k}^1)$, $(\bar{x}^2, \bar{k}^2)$, $\ldots, (\bar{x}^l, \bar{k}^l)$ is known. The training set was selected by an experimenter (may be, when she/he observed a real object) and is regarded as a typical set that summarises quite probable examples of the observed object behaviour. Learning has to create such a model P^* in which none of the probabilities $p(x^j, k^j)$ corresponding to an instance in the given training set is too small. More precisely, such a maximal value ε and such a model P^* are to be found to secure that the probability $p(\bar{x}^j, \bar{k}^j)$ of any example $(\bar{x}^j, \bar{k}^j)$ is not less than ε. This means that

$$\begin{aligned} P^* &= (p_1^*, p_2^*, \ldots, p_n^*) \\ &= \operatorname*{argmax}_{p_1} \max_{p_2} \cdots \max_{p_n} \min_j \left(p_1(k_0^j, x_1^j, k_1^j) \prod_{i=2}^{n} \frac{p_i(k_{i-1}^j, x_i^j, k_i^j)}{\sum\limits_{k \in K} \sum\limits_{x \in X} p_i(k_{i-1}^j, x, k)} \right) . \end{aligned}$$

8.7.3 Tuning of the recognition algorithm

Assume that someone has created an algorithm which recognises the sequence $\bar{k} = (k_0, k_1, \ldots, k_n)$, based on the sequence $\bar{x} = (x_1, x_2, \ldots, x_n)$ in the formulation we have presented in Section 4.2. It is an algorithm of the form

$$\bar{k} = \operatorname*{argmax}_{k_0, k_1, \ldots, k_n} \sum_{i=1}^{n} f_i(k_{i-1}, x_i, k_i) , \tag{8.73}$$

where the functions $f_i \colon K \times X \times K \to \mathbb{R}$ depend, in a certain way, on the statistical model $(p_i \colon K \times X \times K \to \mathbb{R},\ i = 1, \ldots, n)$. It is quite possible that the author had created a program for the calculation of (8.73) before he got the information about the statistical model $P = (p_i,\ i = 1, 2, \ldots, n)$. After the program was written a question arose about what functions f_i are to be included into the program. The functions f_i can be created in two different ways according to the nature of the training set of examples $(\bar{x}^j, \bar{k}^j)$, $j = 1, \ldots, l$. If the pair $(\bar{x}^j, \bar{k}^j)$ can be regarded as an instance of a random pair the probability distribution of which is $p(\bar{x}, \bar{k})$ and if, at the same time, the pairs $(\bar{x}^j, \bar{k}^j)$ at different j are mutually independent then the ensemble P can be built up according to the requirement (8.72). Then, in an appropriate manner according to P, the functions f_i have to be created.

The origin of the set of examples $(\bar{x}^j, \bar{k}^j)$, $j = 1, \ldots, l$, can be quite different. The set might be created not as an outcome of the observation of a real object, but as a training set by which the requirements for the future algorithm are expressed. In this case $\bar{k}^j$ is a sequence which the algorithm has to create as a result of recognition when the sequence $\bar{x}^j$ is brought to the input. Here the functions f_i are no longer considered as statistical parameters of a recognised object, but as parameters of a recognition algorithm. These parameters are to be tuned up so that the algorithm should operate correctly at the examples of the training set. The learning task, i.e., the tuning of the algorithm to the beforehand given training set, lies in seeking the functions $f_1, f_2, \ldots, f_n$ which satisfy the following system of relations

$$(k_0^j, k_1^j, \ldots, k_n^j) = \underset{k_0}{\operatorname{argmax}} \max_{k_1} \cdots \max_{k_n} \sum_{i=1}^{n} f(k_{i-1}, x_i^j, k_i)\,, \quad j = 1, 2, \ldots, l\,.$$

8.7.4 Task of unsupervised learning

The ensemble $P = (p_1, p_2, \ldots, p_n)$ does not only uniquely determine the probabilities $p(\bar{x}, \bar{k})$ of each pair $(\bar{x}, \bar{k})$ which consists of a sequence of observed parameters $\bar{x} = (x_1, x_2, \ldots, x_n)$ and a sequence of hidden parameters $\bar{k} = (k_0, k_1, \ldots, k_n)$, but it also determines the probability of each sequence $\bar{x} = (x_1, x_2, \ldots, x_n)$ of the observed parameters. Its probability is given by the sum $\sum_{\bar{k} \in K^{n+1}} p(\bar{x}, \bar{k})$.

Assume that l experiments were performed with the examined object during which only the values of the observed features were observed. In the unsupervised learning task the point is sought on the basis of experimental data, i.e., the maximum likelihood model of the object

$$P^* = (p_1^*, p_2^*, \ldots, p_n^*) = \underset{P}{\operatorname{argmax}} \prod_{j=1}^{l} \sum_{\bar{k} \in K^{n+1}} p(\bar{x}^j, \bar{k})$$

$$= \underset{p_1}{\operatorname{argmax}} \cdots \max_{p_n} \prod_{j=1}^{l} \sum_{k_0} \sum_{k_1} \cdots \sum_{k_n} p_1(k_0, x_1^j, k_1) \prod_{i=2}^{n} \frac{p_i(k_{i-1}, x_i^j, k_i)}{\sum\limits_{x \in X} \sum\limits_{k \in K} p_i(k_{i-1}, \bar{x}, \bar{k})}\,.$$

8.8 Maximum likelihood estimate of the model

The maximum likelihood estimation of a model is defined by requirement (8.72). A correct estimate can be guessed on the basis of mere common sense intuition. First, we will show this estimate and then demonstrate that it satisfies the requirement (8.72). Assume that on the basis of the experimental data $\big((\bar{x}^j, \bar{k}^j)\,,\; j = 1, 2, \ldots, l\big)$ for the fixed triplet $k' \in K, x' \in X, k'' \in K$ it is necessary to find a joint probability that the value of the $(i-1)$-th hidden parameter will be k', the value of the i-th observed parameter will be x', and the value of the i-th hidden parameter will be k''. Without further specifying this task, we could hardly look for a more natural procedure than simply calculating how many times the case $k_{i-1} = k'$, $x_i = x'$, $x_i = k''$ occurred in the experiments, and then dividing the obtained number by l.

We will prove this quite reasonable recommendation, i.e., we will see that the model created in this way maximises (8.72). We need this proof because maximum likelihood estimation will be used later as an element of more complex algorithms for minimax estimation and unsupervised learning. Analysis of these more complex problems will be based not only on the fact that the above stated way of creating the model is reasonable, but on the property that the created model maximises (8.72).

For the sake of formal analysis the task will be expressed in the following equivalent form

$$P^* = (p_1^*, p_2^*, \dots, p_n^*)$$
$$= \operatorname*{argmax}_{p_1} \cdots \max_{p_n} \sum_{j=1}^{l} \left(\log p_1(k_0^j, x_1^j, k_1^j) + \sum_{i=2}^{n} \log \frac{p_i(k_{i-1}^j, x_i^j, k_i^j)}{\sum_{x' \in X} \sum_{k' \in K} p_i(k_{i-1}^j, x', k')} \right).$$

The experimental data $\left((\bar{x}^j, \bar{k}^j),\ j = 1, 2, \dots, l\right)$ can be expressed by means of the functions $g\colon X^n \times K^{n+1} \to \mathbb{Z}$, where $\mathbb{Z}$ is a set of integers. For each sequence $\bar{x} \in X^n$ and the sequence $\bar{k} \in K^{n+1}$ the number $g(\bar{x}, \bar{k})$ states how many times the pair $(\bar{x}, \bar{k})$ occurred in the experimental data $\left((\bar{x}^j, \bar{k}^j),\ j = 1, 2, \dots, l\right)$. If we use the notation g then the previous expression for P^* assumes the form

$$P^* = \operatorname*{argmax}_{P} \left(\sum_{\bar{x} \in X^n} \sum_{\bar{k} \in K^{n+1}} g(\bar{x}, \bar{k}) \left(\log p_1(k_0, x_1, k_1) + \sum_{i=2}^{n} \log \frac{p_i(k_{i-1}, x_i, k_i)}{\sum_{x' \in X} \sum_{k' \in K} p_i(k_{i-1}, x', k')} \right) \right). \tag{8.74}$$

The task given by the previous relation does not change if the integer function g is substituted by the function α for which there hold $\alpha(\bar{x}, \bar{k}) = g(\bar{x}, \bar{k}) / l$, i.e.,

$$P^* = \operatorname*{argmax}_{P} \left(\sum_{\bar{x} \in X^n} \sum_{\bar{k} \in K^{n+1}} \alpha(\bar{x}, \bar{k}) \left(\log p_1(k_0, x_1, k_1) + \sum_{i=2}^{n} \log \frac{p_i(k_{i-1}, x_i, k_i)}{\sum_{x' \in X} \sum_{k' \in K} p_i(k_{i-1}, x', k')} \right) \right). \tag{8.75}$$

For the function α

$$\sum_{\bar{x} \in X^n} \sum_{\bar{k} \in K^{n+1}} \alpha(\bar{x}, \bar{k}) = 1,$$

holds and therefore α can be understood as the probability distribution on the set of all pairs $(\bar{x}, \bar{k})$, i.e., over the set of all possible outcomes of the experiment.

We will introduce the notation $X_i(x')$, $i = 1, 2, \dots, n$, $x' \in X$, for the set of such sequences $\bar{x} = (x_1, x_2, \dots, x_n) \in X^n$ in which $x_i = x'$. The notation $K_i(k', k'')$, $i = 1, 2, \dots, n$, $k' \in K$, $k'' \in K$, means the set of such sequences $\bar{k} = (k_0, k_1, \dots, k_n) \in K^{n+1}$ for which $k_{i-1} = k'$, $k_i = k''$ holds. By $\alpha_i(k', x', k'')$ we will denote the sum

$$\alpha_i(k', x', k'') = \sum_{\bar{x} \in X_i(x')} \sum_{\bar{k} \in K_i(k', k'')} \alpha(\bar{x}, \bar{k}). \tag{8.76}$$

The previously mentioned recommendation for calculating the numbers $p_i(k_{i-1}, x_i, k_i)$ was stated on the basis of understanding what probability really is. In its precise formulation the recommendation reads that the functions p_i should be identical with the functions α_i defined by the previous formula. The proof that they are the functions p_i created just in this way which solve the optimisation task (8.75), and thus also the task (8.72), is given in the following theorem.

Theorem 8.1 Maximum likelihood estimation of the Markovian model. *Let X and K be two finite sets and let α be a non-negative function of the form $X^n \times K^{n+1} \to \mathbb{R}$ for which there hold*

$$\sum_{\bar{x} \in X^n} \sum_{\bar{k} \in K^{n+1}} \alpha(\bar{x}, \bar{k}) = 1 .$$

Let p_i^*, $i = 1, 2, \ldots, n$, be functions of the form $K \times X \times K \to \mathbb{R}$ for which there hold

$$p_i^*(k', x', k'') = \sum_{\bar{x} \in X_i(x')} \sum_{\bar{k} \in K_i(k', k'')} \alpha(\bar{x}, \bar{k}) ,$$

and p_i, $i = 1, 2, \ldots, n$, be arbitrary non-negative functions $K \times X \times K \to \mathbb{R}$ for which there hold

$$\sum_{k' \in K} \sum_{x' \in X} \sum_{k'' \in K} p_i(k', x', k'') = 1 .$$

In this case the inequality holds

$$\sum_{\bar{x} \in X^n} \sum_{\bar{k} \in K^{n+1}} \alpha(\bar{x}, \bar{k}) \left(\log p_1^*(k_0, x_1, k_1) + \sum_{i=2}^{n} \log \frac{p_i^*(k_{i-1}, x_i, k_i)}{\sum\limits_{x' \in X} \sum\limits_{k' \in K} p_i^*(k_{i-1}, x', k')} \right)$$
$$\geq \sum_{\bar{x} \in X^n} \sum_{\bar{k} \in K^{n+1}} \alpha(\bar{x}, \bar{k}) \left(\log p_1(k_0, x_1, k_1) + \sum_{i=2}^{n} \log \frac{p_i(k_{i-1}, x_i, k_i)}{\sum\limits_{x' \in X} \sum\limits_{k' \in K} p_i(k_{i-1}, x', k')} \right) .$$

▲

Proof. The basis for the proof is Lemma 6.1. First we will make clear the relationship between the sums

$$\sum_{\bar{x} \in X^n} \sum_{\bar{k} \in K^{n+1}} \alpha(\bar{x}, \bar{k}) \log p_1^*(k_0, x_1, k_1) \text{ and } \sum_{\bar{x} \in X^n} \sum_{\bar{k} \in K^{n+1}} \alpha(\bar{x}, \bar{k}) \log p_1(k_0, x_1, k_1). \tag{8.77}$$

For the first of these sums there hold

$$\begin{aligned}
&\sum_{\bar{x}\in X^n} \sum_{\bar{k}\in K^{n+1}} \alpha(\bar{x},\bar{k}) \log p_1^*(k_0,x_1,k_1) \\
&\quad = \sum_{x_1\in X} \sum_{\bar{x}\in X_1(x_1)} \sum_{k_0\in K} \sum_{k_1\in K} \sum_{\bar{k}\in K_1(k_0,k_1)} \alpha(\bar{x},\bar{k}) \log p_1^*(k_0,x_1,k_1) \\
&\quad = \sum_{k_0\in K} \sum_{x_1\in X} \sum_{k_1\in K} \left(\sum_{\bar{x}\in X_1(x_1)} \sum_{\bar{k}\in K_1(k_0,k_1)} \alpha(\bar{x},\bar{k}) \right) \log p_1^*(k_0,x_1,k_1) \\
&\quad = \sum_{k_0\in K} \sum_{x_1\in X} \sum_{k_1\in K} p_1^*(k_0,x_1,k_1) \log p_1^*(k_0,x_1,k_1) \,. \qquad (8.78)
\end{aligned}$$

Similarly we can demonstrate that the second sum in (8.77) is

$$\sum_{k_0\in K} \sum_{x_1\in X} \sum_{k_1\in K} p_1^*(k_0,x_1,k_1) \log p_1(k_0,x_1,k_1) \,. \qquad (8.79)$$

Since both the sum $\sum_{k_0\in K} \sum_{x_1\in X} \sum_{k_1\in K} p_1^*(k_0,x_1,k_1)$ and the sum $\sum_{k_0\in K} \sum_{x_1\in X} \sum_{k_1\in K} p_1(k_0,x_1,k_1)$ are equal to 1, we can find on the basis of Lemma 6.1 that (8.78) is not less than (8.79), and thus

$$\sum_{\bar{x}\in X^n} \sum_{\bar{k}\in K^{n+1}} \alpha(\bar{x},\bar{k}) \log p_1^*(k_0,x_1,k_1) \geq \sum_{\bar{x}\in X^n} \sum_{\bar{k}\in K^{n+1}} \alpha(\bar{x},\bar{k}) \log p_1(k_0,x_1,k_1) \,. \qquad (8.80)$$

Now we will make clear the relationship between the sums

$$\sum_{\bar{x}\in X^n} \sum_{\bar{k}\in K^{n+1}} \alpha(\bar{x},\bar{k}) \log \frac{p_i^*(k_{i-1},x_i,k_i)}{\sum\limits_{x'\in X} \sum\limits_{k'\in K} p_i^*(k_{i-1},x',k')}$$

and

$$\sum_{\bar{x}\in X^n} \sum_{\bar{k}\in K^{n+1}} \alpha(\bar{x},\bar{k}) \log \frac{p_i(k_{i-1},x_i,k_i)}{\sum\limits_{x'\in X} \sum\limits_{k'\in K} p_i(k_{i-1},x',k')} \,. \qquad (8.81)$$

The first of the sums is

$$\begin{aligned}
&\sum_{\bar{x}\in X^n} \sum_{\bar{k}\in K^{n+1}} \alpha(\bar{x},\bar{k}) \log \frac{p_i^*(k_{i-1},x_i,k_i)}{\sum\limits_{x'\in X} \sum\limits_{k'\in K} p_i^*(k_{i-1},x',k')} \\
&= \sum_{k_{i-1}\in K} \sum_{x_i\in X} \sum_{k_i\in K} \sum_{\bar{x}\in X_i(x_i)} \sum_{\bar{k}\in K_i(k_{i-1},k_i)} \alpha(\bar{x},\bar{k}) \log \frac{p_i^*(k_{i-1},x_i,k_i)}{\sum\limits_{x'\in X} \sum\limits_{k'\in K} p_i^*(k_{i-1},x',k')} \\
&= \sum_{k_{i-1}\in K} \sum_{x_i\in X} \sum_{k_i\in K} \left(\sum_{\bar{x}\in X_i(x_i)} \sum_{\bar{k}\in K_i(k_{i-1},k_i)} \alpha(\bar{x},\bar{k}) \right) \log \frac{\alpha_i(k_{i-1},x_i,k_i)}{\sum\limits_{x'\in X} \sum\limits_{k'\in K} \alpha_i(k_{i-1},x',k')} \\
&= \sum_{k_{i-1}\in K} \sum_{x_i\in X} \sum_{k_i\in K} \alpha_i(k_{i-1},x_i,k_i) \log \frac{\alpha_i(k_{i-1},x_i,k_i)}{\sum\limits_{x'\in X} \sum\limits_{k'\in K} \alpha_i(k_{i-1},x',k')} \,. \qquad (8.82)
\end{aligned}$$

With respect to similar considerations we claim that the second sum in (8.81) is

$$\sum_{k_{i-1}\in K}\sum_{x_i\in X}\sum_{k_i\in K}\alpha_i(k_{i-1},x_i,k_i)\log\frac{p_i(k_{i-1},x_i,k_i)}{\sum\limits_{x'\in X}\sum\limits_{k'\in K}p_i(k_{i-1},x',k')}\,. \tag{8.83}$$

Since the sum $\sum\limits_{x_i\in X}\sum\limits_{k_i\in K}\frac{p_i(k_{i-1},x_i,k_i)}{\sum_{x'\in X}\sum_{k'\in K}p_i(k_{i-1},x',k')}$ is equal to 1 at any value k_{i-1} we claim on the basis of Lemma 6.1 that the inequality

$$\begin{aligned}&\sum_{x_i\in X}\sum_{k_i\in K}\alpha_i(k_{i-1},x_i,k_i)\log\frac{\alpha_i(k_{i-1},x_i,k_i)}{\sum\limits_{x'\in X}\sum\limits_{k'\in K}\alpha_i(k_{i-1},x',k')}\\ \geq&\sum_{x_i\in X}\sum_{k_i\in K}\alpha_i(k_{i-1},x_i,k_i)\log\frac{p_i(k_{i-1},x_i,k_i)}{\sum\limits_{x'\in X}\sum\limits_{k'\in K}p_i(k_{i-1},x',k')}\end{aligned}$$

is correct at any value k_{i-1}. If we sum this inequality over all k_{i-1} then we obtain the inequality

$$\begin{aligned}&\sum_{k_{i-1}\in K}\sum_{x_i\in X}\sum_{k_i\in K}\alpha_i(k_{i-1},x_i,k_i)\log\frac{\alpha_i(k_{i-1},x_i,k_i)}{\sum\limits_{x'\in X}\sum\limits_{k'\in K}\alpha_i(k_{i-1},x',k')}\\ \geq&\sum_{k_{i-1}\in K}\sum_{x_i\in X}\sum_{k_i\in K}\alpha_i(k_{i-1},x_i,k_i)\log\frac{p_i(k_{i-1},x_i,k_i)}{\sum\limits_{x'\in X}\sum\limits_{k'\in K}p_i(k_{i-1},x',k')}\,,\end{aligned}$$

and thus owing to (8.82) and (8.83) we also obtain the inequality

$$\begin{aligned}&\sum_{\bar{x}\in X^n}\sum_{\bar{k}\in K^{n+1}}\alpha(\bar{x},\bar{k})\log\frac{p_i^*(k_{i-1},x_i,k_i)}{\sum\limits_{x'\in X}\sum\limits_{k'\in K}p_i^*(k_{i-1},x',k')}\\ \geq&\sum_{\bar{x}\in X^n}\sum_{\bar{k}\in K^{n+1}}\alpha(\bar{x},\bar{k})\log\frac{p_i(k_{i-1},x_i,k_i)}{\sum\limits_{x'\in X}\sum\limits_{k'\in K}p_i(k_{i-1},x',k')}\end{aligned}$$

which is satisfied for any $i=2,3,\ldots,n$. If we sum this inequality over all i

$$\begin{aligned}&\sum_{\bar{x}\in X^n}\sum_{\bar{k}\in K^{n+1}}\alpha(\bar{x},\bar{k})\sum_{i=2}^{n}\log\frac{p_i^*(k_{i-1},x_i,k_i)}{\sum\limits_{x'\in X}\sum\limits_{k'\in K}p_i^*(k_{i-1},x',k')}\\ \geq&\sum_{\bar{x}\in X^n}\sum_{\bar{k}\in K^{n+1}}\alpha(\bar{x},\bar{k})\sum_{i=2}^{n}\log\frac{p_i(k_{i-1},x_i,k_i)}{\sum\limits_{x'\in X}\sum\limits_{k'\in K}p_i(k_{i-1},x',k')}\,.\end{aligned} \tag{8.84}$$

The inequality expressed by Theorem 8.1 being proved, is an evident consequence of the inequality (8.84), and the previously proved inequality (8.80). ∎

If the statistical model of an object is known to be Markovian then Theorem 8.1 proved the following algorithm for the maximum likelihood estimate of the statistical model of an object.

Algorithm 8.1 Maximum likelihood estimation of the Markovian model

1. The outcomes of the experimental examination of an object which were originally expressed by the ensemble $((\bar{x}^j, \bar{k}^j),\ j = 1, 2, \ldots, n)$, $\bar{x}^j \in X^n$, $\bar{k}^j \in K^{n+1}$, are to be expressed as a function $\alpha\colon X^n \times K^{n+1} \to \mathbb{R}$ the value $\alpha(\bar{x}, \bar{k})$, $\bar{x} \in X^n$, $\bar{k} \in K^{n+1}$ of which is the relative frequency of the pair $(\bar{x}, \bar{k})$ in the experiment.
2. For each $i = 1, 2, \ldots, n$ and for each $x' \in X$, $k' \in K$ and $k'' \in K$ the probabilities
$$p_i(k', x', k'') = \sum_{\bar{x} \in X_i(x')} \sum_{\bar{k} \in K_i(k', k'')} \alpha(\bar{x}, \bar{k}) \tag{8.85}$$
are to be calculated which in the experiment means the relative frequency of the value x' in the position i of the sequence x together with the values k' and k'' in the $i-1$-th and i-th positions of the sequence $\bar{k}$.
3. The ensemble of numbers $p_i(k', x', k'')$ expresses the Markovian model of the object examined in the sense that for each pair of sequences $\bar{x} \in X^n$ and $\bar{k} \in K^{n+1}$ it determines their joint probability according to the formula
$$p(\bar{x}, \bar{k}) = p_1(k_0, x_1, k_1) \prod_{i=2}^{n} \frac{p_i(k_{i-1}, x_i, k_i)}{\sum_{x' \in X} \sum_{k' \in K} p_i(k_{i-1}, x', k')} . \tag{8.86}$$

Theorem 8.1 proved claims that the Markovian model of the object obtained is the most likely one in the sense that the probability of the ensemble $((\bar{x}^j, \bar{k}^j),\ j = 1, 2, \ldots, l)$ experimentally observed in this model is not less than the probability of the same experimental outcomes in any other Markovian model of the object.

For further explanation it is helpful to consider Algorithm 8.1, i.e., the formulæ (8.85) and (8.86) as a transformation of the function $\alpha\colon X^n \times K^{n+1} \to \mathbb{R}$ to the function $p\colon X^n \times K^{n+1} \to \mathbb{R}$. The transformation converts the probability distribution α which need not be Markovian and can be of any kind to the probability distribution p which must be Markovian. The probability distribution p which is formed on the basis of the probability distribution α will be referred to as the *Markovian approximation of the function* α and will be denoted as α^M. The index ^{M} in the denotation α^M is understood as an operator which affects the function α and transforms it into the function $p = \alpha^M$. For example, the denotation $(\alpha + \beta)^M$ means the Markovian approximation of the sum of functions α and β. The denotation $\alpha^M(x)$ means the value of the function that is the Markovian approximation of the function α in the point x, etc..

From the definition of the Markovian approximation there immediately follows
$$\sum_{\bar{x} \in X^n} \sum_{\bar{k} \in K^{n+1}} \alpha(\bar{x}, \bar{k}) \log \alpha^M(\bar{x}, \bar{k}) \geq \sum_{\bar{x} \in X^n} \sum_{\bar{k} \in K^{n+1}} \alpha(\bar{x}, \bar{k}) \log p(\bar{x}, \bar{k})$$
where $p\colon X^n \times K^{n+1} \to \mathbb{R}$ is any probability distribution of a Markovian form.

We will call to mind once more the idea that even if the Markovian approximation given in (8.85) and (8.86) seems to be non-constructive because of the multi-dimensional sum on the right-hand side of (8.85), the calculation of (8.85) is actually a quite natural and easily realisable processing of experimental data. Its complexity is proportional to the length of the experiment l, i.e., to the number of non-zero addition terms on the right-hand side of (8.85). In constructing further algorithms the Markovian approximation will be used as a multiple performed operation, and it is necessary therefore to realise that it is a natural and easily calculable kind of operation.

8.9 Minimax estimate of a statistical model

8.9.1 Formulation of an algorithm and its properties

In this subsection we will give a survey of the main results of an analysis of a task without a proof including the algorithm of its solution, and we will prove them later in subsequent subsections. From the algorithm presented we will see that the *minimax estimate is reduced to a sequence of the maximum likelihood estimates, i.e., to a step by step calculation of certain Markovian approximations.*

The input information for the minimax estimate is the finite training set L consisting of the pairs $(\bar{x}, \bar{k})$ which represent a typical (probable) behaviour of the object. The task seeks an ensemble of functions $P^* = (p_1^*, p_2^*, \ldots, p_n^*)$ of the form $p_i^* \colon K \times X \times K \to \mathbb{R}$,

$$\begin{aligned} P^* &= (p_1^*, p_2^*, \ldots, p_n^*) \\ &= \underset{p_1,\ldots,p_n}{\operatorname{argmax}} \min_{(\bar{x},\bar{k})\in L} \left(\log p_1(k_0, x_1, k_1) + \sum_{i=2}^{n} \log \frac{p_i(k_{i-1}, x_i, k_i)}{\sum\limits_{x'\in X} \sum\limits_{k'\in K} p_i(k_{i-1}, x', k')} \right) . \end{aligned}$$

We will introduce an algorithm which solves this task with any predefined accuracy $\varepsilon > 0$.

The algorithm alters stepwise the integer numbers $n(\bar{x}, \bar{k})$ for each pair $\bar{x} = (x_1, \ldots, x_n)$ and $\bar{k} = (k_0, k_1, \ldots, k_n)$ which occurs in the training set L, and the integer numbers $n_i(k', x', k'')$ for each $i = 1, 2, \ldots, n$ and for each triplet $k' \in K$, $x' \in X$, $k'' \in K$. The numbers $n(\bar{x}, \bar{k})$ and $n_i(k', x', k'')$ serve for calculating current values of the probabilities $\alpha(\bar{x}, \bar{k})$ and $p_i(k', x', k'')$ in the following way

$$\alpha(\bar{x}, \bar{k}) = \frac{n(\bar{x}, \bar{k})}{\sum\limits_{(\bar{x},\bar{k})\in L} n(\bar{x}, \bar{k})} , \quad p_i(k', x', k'') = \frac{n_i(k', x', k'')}{\sum\limits_{k'\in K} \sum\limits_{x'\in X} \sum\limits_{k''\in K} n_i(k', x', k'')} .$$

In each step the algorithm creates a Markovian approximation of the current distribution α. The initial values $n^1(\bar{x}, \bar{k})$ and $n_i^1(x', k', x'')$ can be arbitrary(!). For the sake of unambiguity we define them in such a way that $n^1(\bar{x}, \bar{k}) = 1$ for any $(\bar{x}, \bar{k})$ from the training set L. For the automaton states k, k' and the output signal x', let us introduce the number $n_i^1(k', x', k'')$ stating how many times a sequence occurred in the training set such that the state was k' in the instant $i - 1$, the state was k'' in the instant i, and the observed output signal was x' in the instant i. Assume that prior to the step t the numbers $n^t(\bar{x}, \bar{k})$, $(\bar{x}, \bar{k}) \in L$, and the numbers $n_i^t(k', x', k'')$, $i = 1, 2, \ldots, n$, were computed. Next values of these numbers are calculated according to the rules:

Algorithm 8.2 Minimax estimate of the statistical model

1. The probabilities are calculated for $i = 1, 2, \ldots, n$, $k' \in K$, $x' \in X$, $k'' \in K$,

$$\alpha^t(\bar{x}, \bar{k}) = \frac{n^t(\bar{x}, \bar{k})}{\sum\limits_{(\bar{x}, \bar{k}) \in L} n^t(\bar{x}, \bar{k})}, \quad (\bar{x}, \bar{k}) \in L,$$

$$p_i^t(k', x', k'') = \frac{n_i^t(k', x', k'')}{\sum\limits_{k' \in K} \sum\limits_{x' \in X} \sum\limits_{k'' \in K} n_i^t(k', x', k'')}.$$

2. The probabilities $p^t(\bar{x}, \bar{k})$ are calculated according to the formula (8.86), i.e.,

$$p^t(\bar{x}, \bar{k}) = p_1^t(k_0, x_1, k_1) \prod_{i=2}^{n} \frac{p_i^t(k_{i-1}, x_i, k_i)}{\sum\limits_{x' \in X} \sum\limits_{k' \in K} p_i^t(k_{i-1}, x', k')}, \quad (\bar{x}, \bar{k}) \in L. \tag{8.87}$$

3. It is verified if the following inequality is satisfied

$$\sum_{(\bar{x}, \bar{k}) \in L} \alpha^t(\bar{x}, \bar{k}) \log p^t(\bar{x}, \bar{k}) - \min_{(\bar{x}, \bar{k}) \in L} \log p^t(\bar{x}, \bar{k}) \le \varepsilon. \tag{8.88}$$

4. If the inequality (8.88) is satisfied then the algorithm ends, and the current values $p_i^t(k', x', k'')$ form the ε-solution of the task.
5. If the inequality (8.88) is not satisfied then the following calculations are performed.

   (a) Any pair $(\bar{x}^*, \bar{k}^*) \in L$ is found for which there holds

   $$p^t(\bar{x}^*, \bar{k}^*) = \min_{(\bar{x}, \bar{k}) \in L} p^t(\bar{x}, \bar{k}).$$

   (b) New values of the numbers $n(\bar{x}, \bar{k})$ and $n_i(k_{i-1}, x_i, k_i)$ are calculated

   $$\begin{aligned}
   n^{t+1}(\bar{x}, \bar{k}) &= n^t(\bar{x}, \bar{k}) + 1, && \text{if } \bar{x} = \bar{x}^*, \ \bar{k} = \bar{k}^*, \\
   n^{t+1}(\bar{x}, \bar{k}) &= n^t(\bar{x}, \bar{k}), && \text{if } (\bar{x}, \bar{k}) \ne (\bar{x}^*, \bar{k}^*), \\
   n_i^{t+1}(k', x', k'') &= n_i^t(k' x', k'') + 1, && \text{if } k_{i-1}^* = k', \ x_i^* = x', \ k_i^* = k'', \\
   n_i^{t+1}(k', x', k'') &= n_i^t(k' x', k''), && \text{if } (k_{i-1}^*, x_i^*, k_i^*) \ne (k', x', k'').
   \end{aligned}$$

6. It is proceeded to the $(t+1)$-th iteration of the algorithm, starting from the step 1.

For the algorithm formulated in this way the following two theorems are valid.

Theorem 8.2 On convergence of an algorithm in a finite number of steps. *For any predefined positive value ε Algorithm 8.2 gets after a finite number of steps to the state in which the inequality (8.88) is satisfied and the algorithm ends.*

▲

Let $F(P)$ denote the number

$$F(P) = \min_{(x,k)\in L} \left(\log p_1(k_0, x_1, k_1) + \sum_{i=2}^{n} \log \frac{p_i(k_{i-1}, x_i, k_i)}{\sum\limits_{x'\in X} \sum\limits_{k'\in K} p_i(k_{i-1}, x', k')} \right)$$

which is to be maximised in the task.

Theorem 8.3 On achieving an arbitrary predefined accuracy. *Let P' be the result of Algorithm 8.2 and P^* be the solution of a task concerning the minimax estimate of a statistical model. Then there holds*

$$F(P^*) - F(P') \le \varepsilon \,.$$

▲

The remaining part of Section 8.9 will be devoted to the proof of the properties of Algorithm 8.2, and to further important properties of the whole task. We will see that the task of the minimax estimate of a statistical model can be reduced to the sequence of easier tasks of the maximum likelihood estimate. Such reduction is possible not only in the class of Markovian models, but even for a far more general case. Because of their generality further results reach much beyond the scope of problems concerning Markovian sequences. Furthermore, removing dependance on the class of statistical models, we will see that the minimax estimate tasks will be reduced to special tasks of *convex optimisation.* This means that for the minimax estimate not only the algorithm mentioned above can be applied, but an abundance of methods which are provided by the widely developed theory of convex optimisation.

We will quote, without a proof, known mathematical results which will be used by us. Let X be a convex subset of linear space and $f\colon X \to \mathbb{R}$ be a real valued function defined on this set. For this function the *convexity* is defined in the following three equivalent ways.

1. The function f is convex if the following set

$$\{(y, f(x)) \mid y \ge f(x)\}$$

is convex.

2. The function f is convex if for any $x_1 \in X$, $x_2 \in X$ and α, $0 \le \alpha \le 1$, the inequality is satisfied,

$$f\big(\alpha\, x_1 + (1-\alpha)\, x_2\big) \le \alpha\, f(x_1) + (1-\alpha)\, f(x_2)\,.$$

3. The function f is convex if for each point $x_0 \in X$ there exists such a linear function $L_{x_0}\colon X \to \mathbb{R}$, that the inequality

$$L_{x_0}(x - x_0) \le f(x) - f(x_0) \tag{8.89}$$

holds for each $x \in X$.

Let vector $g(x_0) \in X$ correspond to the linear function L_{x_0} in such a way that $L_{x_0}(x)$ is a scalar product $\langle g(x_0), x\rangle$. The vector $g(x_0)$ is referred to as the *generalised gradient* f in the point x_0. The generalised gradient is known to exist in any point of a convex function. If the function f and the definition (8.89) unambiguously determine the linear function L_{x_0}, and thus also the generalised gradient $g(x_0)$, then the function f is called differentiable (or smooth) and the generalised gradient $g(x_0)$ is simply called gradient.

Any convex function on a finitely dimensional set is *continuous*. This means that the difference $f(x) - f(x_0)$ approaches zero if x approaches x_0. A linear function is a particular case of the convex function and therefore $\lim_{x \to x_0} g(x_0)(x - x_0) = 0$.

If, however, the function f is smooth and $g(x_0) \neq 0$ then

$$\lim_{x \to x_0} \frac{f(x) - f(x_0)}{g(x_0)(x - x_0)} = 1 .$$

Such a situation is a special case of a situation in which two infinitely small quantities $f(x) - f(x_0)$ and $g(x_0)(x - x_0)$ are of the same order. Let us recall the precise formulation of this concept. Let $u_1, u_2, \ldots, u_i, \ldots$ and $v_1, v_2, \ldots, v_i, \ldots$ be two infinitely small quantities in the sense that $\lim_{i \to \infty} u_i = \lim_{i \to \infty} v_i = 0$. These two infinitely small quantities are defined as *infinitely small of the same order* if there exists a limit

$$\lim_{i \to \infty} \frac{u_i}{v_i}$$

which is neither zero nor infinite. The infinitely small quantities of the same order have the following important property.

Let u_i and v_i, $i = 1, 2, \ldots, \infty$, be two infinitely small quantities of the same order. In this case if a series of numbers $\sum_{i=1}^{n} u_i$, $n = 1, 2, \ldots, \infty$, converges to a finite value then the series of numbers $\sum_{i=1}^{n} v_i$, $n = 1, 2, \ldots, \infty$, also converges to a finite value (possibly different one). Further on, if $\lim_{n \to \infty} \sum_{i=1}^{n} u_i = \infty$ then also $\lim_{n \to \infty} \sum_{i=1}^{n} v_i = \infty$.

The properties of a minimax task which result in the possibility of its constructive solution are satisfied not only for Markovian models, but it is so even in more general cases. Furthermore, an analysis in the general case is easier since the important properties are not overshadowed by unnecessary details. In the following subsection we will examine a task for the general case, and then transfer the results of the analysis without difficulty to a more specialised case of Markovian models.

8.9.2 Analysis of a minimax estimate

The problem of the minimax model estimation has been formulated in the general form in Lecture 3. With respect to the present explanation we will redefine the task in a somewhat different form.

Let X be a set for which a class $\mathcal{P}$ of functions of the form $p: X \to \mathbb{R}$ is specified. Let $(x^j,\ j = 1, \ldots, n)$ be a multi-set of elements from the set X. The task is to find a function p^* from the class $\mathcal{P}$ which maximises the number $\min_j p(x^j)$, i.e.,

$$p^* = \operatorname*{argmax}_{p \in \mathcal{P}} \min_j p(x^j) . \tag{8.90}$$

Since the solution of the task, i.e., the function p^* does not depend on how many times an element has occurred in the multi-set $(x^j,\ j = 1,\dots,n)$, and depends on a single occurrence in the multi-set at least, the input information can be regarded as a finite subset $L \subset X$, not as a multi-set $(x^j,\ j = 1,\dots,n)$. The task (8.90) assumes the form

$$p^* = \operatorname*{argmax}_{p\in\mathcal{P}} \min_{x\in L} p(x) ,$$

or in the equivalent expression

$$p^* = \operatorname*{argmax}_{p\in\mathcal{P}} \min_{x\in L} \log p(x) . \tag{8.91}$$

Let us recall the maximum likelihood estimate of the model p, i.e., seeking

$$p^* = \operatorname*{argmax}_{p\in\mathcal{P}} \sum_{x\in L} \alpha(x) \log p(x) . \tag{8.92}$$

This means that a function p^* is being sought with which the probability of a multi-set is maximised and in which $\alpha(x)$ is the relative occurrence of the element x. The function p^* according to the relation (8.92) depends on the coefficients $\alpha(x)$, and therefore the corresponding algorithm can be regarded as an operator which transfers the function $\alpha\colon L \to \mathbb{R}$ to the function $\alpha^M\colon X \to \mathbb{R}$ in such a way that

$$\alpha^M = \operatorname*{argmax}_{p\in\mathcal{P}} \sum_{x\in L} \alpha(x) \log p(x) .$$

The number $\max\limits_{p\in\mathcal{P}} \sum\limits_{x\in L} \alpha(x) \log p(x)$ is simply the number

$$\sum_{x\in L} \alpha(x) \log \alpha^M(x)$$

which depends on the function $\alpha\colon L \to \mathbb{R}$ and which will be denoted $Q(\alpha)$. For the number $Q(\alpha)$ according to its definition there holds that the inequality

$$Q(\alpha) = \sum_{x\in X} \alpha(x) \log \alpha^M(x) \geq \sum_{x\in X} \alpha(x) \log p(x)$$

is satisfied for any function $p \in \mathcal{P}$. As before in Section 8.8 the function $\alpha^M\colon X \to \mathbb{R}$ will be referred to as an approximation of the function $\alpha\colon L \to \mathbb{R}$ in the class $\mathcal{P}$, or simply approximation of α. If $\mathcal{P}$ is a Markovian class of models then α^M is a Markovian approximation of α.

The maximum likelihood estimate of the model (8.92) has been well examined. There exist programmed solutions for a number of favoured classes $\mathcal{P}$. Particularly, for the class of Markovian models the analysis of the maximum likelihood estimate has been performed in Section 8.8. Even when the minimax estimate has certain advantages, particularly the independence of the random form of the training set, this type of task has been examined far less. At first glance it may seem much more difficult. It is therefore important that the minimax estimate is reduced to the maximum likelihood estimate in the following sense.

If there is a program for the maximum likelihood estimate (8.92) at our disposal then we can quite formally, that is, in a standard way, also create a program for the minimax estimate (8.91). The part solving (8.92) will be included in it as a subroutine. This trick is made possible because we are able to prove that independently on the sets X and $\mathcal{P}$ the solution of the task (8.91) must have the form α^M for some function α. In other words, the minimax estimate is identical with the maximum likelihood estimate for certain coefficients $\alpha(x)$, $x \in L$. The factors $\alpha(x)$ with which the two estimates become identical are extremal in the sense that they minimise the number $Q(\alpha) = \sum_{x \in L} \alpha(x) \log \alpha^M(x)$. Another important result is that for arbitrary sets X, L and $\mathcal{P}$, the function $Q(\alpha)$ is always convex. It can therefore be minimised in various well known ways.

The so far informally expressed statements will be exactly formulated and proved.

Lemma 8.1 On the upper-bound of the function $\min_{x \in L} \log p(x)$. *Let p be any function $X \to \mathbb{R}$ from the class $\mathcal{P}$, $\alpha\colon L \to \mathbb{R}$ be any function for which there holds*

$$\left.\begin{aligned} \sum_{x \in L} \alpha(x) &= 1\,, \\ \alpha(x) &\geq 0\,, \quad x \in L\,. \end{aligned}\right\} \tag{8.93}$$

In this case there holds

$$\min_{x \in L} \log p(x) \leq \sum_{x \in L} \alpha(x) \log \alpha^M(x)\,. \tag{8.94}$$

▲

Proof. The inequality (8.94) results from the evident inequalities

$$\min_{x \in L} \log p(x) \leq \sum_{x \in L} \alpha(x) \log p(x)\,, \tag{8.95}$$

$$\sum_{x \in L} \alpha(x) \log p(x) \leq \sum_{x \in L} \alpha(x) \log \alpha^M(x)\,. \tag{8.96}$$

The inequality (8.95) holds owing to the condition (8.93). The value on the right-hand side of (8.95) is the weighted arithmetic average of the numbers $\log p(x)$, the number on the left-hand side of (8.95) is the least of them. Of course, the least number is not greater than the average.

The inequality (8.96) immediately results from the definition of the function α^M. From the inequalities (8.95) and (8.96) we obtain (8.94). ■

The symbol A will denote the set of functions $\alpha\colon L \to \mathbb{R}$ which satisfy (8.93).

Lemma 8.2 On convexity of the function Q. *The function $Q(\alpha) = \sum_{x \in L} \alpha(x) \log \alpha^M(x)$ is convex on the set A.* ▲

Proof. The symbol lin A will denote the linear closure of the set A. For an arbitrary point $\alpha_0 \in A$, i.e., for an arbitrary function $\alpha_0 \colon L \to \mathbb{R}$ we will determine the linear function $G_{\alpha_0} \colon \operatorname{lin} A \to \mathbb{R}$

$$G_{\alpha_0}(\alpha) = \sum_{x \in L} \alpha(x) \log \alpha_0^M(x) .$$

For the function G_{α_0} there holds

$$G_{\alpha_0}(\alpha_0) = \sum_{x \in L} \alpha_0(x) \log \alpha_0^M(x)$$

and for any $\alpha \in A$ there holds

$$G_{\alpha_0}(\alpha) \leq \sum_{x \in L} \alpha(x) \log \alpha^M(x) .$$

The function G_{α_0} is consequently just the function the existence of which owing to (8.89) determines the convexity of the function $Q(\alpha)$. The following inequality is satisfied on the set A

$$G_{\alpha_0}(\alpha - \alpha_0) \leq \sum_{x \in L} \alpha(x) \log \alpha^M(x) - \sum_{x \in L} \alpha_0(x) \log \alpha_0^M(x) .$$

■

Theorem 8.4 Necessary and sufficient conditions for the minimax estimate of a model.

1. *If there holds*

$$\min_{x \in L} \log \alpha^{*M}(x) = \sum_{x \in L} \alpha^*(x) \log \alpha^{*M}(x) \tag{8.97}$$

then there also holds

$$\alpha^{*M} = \operatorname*{argmax}_{p \in \mathcal{P}} \min_{x \in L} \log p(x) , \tag{8.98}$$

and

$$\alpha^* = \operatorname*{argmin}_{\alpha \in A} \sum_{x \in L} \alpha(x) \log \alpha^M(x) . \tag{8.99}$$

2. *If the function* $Q(\alpha) = \sum_{x \in L} \alpha(x) \log \alpha^M(x)$ *is smooth and it is satisfied that*

$$\alpha^* = \operatorname*{argmin}_{\alpha \in A} \sum_{x \in L} \alpha(x) \log \alpha^M(x) \tag{8.100}$$

then there also holds

$$\min_{x \in L} \log \alpha^{*M}(x) = \sum_{x \in L} \alpha^*(x) \log \alpha^{*M}(x) . \tag{8.101}$$

▲

Proof. The first part of Theorem 8.4 can be proved quite easily. Thanks to Lemma 8.1 the inequality (8.94) holds for any $p \in \mathcal{P}$ and $\alpha \in A$, and thus for any $p \in \mathcal{P}$ and just for the α^*, that satisfies (8.97). So we can write

$$\min_{x\in L} \log p(x) \le \sum_{x\in L} \alpha^*(x) \log \alpha^{*M}(x)\,, \quad p \in \mathcal{P}\,,$$

which together with the condition (8.97) leads to the inequality

$$\min_{x\in L} \log p(x) \le \min_{x\in L} \log \alpha^{*M}(x)\,, \quad p \in \mathcal{P}\,,$$

which is the relation (8.98) written in another form.

We will write the inequality (8.94) for $p = \alpha^{*M}$ and any $\alpha \in A$,

$$\min_{x\in L} \log \alpha^{*M}(x) \le \sum_{x\in L} \alpha(x) \log \alpha^{M}(x)\,, \quad \alpha \in A\,.$$

This inequality together with the condition (8.97) leads to the inequality

$$\sum_{x\in L} \alpha^*(x) \log \alpha^{*M}(x) \le \sum_{x\in L} \alpha(x) \log \alpha^{M}(x)\,, \quad \alpha \in A\,,$$

which is the relation (8.99) expressed in another way. Therefore the first statement of Theorem 8.4 is proved.

Now we will prove the second part of Theorem 8.4 by contradiction. The inequality

$$\min_{x\in L} \log \alpha^{*M}(x) \le \sum_{x\in L} \alpha^*(x) \log \alpha^{*M}(x)$$

is trivial. Assume that the result (8.101) does not hold, i.e., that a strict inequality occurs

$$\min_{x\in L} \log \alpha^{*M}(x) < \sum_{x\in L} \alpha^*(x) \log \alpha^{*M}(x)\,. \tag{8.102}$$

We will prove that in this case there would exist a function $\alpha \in A$ that

$$\sum_{x\in \alpha} \alpha(x) \log \alpha^{M}(x) < \sum_{x\in \alpha} \alpha^*(x) \log \alpha^{*M}(x)\,,$$

i.e., the relation (8.100) would not be satisfied.

We will denote $x' = \operatorname{argmin}_{x\in L} \alpha^{*M}(x)$ and select the function $\alpha' \colon L \to \mathbb{R}$ such that $\alpha'(x') = 1$ and $\alpha'(x) = 0$ for all $x \in L$ except for x'. We can write

$$\min_{x\in L} \log \alpha^{*M}(x) = \sum_{x\in L} \alpha'(x) \log \alpha^{*M}(x)\,,$$

and owing to (8.102) the following relation holds

$$\sum_{x\in L} \bigl(\alpha'(x) - \alpha^*(x)\bigr) \log \alpha^{*M}(x) < 0\,. \tag{8.103}$$

We will examine how the function $Q(\alpha) = \sum_{x \in L} \alpha(x) \log \alpha^M(x)$ behaves on the abscissa that connects the points α^* and α', i.e., the dependence of the number $Q\big(\alpha^* \, (1-\gamma) + \alpha' \, \gamma\big)$ on the coefficient γ. Let us regard the points α^* and α' to be fixed, and the function $Q\big(\alpha^* \, (1-\gamma) + \alpha' \, \gamma\big)$ to be a function of one variable γ. The derivative of this function according to the variable γ in the point $\gamma = 0$ is

$$\frac{\mathrm{d}Q}{\mathrm{d}\gamma} = \sum_{x \in L} \big(\alpha'(x) - \alpha^*(x)\big) \log \alpha^{*M}(x) \, .$$

The derivative is negative because of (8.103). This means that for small values γ at least the number $Q\big(\alpha^* \, (1-\gamma) + \alpha' \, \gamma\big)$ is less than $Q(\alpha^*)$, and thus

$$\alpha^* \neq \operatorname*{argmin}_{\alpha \in A} Q(\alpha)$$

which is in contradiction with the assumption (8.100). In this way the second part of Theorem 8.4 is proved. ■

Theorem 8.4 that has been proved shows the direction in which the solution of the minimax estimate task concerning the model $p \in \mathcal{P}$ should be sought. It is necessary to find such weights $\alpha^*(x)$, $x \in L$, which minimise the convex function $Q(\alpha) = \sum_{x \in L} \alpha(x) \log \alpha^M(x)$. From the second part of Theorem 8.4 it will result that obtained weights $\alpha^*(x)$ satisfy the Equation (8.101). At the same time from the first part of Theorem 8.4 it results that the solution of the minimax task (8.91) is the approximation $p^* = \alpha^{*M}$, i.e., the maximum likelihood estimate

$$p^* = \operatorname*{argmax}_{p \in \mathcal{P}} \sum_{x \in L} \alpha^*(x) \log p(x) \, .$$

Since the function $Q(\alpha)$ is convex various procedures are at hand for its minimisation. But for this minimisation standard procedures need not be used since from the very proof of Theorem 8.4 the following recommendations for creating minimisation algorithms result.

So far a pair $\alpha \in A$ and $p \in P$ has been found satisfying the relation

$$\left.\begin{aligned} \min_{x \in L} \log p(x) &= \sum_{x \in L} \alpha(x) \log p(x) \, , \\ p &= \operatorname*{argmax}_{p \in \mathcal{P}} \sum_{x \in L} \alpha(x) \log p(x) \, , \end{aligned}\right\} \tag{8.104}$$

and the task is solved. If the relation (8.104) is not satisfied then we can see in which way the weights $\alpha(x)$ should be altered. We have to increase the weight $\alpha(x)$ of such a pattern from the training set $x \in L$ the instantaneous probability $p(x)$ of which is lowest, or one of the lowest.

This intuition can be exactly expressed by means of the following algorithm.

Algorithm 8.3 Minimax estimate of the model $p \in \mathcal{P}$

1. The user assigns the required accuracy of the task solution $\varepsilon > 0$. He or she specifies the numbers $n(x) = 1$ for all patterns from the training set L and starts with the iteration $t = 1$.
2. The following values are calculated

$$\alpha^t(x) = \frac{n^t(x)}{\sum_x n^t(x)} \,.$$

3. The maximum likelihood estimate is calculated

$$p^t = \operatorname*{argmax}_{p \in \mathcal{P}} \sum_{x \in L} \alpha^t(x) \log p(x) \,. \tag{8.105}$$

4. If the inequality

$$\sum_{x \in L} \alpha^t(x) \log p^t(x) - \min_{x \in L} \log p^t(x) < \varepsilon \tag{8.106}$$

is satisfied then the algorithm ends and p^t is the solution of the task.

5. If the inequality (8.106) is not satisfied
   (a) It is denoted

$$x' = \operatorname*{argmin}_{x \in L} p^t(x) \,.$$

   (b) New values of the numbers $n(x)$, $x \in L$, are calculated such that

$$\begin{aligned} n^{t+1}(x') &= n^t(x') + 1 \,, \\ n^{t+1}(x) &= n^t(x) \,, \qquad x \in L \,, \quad x \neq x' \,. \end{aligned}$$

6. It proceeds to the next $(t+1)$-th iteration namely by going to step 2 of the algorithm.

Algorithm 8.3 differs from the mentioned assumptions because the condition for ending the iterations is not Equation (8.104) which thanks to Theorem 8.4 guarantees that the task has been solved. A weaker condition (8.106) was used. Using one's common sense one would assume that since the condition (8.106) is an approximate alternative of the condition (8.104) it could be considered as the condition of an approximate solution of the task. This correct assumption is confirmed by the following theorem.

Theorem 8.5 On approximate solution of a minimax estimate task. *If $\alpha \in A$ and $p \in \mathcal{P}$ satisfy the inequality*

$$\sum_{x \in L} \alpha(x) \log p(x) - \min_{x \in L} \log p(x) < \varepsilon \tag{8.107}$$

and the relation

$$p = \operatorname*{argmax}_{p' \in \mathcal{P}} \sum_{x \in L} \alpha(x) \log p'(x)$$

then there holds

$$\min_{x \in L} \log p^*(x) - \min_{x \in L} \log p(x) < \varepsilon$$

where

$$p^* = \operatorname*{argmax}_{p \in \mathcal{P}} \min_{x \in L} p(x) . \tag{8.108}$$

▲

Proof. The inequality (8.94) which is proved by Lemma 8.1 is valid for any $\alpha \in A$ and $p \in \mathcal{P}$. Therefore this is valid even for α which satisfies the condition (8.107), and for p^* which satisfies (8.108). We can write

$$\min_{x \in L} \log p^*(x) - \sum_{x \in L} \alpha(x) \log p(x) \leq 0 . \tag{8.109}$$

By adding (8.109) and (8.107) we obtain

$$\min_{x \in L} \log p^*(x) - \min_{x \in L} \log p(x) < \varepsilon .$$

■

We will demonstrate now that Algorithm 8.3 with any positive ε will undoubtedly get to a state when the condition (8.106) is satisfied and so the model will be found which solves the task with a predefined precision in the sense of Theorem 8.5. This statement is proved on an additional and not very limiting condition.

We will say that the *training set L is not in contradiction with the class $\mathcal{P}$* if there exists such a model $p \in \mathcal{P}$ for which $p(x) \neq 0$ for any $x \in L$. It is naturally understood that if L is in contradiction with $\mathcal{P}$ then all models are wrong because $\min_{x \in L} p(x) = 0$ for each model $p \in \mathcal{P}$ and no optimisation can be thought of. From the assumption that the training set L is not in contradiction with the $\mathcal{P}$ it directly follows that the function $Q(\alpha) = \sum_{x \in L} \alpha(x) \log \alpha^M(x)$ is bound from below by the set A. As before we assume that the function $Q(\alpha) = \sum_{x \in L} \alpha(x) \log \alpha^M(x)$ is smooth.

Theorem 8.6 On algorithm convergence of the minimax estimate of a model. *If the set $L \subset X$ is not in contradiction with the set $\mathcal{P}$ and the function $Q(\alpha) = \sum_{x \in L} \alpha(x) \cdot \log \alpha^M(x)$ is smooth on the set A then Algorithm 8.3 will converge in a finite number of iterations to the state in which the condition (8.106) is satisfied, and the algorithm finishes.* ▲

Proof.

1. Let us assume that Theorem 8.6 is not valid and Algorithm 8.3 will not finish iterating. This would mean that in each iteration of the algorithm the following inequality was satisfied

$$\sum_{x \in L} \alpha^t(x) \log p^t(x) - \min_{x \in L} \log p^t(x) \geq \varepsilon . \tag{8.110}$$

   We will prove that in this case and at a sufficiently large t the quantity $Q(\alpha) = \sum_{x \in L} \alpha^t(x) \log p^t(x)$ could fall below any value. This is, however, impossible owing to the assumption that the set L is not contradictory in the class $\mathcal{P}$.

2. The proof will be performed according to the following scheme.
   (a) It will be proved that

$$\lim_{t \to \infty} |\alpha_t - \alpha_{t-1}| = 0 . \tag{8.111}$$

(b) A linear function $G_t\colon A \to \mathbb{R}$ is introduced

$$G_t(\alpha) = \sum_{x \in L} \alpha(x) \log p^t(x)$$

which is with respect to the function $Q(\alpha) = \sum_{x\in L} \alpha(x) \log \alpha^M(x)$ in the relation

$$G_t(\alpha - \alpha^t) \le Q(\alpha) - Q(\alpha^t), \quad \alpha \in A.$$

The linear function G_t corresponds to the gradient of the function $Q(\alpha)$ at the point α^t.

(c) A sequence of the numbers $G_t(\alpha^{t+1} - \alpha^t)$, $t = 1, 2, \ldots, \infty$, and $Q(\alpha^{t+1}) - Q(\alpha^t)$, $t = 1, 2, \ldots, \infty$, is formed. If (8.111) is proved then also the following is proved

$$\lim_{t\to\infty} G_t(\alpha^{t+1} - \alpha^t) = 0,$$

$$\lim_{t\to\infty} \left(Q(\alpha^{t+1}) - Q(\alpha^t)\right) = 0,$$

since in the finite-dimensional space any linear function is continuous which holds even for any convex function.

(d) The function Q is according to the assumption smooth, and therefore the convergence of sequences of the quantities $G_t(\alpha^{t+1} - \alpha^t)$ and $Q(\alpha^{t+1}) - Q(\alpha^t)$ toward zero are of the same order. If there holds that

$$\sum_{t=1}^{\infty} G_t(\alpha^{t+1} - \alpha^t) = -\infty, \tag{8.112}$$

then there also holds

$$\sum_{t=1}^{\infty} \left(Q(\alpha^{t+1}) - Q(\alpha^t)\right) = -\infty. \tag{8.113}$$

The number on the left-hand side of the last equality is nothing else than $\lim_{t\to\infty}\left(Q(\alpha^t) - Q(\alpha^1)\right)$. Therefore the relation (8.113) would mean that in spite of the lower-bound of the function $Q(\alpha)$ the number $Q(\alpha^t)$ could fall below any negative number at sufficiently great t. In this way, the theorem would be proved through contradiction.

Thus, it is necessary to prove that the assumption (8.110) results in the relations (8.111) and (8.112) and Theorem 8.6 will be proved.

3. We will denote by $n^t = \sum_{x\in L} n^t(x)$, $x^t = \operatorname{argmin}_{x\in L} p^t(x)$, and by α''^t the function $L \to \mathbb{R}$ for which there holds $\alpha''^t(x) = 1$ if $x = x^t$ and $\alpha''^t(x) = 0$ if $x \ne x^t$. With these notations the inequality (8.110) assumes the form

$$\sum_{x\in L} \left(\alpha^t(x) - \alpha''^t(x)\right) \log p^t(x) \ge \varepsilon, \tag{8.114}$$

and the weights $\alpha^{t+1}(x)$ in the algorithm can be expressed as

$$\alpha^{t+1}(x) = \alpha^t(x)\,\frac{n^t}{n^t+1} + \alpha''^t(x)\,\frac{1}{n^t+1}.$$

The difference between the weights α in the two successive iterations is

$$\alpha^{t+1}(x) - \alpha^t(x) = \frac{\alpha'^t(x) - \alpha^t(x)}{n^t + 1} \,.$$

4. The number n^t in the first iteration is $|L|$ and it is increased by one in each iteration. Therefore there holds that $n^t = |L| + t - 1$, and therefore

$$\alpha^{t+1}(x) - \alpha^t(x) = \frac{\alpha'^t(x) - \alpha^t(x)}{|L| + t} \,. \tag{8.115}$$

The numerator is limited and therefore we can see that for all $x \in L$

$$\lim_{t \to \infty} |\alpha^{t+1}(x) - \alpha^t(x)| = 0$$

holds and the relation (8.111) is proved.

5. The difference $G_t(\alpha^{t+1} - \alpha^t)$ is

$$G_t(\alpha^{t+1}) - G_t(\alpha^t) = \sum_{x \in L} \left(\alpha^{t+1}(x) - \alpha^t(x) \log p^t(x)\right) \,.$$

Owing to (8.115) we can write

$$G_t(\alpha^{t+1}) - G_t(\alpha^t) = \sum_{x \in L} \frac{\alpha'^t(x) - \alpha^t(x)}{|L| + t} \log p^t(x) \,.$$

At last we can see, owing to (8.114), that the difference is negative and, moreover,

$$G_t(\alpha^{t+1}) - G_t(\alpha^t) \leq -\frac{\varepsilon}{|L| + t} \,.$$

6. The sum $\sum_{t=1}^{T} G_t(\alpha^{t+1} - \alpha^t)$ is not greater than $-\varepsilon \sum_{t=1}^{T} \frac{L}{|L|+t}$, and therefore with increasing T the sum can fall below any negative number. The relation (8.112) is proved. ■

Thus we have proved that rather simple algorithm performs the minimax evaluation of the statistical model of the object. Of course, this algorithm can be treated as a simple one only under the condition that a simple algorithm for the maximum likelihood estimate is available. If we already have such a program then the program for the minimax estimate is designed in quite a mechanical way. The existing program is extended by simple operations of adding up a one, and enclosing into a cycle which is not worth mentioning. This superstructure does not depend on a concrete task, i.e., on the form of the observation set X and on the class of models $\mathcal{P}$. In this way a close relationship of two extensive estimation tasks is revealed which would seem, at a cursory glance, to be different.

The algorithm mentioned above is, because of its universal character, suitable for using even when specific properties of the applied problem under consideration are not yet known. With increasing knowledge of the task even other algorithms can be better suited, for example those of the gradient descent, or the methods of reciprocal gradients, and many others. The competence of using all these methods is based on the following two previously proved and universally true properties.

- The minimax estimate of the model for some training set L is identical with maximum likelihood estimate of the model for some training multi-set L^* where every member $x \in L$ occurs with relative frequency $\alpha(x)$.
- The coefficients $\alpha(x)$, $x \in L$, which provide the equivalence of these two estimates, minimise the well defined convex function.

8.9.3 Proof of the minimax estimate algorithm of a Markovian model

The general algorithm for the minimax estimate of a model described in Subsection 8.9.2 is defined except for the operation

$$p^t(x) = \operatorname*{argmax}_{p \in \mathcal{P}} \sum_{x \in L} \alpha^t(x) \log p(x) .$$

The program implementing the operation was assumed to have been available. For the case in which $\mathcal{P}$ is a set of Markovian models, we defined this operation by means of the formulæ (8.85) and (8.86) and called it a Markovian approximation.

It can be noted that including the calculations (8.85), (8.86) from Section 8.8 in the general Algorithm 8.3 will lead to the particular Algorithm 8.2. It is not difficult to make certain that the set of Markovian models $\mathcal{P}$ satisfies the conditions on which Theorems 8.4, 8.5 and 8.6 are valid. These are only two conditions: consistence of the training set L with respect to the set $\mathcal{P}$ and the smoothness of the function $\sum \alpha(\bar{x}, \bar{k}) \log \alpha^M(\bar{x}, \bar{k})$. The first property is satisfied because for any training set $\big((\bar{x}^j, \bar{k}^j)\,,\, j = 1, \dots, l\big)$ such a Markovian model exists in which each pair $(\bar{x}^j, \bar{k}^j)$ occurs with non-zero probability. It can also be noticed that for the set of Markovian models $\mathcal{P}$ the dependence of the value $\max_{p \in \mathcal{P}} \sum_{(\bar{x}, \bar{k}) \in L} \alpha(\bar{x}, \bar{k}) \log p(\bar{x}, \bar{k})$ on coefficients $\alpha(\bar{x}, \bar{k})$ is not only convex but also smooth, i.e., differentiable. Theorems 8.2 and 8.3 are thus special cases of proved Theorems 8.5 and 8.6, respectively. We need not, therefore, prove them.

8.10 Tuning the algorithm that recognises sequences

The task of tuning a pattern recognition algorithm was already formulated in Lecture 4 in the general form, and in the Subsection 8.7.3 particularly for Markovian models. The objective of the task is to create for the training set $L = \big((\bar{x}^j, \bar{k}^j),\ j = 1, \dots, l\big)$ the ensemble of functions $f_i\colon K \times X \times K \to \mathbb{R}$, $i = 1, 2, \dots, n$, which fulfill the relation

$$\bar{k}^j = \operatorname*{argmax}_{\bar{k} \in K^{n+1}} \sum_{i=1}^{n} f_i(k_{i-1}, x_i^j, k_i)\,, \quad j = 1, \dots, l\,. \tag{8.116}$$

Written in another way, the function values $f_i(k_{i-1}, x_i, k_i)$ have to satisfy the system of inequalities

$$\sum_{i=1}^{n} f_i(k_{i-1}^j, x_i^j, k_i^j) > \sum_{i=1}^{n} f_i(k_{i-1}, x_i^j, k_i)\,, \quad \bar{k} \neq \bar{k}^j\,, \quad j = 1, 2, \ldots, l\,. \tag{8.117}$$

The system of inequalities (8.117) consists of an enormous number $(|K|^{n+1} - 1)l$ of linear inequalities which restrict $n \times |K|^2 \times |X|$ variables $f_i(k', x', k'')$. In spite of an enormously great number of equalities, a solution of the system (8.117) can be found with the methods explained in Lecture 4. Naturally, the solution can be found by these methods only if such a solution exists at all. The main advantage of those methods, be it the Perceptron method, Kozinec's or similar ones, is that they do not require a check of all the inequalities in the system. It suffices to have at disposal a constructive way for checking whether the given ensemble of numbers $f_i(k', x', k'')$ satisfies the system (8.117). If the system is not satisfied then it is sufficient to find a single inequality which is not satisfied. The current modification of numbers $f_i(k', x', k'')$ depends only on this single found inequality. Such a constructive method exists for a system of linear inequalities of the form (8.117) and the procedure is represented by the system of relations (8.116). For each $j = 1, \ldots, l$, it suffices to find out the sequence

$$\bar{k}^{*j} = \operatorname*{argmax}_{\bar{k}} \sum_{i=1}^{n} f_i(k_{i-1}, x_i^j, k_i)\,, \tag{8.118}$$

and then check if the following holds

$$\bar{k}^{*j} = \bar{k}^j\,, \quad j = 1, 2, \ldots, n\,. \tag{8.119}$$

The sequence (8.118) is obtained with the help of constructive algorithms seeking the shortest path in the graph which was introduced in Subsection 8.4.4. If the equation (8.119) is not satisfied for some j then it also determines the inequality from the system (8.117) which is not satisfied. Namely it is that inequality the left-hand side of which corresponds to the sequence $k_0^j, k_1^j, \ldots, k_n^j$ and its right-hand side to the sequence $k_0^{*j}, k_1^{*j}, \ldots, k_n^{*j}$.

The advantage of perceptron and Kozinec algorithms is that it is not required to know all unsatisfied inequalities from the system to change values $f_i(k', x', k'')$. It suffices to know just a single inequality out of them. If for some j the equation (8.119) is not satisfied then it leads immediately to modification of values $f_i(k', x', k'')$ for which the selected algorithm is used. If the perceptron algorithm is used, and it is the most easy one to formulate, then the rule performing the modification has very simple form. Numbers $f_i(k', x', k'')$ are to be increased by one if $k_{i-1}^j = k', x_i^j = x', k_i^j = k''$, and to be decreased by one if $k_{i-1}^{*j} = k', x_i^j = x', k_i^{*j} = k''$. Otherwise the values remain unchanged. In the case if $k_{i-1}^{*j} = k_{i-1}^j = k'$ and simultaneously $k_i^{*j} = k_i^j = k''$, the mentioned modification rule has to be understood in such a way that the value

$f_i(k', x', k'')$ is increased by one at first and decreased by one afterwards. The result is the same as there were no modification.

The Novikoff Theorem 5.5 is valid for the algorithm formulated in a described way because it is not required that the number of inequalities were small. It is required only that the system has a solution. It follows from the Novikoff theorem that if such an ensemble of numbers exists that satisfies the system of relations (8.116) then in a finite number of steps the variables $f_i(k', x', k'')$ assume values that fulfill the relation (8.116).

8.11 The maximum likelihood estimate of statistical model in unsupervised learning

Let X^n be a set of sequences of the form $\bar{x} = (x_1, x_2, \dots, x_n)$, $x_i \in X$, of the length n. Let K^{n+1} be a set of sequences of the form $\bar{k} = (k_0, k_1, \dots, k_n)$, $k_i \in K$, of the length $n+1$. The function $p\colon X^n \times K^{n+1} \to \mathbb{R}$ gives for each sequence $x \in X^n$ and $k \in K^{n+1}$ its joint probability $p(\bar{x}, \bar{k})$. We know that the function p has the form

$$p(\bar{x}, \bar{k}) = p_1(k_0, x_1, k_1) \prod_{i=2}^{n} \frac{p_i(k_{i-1}, x_i, k_i)}{\sum_{k' \in K} \sum_{x' \in X} p_i(k_{i-1}, x', k')} \,. \tag{8.120}$$

This means that the function $p\colon X^n \times K^{n+1} \to \mathbb{R}$ is uniquely determined by n functions p_i, $i = 1, \dots, n$, of the form $K \times X \times K \to \mathbb{R}$. The set of functions $p\colon X^n \times K^{n+1} \to \mathbb{R}$ of the form (8.120) will be denoted by $\mathcal{P}$ and the ensemble $(p_i\,,\; i = 1, \dots, n)$ by P.

Let $(\bar{x}^1, \bar{x}^2, \dots, \bar{x}^l)$ be an ensemble of mutually independent sequences and let each of them be of the length n, i.e., $\bar{x}^j \in X^n$, $j = 1, \dots, l$. In addition, let us assume that each sequence $\bar{x}^j$ is an instance of a random sequence the probability distribution of which is $\sum_{\bar{k} \in K^{n+1}} p(\bar{x}, \bar{k})$. In this case, the probability of the ensemble $(\bar{x}^1, \bar{x}^2, \dots, \bar{x}^l)$ is given as $\prod_{j=1}^{l} \sum_{\bar{k} \in K^{n+1}} p(\bar{x}^j, \bar{k})$, and this probability depends on what the function $p \in \mathcal{P}$ is like. The task of the maximum likelihood estimate of a model in the context of unsupervised learning is defined as seeking the model $p^* \in \mathcal{P}$ which maximises this function, i.e.,

$$\begin{aligned} p^* &= \operatorname*{argmax}_{p \in \mathcal{P}} \prod_{j=1}^{l} \sum_{\bar{k} \in K^{n+1}} p(\bar{x}^j, \bar{k}) \\ &= \operatorname*{argmax}_{p \in \mathcal{P}} \sum_{j=1}^{l} \log \sum_{\bar{k} \in K^{n+1}} p(\bar{x}^j, \bar{k}) \,. \end{aligned} \tag{8.121}$$

The similarity of this task to the task of unsupervised learning which was discussed in Lecture 6 is quite evident even if, strictly speaking, the tasks are different. Formerly we analyzed a case of supervised learning in which the sought model $p(x, k)$ is decomposed into $|K| + 1$ independent tasks. The first of them is seeking *a priori* probabilities $p_K(k)$ for each value of the hidden parameter k. Other $|K|$ tasks seek the distribution of conditional probabilities $p_{X|k}(x)$ under the condition k for each value $k \in K$. It was assumed that the choice of functions $p_{X|k'}$ from the known set $\mathcal{P}$ does not, in any way, affect the choice of another function $p_{X|k''}$. In other words it referred to a situation in which each function $p_{X|k}$, $k \in K$, was completely determined by its own value a_k of the parameter a. The choice of the value $a_{k'}$ for some value k' affected in no way the choice of the value $a_{k''}$ for any other value k''. The case we are dealing with now is different. If $\bar{k}'$ and $\bar{k}''$ are two fixed sequences then

$$p_{X^n|\bar{k}'}(\bar{x}) = \prod_{i=1}^{n} \frac{p_i(k'_{i-1}, x_i, k'_i)}{\sum\limits_{x' \in X} p_i(k'_{i-1}, x', k'_i)}$$

and

$$p_{X^n|\bar{k}''}(\bar{x}) = \prod_{i=1}^{n} \frac{p_i(k''_{i-1}, x_i, k''_i)}{\sum\limits_{x' \in X} p_i(k''_{i-1}, x', k''_i)}.$$

The functions $p_i \colon K \times X \times K \to \mathbb{R}$, $i = 1, \ldots, n$, which determine these two probability distributions are to be the same in both the expressions. The question therefore is that the parameters of the model are given by the ensemble $(p_1, p_2, \ldots, p_n)$ of the functions of three variables that affect simultaneously the probability distributions $p_{X^n|\bar{k}}$ for all $\bar{k} \in K^{n+1}$. In addition, they also affect the *a priori* probability $p_{K^{n+1}}(\bar{k})$ for each sequence $\bar{k} \in K^{n+1}$.

In spite of this formal difference of the task (8.121) from the unsupervised tasks analysed in Lecture 6, these two tasks have a kind of affinity of thought. The task (8.121) can be solved through a slight modification of all considerations that we quoted for the unsupervised learning task before. We will briefly recall the considerations, but now it will be within the examined Markovian model.

For $t = 1, 2, \ldots, \infty$ we will create a sequence of models $p^t \in \mathcal{P}$, i.e., the sequence of ensembles of the numbers $p_i^t(k', x', k'')$, $i = 1, \ldots, n$, $k' \in K$, $x' \in X$, and $k'' \in K$. Each ensemble determines the following probability for each pair $(\bar{x}, \bar{k}) \in X^n \times K^{n+1}$,

$$p^t(\bar{x}, \bar{k}) = p_1^t(x_0, x_1, k_1) \prod_{i=2}^{n} \frac{p_i^t(k_{i-1}, x_i, k_i)}{\sum\limits_{x' \in X} \sum\limits_{k' \in K} p_i^t(k_{i-1}, x', k')}. \tag{8.122}$$

We denote by $J_i(x')$ the subset of those indices j for which $x_i^j = x'$ holds, and $K_i(k', k'')$ the set of those sequences k for which $k_{i-1} = k'$, $k_i = k''$ holds.

Assume that we know the ensemble of numbers $p_i^t(k',\ x',\ k'')$. A new ensemble of numbers $p_i^{t+1}(k', x', k'')$ is being built in the following two steps.

Recognition. For each sequence $\bar{x}^j$ and each sequence $\bar{k} \in K^{n+1}$ we calculate the number

$$\alpha^t(\bar{x}^j, \bar{k}) = \frac{p^t(\bar{x}^j, \bar{k})}{\sum_{\bar{k}' \in K^{n+1}} p^t(\bar{x}^j, \bar{k}')}, \tag{8.123}$$

where the numbers $p^t(\bar{x}^j, \bar{k})$ are calculated according to the formula (8.122).

Learning. For each $i = 1, 2, \ldots, n$, $k' \in K$, $x' \in X$, $k'' \in K$, the following numbers are calculated

$$p_i^{t+1}(k', x', k'') = \frac{1}{l} \sum_{j \in J_i(x')} \sum_{\bar{k} \in K_i(k', k'')} \alpha^t(\bar{x}^j, \bar{k}) . \tag{8.124}$$

The given algorithm is, of course, not suitable for use since the set K^{n+1} is extremely extensive, and therefore the numbers $\alpha(\bar{x}^j, \bar{k})$ cannot be calculated for each sequence $\bar{k} \in K^{n+1}$. Similarly the sum over all sequences $\bar{k} \in K_i(k', k'')$ in the formula (8.124) cannot be computed. Later we will show how an algorithm is to be formed which is equivalent to the above quoted algorithm, but can be built in a constructive way. But now we will use the definition of the algorithm in the given non-constructive form to prove the following theorem.

Theorem 8.7 On unsupervised learning in Markovian sequences. *Let $p_i^t(k', x', k'')$ and $p_i^{t+1}(k', x', k'')$, $i = 1, 2, \ldots, n$, $k' \in K$, $x' \in X$, $k'' \in K$, be two ensembles calculated according to the relations (8.123) and (8.124). Let p^t, p^{t+1} be two successive models, i.e., two functions of the form $X^n \times K^{n+1} \to \mathbb{R}$ defined by the formula (8.122). In this case there holds that*

$$\prod_{j=1}^{l} \sum_{\bar{k} \in K^{n+1}} p^{t+1}(\bar{x}^j, \bar{k}) \geq \prod_{j=1}^{l} \sum_{\bar{k} \in K^{n+1}} p^t(\bar{x}^j, \bar{k}) .$$

▲

Proof. Since there holds that

$$\sum_{\bar{k} \in K^{n+1}} \alpha^t(\bar{x}^j, \bar{k}) = 1 , \quad j = 1, \ldots, l ,$$

also the two following equations hold

$$\sum_{j=1}^{l} \log \sum_{\bar{k} \in K^{n+1}} p^t(\bar{x}^j, \bar{k})$$

$$= \sum_{j=1}^{l} \sum_{\bar{k} \in K^{n+1}} \alpha^t(\bar{x}^j, \bar{k}) \log p^t(\bar{x}^j, \bar{k}) - \sum_{j=1}^{l} \left(\sum_{\bar{k} \in K^{n+1}} \alpha^t(\bar{x}^j, \bar{k}) \log \frac{p^t(\bar{x}^j, \bar{k})}{\sum_{\bar{k} \in K^{n+1}} p^t(\bar{x}^j, \bar{k})} \right) ,$$

$$\sum_{j=1}^{l} \log \sum_{\bar{k} \in K^{n+1}} p^{t+1}(\bar{x}^j, \bar{k})$$

$$= \sum_{j=1}^{l} \sum_{\bar{k} \in K^{n+1}} \alpha^t(\bar{x}^j, \bar{k}) \log p^{t+1}(\bar{x}^j, \bar{k}) - \sum_{j=1}^{l} \left(\sum_{\bar{k} \in K^{n+1}} \alpha^t(\bar{x}^j, \bar{k}) \log \frac{p^{t+1}(\bar{x}^j, \bar{k})}{\sum_{\bar{k} \in K^{n+1}} p^{t+1}(\bar{x}^j, \bar{k})} \right) .$$

According to (8.123)

$$\alpha^t(\bar{x}^j, \bar{k}) = \frac{p^t(\bar{x}^j, \bar{k})}{\sum_{\bar{k} \in K^{n+1}} p^t(\bar{x}^j, \bar{k})}$$

holds for each $j = 1, 2, \ldots, l$. Consequently owing to Lemma 6.1 the inequality

$$\sum_{\bar{k} \in K^{n+1}} \alpha^t(\bar{x}^j, \bar{k}) \log \frac{p^t(\bar{x}^j, \bar{k})}{\sum\limits_{\bar{k} \in K^{n+1}} p^t(\bar{x}^j, \bar{k})} \geq \sum_{\bar{k} \in K^{n+1}} \alpha^t(\bar{x}^j, \bar{k}) \log \frac{p^{t+1}(\bar{x}^j, \bar{k})}{\sum\limits_{\bar{k} \in K^{n+1}} p^{t+1}(\bar{x}^j, \bar{k})}$$

holds also for each $j = 1, 2, \ldots, l$. So we write

$$\sum_{j=1}^{l} \sum_{\bar{k} \in K^{n+1}} \alpha^t(\bar{x}^j, \bar{k}) \log \frac{p^t(\bar{x}^j, \bar{k})}{\sum\limits_{\bar{k} \in K^{n+1}} p^t(\bar{x}^j, \bar{k})} \geq \sum_{j=1}^{l} \sum_{\bar{k} \in K^{n+1}} \alpha^t(\bar{x}^j, \bar{k}) \log \frac{p^{t+1}(\bar{x}^j, \bar{k})}{\sum\limits_{\bar{k} \in K^{n+1}} p^{t+1}(\bar{x}^j, \bar{k})} . \tag{8.125}$$

The numbers $p_i^{t+1}(k', x', k'')$ calculated according to (8.124) satisfy the conditions of Theorem 8.1 from which it follows that

$$\sum_{j=1}^{l} \sum_{\bar{k} \in K^{n+1}} \alpha^t(\bar{x}^j, \bar{k}) \log p^{t+1}(\bar{x}^j, \bar{k}) \geq \sum_{j=1}^{l} \sum_{\bar{k} \in K^{n+1}} \alpha^t(\bar{x}^j, \bar{k}) \log p^t(\bar{x}^j, \bar{k}) . \tag{8.126}$$

The following inequality follows from the inequalities (8.125) and (8.126)

$$\sum_{j=1}^{l} \log \sum_{\bar{k} \in K^{n+1}} p^{t+1}(\bar{x}^j, \bar{k}) \geq \sum_{j=1}^{l} \log \sum_{\bar{k} \in K^{n+1}} p^t(\bar{x}^j, \bar{k}) ,$$

and the theorem is proved.

■

We will show how the above algorithm can be transformed to an equivalent algorithm which can be realised in a constructive way. The numbers $\alpha^t(\bar{x}^j, \bar{k})$ and the probabilities $p^t(\bar{x}^j, \bar{k})$ in the algorithm are only auxiliary data which facilitate the proof of Theorem 8.7. In the constructive application of the algorithm these numbers can be excluded and the algorithm can be expressed in such a way that it will operate only with parameters $p_i^t(k', x', k'')$ and numbers $\alpha_i^t(k', x^j, k'')$ which are defined as

$$\alpha_i^t(k', x_i^j, k'') = \sum_{\bar{k} \in K_i(k', k'')} \alpha^t(\bar{x}^j, \bar{k}) .$$

The formula (8.124) assumes a quite simple form

$$p_i^{t+1}(k', x', k'') = \frac{1}{l} \sum_{j \in J_i(x')} \alpha_i^t(k', \bar{x}^j, k'')$$

the calculation of which does not pose any unsurmountable obstacles because it requires only addition over the sequences which are present in the training multi-set.

The calculation according to the formula (8.123) is immediately replaced by the calculation of the sum $\alpha_i^t(k', \bar{x}^j, k'')$ on the basis of the numbers $p_i^{t-1}(k', x', k'')$. The mechanism of this calculation was analysed in detail in Section 8.5 since the number $\alpha_i^t(k', \bar{x}^j, k'')$ is nothing else than the joint *a posteriori* probability of the event $k_{i-1} = k'$, $k_i = k''$ for an observed sequence $\bar{x}^j$.

8.12 Discussion

I have had an ambiguous impression from your lecture. On the one hand, I noticed that the Markovian model of an object, which you examined in detail in your lecture, allows us to solve precisely a number of pattern recognition tasks. I seem to understand the lecture to such an extent that I could solve the tasks by myself. And so I gather a self-confident feeling that in the sphere of structural recognition of sequences I can master many things. At the same time I assume that there exist treacherous pitfalls, into which I can fall more easily when I do not know about their existence. I would rather know them beforehand. I well remember Example 7.1 on recognising vertical and horizontal lines. The task seemed to be quite easy, but in fact it happened to be fantastically complicated. I would like to find a similar task even among Markovian models.

You would naturally come across such a task if you had enough time for it. You would master one task after another and we estimate that after the tenth task at the latest you would come across what you are looking for. We will try to make the job easier for you by seeking it together with you. But anyhow, let us start from a simple task.

Let us assume, as we did in the lecture, that $\bar{x} \in X^n$ and $\bar{k} \in K^{n+1}$ are two random sequences the joint probability distribution $p(\bar{x}, \bar{k})$ of which is Markovian. We already know that a sequence $\bar{k}^* = \operatorname{argmax}_{\bar{k} \in K^{n+1}} p(\bar{x}, \bar{k})$ can be constructively created for any sequence $\bar{x}$. How would you design an algorithm if you did not need to know the whole sequence $\bar{k}^*$, but only to find out how many times the certain value $\sigma \in K$ occurred in the sequence?

Here a kind of trouble is hidden and I did not reveal it. When I already have the sequence $\bar{k}^*$ it is then easy to count how many times σ occurs in it. I wonder what I passed over when I cannot see why it should not be a correct solution.

Your suggestion is not incorrect. But a trouble is there in spite of all that. You think it self-evident that for solving the task it is necessary to find the whole sequence $\bar{k}^*$. But in formulating the task we pointed out that a whole sequence was not needed in your application. In your algorithm the whole sequence is only an auxiliary piece of information, from which you will select the final result. You have not suggested the best way.

Yes, it was said that I did not have to create the whole sequence $\bar{k}^*$, but it was not said that I was not allowed to create it. Why would not I be able to find it as an auxiliary sequence?

Since you would unnecessarily waste the memory and you may be later short of it. In creating this auxiliary information you must have enough memory for quantities $\mathrm{ind}_i(k)$ for each $i = 1, 2, \ldots, n$, and for each $k \in K$, i.e., the memory of $n\,|K|\,\log|K|$ bits. Remember the procedure presented in Subsection 8.4.4. Do not forget that the length n of the sequence can be so large that you may not have the necessary memory any more. In this case you were not able to realise your procedure. It does not mean, however, that the task cannot be solved by another algorithm which can find out how many times the value σ has occurred in the most probable sequence without your having built up the most probable sequence. What should such an algorithm look like?

I understand it now. Let the numbers $\mathrm{ind}_i(k)$ *have the same sense as defined in the lecture. I will introduce other numbers* $h_i(k)$, $i = 0, 1, \ldots, n$, $k \in K$, *which mean how many times the value* σ *occurred in the sequence* $k'_0, k'_1, \ldots, k'_i$, *which maximises the probability* $p(x_1^i, k_0^i)$ *and in which* $k'_i = k$. *The numbers* $h_i(k)$ *are calculated according to the following procedure:*

$$
\begin{aligned}
h_0(k) &= \begin{cases} 1, & \text{if} \quad k = \sigma\,, \\ 0, & \text{if} \quad k \neq \sigma\,. \end{cases} \\
h_i(k) &= \begin{cases} h_{i-1}\big(\mathrm{ind}_i(k)\big) + 1\,, & \text{if} \quad k = \sigma\,, \\ h_{i-1}\big(\mathrm{ind}_i(k)\big)\,, & \text{if} \quad k \neq \sigma\,. \end{cases}
\end{aligned}
\tag{8.127}
$$

If $k^* = (k_0^*, k_1^*, \ldots, k_n^*)$ *is the most probable sequence then the number* $h_n(k_n^*)$ *is the solution of the task. To find it we need not know the whole sequence* k^*, *but it is sufficient to know only its last element. To find that last element I need not know the numbers* $\mathrm{ind}_i(k)$, *and therefore neither the memory for them is needed. The number* $\mathrm{ind}_i(k)$, *which occurs in the relation (8.127), is used for each* i *and* k *only once, immediately after its being calculated and therefore to remember it, only one* $\log|K|$*-bit cell is sufficient. The numbers* $h_i(k)$, $k \in K$, *which the algorithm has calculated, are also used only once, namely in calculating the numbers* $h_{i+1}(k)$. *To store the numbers* $h_i(k)$, $k \in K$, *the memory for* $2\,|K| \log n$ *bits is sufficient, which is far less than* $n\,|K| \log|K|$ *bits that would be necessary if I wanted to reconstruct the whole sequence* k^*.

Now try to create an algorithm which, for a known stochastic automaton, finds out how many times the automaton got into the particular state σ while it was generating the known output sequence $(x_1, x_2, \ldots, x_n)$.

Have not I done that just now?

Of course not. The most probable sequence need not be the real one.

But, I do not know any real sequence. I cannot say that a particular sequence $\bar{k}$ *is or is not the real one. At most, I can calculate the* a posteriori *probability of it being real. And so the question you are asking is similar to that which follows. There exists a random quantity with a known probability distribution. On this basis it is to find out what this quantity is equal to. Well, it is nonsense. A random quantity is not identical with any fixed quantity.*

We are glad that your criticism of us is so sharp. It is actually nonsense. Let us formulate the question in a proper way. For a known stochastic automaton, for a given state σ, and a given number l it is necessary to find the probability that the automaton passed the state σ l times when generating the observed sequence of symbols $(x_1, x_2, \ldots, x_n)$.

I have understood the assignment. I denote the sequence $(x_1, x_2, \ldots, x_i)$ *by* x_1^i *and assume that* x_1^0 *is an empty sequence. I denote by the symbols* $g_i(x_1^i, l, k)$, $i = 0, 1, \ldots, n$, $l = 0, 1, 2, \ldots, i+1$, $k \in K$, *the joint probability of the following three events:*

1. *The first* i *elements in the output sequence which the automaton generates are the values* x_1^i.
2. *In the sequence of states* $k_0, k_1, \ldots, k_i$ *through which the automaton passes the state* σ *occurs* l *times.*
3. *The state* k_i *is* k.

According to this definition there holds

$$g_0(x_1^0, l, k_0) = \begin{cases} p_0(k_0)\,, & \text{if } \; l = 1 \; \text{ and } \; k_0 = \sigma\,, \\ p_0(k_0)\,, & \text{if } \; l = 0 \; \text{ and } \; k_0 \neq \sigma\,, \\ 0\,, & \text{in other cases}\,. \end{cases} \tag{8.128}$$

I denote by the symbol $g_i'(x_1^i, l, k)$, $i = 1, 2, \ldots, n$, $l = 0, 1, \ldots, i$, $k \in K$, *the joint probability of somewhat different events than those whose probability is denoted by* g.

1. *The first* i *elements of the output sequence which are generated by the automaton are* x_1^i.
2. *In the sequence of states* $(k_0, k_1, \ldots, k_{i-1})$ *through which the automaton passes the state* σ *occurs* l*-times.*
3. *The state* k_i *is* k.

From the definitions of g_i *and* g_i' *their mutual relation follows*

$$g_i(x_1^i, l, k) = \begin{cases} g_i'(x_1^i, l, k)\,, & \text{if } \; k \neq \sigma\,, \\ g_i'(x_1^i, l-1, k)\,, & \text{if } \; k = \sigma\,. \end{cases} \tag{8.129}$$

and the fraction

$$\frac{\sum_{k \in K} g_n(x_1^n, l, k)}{\sum_{l=0}^{n} \sum_{k \in K} g_n(x_1^n, l, k)}$$

is the conditional probability sought that the automaton passed l*-times through the state* σ *during the generation of the assigned sequence* x_1^n. *The algorithm which I am expected to create should gradually calculate the numbers* $g_0(), g_1(), \ldots, g_n()$, *starting from the numbers* g_0 *which are defined by (8.128). For the formulation of the algorithm I will further introduce auxiliary numbers* $g_i''(x_1^i, l, k', k'')$ *which mean the joint probability of the following four events.*

1. *The first* i *elements generated by the automaton into the output sequence are the values* x_1^i.

2. *In the sequence of states* $k_0, k_1, \ldots, k_{i-1}$ *through which the automaton passes the state* σ *occurs* l*-times.*
3. *The state* k_{i-1} *is* k'*.*
4. *The state* k_i *is* k''*.*

The probabilities g' *and* g'' *must satisfy the following relation which holds for any probabilities:*

$$g'_i(x_1^i, l, k) = \sum_{k' \in K} g''_i(x_1^i, l, k', k) . \tag{8.130}$$

For the probabilities g''_i *there holds, in addition,*

$$g''_i(x_1^i, l, k', k) = g''_i\big((x_1^{i-1}, x_i), l, k', k\big) = g_{i-1}(x_1^{i-1}, l, k')\, p_i(x_i, k \mid k') , \tag{8.131}$$

where $p_i(x_i, k \mid k')$ *is the known probability which characterises the automaton. The relation (8.131) is correct because with the fixed state* k' *in the* $(i-1)$*-th instant the random quantity* x_i *and* k*, which are realised subsequent to this instant, do not depend on the random events* x_1^{i-1} *and* l*, which precede this instant. After including (8.131) into (8.130) I obtain*

$$g'_i(x_0^i, l, k) = \sum_{k' \in K} g_{i-1}(x_1^{i-1}, l, k')\, p_i(x_i, k \mid k') .$$

I will rewrite the same using (8.129)

$$g_i(x_1^i, l, k) = \begin{cases} \sum\limits_{k' \in K} g_{i-1}(x_1^{i-1}, l, k')\, p_i(x_i, k \mid k') , & \text{if } k \neq \sigma , \\ \sum\limits_{k' \in K} g_{i-1}(x_1^{i-1}, l-1, k')\, p_i(x_i, k \mid k') , & \text{if } k = \sigma . \end{cases} \tag{8.132}$$

One calculation according to the formula (8.132) assumes the complexity $\mathcal{O}(l\,|K|^2)$ *and the whole algorithm has the complexity* $\mathcal{O}(l^2\,|K|^2)$*. This complexity seems to me too high. Cannot it be made lower in some way?*

You have managed the task quite well. It does not seem to us that the algorithm could be made substantially faster. Notice, however, that if you did not have to calculate the whole probability distribution of the quantity l, but only some characteristics of the random quantity l, such as the mathematical expectation or variance, then the calculation could be made faster. If you were to calculate the probability of each value l when only the mathematical expectation of this value is needed you would make many superfluous calculations.

Let us direct your attention to a more difficult task because the limit of your resources does not seem to have been attained yet.

Let us assume that you are interested not only in the total number of states σ in the sequence $(k_0, k_1, \ldots, k_n)$, but also in their positions. This means that you are interested in the set I' of all instances in which $k_i = \sigma$ has occurred. This task may be one of the simplest out of the class of tasks referred to as *segmentation*. In the given case it is the *segmentation of the time interval* $(0, 1, 2, \ldots, n)$ into *subintervals* which are separated from each other by the state σ and inside them the state σ does not occur. It is a task which, because of its treacherous character, is quite near to the task concerning vertical and horizontal lines you have remembered. At first glances it seems that it is a simple and common task, such as finding locations in a text document in which a certain letter is placed. If the character sought is a space then it means a segmentation of the text into individual words.

We will formulate the task exactly and you will see that it can be solved, but its solution will require some mental effort. The set $I' \subset \{0, 1, \ldots, n\}$ will be called segmentation. Let us define a set $K(I')$ of sequences $\bar{k} = (k_0, k_1, \ldots, k_n)$ for each segmentation I'. The sequence $\bar{k} = (k_0, k_1, \ldots, k_n)$ belongs to $K(I')$, if $k_i = \sigma$ for all $i \in I'$ and, at the same time, $k_i \neq \sigma$ for all $i \notin I'$. The probability that the actual segmentation is I' is equal to the probability that the actual sequence $\bar{k}$ belongs to the set $K(I')$. Under the condition that the sequence $\bar{x}$ is known, the probability of the segmentation I' is given by the sum $\sum_{\bar{k} \in K(I')} p(\bar{k} \mid \bar{x})$. The most probable segmentation I^* is

$$I^* = \operatorname*{argmax}_{I'} \sum_{\bar{k} \in K(\bar{I}')} p(\bar{k} \mid \bar{x}) = \operatorname*{argmax}_{I'} \sum_{\bar{k} \in K(I')} p(\bar{x}, \bar{k}) \,. \tag{8.133}$$

Try to design an algorithm which for each given sequence $\bar{x}$ yields the segmentation (8.133).

I have mastered the task, but it may be the most difficult task I can still manage. I have found that the task (8.133) can be transformed to a form in which it can be solved with dynamic programming. But in setting up the particular task of dynamic programming a quite complicated algorithm is needed. This algorithm calculates the data which are the input for the task of dynamic programming itself.

First, I made clear what the function

$$F(I') = \sum_{\bar{k} \in K(I')} p(\bar{x}, \bar{k})$$

looks like which, according to the task (8.133), is to be maximised.

Since I' *is a subsequence of the ordered sequence* $I = (0, 1, \ldots, n)$, *the subsequence* I' *is also ordered. Thus it can be expressed as a sequence of indices* $(i_0, i_1, \ldots, i_q, \ldots, i_Q)$, $i_q > i_{q-1}$, $q = 1, 2, \ldots, Q$, $Q = |I'| - 1$. *Furthermore, I assumed that indices* i_0 *and* i_Q *were known, i.e.,* $i_0 = 0$ *and* $i_Q = n$. *This means that the automaton began and finished the generating of the observed sequence in the state* σ. *I do not intend to hold you back by explaining why I am entitled to have such an assumption. I claim that in this way I do not make the task narrower.*

For some given segmentation $I' = (i_0, i_1, \ldots, i_Q)$ the pair $(\bar{x}, \bar{k})$ can be expressed as a concatenation of a certain number of subsequences in the form

$$
\begin{array}{rccc}
(\bar{x}, \bar{k}) = & & & k_0\,, \\
& x_1^{i_1}\,, & k_1^{i_1-1}\,, & k_{i_1}\,, \\
& x_{i_1+1}^{i_2}\,, & k_{i_1+1}^{i_2-1}\,, & k_{i_2}\,, \\
& & \vdots & \\
& x_{i_{q-1}+1}^{i_q}\,, & k_{i_{q-1}+1}^{i_q-1}\,, & k_{i_q}\,, \\
& & \vdots & \\
& x_{i_{Q-1}+1}^{n}\,, & k_{i_{Q-1}+1}^{n-1}\,, & k_n\,.
\end{array}
$$

Because the recognised object is Markovian and because a stochastic automaton is being analysed, it follows that at a fixed state k_{i_q} the variables x_i, k_i, $i < i_q$, do not depend on the variables x_i, k_i, $i > i_q$. Therefore the joint probability of the pair $(\bar{x}, \bar{k})$ has the form of the product

$$
p(\bar{x}, \bar{k}) = p_0(k_0) \prod_{q=1}^{Q} p'_q\left(x_{i_{q-1}+1}^{i_q}\,,\ k_{i_{q-1}+1}^{i_q-1}\,,\ k_{i_q} \mid k_{i_{q-1}}\right) . \tag{8.134}
$$

In this product the number $p_0(k)$, $k \in K$, means the probability that the initial state of the automaton is k. The number $p'_q(x_{i_{q-1}+1}^{i_q}, k_{i_{q-1}+1}^{i_q-1}, k_{i_q} \,|\, k_{i_{q-1}})$ means a conditional probability of the event which under the condition that the automaton was in the state $k_{i_{q-1}}$ in the (i_{q-1})-th instant, the further $i_q - i_{q-1}$ output symbols will be $x_{i_{q-1}+1}^{i_q}$ and the automaton will pass through $i_q - i_{q-1}$ states $k_{i_{q-1}+1}^{i_q}$.

In the sum $\sum_{\bar{k} \in K(I')} p(\bar{x}, \bar{k})$, which depends on the segmentation $I' = (i_0, i_1, \ldots, i_Q)$, only such sequences $\bar{k}$ occur in which $k_{i_q} = \sigma$, $q = 0, 1, \ldots, Q$, and $k_i \neq \sigma$ for other indices i. Therefore the product (8.134) assumes the form

$$
p(\bar{x}, \bar{k}) = p_0(\sigma) \prod_{q=1}^{Q} p'_q\left(x_{i_{q-1}+1}^{i_q}\,,\ k_{i_{q-1}+1}^{i_q-1}\,,\ \sigma \mid \sigma\right) . \tag{8.135}
$$

This product is to be summed over all sequences of the set $K(I')$. This means that the summation $\sum_{k \in K(I')}$ must be performed, i.e., the multi-dimensional sum

$$
\sum_{k_1^{i_1-1}} \sum_{k_{i_1+1}^{i_2-1}} \cdots \sum_{k_{i_{q-1}+1}^{i_q-1}} \cdots \sum_{k_{i_{Q-1}+1}^{n-1}} . \tag{8.136}
$$

This summation of the products (8.135) has the same form as, e.g., the sum

$$
\sum_{z_1} \sum_{z_2} \cdots \sum_{z_m} \prod_{i=1}^{m} \varphi_i(z_i)
$$

of products $\prod_{i=1}^{m} \varphi_i(z_i)$ *which expresses the same as the product* $\prod_{i=1}^{m} \sum_{z_i} \varphi_i(z_i)$ *of the sums* $\sum_{z_i} \varphi_i(z_i)$. *The equality*

$$\sum_{z_1} \sum_{z_2} \cdots \sum_{z_m} \prod_{i=1}^{m} \varphi_i(z_i) = \prod_{i=1}^{m} \sum_{z_i} \varphi_i(z_i)$$

is universally correct for any functions φ_i, $i = 1, 2, \dots, m$. *Therefore the function* $F(I')$ *which is expressed as the sum (8.136) of the products (8.135) will assume the form*

$$\begin{aligned} F(I') &= \sum_{\bar{k} \in K(\bar{I}')} p(\bar{x}, \bar{k}) \\ &= p_0(\sigma) \prod_{q=1}^{Q} \sum_{k_{i_{q-1}+1}^{i_q - 1}} p'_q \left(x_{i_{q-1}+1}^{i_q},\, k_{i_{q-1}+1}^{i_q - 1},\, \sigma \,|\, \sigma \right) . \end{aligned} \tag{8.137}$$

We will introduce a more general denotation $p'_{ij}(x_{i+1}^{j}, k_{i+1}^{j-1}, \sigma \,|\, \sigma)$ *for the probability that under the condition* $k_i = \sigma$ *the automaton will generate the sequence* x_{i+1}^{j} *and will pass through the sequence of the states* (k_{i+1}^{j-1}, σ). *Therefore the denotation* p'_q *used will now be written as* p'_{i_{q-1}, i_q}. *Each factor in the product (8.137) has now the form*

$$\sum_{k_{i+1}^{j-1}} p'_{ij} \left(x_{i+1}^{j},\, k_{i+1}^{j-1},\, \sigma \,|\, \sigma \right) \tag{8.138}$$

and depends only on indices i *and* j. *The sum (8.138) evidently does not depend on the subsequence* k_{i+1}^{j-1} *because the sum is taken over the set of these subsequences. The sum (8.138) does not depend on the subsequence* x_{i+1}^{j} *because in each calculation this subsequence is fixed. The denotation* $\Phi(i,j)$ *will be introduced in (8.138), i.e.,*

$$\Phi(i,j) = \sum_{k_{i+1}^{j-1}} p'_{ij} \left(x_{i+1}^{j},\, k_{i+1}^{j-1},\, \sigma \,|\, \sigma \right) . \tag{8.139}$$

In the previous sum the summation is taken over all sequences k_{i+1}^{j-1}, *in which* σ *does not occur. The expression (8.137) for* $F(I')$ *can be written more concisely by using the introduced symbol* $\Phi(i,j)$,

$$F(I') = p_0(\sigma) \prod_{q=1}^{Q} \Phi(i_{q-1}, i_q) . \tag{8.140}$$

In this way I have succeeded in decomposing the original task into two separate tasks. In the former the value $\Phi(i,j)$ *for each pair of indices* (i,j), $i = 0, 1, \dots, n-1$, $j = 1, 2, \dots, n$, $i < j$, *is calculated according to the definition*

(8.139). In the latter task the segmentation I^* is found, i.e., the sequence $i_0^*, i_1^*, \ldots, i_Q^*$ (with Q not known beforehand) which minimises (8.140),

$$\begin{aligned} I^* &= (i_0^*, i_1^*, i_2^*, \ldots, i_{Q-1}^*, i_Q^*) \\ &= \operatorname*{argmax}_{i_1, i_2, \ldots, i_{Q-1}} \prod_{q=1}^{Q} \Phi(i_{q-1}, i_q) \end{aligned} \tag{8.141}$$

under the condition $0 = i_0 < i_1 < i_2 < \cdots < i_{Q-1} < i_Q = n$.

The calculation procedure for the first task is similar to the procedures you quoted in the lecture. Let $\Phi'(i,j,k)$ be auxiliary values defined by the expression

$$\Phi'(i,j,k) = \sum_{k_{i+1}^{j-1}} p'_{ij}(x_{i+1}^j, k_{i+1}^{j-1}, k \mid \sigma)\,.$$

The values $\Phi(i,j)$ sought are $\Phi'(i,j,\sigma)$. The values $\Phi'(i-1,i,k)$ are calculated according to the following recursive formula

$$\Phi'(i,j,k) = \sum_{k' \neq \sigma} \Phi'(i,j-1,k')\, p_j(x_j, k \mid k')\,. \tag{8.142}$$

The calculation begins with the values $\Phi'(i-1,i,k)$ which for each i are the known probabilities $p_i(x_i, k \mid \sigma)$ characterising the automaton. The calculation of the collection of numbers $\Phi'(i,j,k)$ for all triplets (i,j,k) according to the formula (8.142) has a complexity $\mathcal{O}(n^2\,|K|^2)$. The values $\Phi'(i,j,k)$ have been calculated, thus we know the numbers $\Phi(i,j) = \Phi'(i,j,\sigma)$.

Now I will search for the segmentation I^* satisfying the requirement (8.141). Let $i^* > 0$ be the chosen index and $J(i^*)$ be a set that contains all sequences of the form

$$i_0, i_1, i_2, \cdots, i_Q\,, \quad 0 = i_0 < i_1 < i_2 < \cdots < i_{Q-1} < i_Q = i^*\,.$$

Each sequence of this type is characterised by the number $\prod_{q=1}^{Q} \Phi(i_{q-1}, i_q)$. I will denote the largest of them $F^*(i^*)$,

$$F^*(i^*) = \max_{(i_0, i_1, \ldots, i_Q) \in J(i^*)} \prod_{q=1}^{Q} \Phi(i_{q-1}, i_q)\,. \tag{8.143}$$

Let $i_0^*, i_1^*, \ldots, i_{Q-1}^*, i_Q^*$ be a sequence which maximises (8.143). I will denote the last but one index i_{Q-1}^* in the sequence by the symbol $\operatorname{ind}(i^*)$. The last index i_Q^* is, as was said, i^*. I define $F^*(i^*) = 1$ for $i^* = 0$. The following holds *for $i^* > 0$ in the general case,*

$$\begin{aligned} F^*(i^*) &= \underbrace{\max_{i_1} \cdots \max_{i_{Q-1}}}_{0<i_1<\ldots<i_{Q-1}<i^*} \prod_{q=1}^{Q} \Phi(i_{q-1}, i_q) \\ &= \max_{i_{Q-1}<i^*} \underbrace{\max_{i_1} \max_{i_2} \cdots \max_{i_{Q-2}}}_{0<i_1<i_2<\cdots<i_{Q-2}<i_{Q-1}} \Phi(i_{Q-1}, i^*) \prod_{q=1}^{Q-1} \Phi(i_{q-1}, i_q) \\ &= \max_{i_{Q-1}<i^*} \Phi(i_{Q-1}, i^*)\, F^*(i_{Q-1})\,. \end{aligned}$$

The previous expression will be written in a briefer manner

$$F^*(i^*) = \max_{i<i^*} \Phi(i, i^*)\, F^*(i)\,. \tag{8.144}$$

I will define the index $\mathrm{ind}(i^*)$ *as*

$$\mathrm{ind}(i^*) = \operatorname*{argmax}_{i<i^*} \Phi(i, i^*)\, F^*(i)\,. \tag{8.145}$$

By means of expressions (8.144) and (8.145) I gradually calculate $F^*(1)$, $F^*(2)$, $\ldots$, $F^*(n)$ *and* $\mathrm{ind}(1)$, $\mathrm{ind}(2)$, $\ldots$, $\mathrm{ind}(n)$ *with complexity* $\mathcal{O}(n^2)$. *The index* $\mathrm{ind}(n)$ *will be the last but one index* i_{Q-1} *in the sought result of the segmentation. The index* $\mathrm{ind}(i_{Q-1})$ *determines* i_{Q-2}, *and so by means of indices in the array* ind *I am passing to smaller indices, and this I am doing until I come across the index* 0.

I believe that I have managed the task, at last. But I am sure that I have now really reached the limits of my capabilities.

Well, you wished so yourself. We wonder if you are able to formulate a task that you probably will not be able to master.

I realise quite clearly now that we dealt with the most primitive variant of the segmentation task. For actual applications the tasks should be formulated with far greater care. From these lectures I have learned that seeking the most probable value of a hidden parameter of an object seems to be natural only at first glance. In the first, usually evident, serious plunge into the application task the roughness of such an attitude is already obvious.

Seeking the most probable segmentation would result from an unmentioned assumption that all deviations of the estimated segmentation from the actual one are equally significant. This is, however, an unforgivable simplification of a task. For example, if the algorithm has wrongly decided that the automaton, at a certain instant, has passed through the state σ, *this error has smaller or greater significance according to what the actual state of the automaton was, or whether the automaton was in the state* σ *at a not distant time, etc..*

So the solution of an applied problem must begin with the careful definition of penalties $d(I', I'')$, *which estimate how dangerous the situation is when the* *segmentation I'' is assumed instead of actual segmentation I'. I can guess that a mere calculation of the penalty for the given pair of the segmentations I', I'' will be devilishly difficult. The mathematical expectation of such reasonably defined penalties, i.e., the risk, must be thoroughly analysed. Even this may mean considerable effort. When it is done then the optimisation problem that seeksthe segmentation minimising risk can be solved. The algorithm obtained in such a way could be considered as quite a good achievement.*

At last we can see you to be such as we were used to. Before, it seemed to us for a while that somebody else was discussing instead of you. You may, sometime, manage to master even the tasks you see now.

I doubt it. Not because I would underestimate myself. There is a more serious reason here, which lies in my respect for the community who are engaged in pattern recognition (especially in image segmentation) and who do not go into these highly interesting tasks. This may not be just by chance.

In the core of all the difficult tasks you quoted in the lectures, as well as of those we came across together in our discussions, there appears to be an irrefutable fact that the most probable value of a random variable is not identical with its actual value. *Should a certain feature of the random object be estimated, not only the most probable object is to be taken into consideration but also other objects having smaller probabilities. Even when you have been repeating this idea since the first lecture, I have actually understood it only now.*

Why, in the great majority of research work and application tasks, is nothing else done than seeking the most probable sequences of hidden parameters? The sequences found are then manipulated as if they were real ones. In the case in which this starting point was correct, all the tasks we have analysed, including the difficult segmentation task, would then be reduced to a single task seeking the most probable sequence. But such a procedure is erroneous. But when nearly everybody does so there must be a more serious explanation than merely stating that it is a wrong procedure. Lacking a proper explanation, I do not dare to leave the smooth path along which everybody walks, and take up other paths, along which nobody has walked so far. You may think me to be too conservative, but anyhow, conservatism is not always the worst virtue. In the present case it is respect for the well known ways in pattern recognition and a fear of that being destroyed which has already been achieved.

Your respect for the established views impresses us, and therefore we were thinking about your question for rather a long time, but we have not arrived at anything convincing. But we recalled an old joke which comes from the Ukrainian city of Odessa. In addition to its many beautiful sights, this city is known for the famous brilliance of spirit of its inhabitants. Among many stories coming from Odessa you could find this one also.

A person had a pair of trousers made at a tailor. The tailor finished the work ordered only after a week's time. It seemed to be too long for the customer and he reproached the tailor saying that one week had been enough for Almighty God to create the whole world. The tailor could only defend himself by saying: 'But, look at the world and look at these trousers'.

Well, now look at the lot of algorithms for segmentation and...

Thank you. I would also like to discuss with you the second part of the lecture devoted to learning. I am fascinated by the level of the universality and abstractness at which the tasks were examined, and by the relation between the plentiful classes of tasks being successfully revealed without creating algorithms for solving them. My attention was attracted by the relation between the minimax and the maximum likelihood estimate of the statistical model. Later, I was captivated by the quite unexpected relation between tuning the algorithm recognising Markovian sequences, on the one hand, and the perceptron or Kozinec algorithms on the other.

I do not even mention the relation between learning and unsupervised learning which I had already understood in Lecture 6. Now I have again made sure of its fruitfulness when I saw how the task estimating the Markovian object in unsupervised learning can be reduced from astonishing complexity to a task of supervised learning the complexity of which is not worth mentioning. When I see how the extensive classes of tasks start cooperating and fusing into one river, I start imagining that, at last, we hold a pneumatic drill in our hands to cope with the rock representing pattern recognition. Certainly, it is not dynamite yet, but it is already not a nail file, with which we jabbed at the rock before.

We are pleased at your enthusiasm. As we know you, we expect that now a kind of damned 'but' must follow.

Yes, you are right. It seems to me that you have thoroughly discussed a certain aspect of building up the statistical model of an object, but from my point of view, it is not the most important one. The question has remained aside of how to find the structure of a complex object. It seems to me that it is the most significant question in structural pattern recognition. I am going to explain what I mean. When we know that an object is composed of parts and we want to apply a method for recognition which was explained in the lecture, we have to order the known set of the parts so that it may become a sequence. Such an ordering is sometimes not known beforehand.

The ordering may be unknown even in the case if the sequence $k_0, k_1, k_2, \ldots, k_i, \ldots$ is a process which develops in time. The index i then represents time. For example, let k_i denote the behaviour of a person on the i-th day. Only at a rough glance one can assume that the behaviour of a person today is entirely dependent on his behaviour of yesterday. Let us imagine that a person has two ranges of interest. On week days he is at his place of work and on Saturdays and Sundays he is at his holiday home. In this case his behaviour at his holiday home on Saturday will be less affected by what he did at his work on Friday, but rather by what he had done on the previous Sunday. In *addition to the natural ordering according to time known beforehand, there is another ordering, which represents the dependence between parts of the object and which can be unknown beforehand. In the learning theory in the lecture, you stressed too strongly the estimate of numeral parameters of the statistical model of the object and entirely ignored the estimate of mutual dependencies between parts of the object. It seems to me that the structure, which has not been considered, is the most significant matter in structural pattern recognition.*

Naturally, I tried to fill the hiatus in your explanation by myself, but I arrived at pessimistic conclusions. When formulating the task I learned that the task is equivalent to the travelling salesman problem. The number of destinations the travelling salesman is to visit is equal to the number of parts of a complex object which I would like to order into a sequence. This classic task from the theory of algorithms is known to be NP hard and so hopeless to solve exactly. The total outcome of my effort is apparently negative. I would like to ask you if the situation is really so hopeless or whether I can still hope that our tasks are specific in some aspect so that they do not correspond to the general travelling salesman problem, but to its particular case which can be mastered in practice.

We are afraid that the situation is quite hopeless. We will not even ask you how you formulated the task that you have arrived at in the travelling salesman problem. We have tried ourselves to solve the task many times, but not a single time did we manage to avoid either the travelling salesman problem or the task seeking the Hamiltonian cycle in the graph which is, from the point of view of its solution, also hopeless.

Forgive me, I do not know what the Hamilton cycle is. Could you explain it to me?

A task is usually formulated by a simple example. Assume that you are inviting a set I of people to a banquet. You know about each pair of guests whether they are acquainted with one another. This knowledge can be expressed by an unoriented graph whose vertices are formed by the set I. Two vertices in a graph are connected by an edge if and only if they represent a pair of guests who are acquainted with each other. You are to seat the guests around a round table so that each guest is acquainted with the guest on left-hand, as well as with his or her right-hand neighbour. In terms of graphs, you are to find if the graph contains a cycle which passes through each vertex just once, and along each edge at most once. The graphs containing such a cycle are called Hamiltonian graphs.

Such a trifle cannot be solved?

It is even worse. So far, nobody has solved the task yet in polynomial time, but nobody has proved that it is not solvable in the polynomial time either. But we frankly advise you, do not try to solve it. It is an abyss similar to Fermat's last theorem.

With the important difference that Fermat's last theorem has already been solved.

We did not even know that. Well, let us wait about three hundred years until the situation with NP-complete problems is cleared up.

And what connects the travelling salesman problem with Hamilton cycles?

Let us assume that a non-negative real function is defined on a set of edges of the graph. The value of that function indicates the length of the particular edge. The travelling salesman problem represents seeking the Hamilton cycle which minimises the sum of the lengths of the graph edges from which it is set up. At first glance the travelling salesman task seems to be far more difficult than the Hamilton cycle task, but in fact they are tasks of the same order of complexity.

But I have come across a quite different task. My task is not reduced to seeking a cycle but to seeking for a chain which passes through all graph vertices. Maybe, in this case, the task is not hopeless from the point of view of complexity.

These two tasks are, from a complexity point of view, equivalent.

I resent it extremely. The reduction of the original task concerning the structure of a complex object to the travelling salesman problem was not simple. The blown bubble has at last burst.

You would resent it even more if you learned that you had been within reach of a very beautiful task concerning an estimate of the complex object structure which is solvable. From the beginning you have been tied up to an idea that the structure sought must be a chain. Recall that in the lecture a whole section was purposely devoted to a more general structure...

I have got it now! All results of the lecture come in useful even in the case in which the object sought has the structure of an acyclic graph. Now, I should generalise the tasks even more, so that with respect to the training multi-set not only numerical parameters of the statistical model of the object, but also its structure, i.e., the mutual relations of the parts, may be estimated. When I do not require that the structure should correspond to the chain then I probably may come to a solvable task.

Yes, that is right. Now, do not hurry because you have come across a task the solution of which is made possible thanks to a nearly hundred years effort. In 1968 the American Chow [Chow, 1965; Chow and Liu, 1968] formulated a task which we now understand as an estimate of the structure of mutual dependences between parts of a complex object. He also demonstrated that the task was equivalent to a task formulated in 1889 by the mathematician Cayley in graph theory [Cayley, 1889]. Cayley task was, for a long time, considered a difficult one. Nearly forty years later it was cleverly solved by the Moravian mathematician Otakar Borůvka [Borůvka, 1926]. Unfortunately his solution did not become generally known, and even long after the appearance of Borůvka's article other algorithms for solving Cayley task were published. Today, the three streams which are formed by the research work of Cayley, Borůvka, and Chow, are fortunately united. But yet, not before long, we could see articles which showed that some authors did not know the results mentioned above.

And now, carefully formulate and examine the proposed task, i.e., repeat what Chow had done.

Let I be a finite set of indices by which individual parts of the complex object examined are indexed. The index $i \in I$ is understood as the number of a particular part. Each part is described by two parameters: k_i is the hidden parameter and x_i is the observable parameter of the i-th part of the complex object. The ensemble $\bar{k} = (k_i\,,\; i \in I)$ is the unobservable state of the object and the ensemble $\bar{x} = (x_i\,,\; i \in I)$ is the result of its observation. As before, we assume that the parameters k_i, $i \in I$, take the value from the finite set K, and the parameters x_i, $i \in I$, do so from the finite set X.

The set I is understood as a set of vertices on which an unoriented acyclic continuous graph is created. The set of the graph edges will be denoted G. The notation $(i,j) \in G$ means that the vertices i and j, $i \in I$, $j \in I$, are connected by the edge from G. We assume that with a fixed parameter k_i, corresponding to the i-th part, the observable parameter does not depend on any other parameter of the object. This means that x_i is conditionally independent (at fixed value k_i) on any other parameter of the object. So, $p(\bar{x} \mid \bar{k}) = \prod_{i \in I} p_i(x_i \mid k_i)$. The probability distribution $p(\bar{k})$ is assumed to be Markovian with respect to the graph the edges of which are formed by the set G,

$$p(\bar{k}) = \frac{\prod_{(i,j)\in G} g_{ij}(k_i, k_j)}{\prod_{i\in I} (g_i(k_i))^{h_i - 1}}\,, \tag{8.146}$$

where h_i denotes the number of edges that pass through the vertex i. The joint probability $p(\bar{x}, \bar{k})$ is

$$\begin{aligned} p(\bar{x}, \bar{k}) &= \frac{\prod_{(i,j)\in G}\; g_{ij}(k_i, k_j)\, \prod_{i\in I} p_i(x_i \mid k_i)}{\prod_{i\in I} (g_i(k_i))^{h_i - 1}} \\ &= \frac{\prod_{(i,j)\in G}\; g_{ij}(k_i, k_j)\, \prod_{i\in I} p_i(x_i \mid k_i)\; g_i(k_i)}{\prod_{i\in I} (g_i(k_i))^{h_i}} \\ &= \frac{\prod_{(i,j)\in G}\; g_{ij}(k_i, k_j)\, \prod_{i\in I} p_i(x_i, k_i)}{\prod_{i\in I} (g_i(k_i))^{h_i}}\,. \end{aligned} \tag{8.147}$$

In the formulæ (8.146) and (8.147) the value $g_{ij}(k, k')$, $k \in K$, $k' \in K$, means the joint probability that the i-th hidden parameter will assume the value k

and the j-th parameter the value k'. The probability $p_i(x,k)$, $x \in X$, $k \in K$, represents the joint probability of the value x of the i-th observable parameter and the value k for the i-th hidden parameter. And eventually, $g_i(k)$, $k \in K$, is the probability that the i-th hidden parameter will assume the value k. This means that

$$g_i(k) = \sum_{k' \in K} g_{ij}(k,k')$$

for all such j, that $(i,j) \in G$, and

$$g_i(k) = \sum_{x \in X} p_i(x,k)\,.$$

The statistical model of the object examined is characterised by the ensemble of functions $(p_i,\ i \in I)$, the set of edges G forming the acyclic structure, and the ensemble of functions $\big(g_{ij},\ (i,j) \in G\big)$. If the statistical model is not known then it is to be estimated on the basis of experimental examining of the object. I will be concerned with the task of the maximum likelihood estimate of the statistical model on the basis of experiments with the object, i.e., on the basis of random data $(\bar{x}^j, \bar{k}^j)$, $j = 1,\ldots,l$, the probability distribution of which is $p(\bar{x},\bar{k})$. This means that I am solving the easiest task of those which were formulated in the lecture. As soon as I create an algorithm for solving this easiest task I can then, quite formally, build algorithms for the minimax estimate and for unsupervised learning.

Let the number $n(\bar{x},\bar{y})$ denote how many times the pair $(\bar{x},\bar{k})$ occurs in the experiment, i.e., in the training multi-set. Let

$$\alpha(\bar{x},\bar{y}) = \frac{n(\bar{x},\bar{y})}{l}\,.$$

The task solved requires me to find a set of edges of the graph G, which forms an acyclic structure, and the ensembles of functions $(g_{ij}\,,\ (i,j) \in G)$, $(p_i\,,\ i \in I)$ which maximise the probability of the results of the experiment given by the multi-set $\big((\bar{x}^j, \bar{k}^j),\ j = 1,2,\ldots,l\big)$, i.e.,

$$\big(G^*, (p_i^*, i \in I), (g_{ij}^*, (i,j) \in G^*)\big) \tag{8.148}$$

$$= \operatorname*{argmax}_{G} \max_{(p_i, i \in I)} \max_{(g_{(ij)}, (i,j) \in G)} \sum_{\bar{x},\bar{k}} \alpha(\bar{x},\bar{k}) \log p(\bar{x},\bar{k})$$

$$= \operatorname*{argmax}_{G} \max_{(p_i, i \in I)} \max_{(g_{ij}, (i,j) \in G)} \sum_{\bar{x},\bar{k}} \alpha(\bar{x},\bar{k}) \log \frac{\prod\limits_{(i,j) \in G} g_{ij}(k_i,k_j) \prod\limits_{i \in I} p_i(x_i,k_i)}{\prod\limits_{i \in I} \big(g_i(k_i)\big)^{h_i}}\,.$$

I will make several equivalent modifications of the function which is to be maximised in the formulated task (8.148)

$$\begin{aligned}
&\sum_{\bar{x},\bar{k}} \alpha(\bar{x},\bar{k}) \log \frac{\prod_{(i,j)\in G} g_{ij}(k_i,k_j) \prod_{i\in I} p_i(x_i,k_i)}{\prod_{i\in I} (g_i(k_i))^{h_i}} \\
&= \sum_{\bar{x},\bar{k}} \alpha(\bar{x},\bar{k}) \log \prod_{(i,j)\in G} g_{ij}(k_i,k_j) \\
&\quad + \sum_{\bar{x},\bar{k}} \alpha(\bar{x},\bar{k}) \log \prod_{i\in I} p_i(x_i,k_i) - \sum_{\bar{x},\bar{k}} \alpha(\bar{x},\bar{k}) \log \prod_{i\in I} (g_i(k_i))^{h_i} \\
&= \sum_{(i,j)\in G} \sum_{k_i\in K} \sum_{k_j\in K} \alpha_{ij}(k_i,k_j) \log g_{ij}(k_i,k_j) \\
&\quad + \sum_{i\in I} \sum_{x_i\in X} \sum_{k_i\in K} \beta_i(x_i,k_i) \log p_i(x_i,k_i) - \sum_{i\in I} h_i \sum_{k_i\in K} \alpha_i(k_i) \log g_i(k_i) \,,
\end{aligned} \tag{8.149}$$

where

$$\begin{aligned}
\alpha_{ij}(k_i,k_j) &= \sum_{(x_{i'},\, i'\in I)} \; \sum_{(k_{i'},\, i'\in I\setminus\{i,j\})} \alpha(\bar{x},\bar{k}) \,, \\
\beta_i(x_i,k_i) &= \sum_{(x_{i'},\, i'\in I\setminus\{i\})} \; \sum_{(k_{i'},\, i'\in I\setminus\{i\})} \alpha(\bar{x},\bar{k}) \,, \\
\alpha_i(k_i) &= \sum_{(x_{i'},\, i'\in I)} \; \sum_{(k_{i'},\, i'\in I\setminus\{i\})} \alpha(\bar{x},\bar{k}) \,.
\end{aligned}$$

The last three formulæ are presented only to demonstrate the rightfulness of the last step in deriving the (8.149). In fact, the numbers α_{ij}, α_i and β_i are not calculated according to the formulæ quoted, but it is done in a far simpler way. The number $\alpha_{ij}(k,k')$ means merely the relative frequency of the situation occurring in the experiment when the i-th hidden parameter assumed the value k and the j-th one the value k'. A similar sense is assigned also to the numbers α_i and β_i.

On the basis of similar considerations as those in Theorem 8.1 from the lecture, it can be proved that the numbers $g_{ij}(k_i,k_j)$, $p_i(x_i,k_i)$ and $g_i(k_i)$, which maximise (8.149), are to be equal to the corresponding numbers $\alpha_{ij}(k_i,k_j)$, $\beta_i(x_i,k_i)$ and $\alpha_i(k_i)$. This means that the expression (8.148) assumes the form

$$\begin{aligned}
G^* = \operatorname*{argmax}_G \Bigg(&\sum_{(i,j)\in G} \sum_{k_i\in K} \sum_{k_j\in K} \alpha_{ij}(k_i,k_j) \log \alpha_{ij}(k_i,k_j) \\
&+ \sum_{i\in I} \sum_{x_i\in X} \sum_{k_i\in K} \beta_i(x_i,k_i) \log \beta_i(x_i,k_i) - \sum_{i\in I} h_i \sum_{k_i\in K} \alpha_i(k_i) \log \alpha_i(k_i) \Bigg) .
\end{aligned} \tag{8.150}$$

In the preceding expression the first summand depends on the set G since the addition is done according to those pairs (i,j) which belong to G. So does the third summand, too, since the numbers h_i depend on the set G. The second summand does not depend on the set G, and thus the expression (8.150) can be rewritten to

$$G^* = \operatorname*{argmax}_G \left(\sum_{(i,j)\in G} \sum_{k_i\in K} \sum_{k_j\in K} \alpha_{ij}(k_i,k_j) \log \alpha_{ij}(k_i,k_j) - \sum_{i\in I} h_i \sum_{k_i\in K} \alpha_i(k_i) \log \alpha_i(k_i) \right) .$$

If I introduce the entropies

$$H_{ij} = -\sum_{k_i\in K} \sum_{k_j\in K} \alpha_{ij}(k_i,k_j) \log \alpha_{ij}(k_i,k_j) \tag{8.151}$$

and

$$H_i = -\sum_{k_i\in K} \alpha_i(k_i) \log \alpha_i(k_i) \tag{8.152}$$

then I obtain

$$\begin{aligned} G^* &= \operatorname*{argmax}_G \left(\sum_{i\in I} h_i\, H_i - \sum_{(i,j)\in G} H_{ij} \right) \\ &= \operatorname*{argmax}_G \sum_{(i,j)\in G} (H_i + H_j - H_{ij}) . \end{aligned}$$

At last, I arrived at the following procedure by which the set G is created, i.e., the structure of mutual dependencies between the parts of a complex object.

1. *With respect to the training multi-set $(k^j\,, j = 1,\ldots,l)$ (the data x^j are not used at all) the numbers $\alpha_{ij}(k_i,k_j)$ and $\alpha_i(k_i)$, $i \in I$, $j \in I$, $i \neq j$, $k_i \in K$, $k_j \in K$, are calculated.*
2. *For each pair (i,j), $i \in I$, $j \in I$, $i \neq j$, the entropy H_{ij} is calculated according to the formula (8.151) and for each $i \in I$ the entropy H_i is calculated according to the formula (8.152).*
3. *The set I is understood as a set of graph vertices on which a complete graph is created. In this graph each vertex is connected to each vertex by an edge. The length of the edge connecting the vertices i and j is given as $H_i + H_j - H_{ij}$.*
4. *In the graph obtained a connected acyclic subgraph (i.e., a tree) is to be found which contains all the vertices of the original graph and which maximises the sum of the graph edges that belong to the subgraph.*

And now I would like to know your opinion on whether this is Cayley task or not.

Yes, it is.

Could you, please, explain to me the Borůvka algorithm for solving Cayley task?

It is so simple that we can but admire its cleverness and we are astonished that people were not able to fall upon that simple solution for such a long time. In addition, we are surprised at how it may have happened that even after Borůvka's article it was not known for quite long a time that Cayley task had been solved.

Let M be a set of graph edges, i.e., pairs (i,j) of the form $i \in I$, $j \in I$, $i \neq j$. Let us order the set M according to the edge lengths; if the edge (i',j') occurs in the array M prior to the edge (i'',j'') then the length of the edge (i',j') is greater than or equal to the length of the edge (i'',j''). We seek a subset of edges which forms a connected subgraph G^* containing all the vertices I and maximises the sum of the edge lengths. The subgraph G^* is formed in the following way. One edge after another is examined in the order given by the array M and a decision is made about each current edge whether it belongs to the subgraph G^* or not.

Let the edge (i',j') from the array M be examined at a certain instant and let the subset of edges G' to have been constructed at the preceding instant. They are the edges about which the decision has already been made that they belong to G^*. If there exists a path in the subset G' from the vertex i' to the vertex j' then the decision is made that the edge does not belong to G^*, i.e., the subset G' does not change. If the subset G' contains no path from the vertex i' to the vertex j' then the decision is made that the edge (i',j') belongs to the set G^*, and so the set G' is changed into the set $G' \cup \{(i',j')\}$. The array M is examined as long as there is less than $|I| - 1$ edges in the subset G'.

The simplicity of the algorithm is incredible. Could not you, please, explain the most important ideas of Borůvka's proof to me?

We could, but we do not want to. You should master the proof on your own. Of course, it does not mean that we can equal Borůvka. Imagine it so: we are dwarfs who have climbed onto a giant's shoulders. This statement was neither invented by us nor do we remember who said it, but he must have been a clever man.

For simplicity, I dealt with the task only for the case in which the lengths of graph edges differed from each other. The proof of Borůvka algorithm is based only on two rather obvious statements.

Assertion 8.1 *Let (i_0, j_0) be the greatest edge and G^* be the set of edges sought which is the solution of Cayley task. Then $(i_0, j_0) \in G^*$ holds.* ▲

Proof. *I will prove the assertion by contradiction. I will prove that if $(i_0, j_0) \notin G^*$ then the set G^* is not a solution of Cayley task. Since the set of edges G^* forms a connected graph there exists a path from the vertex i_0 to the vertex j_0. Each edge within this path is shorter than is the length of the edge (i_0, j_0). Let (i', j') be an arbitrary edge within this path. The path from i_0 to j_0 together with the edge (i_0, j_0) creates a cycle. If I take out the edge (i', j') from the set G^* and include the edge (i_0, j_0) in it then the new graph remains to be connected and will, as it was before, contain all vertices. But the total length of its edges will increase since the length of the edge (i', j') is less than the length of the edge (i_0, j_0). From that it follows that the set G^* was not the solution of Cayley task.* ■

Assertion 8.2 *Let G^* be a set of edges that solves Cayley task and G' be its subset. Let (i_0, j_0) be an edge that satisfies two conditions:*

1. *There is no path in the set G' from the vertex i_0 to the vertex j_0.*
2. *Among all edges that satisfy the first condition, the edge (i_0, j_0) has the greatest length.*

In this case $(i_0, j_0) \in G^*$ holds. ▲

Proof. *Assume that $(i_0, j_0) \notin G^*$ (I prove it by contradiction again). Since the edges G^* constitute a connected graph there is a path from the vertex i_0 to the vertex j_0. Because of the first condition at least one edge within the path does not belong to G'. Owing to the second condition this edge is less than the edge (i_0, j_0). All other considerations are the same as those in Assertion 8.1.* ■

I believe that Borůvka algorithm itself is an interesting object for studying regardless that is solves Cayley task. How should it be arranged from the computational point of view? The question about its complexity should be asked, etc., should it not?

Borůvka algorithm has been thoroughly examined, particularly within graph theory and computational geometry.

We are pleased that you have mastered this not very easy lecture.

But still I have not yet made clear for myself a question that is important for me. I am not sure if I have understood the importance of the matrix notation which is the main thread of all the first part of the lecture. According to how you stress the application of the matrix notation, I am afraid that you see something in it I have not noticed. On account of the lecture I have not been convinced that the matrix notation is indispensable for solving the tasks. I understand and I am capable of solving all tasks analysed in the lecture without knowing anything about the matrix notation. Am I right that it is only the matter of concisely expressing something that is clear anyhow? And is it so essential what language or what mathematical symbolism is used for expressing knowledge? It is the knowledge itself that is important, i.e., that invariant something, which does not depend on the form in which it is written. For what do I need something new when I understand all I need even without it?

Your question can be understood both in a narrower and a broader sense. The matrix expression of tasks and algorithms of their solution was used by us as one of the possible ways. After all, we write our lectures not only for you, even though we are very pleased by the collaboration with you. We are going to publish the lectures and we would like them to be understood by the largest possible body of readers. Some one better understands a graph interpretation of the task, another is fond of a statistical interpretation, the third may think the matrix notation to be most understandable. And it is the matrix expression in which we can see at least something common to all the tasks analysed.

Unlike you, we consider it very important in what language the knowledge is expressed. Perhaps, the main attribute of the immaturity of today's pattern recognition, and perhaps even the cause of the immaturity, is that recognition has not yet created a language of its own. We are getting now to your question in the broader sense, whether it is important at all what language the knowledge is expressed in. You yourself answer this question definitively in negative. We have long thought about it and the result is that we do not agree with you. We would be pleased if we could convince you, and you moved your views closer to ours.

Let us have an example which is quite closely related to us now. For writing Borůvka algorithm to solve Cayley task we needed not even one half of a page. Your correct proof of their algorithm took less than a page. Approximately as briefly as that, this subject matter is explained in modern textbooks on graph theory as well. And now an interesting question arises. What is it in the article by Borůvka that he needed 16 pages for interpreting the results? We have read his article and we also advise you to do so, as it is instructive. First, you will find that there is nothing more in the article than the algorithm described. Second, and this is most important, when you start reading you will get the first impression that we were mistaken and referred you to another article, which has nothing in common with Cayley task and the beautiful algorithm for its solution which you are acquainted with now. The reason for it is that in not a single case in the whole article is the word graph used, and entirely different apparatus is applied in writing it behind which it is not easy to see the results that are essential for the article.

It is important that it might have been because of that particular fact that the results had been written in an inconvenient language (of course, we do not mean by it Czech or any other natural language) that the results remained hidden for such a long time. These results recurred, eventually, with other scientists. When Borůvka's article was rediscovered its importance was more historical than scientific. It is a situation similar to that in discovering America, which was mentioned in Lecture 6.

In the development of any scientific branch a period sometimes appears when acquiring new knowledge is quite impossible. The reason for it is that no corresponding symbolics is created, or, if you choose, an adequate language is still missing for a concise interpretation of the knowledge already available. At this stage, creating a new language tool for the interpretation of the old knowledge may be more important than acquiring new knowledge. Only when the old knowledge is expressed in a new terminology and people get used to the new terms do they begin wondering why such simple things had seemed so complicated before.

Your negative attitude to matrix notation reminds us of another circumstance which necessarily accompanies the creation of new tools for interpreting old knowledge. The mere existence of the language might affect the range of knowledge which becomes generally known. Before new language tools are created the greatest popularity is attained by that knowledge which had proved successful in being expressed in the old language. All of that shows a certain infirmity of new language tools, since what has been known best does not sound very familiar in the new language. And, in addition, we have to take into consideration the effort which is needed for mastering the new language. You have nicely expressed this situation in your words 'why do I need anything new when I understand everything I need even without that'.

We have also thought about how it could happen that only not very long ago, perhaps about a thousand years ago, that people were not able to add and multiply arbitrary integer numbers. It was not because the overall level of education would have been low and therefore the majority of population would not have known how to multiply. The situation was more complex. Nobody could operate with arbitrary integer numbers. The capability of adding some large numbers was regarded as an attribute of the highest intellect. The tasks concerning the addition and multiplication of individual numbers or of some number classes became respected scientific research. From the present day point of view it can hardly be understood how it could happen that people were capable of adding and multiplying some numbers but others not. Though, in fact, the procedure through which these operations are performed is the same for all numbers. The explanation that our ancestors were less intelligent than we are would be wrong. Even though the human society is, as a whole, a bit more educated than a thousand years ago, there is no evidence that every individual today would be more capable of mental activity than were his grand or great grandparents. Let us recall the brilliant outcomes of European ancient mathematics from which such concepts as the prime number, the highest common devisor, the lowest common denominator, etc. have their origin. All this gives tribute to a deep understanding for the nature of the integer number. And in spite of all that, people did not know how to add and multiply for long after the time when the concept of the number had been understood rather clearly.

It was because that people did not represent numbers in the unified form to which we are accustomed now and know as the Arabic notation of numbers. The mode of notation was different with different nations, and even with different classes of numbers. Of the earlier forms, Roman numerals are used up to today. In such a mess and disorder in the number representation itself, hardly anybody got the idea that there might be a universal way of manipulating numbers. Well, all the numbers seemed to bear no resemblance to one another. Does not this situation remind you of the present day state of affairs in pattern recognition?

It does, a little, but I wonder what all these historical considerations lead to, all the more so that they are not quite correct. As early as many thousands of years ago, in ancient Mesopotamia, people could manipulate integer numbers quite correctly. It was far earlier than you say.

Do not be very strict this time. The matter is that you and we are not doing historical research, but something quite different. It is important that only not very long ago, about one thousand years ago, people could not manipulate integer numbers in quite extensive territories (when once you are so strict to us).

The mess concerning integer numbers prevailed even in the Central Asian science center, Samarkand, until about a thousand years ago Muhammad from the neighbouring Khwarizmi came to Samarkand and notified them that a new way of number representation was used in Khwarizmi. Thanks to it, the capability of performing mathematical operations ceased to be regarded as an exceptional gift of Nature for some intellectuals and became accessible for any young boy of the street. Muhammad ibn Musa al-Khwarizmi explained the way of representing numbers and calculating with them which has been used in the world up to now. In addition to the facility to calculate, which affected the development of science all over the world for centuries, there was another outcome. The manner started by Muhammad al-Khwarizmi, by which ingenious intellectual inventions can be replaced by disciplined executing of unambiguously formulated regulations, was quite new. In honour of al-Khwarizmi (perhaps also in honour of the country where he came from) the formulated rules began to be called the Khwarizmian way, or the al-Khwarizmi method. Owing to later distortions of the expression the word algorithm originated. You can see in what way significant outcomes in the history of science can originate even because the objects known before were newly expressed or given a new name. And therefore we cannot agree with you that it is of no significance in which formalism the new, as well as the old, knowledge is expressed.

And there is something else we would like to add. The personality of Muhammad al-Khwarizmi is so great that a mere ambition to make him one's example could be regarded as an unforgivable immodesty. In spite of that, try to imagine yourself in his place. If you put yourself in his situation in a quite realistic way, and in all inevitable details then you will find that from the standpoint of al-Khwarizmi, his situation appeared more than ugly. The poor al-Khwarizmi must have listened to a pretty large amount of foolishness in his life.

Someone may have disliked the representation of numbers. For example, the representation of the number 247 seemed to be far less illustrative than CCXLVII. Well, even from the notation CCXLVII one can see that it is a sum C+C+(L−X)+V+I+I, but the number 247 does not say anything as that. To find out what it means, it is necessary to calculate $2 \cdot 10^2 + 4 \cdot 10^1 + 7 \cdot 10^0$. Well now, our colleague al-Khwarizmi, instead of multiplying only when we need it, we will now have to multiply every time when we want only to know what the number means! To some other person, the new way of adding seemed to be far more complicated than the previous one. The fact that the sum of the numbers

V and II is equal to VII is far more understandable than stating that the sum of 5 and 2 is 7. The numeral 7 itself does in no way include the numerals 5 and 2. Yet another person in turn criticised the new way because it needed 10 numerals, 0, 1, 2, 3, 4, 5, 6, 7, 8, 9, whereas for the old way only VII numerals I, V, X, L, C, D, M had been sufficient. It is true, though, that by means of Roman numerals the numbers greater than, e.g., MMDCCCLXXXVIII are difficult to express, but for a great majority of practical applications it is sufficient.

Furthermore, nobody can remember the rules of multiplying numbers. To multiply, one has to know by heart about 100 rules, i.e., the multiplication tables. When it is necessary to know 100 rules then any additional rules seem to be useless since the products of numbers which occur in practice can be easily calculated by common sense. If the poor al-Khwarizmi asked in a shy manner how much to pay for CCLXXIV rams when each costs XLIX ducats then everybody would wonder how it was that al-Khwarizmi was so silly and out of touch of the real life. Has anybody seen CCLXXIV rams for sale? In practice there are always either CC or CCC, or CCL rams, at worst. I wonder where you saw such a silly price, XLIX ducats for one ram? In practice it is always L ducats, and therefore these fine and rather complicated considerations are unworkable. In practice, I am to multiply CC times L, and without all wisdom of yours, my colleague al-Khwarizmi, I know that it will be C times C ducats. At last, colleague al-Khwarizmi, come to see that we cannot reckon on being able to retrain all the merchants in Samarkand to use the new way of writing numbers only for the purpose of making their addition and multiplication easier for you. The whole scientific community of ours is kept by virtue of taxes and charity benefits from the merchants. They are here not for the sake of us, but we are here for the sake of them...

We could continue ad infinitum. We need not even invent silly stories like that. Each of us can hear a lot of them around.

I admit that I was wrong saying that it was of no importance in which formalism knowledge was expressed.

But I would like, in a similar way like you, to add something. The first part of your answer sounded like a beautiful poem. But I understood the second part of it as an irony directed towards me. My attitude is, in fact, nearer to those imaginary blockheads, who did not understand al-Khwarizmi, than to his views. I am not very proud of it, but I do not feel like opposing it either. Therefore, I am saying again that all tasks the solution of which was expressed by you by matrix products can be proved by me even without using them. Could you quote an example of a task that is solved within the formalism mentioned, but is difficult to solve without it?

We know one of such tasks. You yourself may guess that it must be a pretty difficult task. We are planning it for our next lecture.

April 1998

8.13 Link to a toolbox

The public domain Discrete Hidden Markov Models Toolbox was written by J. Dupač as a diploma thesis in Spring 2000. It can be downloaded from the website `http://cmp.felk.cvut.cz/cmp/cmp_software.html`. The toolbox is built on top of Matlab version 5.3 and higher. The source code of the algorithms is available.

The part of the toolbox which is related to this lecture implements algorithms for recognition, supervised and unsupervised learning, which work with Discrete Hidden Markovian Models of sequences and acyclic structures.

8.14 Bibliographical notes

The lecture has dealt with hidden Markovian processes. The idea of this simplified statistical model was introduced by Markov [Markov, 1916]. He first analysed 20,000 words of Pushkin's novel in verse Eugen Onegin. The basic tool used in the lecture was dynamic programming [Bellman and Dreyfus, 1962]. Kovalevski formulated and solved a task of recognising a non-segmented row of letters and used dynamic programming for recognising image sequences [Kovalevski, 1967], [Kovalevski, 1969]. Chazen (she) [Chazen, 1968] was engaged, in the general form, in recognising Markovian sequences by means of dynamic programming. Vincjuk [Vincjuk,] used the Markovian sequence for speech recognition.

The recognition algorithm yielding the sequence of most likely states for a sequence of observation assuming a Markovian statistical model is also known under the name of Viterbi algorithm [Viterbi, 1967]. The training of the statistical model is known as Baum–Welsh algorithm [Baum et al., 1970].

The formalism of generalized matrix products was used by [Aho et al., 1975] for a unified analysis of algorithms. Seeking the best approximation of a multi-dimensional random variable by an acyclic graph was published by Chow [Chow, 1965; Chow and Liu, 1968]. He took into consideration common dependencies of an object, he posed the questions of what the tree should look like and how to find it. Chow reduced this task to the known task of the least continuous subgraph of a graph, i.e., to Cayley task [Cayley, 1889]. The solution of this task belongs to the mathematician Borůvka [Borůvka, 1926] from Brno. Later this solution was many times repeated by other authors, probably the best known work is the work by Kruskal [Kruskal, 1956].

We are not aware of studies which would bring statistical and structural pattern recognition methods together. A certain space is devoted to these problems in Chapters 5, 6 and 7 in the book by Fu [Fu, 1974].

Lecture 9

Regular languages and corresponding pattern recognition tasks

9.1 Regular languages

The previous lecture has been devoted to the analysis of the following model of a recognised object.

Let X and K be two finite sets. The first of them is an alphabet of output symbols and the second is a set of states of an autonomous stochastic finite automaton. The automaton is characterised by the function $p\colon K \times X \times K \to \mathbb{R}$, which describes its behaviour in the following way. The automaton passes through a sequence of states $k_0, k_1, \ldots, k_i, \ldots$ and it produces a sequence of symbols $x_1, x_2, \ldots, x_i, \ldots$ at the output. The initial state of the automaton k_0 is random and in agreement with the probability distribution

$$\sum_{k \in K} \sum_{x \in X} p(k_0, x, k) \, .$$

If the automaton is in the state k_{i-1} in the $(i-1)$-th instant, $i = 1, 2, \ldots$, then a random pair (x_i, k_i) is generated in agreement with the probability distribution

$$\frac{p(k_{i-1}, x_i, k_i)}{\sum\limits_{x \in X} \sum\limits_{k \in K} p(k_{i-1}, x, k)} \, ,$$

the symbol x_i appears at the output, and the automaton passes to the state k_i. The function p characterises the stochastic automaton and it defines the probability distribution on the set of the output sequences of the automaton. Among all possible sequences of the length n, the probability of the sequence $x_1, x_2, \ldots, x_n$ is given by the sum

$$\sum_{k_0} \sum_{k_1} \cdots \sum_{k_n} \left(p(k_0, x_1, k_1) \prod_{i=2}^{n} \frac{p(k_{i-1}, x_i, k_i)}{\sum\limits_{x \in X} \sum\limits_{k \in K} p(k_{i-1}, x, k)} \right) .$$

At the same time the automaton described defines a more rough characteristic of the set of sequences which may occur at its output, i.e., the set of the sequences with non-zero probability. If we are interested only whether the sequence $x_1, x_2, \ldots, x_n$ may occur at the automaton output, and the probability of that sequence is of no interest, such a detailed characteristic, as that described by the function $p\colon K \times X \times K \to \mathbb{R}$, is not needed. It is sufficient to know if the probability $p(k', x, k'')$ of the triplet $k' \in K$, $x \in X$, $k'' \in K$ is zero. The automaton can be described in less detail by the function P of the form $K \times X \times K \to \{0, 1\}$ the value $P(k', x, k'')$ of which is 0 if $p(k', x, k'') = 0$, and 1 if $p(k', x, k'') \neq 0$. Thus the subset of the sequences is determined by a simple binary function of three variables. Naturally, some sequence subsets cannot be defined in this way. It only concerns subsets of a certain form known as *regular languages*. We will introduce this concept more precisely.

Let X be a finite set which will be called an *alphabet*. Its elements are *symbols*. The finite sequence of symbols from the alphabet X is referred to as a *sentence* in the alphabet X. The set of all possible sentences in alphabet X will be denoted X^*. The subset of sentences $L \subset X^*$ is called a *language* in the alphabet X.

Let K be a finite set which will be called the *alphabet of the automaton states*. Let the function $\varphi\colon K \to \{0, 1\}$ define the subset of states which are regarded as the *initial states* of the automaton. If $\varphi(k) = 1$ then it means that the state k is one of the initial states (it can be a single one). Let the function $\psi\colon K \to \{0, 1\}$ define a set of *target states* in a similar way. Let the function $P\colon K \times X \times K \to \{0, 1\}$ be the state transition function of an automaton that has the following sense. If the automaton in the instant $(i - 1)$ occurred in the state k_{i-1} then in the succeeding i-th instant it can produce only such a symbol x_i at its output, and be in such a state k_i for which $P(k_{i-1}, x_i, k_i) = 1$ holds.

The automaton is given by the five-tuplet $A = \langle X, K, \varphi, P, \psi \rangle$. This five-tuplet unambiguously determines the sets of sequences which may occur at the automaton output. This sequence set will be called the language of the automaton A and denoted $L(A)$. A sentence $x_1, x_2, \ldots, x_n$ belongs to the language $L(\langle X, K, \varphi, P, \psi \rangle)$ if

1. $x_i \in X$, $i = 1, 2, \ldots, n$;
2. a sequence $k_0, k_1, \ldots, k_n$ exists for which the following holds:
   (a) $k_i \in K$, $i = 0, 1, \ldots, n$;
   (b) $\varphi(k_0) = 1$;
   (c) $P(k_{i-1}, x_i, k_i) = 1$, $i = 1, 2, \ldots, n$;
   (d) $\psi(k_n) = 1$.

The following equation defines the membership of the sequence to a language in a brief form

$$F(x_1, x_2, \ldots, x_n) = \bigvee_{k_0 \in K} \bigvee_{k_1 \in K} \cdots \bigvee_{k_n \in K} \left(\varphi(k_0) \wedge \left(\bigwedge_{i=1}^{n} P(k_{i-1}, x_i, k_i) \right) \wedge \psi(k_n) \right), \tag{9.1}$$

where $F(x_1, x_2, \ldots, x_n)$ is the statement 'The sequence $\bar{x} = (x_1, x_2, \ldots, x_n)$ belongs to the language $L(\langle X, K, \varphi, P, \psi \rangle)$'.

The definition (9.1) can be written in a briefer way if we accept the following interpretation of the functions φ, P and ψ. The function φ will be understood as a row vector of dimension $|K|$ the k-th component of which is $\varphi(k)$, $k \in K$. The function ψ is a column vector of dimension $|K|$ the k-th component of which is $\psi(k)$. For a given sequence $x_1, x_2, \ldots, x_n$, the matrix P_i, $i = 1, 2, \ldots, n$, will be introduced. The element in its k'-th row and k''-th column is $P(k', x_i, k'')$. Using this denotation we can express the definition (9.1) as the *matrix product*

$$F(x_1, x_2, \ldots, x_n) = \varphi \odot \left(\bigodot_{i=1}^{n} P_i \right) \odot \psi \tag{9.2}$$

on a semi-ring on the set $\{0, 1\}$ with operations $\bigvee$ (it corresponds to addition) and $\bigwedge$ (it corresponds to multiplication).

Furthermore the definition of the regular language by Equation (9.1) there is a number of other equivalent definitions which will be presented later to complete our explanation. Equation (9.1) and particularly its form (9.2) are preferred for pattern recognition purposes because they immediately give rise to an algorithm which recognises if the sentence $x_1, x_2, \ldots, x_n$ belongs to the given language $L(\langle X, K, \varphi, P, \psi \rangle)$.

The language is defined as a set of sentences for which the product (9.2) assumes the value 1. The recognition algorithm will simply calculate this product. The complexity of the calculation is $\mathcal{O}(|K|^2 \, n)$.

9.2 Other ways to express regular languages

9.2.1 Regular languages and automata

Let X be an alphabet of input symbols (not of the output ones as in the previous paragraph) which are led to the input of a finite state automaton. Let K be a set of states of this automaton and k_0 be one of the initial states. Let $q \colon K \times X \to K$ be a transition function which determines behaviour of the automaton in the following manner. If the automaton was in the $(i-1)$-th instant in the state k and in the i-th instant the symbol x appeared at its input then the state of the automaton in the i-th instant will be $q(k, x)$. Therefore if an initial state of the automaton was k_0 and the sentence $x_1, x_2, \ldots, x_n$ appeared at its input then the sequence of states $k_1 = q(k_0, x_1)$, $k_2 = q(k_1, x_2)$, $\ldots$, $k_n = q(k_{n-1}, x_n)$ is unambiguously determined.

In this way the automaton implements the mapping of the set X^* of sequences to the set of states K so that only a single state corresponds to each sequence $\bar{x} \in X^*$. The automaton traverses to this state because the sequence $\bar{x}$ appeared at its input. The mapping will be denoted $Q \colon X^* \to K$. The mapping Q is uniquely determined by the sets X, K, by the initial state k_0, and by the state transition function $q \colon K \times X \to K$.

Let $K' \subset K$ be a subset of states. The subset of sentences $\bar{x} \in X^*$ for which $Q(\bar{x}) \in K'$ is called the language which is accepted by the given automaton.

The language is given by the automaton, i.e., by the five-tuplet $\langle X, K, k_0 \in K, q\colon K \times X \to K, K' \subset K \rangle$.

Only the languages of a certain kind can be defined in this way, not arbitrary ones. It could be proved that a class of all languages which can be expressed in the way mentioned includes all regular languages, such as were defined in the previous paragraph, and does not include any language that is not regular. We do not intend to prove this assertion because our lecture should be mainly based on the definitions (9.1) and (9.2).

Let us point out an important circumstance. We have defined the regular languages in two different ways and both definitions have been based on the automaton. Even if both definitions are equivalent then the automata corresponding to them are quite different. In the first case it is an *autonomous non-deterministic automaton* which generates sentences belonging to the selected language. In the second case it is a *deterministic automaton* to the input of which arbitrary sentences are led. The automaton then separates the sentences into two classes: (a) the sentences belonging to the selected language and (b) all other sentences.

Both definitions have become a basis of extensive research. In this explanation, we exclusively use the first definition in which the language is determined by the automaton generating sentences, and not by the automaton recognising them. If we did not clearly realised this property then it could bring a lot of misunderstandings.

9.2.2 Regular languages and grammars

Let us consider an automaton $A = \langle X, K, \varphi, P, \psi \rangle$ and the corresponding regular language $L(A)$. The five-tuplet $\langle X, K, \varphi, P, \psi \rangle$ need not be understood as an automaton and the elements X, K, φ, P, ψ can have other names and other interpretations than that in Subsection 9.2.1. This interpretation corresponds to *regular grammars*. We will now present the concept of the regular language by means of different terminology and see that the equivalence of the new and the previous definitions of the regular language are almost evident.

The set X and the set K will be called a *terminal alphabet* and a *non-terminal alphabet*, respectively. The function φ will be expressed by a subset $K^0 = \{k \in K \mid \varphi(k) = 1\}$ which is called a *set of axioms*. The functions P and ψ will be expressed by the subset of triplets of the form (k', x, k''), $k' \in K$, $x \in X$, $k'' \in K$, and pairs of the form (k, x), $k \in K$, $x \in X$. These triplets and pairs are usually called substitution, assignment, grammar rules, etc.. We will use the term *rule*.

A set of rules is created according to the automaton by which the language was originally defined. If for some triplet (k', x, k'') the equality $P(k', x, k'') = 1$ holds then the triplet (k', x, k'') becomes one of the rules which is written in the form $k' \to xk''$. Furthermore if in addition to $P(k', x, k'') = 1$ the equation $\psi(k'') = 1$ is also satisfied then along with the rule $k' \to xk''$ a further rule $k' \to x$ is introduced. The set of rules obtained in this way will be denoted R. Therefore the five-tuplet $\langle X, K, \varphi, P, \psi \rangle$ which was called an automaton before has been now expressed by means of a regular grammar given by the quadruplet $\langle X, K, K^0, R\rangle$, where

X is a terminal alphabet,

K is a non-terminal alphabet,

K^0 is a set of axioms,

R is a set of rules.

The regular grammar determines the *language*, i.e., the subset $L \subset X^*$ in the following manner.

1. The sentence consisting of one single symbol which corresponds to one of the axioms is considered to be proved in the given grammar.
2. If some sentence $\bar{x}k'$, $\bar{x} \in X^*$, $k' \in K$, is proved in the grammar and the set of rules contains the rule $k' \rightarrow x'k''$ then the sentence $\bar{x}x'k''$ is considered to be proved.
3. If a sentence $\bar{x}k'$, $\bar{x} \in X^*$, $k' \in K$, is proved in the grammar and the set of rules contains the rule $k' \rightarrow x'$ then the sentence $\bar{x}x'$ belongs to the language defined by the given grammar.

Once we have obtained the grammar $\langle X,\ K,\ K^0,\ R\rangle$ from the automaton $\langle X,\ K,\ \varphi,\ P,\ \psi\rangle$ in the above manner then the language defined by the grammar is just the set of sentences that can occur at the output of the automaton. We will not prove this assertion because it is almost obvious. Also a converse transition is possible, i.e., from a regular grammar to an automaton which will generate all sentences belonging to the language of the given grammar and only them. The autonomous finite automaton and regular grammars are two equivalent means for defining sets of a certain kind, namely, regular languages.

9.2.3 Regular languages and regular expressions

We will introduce the following three operations on a set of languages.

1. *Iteration of the language* L is the language denoted L^* which originates as follows.
   (a) An empty sentence, i.e., a sentence of zero length belongs to L^*.
   (b) If the sentence $\bar{x}$ belongs to L^* and the sentence $\bar{y}$ belongs to L then the sentence $\bar{x}\bar{y}$ belongs to L^* as well.
2. *Concatenation of languages* L_1 and L_2 is a language which will be denoted L_1L_2 and which contains all the sentences of the form $\bar{x}\bar{y}$, $\bar{x} \in L_1$, $\bar{y} \in L_2$, and no other sentence.
3. *Union of languages* L_1 and L_2 is the language $L_1 \cup L_2$.

Let X be a finite set of symbols. Some languages in the alphabet X can be written by means of *regular expressions* in accordance with the following rules.

1. The symbol $\emptyset$ is one of the regular expressions and denotes an *empty set of sentences.*
2. The symbol # is a regular expression for a set that contains one sentence only, which is an empty sentence, i.e., a *sentence of zero length.*

3. For each symbol $x \in X$ the regular expression x means a language which consists of a single sentence containing a single symbol x.
4. Let α be a regular expression of the language L. Then $(\alpha)^*$ is a regular expression for the iteration of the language L.
5. Let α_1 and α_2 be regular expressions of the languages L_1 and L_2. Then the expression $\alpha_1\alpha_2$ denotes the concatenation $L_1 L_2$ of these languages and the expression α_1, α_2 denotes the union $L_1 \cup L_2$ of these languages.

For example, the expression $a(b,c)^*$ is a regular expression of a set of sentences which begins with the symbol a followed by a sequence composed of the symbols b and c of any length (it can even be an empty one).

A basic well known fact about the relation between regular expressions and regular languages is that any regular language can be expressed by means of a regular expression. Similarly, any language expressed by means of a regular expression is regular.

9.2.4 Example of a regular language expressed in different ways

We have presented four ways of how to express regular languages. We will now discuss an example which illustrates these four ways. First we will define in an informal and perhaps easy-to-follow form the language we intend to deal with. Let an alphabet X consist of symbols a, b, c, $+$, $\times$, $=$. We want to express a set of sequences which can be understood as commands in a program written in a programming language. The sequence (i.e., the command) is to be composed of two parts, the left one and the right one, which are separated by the symbol $=$. The left part is to be an identifier, i.e., a nonempty sequence formed by the symbols a, b or c. The right part is to be composed of a sequence of identifiers which are separated by the symbols $+$ or $\times$. For example, the sentence $a = ab + c \times a + aba$ belongs to the language we want to express. The sentences $bc =$ or $a + b$ or $a = a + \times b$ do not belong to the language.

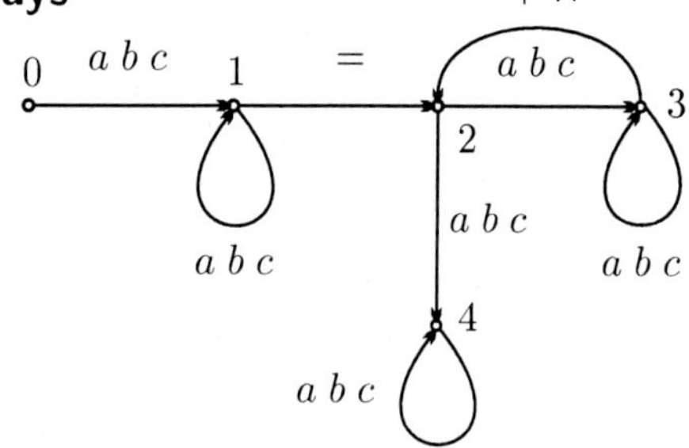

Figure 9.1 Nondeterministic autonomous automaton corresponding to the regular expression $(a,b,c)(a,b,c)^* = ((a,b,c)\ (a,b,c)^*\ (+,\times))^*\ (a,b,c)(a,b,c)^*$.

A regular expression of the language is

$$(a,b,c)(a,b,c)^* = \big((a,b,c)(a,b,c)^*(+,\times)\big)^*(a,b,c)(a,b,c)^* \,.$$

The same language can be expressed by means of a non-deterministic autonomous automaton shown in Fig. 9.1.

The output alphabet consists of the symbols $a, b, c, =, +, \times$. The alphabet of states is the set $\{0,1,2,3,4\}$ which is expressed by a set of graph vertices. The initial state is 0 and the target state is 4. The function P is represented in the

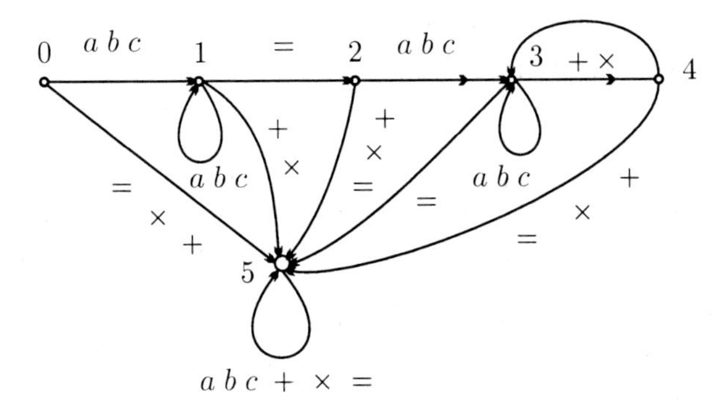

Figure 9.2 Deterministic automaton recognising the language from the example.

graph by means of edges and symbols assigned to the edges. If an edge starts from the vertex k' and points toward the vertex k'' and a symbol x labels the edge then it means that $P(k', x, k'') = 1$. For all other triplets the function P assumes the value 0. This means that the function P assumes the value 1 with the following triplets:

$$
\begin{array}{cc}
(0,a,1),\ (0,b,1),\ (0,c,1)\,, & (3,a,3),\ (3,b,3),\ (3,c,3)\,,\\
(1,a,1),\ (1,b,1),\ (1,c,1)\,, & (3,+,2),\ (3,\times,2)\,,\\
(1,=,2)\,, & (2,a,4),\ (2,b,4),\ (2,c,4)\,,\\
(2,a,3),\ (2,b,3),\ (2,c,3)\,, & (4,a,4),\ (4,b,4),\ (4,c,4)\,.
\end{array}
$$

The table and graph can be understood as grammar of the following form. Its terminal alphabet is $\{a, b, c, =, +, \times\}$, the nonterminal alphabet is $\{0, 1, 2, 3, 4\}$, the axiom is 0, and the set of rules is as follows

$$
\begin{array}{cc}
0 \to a1,\ 0 \to b1,\ 0 \to c1\,, & 3 \to +2,\ 3 \to \times 2\,,\\
1 \to a1,\ 1 \to b1,\ 1 \to c1\,, & 2 \to a4,\ 2 \to b4,\ 2 \to c4\,,\\
1 \to =2\,, & 4 \to a4,\ 4 \to b4,\ 4 \to c4\,,\\
2 \to a3,\ 2 \to b3,\ 2 \to c3\,, & 4 \to a,\ 4 \to b,\ 4 \to c\,.\\
3 \to a3,\ 3 \to b3,\ 3 \to c3\,, &
\end{array}
$$

And finally, the automaton which recognises sentences of the given language is shown by the graph in Fig. 9.2. The input symbol alphabet is $\{a, b, c, =, +, \times\}$. The set of states K is $\{0,\ 1,\ 2,\ 3,\ 4,\ 5\}$. The initial state is 0 and the target state is 3. The automaton recognises if a sentence is correctly formed. Each correct sentence traverses the automaton to the state 3. Each incorrect sentence takes it to some other state. It is achieved by a proper choice of the transition function q which controls transitions of the automaton from state to state according to the input symbol. The transitions between the states are marked by an arrow in Fig. 9.2. If the arrow starts from the vertex k' and points toward the vertex k'' and the symbol x is attached to it then the transition $q(k', x) = k''$ is expressed with it.

9.3 Regular languages respecting faults; best and exact matching

In section 9.1 we have introduced regular languages as a way to define sets of sentences of a certain form. If we look at a problem from a pattern recognition point of view then an algorithm is to be created that will decide for each sentence and for each regular language if the sentence belongs to the given language. The solution of the problem is almost trivial, especially if the regular language is expressed by an automaton generating sentences of the given language. In this case the language is defined as a set of sentences for which the matrix product (9.2) yields the number 1. Such a way of expressing a language immediately provides an explicit recognition algorithm as well.

In addition to the positive features mentioned previously another disadvantage of regular languages is obvious for practical tasks. We are intuitively aware of difficulties in assigning an object to beforehand given classes in the recognition procedure if our decision has to be exact, i.e., yes or no. Such recognition seems to be an overly great simplification for common tasks, such as asking if a person is in good health or if the weather is going to be OK on the following day. We expect that the recognition outcome should have a more complicated form than that of one single bit. The outcome should be a number, at least, providing the measure of certainty with respect to the statement that the recognised object possesses the property examined. Many tasks in applied pattern recognition should not be formulated on the basis of the set of correctly formed sentences, but on the basis of a proper real-valued function defined on the set of all possible sentences.

The first idea might be that the real function sought should be understood as a probability distribution on the set of all possible sentences X^*. That would, of course, mean going back to our previous lecture where alternative tasks were discussed which were based on the assumption of a stochastic mechanism for generating random sentences.

This approach, however, is not the only possible way beyond the scope of a discrete two-valued model of a recognised object. In pattern recognition other methods have originated, primarily the two following classes.

The first class is based on direct generalisation of the regular language. Formal constructions do not determine a subset L of acceptable sentences, but a non-negatively defined function $F\colon X^* \to \mathbb{R}$. Its value $F(\bar{x})$ determines for each sentence $\bar{x} \in X^*$ the measure of acceptability of the particular sentence. The recognition of the sentence $\bar{x}$ is understood in this case as the computation of the number $F(\bar{x})$.

The second class of tasks beyond the scope of pure discrete models defines the function $d\colon X^* \times X^* \to \mathbb{R}$ (usually also a non-negatively defined one) which for each two sequences $\bar{x}_1$ and $\bar{x}_2$ determines to what extent these two sequences differ. The actual concept of the regular language is by no means modified. Pattern recognition tasks are formulated on the basis of two concepts: the regular language L and the function d. Recognition is then understood as a computation of a 'distance' $\bar{x}$ of the observed sentence from the language L, i.e., the computation of the number

$$\min_{\bar{y} \in L} d(\bar{x}, \bar{y}) .$$

Such tasks are called *best matching problems.* The tasks checking the validity of the relation $\bar{x} \in L$ are called *exact matching problems.*

The following subsection is devoted to pattern recognition tasks in which the fundamental concept is not the language as a subset of a sequence, but as a function defined on a set of sequences.

9.3.1 Fuzzy automata and languages

The concepts of fuzzy automata and languages are constructed on the basis of fuzzy sets and operations of their union and intersection, and are defined in the following way.

Let X be a set, it need not be only an alphabet of symbols. The subset $X' \subset X$ can be considered as a function $f \colon X \to \{0, 1\}$ which assumes the value 1 for the subset X' and assumes the value 0 outside the subset X'. The fuzzy subset of the set X is regarded as the function $f \colon X \to \mathbb{R}$ which assumes values in the interval 0 to 1. Let f_1 and f_2 be two fuzzy subsets of the set X. Their intersection is defined as the function $f \colon X \to \mathbb{R}$ the value of which at the point x is $f(x) = \min\big(f_1(x), f_2(x)\big)$. The union of f_1 and f_2 is defined as the function $f \colon X \to \mathbb{R}$ the value at the point x is $f(x) = \max\big(f_1(x), f_2(x)\big)$.

In section 9.1 the automaton has been defined as a five-tuplet $\langle X, K, \varphi, P, \psi \rangle$, where X and K are two finite sets. The function φ determines the subset of initial states, the function P denotes the subset of triplets (k', x, k''), $k' \in K$, $x \in X$, $k'' \in K$, and ψ defines the subset of target states. A fuzzy automaton is defined by a similar five-tuplet. The difference lies in considering φ, P and ψ as fuzzy sets.

The automaton $\langle X, K, \varphi, P, \psi \rangle$ (which is not a fuzzy one) determines the language as a subset of sentences in the form $x_1, x_2, \ldots, x_n$, $n = 1, 2, 3, \ldots$ for which the following equation holds,

$$\bigvee_{k_0 \in K} \bigvee_{k_1 \in K} \cdots \bigvee_{k_n \in K} \left(\varphi(k_0) \wedge \left(\bigwedge_{i=1}^{n} P(k_{i-1}, x_i, k_i) \right) \wedge \psi(k_n) \right) = 1 . \qquad (9.3)$$

The fuzzy automaton $\langle X, K, \varphi, P, \psi \rangle$ defines the fuzzy language as the function $X^* \to \mathbb{R}$ which is determined by the left-hand side of Equation (9.3) with the operation $\bigvee$ understood as a union of fuzzy sets, and the operation $\bigwedge$ as an intersection of fuzzy sets. Thus, the language of the fuzzy automaton $\langle X, K, \varphi, P, \psi \rangle$ is defined as a fuzzy subset defined by the function

$$\max_{k_0 \in K} \max_{k_1 \in K} \cdots \max_{k_n \in K} \min \Big(\varphi(k_0), \min_i P(k_{i-1}, x_i, k_i), \psi(k_n) \Big) . \qquad (9.4)$$

The pair of operations (max, min) constitutes a semi-ring on a set of real numbers from 0 to 1, where max is considered as addition, and min as multiplication. Thus the expression (9.4) can be written as a matrix product

$$F(x_1, x_2, \ldots, x_n) = \varphi \odot \left(\bigodot_{i=1}^{n} P_i \right) \odot \psi\,, \tag{9.5}$$

where φ is a row vector, P_i, $i = 1, \ldots, n$, are matrices, and ψ is a column vector. The above quantities are constructed in a similar way as those in the analysis of earlier problems, e.g., in writing Equation (9.2). Matrix multiplication in the formula (9.5) is to be performed in a relevant semi-ring, i.e., with operation max as an addition and operation min as a multiplication. In calculating the product in (9.5) keep in mind that matrix multiplication is not commutative.

We can see that fuzzyfication of the concepts—the automaton and the regular language—has kept the pattern recognition task on a trivial level. The expression (9.5) formulates the task as a calculation to what extent the sentence $x_1, x_2, \ldots, x_n$ belongs to the given fuzzy language. At the same time, an algorithm for this calculation is defined by the same expression (9.5). Complexity of this calculation is evidently $\mathcal{O}(|K|^2 n)$.

9.3.2 Penalised automata and corresponding languages

Let X and K be two finite sets and $\varphi\colon K \to \mathbb{R}$, $P\colon K \times X \times K \to \mathbb{R}$, $\psi\colon K \to \mathbb{R}$ be three functions determining the behaviour of a finite automaton. For the initial state of the automaton any state $k \in K$ can be chosen, but any such choice is penalised by $\varphi(k)$. If the automaton was in the $(i-1)$-th instant in the state k_{i-1} then it can generate the symbol x_i, traverse to the state k_i and pay a penalty $P(k_{i-1}, x_i, k_i)$ for this step. And eventually, the automaton can interrupt its operation at any instant i and declare the generated sentence as finished. For finishing it the automaton will pay a penalty $\psi(k_i)$ which depends on the state k_i in which the operation was finished.

The five-tuplet $\langle X, K, \varphi, P, \psi\rangle$ and the number ε define the language as a set of sentences which can be generated with the total penalty being less than or equal to ε. The set of languages which can be defined in this way contains all regular languages, but it contains the languages that are not regular too. They will be called *penalised languages*.

The pattern recognition task questioning if the given sentence $x_1, x_2, \ldots, x_n$ belongs to a penalised language can be reduced to calculating the number

$$\min_{k_0} \min_{k_1} \cdots \min_{k_n} \left(\varphi(k_0) + \sum_i P(k_{i-1}, x_i, k_i) + \psi(k_n) \right) \tag{9.6}$$

and comparing it with the number ε. The calculation (9.6) can be expressed as the matrix product

$$\varphi \odot \left(\bigodot_{i=1}^{n} P_i \right) \odot \psi$$

in which the matrix is multiplied in a semi-ring with the operation min as an addition and the operation + as a multiplication.

We can see that even in this formulation the pattern recognition task is trivial because its formulation directly results in the algorithm of its calculation. The complexity of the algorithm remains $\mathcal{O}(|K|^2 n)$.

9.3.3 Simple best matching problem

Let $d\colon X \times X \to \mathbb{R}$ be a function the value $d(x', x'')$, $x' \in X$, $x'' \in X$, of which means the penalty for replacing the symbol x' for x''. This function also determines the value for the pair of sequences $x'_1, x'_2, \dots, x'_n$ and $x''_1, x''_2, \dots, x''_n$, as the sum

$$\sum_{i=1}^{n} d(x'_i, x''_i)\,.$$

In a special case in which the variables x_i assume only two values, and with a relevant choice of the function d, the given sum is the *Hamming distance* known in the theory of coding. Let $L \subset X^*$ be a regular language (neither fuzzy, nor penalised) which corresponds to the automaton $\langle X, K, \varphi, P, \psi\rangle$. In the *simple best matching problem* the given sequence $x_1, x_2, \dots, x_n$ is to be substituted by the sequence $y_1, y_2, \dots, y_n$ from the language L to obtain the minimal distance $\sum_{i=1}^{n} d(y_i, x_i)$. This means that it is necessary to solve the minimisation task

$$\left.\begin{aligned} \min_{y_1}\min_{y_2}\cdots\min_{y_n}\sum_{i=1}^{n} d(y_i, x_i)\,, &\quad \text{under the conditions:} \\ \varphi(k_0) = 1\,, & \\ P(k_{i-1}, y_i, k_i) = 1\,, &\quad i = 1, \dots, n\,, \\ \psi(k_n) = 1\,, & \end{aligned}\right\} \tag{9.7}$$

with the known sequence $x_1, x_2, \dots, x_n$ and given functions d, φ, P and ψ. This problem can be transformed on the basis of the following considerations. Let

$$M = \max_{y \in X}\max_{x \in X} d(y, x)\,.$$

We will introduce new functions φ', P' and f' which are defined as follows

$$\begin{aligned} \varphi'(k) &= Mn + 1\,, &\text{if}\quad \varphi(k) &= 0\,, \\ \varphi'(k) &= 0\,, &\text{if}\quad \varphi(k) &= 1\,, \\ \psi'(k) &= Mn + 1\,, &\text{if}\quad \psi(k) &= 0\,, \\ \psi'(k) &= 0\,, &\text{if}\quad \psi(k) &= 1\,, \\ P'(k', y, k'') &= Mn + 1\,, &\text{if}\quad P(k', y, k'') &= 0\,, \\ P'(k', y, k'') &= 0\,, &\text{if}\quad P(k', y, k'') &= 1\,. \end{aligned}$$

Using the new functions φ', P' and f' we will rewrite the task (9.7) as

$$\min_{k_0} \cdots \min_{k_n} \min_{y_1} \cdots \min_{y_n} \left(\varphi'(k_0) + \sum_{i=1}^{n} \Big(P'(k_{i-1}, y_i, k_i) + d(y_i, x_i) \Big) + \psi'(k_n) \right) . \tag{9.8}$$

For arbitrary values $k_0, \ldots, k_n$, $y_1, \ldots, y_n$ which satisfy the conditions of the problem (9.7) the sum minimised will not be greater than Mn. For values which do not satisfy the conditions of the problem (9.7) the sum will not be less than $Mn + 1$. The result is that the minimum in the problem (9.8) can occur only in the points $k_0, \ldots, k_n$, $y_1, \ldots, y_n$ which satisfy the conditions (9.7). For points satisfying the conditions (9.7) the functions which are minimised in the problems (9.8) and (9.7) assume the same values.

The number (9.8) obviously is

$$\min_{k_0} \min_{k_1} \cdots \min_{k_n} \left(\varphi'(k_0) + \sum_{i=1}^{n} \min_{y_i} \Big(P'(k_{i-1}, y_i, k_i) + d(y_i, x_i) \Big) + \psi'(k_n) \right) .$$

If the notation

$$P_i''(k_{i-1}, x_i, k_i) = \min_{y \in X} \Big(P'(k_{i-1}, y, k_i) + d(y, x_i) \Big) \tag{9.9}$$

is introduced then the value (9.8) can be written as

$$\min_{k_0} \min_{k_1} \cdots \min_{k_n} \left(\varphi'(k_0) + \sum_{i=1}^{n} P_i''(k_{i-1}, x_i, k_i) + \psi'(k_n) \right) .$$

Note that the preceding expression is of the same form as that of the expression (9.6) which had to be calculated in the recognition task based on the penalised language. It follows that even this simple best matching problem can be reduced to a calculation of the matrix product

$$\varphi' \odot \left(\bigodot_{i=1}^{n} P_i'' \right) \odot \psi' . \tag{9.10}$$

In the previous tasks the relevant matrices and vectors were taken directly from input data. The examined best matching problem differs from the preceding tasks in creating the matrices P_i'' as several, not very difficult, optimisation tasks (9.9). The computational complexity of calculating matrices P_i'' is $\mathcal{O}(|K|^2\, |X|\, n)$. After the matrices have been computed the product (9.10) is to be calculated the computational complexity of which is $\mathcal{O}(|K|^2\, n)$. It is clear that the expression (9.9) need not be computed only if the sequence $x_1, x_2, \ldots, x_n$ is already known. The numbers $P''(k', x, k'')$ can be computed in advance for all $k' \in K$, $x \in X$, $k'' \in K$ according to the formula

$$P''(k', x, k'') = \min_{y \in X} \Big(P'(k', y, k) + d(y, x) \Big) . \tag{9.11}$$

In this way the analysed best matching problem (9.7) is directly reduced to a task for recognising a language defined by the penalised automaton $\langle X, K, \varphi', P'', \psi' \rangle$. Then for each sequence $x_1, x_2, \ldots, x_n$ preliminary calculations of (9.9) need not be performed since they are substituted by other preliminary calculations of (9.11) with calculation complexity $\mathcal{O}(|K|^2\,|X|^2)$.

9.4 Partial conclusion after one part of the lecture and introduction to further explanation

We have discussed a quite extensive group of problems connected with recognising sequences. We have found that computational procedures for their solutions are similar and lead to the computation of matrix products in some semi-rings. Computational algorithms for various problems differ only in the semi-ring needed for the given problem.

1. The calculation of a *probability that a stochastic automaton will generate a given sequence* requires multiplication of matrices in a semi-ring $(+, \text{product})$ i.e., the matrix is multiplied in a usual sense.
2. The search for the *most probable sequence of states through which the stochastic automaton has passed* requires multiplication of matrices in a semi-ring $(\max, \text{product})$.
3. The decision *if the sentence belongs to the given regular language* requires matrix multiplication in a semi-ring $(\vee, \wedge)$.
4. The decision on the extent of *how much the given sentence belongs to a fuzzy regular language* can be transposed to multiplication of matrices in a semi-ring $(\max, \min)$.
5. The decision *if the sentence belongs to the given penalised language* requires multiplication of matrices in a semi-ring $(\min, +)$.
6. The solution of the *simple best matching problem* with the given regular language requires multiplication of matrices in a semi-ring $(\min, +)$ similarly as in previous item.

The uniformity of algorithms for solving problems the original formulations of which seemed to differ should attract our attention for several reasons. First, algorithms for solving diverse tasks are easier to remember because they are similar. Second, the uniformity of algorithms offers a purely pragmatic advantage of saving effort in writing problem-solving programs. The algorithms differ in several parameters only, i.e., in the selected operations from the respective semi-rings.

Last but not least by understanding the common features of the given problems, a subjective feeling is achieved that all the mentioned tasks are easy. Some tasks seemed to be far more difficult than some others at first glance in their original varied formulations. The task answering the question if the given sentence belongs to a certain regular language, i.e., an exact matching problem, seems to be much easier at first glance than seeking a sentence which is most similar to the given one. But if it is found that both tasks are solved by the same algorithm then the second task does not appear to be in any way so fantastically complicated either.

Let us give another example concerning a pair of problems of seemingly different complexity, and being closely related to our explanation. Seeking a path between two vertices of a graph does not substantially differ from a seemingly more complicated task which seeks the shortest path between these two vertices.

We have seen that transformation of the original problems to a matrix form has not, so far, required complicated calculations. Only with the best matching problem the transition from the original formulation to a matrix form requires a certain, but not very complicated transformation of input data. In all other tasks no transformation of input data has been needed. The matrices to be multiplied are stored in the initial data. Extracting matrices from the initial data does not require any transformation of the initial data. They only have to be interpreted in another way. If we use the chess terminology then the problem is in winning position from the very beginning.

We also know how to solve the simplest best matching problem in which a simple transformation of initial data is required to convert the problem in a matrix form which makes the solution evident. If we continue with the chess analogy then the winning position can be achieved in a single move. In our explanation we have now worked our way to a task in which we will need multiple modifications to transpose the task to the form of a matrix product. The winning position will be achieved in several moves and a certain auxiliary task is solved with every move. This more complicated problem is Levenstein's well known approximation of the assigned sentence by means of a sentence of the given regular language. The analysis of this task will be dealt with in the remaining part of Lecture 9.

9.5 Levenstein approximation of a sentence by a regular-language sentence

9.5.1 Preliminary formulation of the task

We are going to analyse a task belonging to the class of the best matching problems which has the following form. Let L be a regular language determined by the automaton $\langle X, K, \varphi, P, \psi \rangle$, and let the function $d\colon X^* \times X^* \to \mathbb{R}$ be given which for each pair of sentences $\bar{x}_1 \in X^*$, $\bar{x}_2 \in X^*$ provides the number $d(\bar{x}_1, \bar{x}_2)$ called the *dissimilarity between a sentence* $\bar{x}_2$ *and a sentence* $\bar{x}_1$. The dissimilarity is not necessarily symmetrical and need not have other features of metrics either. The objective is to create an algorithm which will compute the number

$$D(\bar{x}) = \min_{\bar{y} \in L} d(\bar{y}, \bar{x}) , \tag{9.12}$$

for each sentence $\bar{x} \in X^*$ and for each regular language $L \subset X^*$. The number $D(\bar{x})$ will be called the *dissimilarity between the sentence* $\bar{x}$ *and the language* L.

We will deal with the problem (9.12) for the case in which the function d is defined in a special way known as the Levenstein function.

9.5.2 Levenstein dissimilarity

We will introduce three edit operations by which a sentence can be rewritten to another sentence, and show how to penalise these operations. The operations will be denoted by two-letter abbreviations which represent the first two letters of their names.

INsert transforms the sentence $\bar{x}_1\bar{x}_2$, $\bar{x}_1 \in X^*$, $\bar{x}_2 \in X^*$, to the sentence $\bar{x}_1 x \bar{x}_2$, $x \in X$. The penalty for this operation is determined by the function $in\colon X \to \mathbb{R}$ the value $in(x)$, $x \in X$, of which corresponds to the penalty for inserting the symbol x to the sentence.

CHange (also replace) transforms the sentence $\bar{x}_1 x \bar{x}_2$, $\bar{x}_1 \in X^*$, $x \in X$, $\bar{x}_2 \in X^*$, to the sentence $\bar{x}_1 x' \bar{x}_2$, $x' \in X$. The transformation is penalised according to the function $ch\colon X \times X \to \mathbb{R}$. The number $ch(x, x')$, $x \in X$, $x' \in X$, is the penalty for changing the symbol x for the symbol x' in the sentence.

DElete transforms the sentence $\bar{x}_1 x \bar{x}_2$, $\bar{x}_1 \in X^*$, $x \in X$, $\bar{x}_2 \in X^*$, to the sentence $\bar{x}_1\bar{x}_2$. The delete transformation is penalised by the function $de\colon X \to \mathbb{R}$. The number $de(x)$, $x \in X$, is the penalty for deleting the symbol x from the sentence.

The penalty for the sequence of the above mentioned edit operations is defined as a sum of penalties for individual operations. The penalty for an empty sequence of edit operations is zero by definition.

Let us have two sentences, $\bar{x}_1$ and $\bar{x}_2$. There is an infinite number of sequences of edit operations transforming the sentence $\bar{x}_1$ to the sentence $\bar{x}_2$. The price of the cheapest sequence of edit operations transforming the sentence $\bar{x}_1$ to the sentence $\bar{x}_2$ defines the Levenstein dissimilarity between the sentences $\bar{x}_2$ and $\bar{x}_1$. The *Levenstein dissimilarity* is denoted by $d(\bar{x}_1, \bar{x}_2)$. Sometimes the same concept is called *Levenstein function* or edit distance or Levenstein deviation, too.

Levenstein dissimilarity is determined by the triplet of functions $in\colon X \to \mathbb{R}$, $ch\colon X \times X \to \mathbb{R}$, and $de\colon X \to \mathbb{R}$. The set of such functions states the class of functions $X^* \times X^* \to \mathbb{R}$ which will be dealt with in this lecture. Some functions of this class can be understood as definitions of metric relations on the set X^*, but in the general case Levenstein dissimilarity need not possess properties of a distance. We will analyse tasks which are based on Levenstein dissimilarity in their general form. We will be supported by one single assumption that the functions in, ch and de are positive-semidefinite.

The presented definition of Levenstein dissimilarity is somewhat treacherous. Its calculation induces us to use additional features of Levenstein dissimilarity which seem to be natural but actually do not result from its definition. Algorithms based on such seemingly self-evident assumptions (which are in fact additional) solve the task only in particular cases. In the general case their correct performance cannot be guaranteed.

Nothing serious happens when some algorithms fail to solve a task in its complete generality, but solve only a subset of the possible tasks. But it can be disastrous when this subset is not precisely determined and the algorithm starts

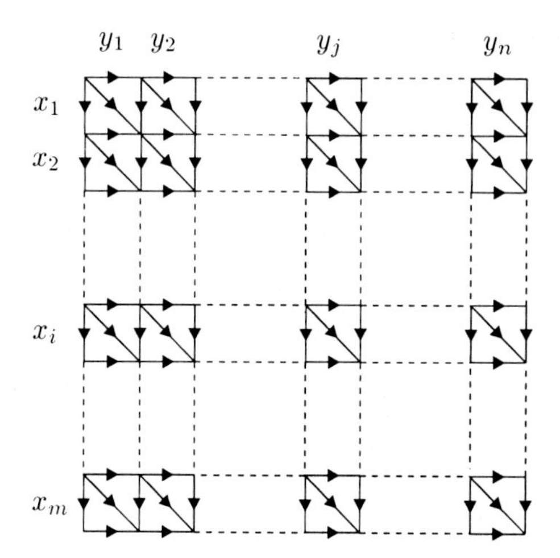

Figure 9.3 Graph of transitions illustrating the performance of an algorithm for calculating Levenstein dissimilarity.

to be used even for tasks for which its correct performance is not guaranteed. The well known and favoured algorithm for calculating Levenstein similarity is a glowing example of such a pseudo-solution of tasks. We intend to mention it in our explanation for completeness and will show where it is in error.

9.5.3 Known algorithm calculating Levenstein dissimilarity

Let us have two sentences from X^*, namely $\bar{x} = (x_1, x_2, \ldots, x_i, \ldots, x_m)$ and $\bar{y} = (y_1, y_2, \ldots, y_j, \ldots \ldots, y_n)$. A commonly used algorithm for calculating Levenstein dissimilarity $d(\bar{y}, \bar{x})$ between the sentences $\bar{x}$ and $\bar{y}$ is represented by a graph in Fig. 9.3.

Fig. 9.3 shows a rectangle composed of $m \times n$ squares which are arranged in m rows and n columns. To the left of the i-th row the i-th symbol of the sentence $\bar{x}$ is written. Above the j-th column the j-th symbol of the sentence $\bar{y}$ is written. Horizontal, vertical and diagonal abscissas in the figure represent graph edges. The set of graph edges defines the set of admissible paths from the left-hand upper corner of the rectangle to its right-hand bottom corner. The edges are so oriented that the admissible motion along different edges is only downward in the vertical direction, to the right in the horizontal direction, and from Northwest to Southeast in the diagonal direction.

To each path in the graph a sequence of edit operations corresponds which transforms the sentence $\bar{y}$ to the sentence $\bar{x}$. The path along the vertical edge in the i-th row corresponds to insertion of the symbol x_i into the sentence. The path along the horizontal edge in the j-th columns of the graph corresponds to deletion of the symbol y_j from the sentence. At last, the path along the diagonal edge in the i-th row and j-th column expresses the change of the symbol y_j for the symbol x_i in the sentence.

Let each edge be weighted by the length which will correspond to the penalty for the respective edit operation in the sentence. The length of each vertical edge in the i-th row represents the penalty $in(x_i)$ for inserting the symbol x_i. The length of the horizontal edge in the j-th column is the penalty $de(y_j)$ for deleting the symbol y_i. Finally, the length of the diagonal edge in the i-th row and j-th column is the penalty $ch(y_j, x_i)$ for changing the symbol y_j to the symbol x_i. In so defined lengths of the edges, the length of each path in the graph will be equal to the total penalty for using the respective edit operations. Therefore (watch out, an error will follow!) the search for the best sequence of edit operations which transform $\bar{y}$ to the sentence $\bar{x}$ is reduced to the known task of seeking the shortest path in the graph from the top left vertex to the bottom right vertex. The error in reasoning mentioned above will be seen from the following counterexample.

Example 9.1 Undermining the often used algorithm for calculating Levenstein dissimilarity. *Let the alphabet X be $\{a, b, c, d\}$, let the sentence $\bar{y}$ consist of one single symbol a and let the sentence $\bar{x}$ consist of one single symbol b. Levenstein dissimilarity between the sentence b and the sentence a is to be calculated. The corresponding graph is shown in Fig. 9.4. Only three paths exist from the top left vertex to the bottom right vertex in the graph. These three paths correspond to three possible transformations of the sentence a to the sentence b.*

1. *possibility.*
   - *Changing a to b.*
2. *possibility.*
   - *Deleting a.*
   - *Inserting b.*
3. *possibility.*
   - *Inserting b.*
   - *Deleting a.*

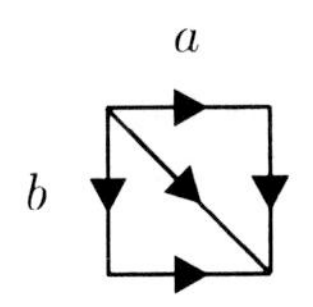

***Figure 9.4** Counterexample. Computing Levenstein dissimilarity.*

In seeking the shortest path, i.e., in seeking the best of these three alternatives, we will arrive at the result that Levenstein dissimilarity between b and a is $\min\big(ch(a,b), de(a) + in(b)\big)$. *However, the factual mismatch of b and a can be far less because the graph 9.4 has shown only three possible ways of transforming the sentence a to the sentence b. The factual number of possibilities is much larger. For example, consider the following.*

- *Changing a to c.*
- *Deleting c.*
- *Inserting d.*
- *Changing d to b.*

There are many other possibilities. The graph in Fig. 9.4 shows only a small portion of possible sequences of edit operations of a sentence. Therefore the quoted procedure can solve the task only if there is an a priori certainty that the cheapest sequence of the edit operations belongs to just that small portion. There is, of course, no such certainty in the general case. ▲

Our ascertainment that the generally applied procedure seeking Levenstein dissimilarity solves the task only in particular cases would not necessarily have disastrous consequences. As soon as the hitch has been discovered and understood, an algorithm for correct calculation of Levenstein dissimilarity can be immediately created. We have dwelt on this example just to show one of the many treacherous traps in which a person handling the task without due care can be easily trapped. We note at the same time that we are not interested in the calculation of Levenstein dissimilarity between a given sentence and some other sentence which is also given. We are interested in the dissimilarity between a given sentence and an extensive, actually infinite, set of sentences, i.e., between a sentence and a regular language. Even at first glance this is a much more difficult task. There are far more such hidden traps as we will see later.

9.5.4 Modified definition of Levenstein dissimilarity and its properties

We will offer another definition of Levenstein dissimilarity which seems more complicated only at first glance. Let X be a finite alphabet and X^* be a set of all sentences composed of symbols of the alphabet. We will regard the set X^* to be vertices of a graph. In a strict sense we should not speak about a graph since it consists of an infinite number of vertices. But we will not take into consideration the non-constructiveness of the mentioned graph because it has been introduced for another purpose than for computational manipulations.

Let us consider three functions $in\colon X \to \mathbb{R}$, $ch\colon X \times X \to \mathbb{R}$ and $de\colon X \to \mathbb{R}$. In the graph *arrows* (transitions) of three types will be introduced, *in*, *ch* and *de*. Let two vertices of the graph correspond to two sequences $\bar{x}_1\bar{x}_2$, $\bar{x}_1 x \bar{x}_2$, where $\bar{x}_1 \in X^*$, $\bar{x}_2 \in X^*$, $x \in X$. This means that these two sequences can be obtained from one another with insertion or deletion of the symbol x. We will introduce an arrow *in* which leads from the vertex $\bar{x}_1\bar{x}_2$ to the vertex $\bar{x}_1 x \bar{x}_2$, and declare its length to be $in(x)$. Similarly we will introduce an arrow *de* which leads from the vertex $\bar{x}_1 x \bar{x}_2$ to the vertex $\bar{x}_1\bar{x}_2$, and state its length to be $de(x)$.

Let two vertices of the graph correspond to two sequences $\bar{x}_1 y \bar{x}_2$ and $\bar{x}_1 x \bar{x}_2$ in which $\bar{x}_1 \in X^*$, $\bar{x}_2 \in X^*$, $x \in X$, $y \in X$. We will introduce an arrow *ch* which begins in the vertex $\bar{x}_1 y \bar{x}_2$ and passes toward the vertex $\bar{x}_1 x \bar{x}_2$, and state its length to be $ch(y, x)$. We will also introduce an arrow in the opposite direction, i.e., from the vertex $\bar{x}_1 x \bar{x}_2$ to the vertex $\bar{x}_1 y \bar{x}_2$, and declare its length to be $ch(x, y)$.

Levenstein dissimilarity is defined as a function $d\colon X^* \times X^* \to \mathbb{R}$ the value $d(\bar{y}, \bar{x})$ of which is given by the length of the shortest path in the created graph from the vertex that represents the sentence $\bar{y}$ to the vertex corresponding to the sentence $\bar{x}$. The following lemma states an important property of Levenstein dissimilarity.

Lemma 9.1 On the ordering of edit operations within the shortest path. *For any two sentences $\bar{y} \in X^*$ and $\bar{x} \in X^*$ one of the shortest paths from the vertex $\bar{y}$ to the vertex $\bar{x}$ has the following form. It begins with a sequence (it* *can even be an empty one) of arrows of the type in. It is followed by a sequence (it can even be an empty one) of the type ch. Finally it ends with a sequence (it can even be an empty one) of arrows of the type de.* ▲

Proof. Let $\bar{y}$, $\bar{x}_1$, $\bar{x}_2$, ..., $\bar{x}_n$, $\bar{x}$ be the shortest paths from the vertex $\bar{y}$ to the vertex $\bar{x}$ which does not possess the property stated by the lemma to be proved. Failure in this property can be revealed in one of the triplet of vertices $\bar{x}_{i-1}$, $\bar{x}_i$, $\bar{x}_{i+1}$, i.e., in a pair of arrows $\bar{x}_{i-1}, \bar{x}_i$ and $\bar{x}_i, \bar{x}_{i+1}$. There are only three cases in which the failure in the property being proved can occur. Each of them will be discussed separately, and the causes of the failure will be examined.

1. The arrow $(\bar{x}_{i-1}, \bar{x}_i)$ is of the type *de* and the arrow $(\bar{x}_i, \bar{x}_{i+1})$ is of the type *in*. This means that the sentences $\bar{x}_{i-1}, \bar{x}_i, \bar{x}_{i+1}$ can assume one of the following forms,

$$\text{either} \quad \left.\begin{aligned} \bar{x}_{i-1} &= \bar{x}'y\bar{x}''\bar{x}''' , \\ \bar{x}_i &= \bar{x}'\bar{x}''\bar{x}''' , \\ \bar{x}_{i+1} &= \bar{x}'\bar{x}''x\bar{x}''' , \end{aligned}\right\} \quad \text{or} \quad \left.\begin{aligned} \bar{x}_{i-1} &= \bar{x}'\bar{x}''y\bar{x}''' , \\ \bar{x}_i &= \bar{x}'\bar{x}''\bar{x}''' , \\ \bar{x}_{i+1} &= \bar{x}'x\bar{x}''\bar{x}''' , \end{aligned}\right\}$$

where $\bar{x}' \in X^*$, $\bar{x}'' \in X^*$, $\bar{x}''' \in X^*$, $x \in X$, $y \in X$.

In the first case the sentence $\bar{x}_i = \bar{x}'\bar{x}''\bar{x}'''$ will be changed to the sentence $\bar{x}'_i = \bar{x}'y\bar{x}''x\bar{x}'''$. In the second case it will be changed to the sentence $\bar{x}'_i = \bar{x}'x\bar{x}''y\bar{x}'''$. In both cases we will obtain a new path from the sentence $\bar{y}$ to the sentence $\bar{x}$ which has exactly the same length as that of the original path. With the new path the arrow $(\bar{x}_{i-1}, \bar{x}'_i)$ will be of the type *in* and the arrow $(\bar{x}'_i, \bar{x}_{i+1})$ will be of the type *de*.

2. The arrow $(\bar{x}_{i-1}, \bar{x}_i)$ is of the type *ch* and the arrow $(\bar{x}_i, \bar{x}_{i+1})$ is of the type *in*. Two alternatives can occur,

$$\text{either} \quad \left.\begin{aligned} \bar{x}_{i-1} &= \bar{x}'y\bar{x}''\bar{x}''' , \\ \bar{x}_i &= \bar{x}'x\bar{x}''\bar{x}''' , \\ \bar{x}_{i+1} &= \bar{x}'x\bar{x}''z\,\bar{x}''' , \end{aligned}\right\} \quad \text{or} \quad \left.\begin{aligned} \bar{x}_{i-1} &= \bar{x}'\bar{x}''y\bar{x}''' , \\ \bar{x}_i &= \bar{x}'\bar{x}''x\bar{x}''' , \\ \bar{x}_{i+1} &= \bar{x}'z\,\bar{x}''x\bar{x}''' . \end{aligned}\right\}$$

In the first case the sentence $\bar{x}_i = \bar{x}'x\bar{x}''\bar{x}'''$ will be changed to the sentence $\bar{x}'_i = \bar{x}'y\bar{x}''z\bar{x}'''$. In the second case the sentence $\bar{x}_i$ will be changed to the sentence $\bar{x}'_i = \bar{x}'z\bar{x}''y\bar{x}'''$. In both cases we will obtain a new path from the sentence $\bar{y}$ to the sentence $\bar{x}$ which has exactly the same length as that of the original path. With the new path the arrow $(\bar{x}_{i-1}, \bar{x}'_i)$ will be of the type *in* and the arrow $(\bar{x}'_i, \bar{x}_{i+1})$ will be of the type *ch*.

3. The arrow $(\bar{x}_{i-1}, \bar{x}_i)$ is of the type *de*, and the arrow $(\bar{x}_i, \bar{x}_{i+1})$ is of the type *ch*. Two alternatives can occur again,

$$\text{either} \quad \left.\begin{aligned} \bar{x}_{i-1} &= \bar{x}'x\bar{x}''y\bar{x}''' , \\ \bar{x}_i &= \bar{x}'\bar{x}''y\bar{x}''' , \\ \bar{x}_{i+1} &= \bar{x}'\bar{x}''z\,\bar{x}''' , \end{aligned}\right\} \quad \text{or} \quad \left.\begin{aligned} \bar{x}_{i-1} &= \bar{x}'y\bar{x}''x\bar{x}''' , \\ \bar{x}_i &= \bar{x}'y\bar{x}''\bar{x}''' , \\ \bar{x}_{i+1} &= \bar{x}'z\,\bar{x}''\bar{x}''' . \end{aligned}\right\}$$

In the first case the sentence $\bar{x}_i = \bar{x}'\bar{x}''y\bar{x}'''$ will be changed to the sentence $\bar{x}'_i = \bar{x}'x\bar{x}''z\bar{x}'''$. In the second case the sentence $\bar{x}_i = \bar{x}'y\bar{x}''\bar{x}'''$ will be changed to the sentence $\bar{x}'_i = \bar{x}'z\ \bar{x}''x\bar{x}'''$. The obtained new path from $\bar{y}$ to $\bar{x}$ will have exactly the same length as that of the original path. The arrow $(\bar{x}_{i-1}, \bar{x}'_i)$ will be of the type *ch* and the arrow $(\bar{x}'_i, \bar{x}_{i+1})$ will be of the type *de*.

We have seen that by gradually changing the original path from $\bar{y}$ to $\bar{x}$ we will find a path where none of the three quoted situations will occur. For the resulting paths, the property stated in the Lemma being proved will be satisfied. ■

Thanks to the proved lemma we can see that one of the shortest paths from the sentence $\bar{y}$ to the sentence $\bar{x}$ consists of three sections. Note that each of them can be empty. The first section of the path passes through the arrows *in*, the second section is formed by the arrows *ch*, and finally the third section consists of the arrows *de*.

If we take this property into consideration then we can define Levenstein dissimilarity in one more way. We will introduce three partial Levenstein dissimilarities which will be denoted d_{in}, d_{ch} and d_{de}. The number $d_{in}(\bar{y}, \bar{x})$ is defined as the length (the shortest one) of the path from the vertex $\bar{y}$ to the vertex $\bar{x}$ which passes through the arrows *in*. In this definition, the word 'shortest' is redundant. The path from $\bar{y}$ to $\bar{x}$ exists only if the sequence $\bar{y}$ is a subsequence in the sequence $\bar{x}$. If such a path exists then the length of all the paths from $\bar{y}$ to $\bar{x}$ are the same. If such a path does not exist then we consider $d_{in}(\bar{y}, \bar{x}) = \infty$.

The number $d_{de}(\bar{y}, \bar{x})$ determines the length (the shortest one) of the path from the vertex $\bar{y}$ to the vertex $\bar{x}$ which passes arrows *de*. If such a path does not exist then we declare $d_{de}(\bar{y}, \bar{x}) = \infty$. The adjective 'shortest' is redundant here too as the lengths of all possible paths from $\bar{y}$ to $\bar{x}$ are the same.

Similarly $d_{ch}(\bar{y}, \bar{x})$ is the length of the shortest (the adjective is needed here) path from $\bar{y}$ to $\bar{x}$, that passes through arrows *ch*. If such a path does not exist then we define $d_{ch}(\bar{y}, \bar{x}) = \infty$. The previous case occurs if the lengths of the sentences $\bar{y}$ and $\bar{x}$ are not the same. If the sentences are identical, i.e., $\bar{y} = \bar{x}$ then we define $d_{in}(\bar{y}, \bar{x}) = d_{ch}(\bar{y}, \bar{x}) = d_{de}(\bar{y}, \bar{x}) = 0$.

The concept of the arrow has been used so far for representing admissible edit operations. Now we will supply additional arrows which will denote a repeated application of the same edit operation, and will be called *long arrows*. We intend to distinguish them from the hitherto used arrows which will be called *short arrows*. Let $\bar{y}$ and $\bar{x}$ be two sentences and let the sentence $\bar{x}$ be obtained from the sentence $\bar{y}$ by inserting symbols. We will introduce a long arrow *in* which starts from the vertex $\bar{y}$ and leads toward the vertex $\bar{x}$. Its length will be stated as $d_{in}(\bar{y}, \bar{x})$. Let $\bar{y}$ and $\bar{x}$ be two sentences of the same length. We will introduce a long arrow *ch* which leads from $\bar{y}$ towards $\bar{x}$ and be of the length $d_{ch}(\bar{y}, \bar{x})$. Let $\bar{y}$ and $\bar{x}$ be two such sentences, the sentence $\bar{x}$ can be obtained from the sentence $\bar{y}$ by deleting symbols from it. We will introduce a long arrow *de* which leads from $\bar{y}$ toward $\bar{x}$ and be of the length $d_{de}(\bar{y}, \bar{x})$.

After the concepts have been introduced it is clear that the length of the shortest path from $\bar{y}$ to $\bar{x}$ along the short arrows, i.e., *Levenstein dissimilarity* between the sentences $\bar{x}$ and $\bar{y}$, is *given by the length of the shortest path* from $\bar{y}$ to $\bar{x}$ *along the long arrows*. This path consists of three long arrows *in*, *ch* and *de* at most, in the respective order. Some arrows can also be missing. The mathematical representation of the previous assertions is the relation

$$d(\bar{y}, \bar{x}) = \min_{\bar{z}_1 \in X^*} \min_{\bar{z}_2 \in X^*} \Big(d_{in}(\bar{y}, \bar{z}_1) + d_{ch}(\bar{z}_1, \bar{z}_2) + d_{de}(\bar{z}_2, \bar{x}) \Big) . \tag{9.13}$$

Though the preceding relation is not suitable to a constructive calculation of Levenstein dissimilarity $d(\bar{y}, \bar{x})$ between the sentences $\bar{x}$ and $\bar{y}$, it decomposes Levenstein dissimilarity into three partial, more understandable concepts. It can be seen that in the calculation two auxiliary sentences $\bar{z}_1$ and $\bar{z}_2$ are used on which three summands depend. The sum is to be minimised by these two sentences. Each of the summands can be found by other, more detailed concepts. We will do it later at a more suitable time.

9.5.5 Formulation of the problem and comments to it

Let us have two finite sets X, K, three functions $\varphi\colon K \to \{0,1\}$, $P\colon K \times X \times K \to \{0,1\}$, $\psi\colon K \to \{0,1\}$ which determine the regular language $L \subset X^*$ as a set of sequences $x_1, x_2, \ldots, x_n$ for which the following logical proposition holds

$$\bigvee_{k_0 \in K} \bigvee_{k_1 \in K} \cdots \bigvee_{k_n \in K} \left(\varphi(k_0) \wedge \left(\bigwedge_{i=1}^{n} P(k_{i-1}, x_i, k_i) \right) \wedge \psi(k_n) \right) .$$

Let us also have three functions $in\colon X \to \mathbb{R}$, $ch\colon X \times X \to \mathbb{R}$ and $de\colon X \to \mathbb{R}$ which define Levenstein dissimilarity $d\colon X^* \times X^* \to \mathbb{R}$.

The task is to create an algorithm which for each sequence $\bar{x} \in X^*$ and for each six-tuplet of functions (φ, P, ψ, in, ch, de) calculates the number

$$D(\bar{x}) = \min_{\bar{y} \in L} d(\bar{y}, \bar{x}) . \tag{9.14}$$

This task substantially differs from all previous tasks recognising sequences. The previous tasks were stated as seeking the optimum on a set of sequences. Roughly speaking the best sequence was to be found. From the nature of the task the length of the sequence was known beforehand. Even if such a defined set was extremely extensive then it was only a finite set after all. The task (9.14) which has now been formulated requires seeking the best sentence in the whole regular language. The length of the sequence sought is not limited beforehand and the respective regular language is infinite. Therefore it is a type of optimisation in an infinite domain.

This specificity of our task can be expressed in another way. In the previous cases we were dealing with multi-dimensional optimisation tasks. The dimension of the optimisation space was usually quite large, but it was known beforehand at least. The number of variables according to which optimisation was performed was known. The task (9.14) is more complicated than the multi-dimensional optimisation task because the number of variables is not predetermined and moreover it can be arbitrarily large.

We will see another specificity of the task analysed when compared with the previous ones. The criterion functions (qualities) which were optimised in previous tasks were simple in a sense. We wish to say that if the sequence sought was already known then its calculation would not pose any computational restraints. This is because the quality was explicitly expressed by a mathematical formula which was to be simply calculated. In other words, $\min_x f(x)$ was to be found in a situation in which the procedure computing the value $f(x)$ for each x was known. The task (9.14) is substantially different. The minimised function $d(\bar{y}, \bar{x})$ is unambiguously determined, but is not explicitly expressed. Moreover, strictly speaking, we have not known the algorithm so far which would calculate the number $d(\bar{y}, \bar{x})$ for each pair $(\bar{y}, \bar{x})$ and for each Levenstein dissimilarity. The algorithm quoted before does not calculate that number correctly. However, its complexity is $\mathcal{O}(n_y n_x)$ where n_y and n_x are the lengths of the sentences $\bar{y}$ and $\bar{x}$, respectively. The complexity of the correct algorithm will hardly be less. The task (9.14) which we try to solve requires finding the smallest number out of infinitely many numbers. This is to be done in a situation in which a mere calculation of some numbers may take an arbitrarily long time because the respective regular language includes arbitrarily long sentences.

We erect a pedestal for the formulated problem not only because it deserves it, but also to point out beforehand that its analysis may not be very simple if easy-going negligence is excluded. To simplify our further explanation in the following subsection we present without proof the basic results and assertions on the solvability of the task. Proofs will be quoted in the next subsection. At the end of the lecture, explicit formulæ for solving the task will be given. We do not recommend using formulæ without knowing what they mean.

9.5.6 Formulation of main results and comments to them

Let $\langle X, K, \varphi, P, \psi \rangle$ be a penalised automaton, i.e., X and K be two finite sets, $\varphi\colon K \to \mathbb{R}$, $P\colon K \times X \times K \to \mathbb{R}$, $\psi\colon K \to \mathbb{R}$ be three non-negatively defined functions. The quoted five-tuplet determines the function $F\colon X^* \to \mathbb{R}$ the value $F(\bar{x})$, $\bar{x} \in X^*$, of which means the minimal penalty for generating the sequence $\bar{x}$,

$$F(\bar{x}) = \min_{k_0} \min_{k_1} \cdots \min_{k_n} \left(\varphi(k_0) + \sum_{i=1}^{n} P(k_{i-1}, x_i, k_i) + \psi(k_n) \right) . \qquad (9.15)$$

The function F is simple in the sense that its calculation has a complexity $\mathcal{O}(|K|^2\, n)$, where n is the length of the sequence $\bar{x}$. In the same sense all tasks analysed so far have been simple ones.

The most important outcome will be that for any function $D\colon X^* \to \mathbb{R}$ of the form (9.14) there exists its equivalent expression in the form (9.15). The calculation of the function D has therefore the same complexity as all the previous tasks have. Let us put aside for the time being an important question about complexity of the transition from the expression (9.14) to the expression (9.15). Let this outcome be formulated in a precise way.

Theorem 9.1 Equivalent expression of Levenstein dissimilarity. *Let X and K be two finite sets, $\varphi\colon K \to \{0,1\}$, $P\colon K \times X \times K \to \{0,1\}$, $\psi\colon K \to \{0,1\}$ be three functions which define a regular language L containing the sentences $x_1, x_2, \ldots, x_n$ for which there holds*

$$\bigvee_{k_0 \in K} \bigvee_{k_1 \in K} \cdots \bigvee_{k_n \in X} \left(\varphi(k_0) \wedge \left(\bigwedge_{i=1}^{n} P(k_{i-1}, x_i, k_i) \right) \wedge \psi(k_n) \right) = 1 . \qquad (9.16)$$

Let us consider three non-negatively defined functions $in\colon X \to \mathbb{R}$, $ch\colon X \times X \to \mathbb{R}$, $de\colon X \to \mathbb{R}$ which determine Levenstein dissimilarity $d\colon X^* \times X^* \to \mathbb{R}$ and Levenstein dissimilarity $D\colon X^* \to \mathbb{R}$,

$$D(\bar{x}) = \min_{\bar{y} \in L} d(\bar{y}, \bar{x}) . \qquad (9.17)$$

For each six-tuplet $(\varphi, P, \psi, in, ch, de)$ which determines the function D with respect to (9.17) there is a pair of functions P', ψ' such that the equality

$$D(\bar{x}) = \min_{k_0} \min_{k_1} \cdots \min_{k_n} \left(\varphi(k_0) + \sum_{i=1}^{n} P'(k_{i-1}, x_i, k_i) + \psi'(k_n) \right) \qquad (9.18)$$

is satisfied for each sentence $\bar{x} = (x_1, x_2, \ldots, x_n) \in X^*$. ▲

The above stated theorem claims that the analysed task is in the same complexity class as the preceding tasks. The computation complexity of the number $D(\bar{x})$ is, despite all its complicated expression, $\mathcal{O}(|K|^2 n)$, where n is the length of the sentence $\bar{x}$. It is, therefore, of the same complexity order as determining if the sentence $\bar{x}$ belongs to a regular language (an ordinary one, neither fuzzy, nor penalised). Thus Theorem 9.1 has expressed such an incredibly convenient property that it deserves to be more thoroughly thought over ...

The same attention should be paid to the following property. Assume that we are now not interested in the calculation of $D(\bar{x})$ according to (9.18), but in manipulating a pair of fixed sentences $\bar{y}, \bar{x}$, i.e., (a) in making clear whether $\bar{y}$ belongs to the language L, and (b) in calculating $d(\bar{y}, \bar{x})$. The answer to the first question is provided by the relation (9.16) verification of which is of complexity $\mathcal{O}(|K|^2 n_y)$, where n_y is the length of the sentence $\bar{y}$. Even if we have not yet formulated the algorithm for calculating $d(\bar{y}, \bar{x})$, we can assert that its complexity is not less than $\mathcal{O}(n_y n_x)$. The total complexity is then $\mathcal{O}(|K|^2 n_y + n_y n_x)$. Theorem 9.1 actually asserts that a complexity of the calculation $\min_{\bar{y} \in L} d(\bar{y}, \bar{x})$ is $\mathcal{O}(|K|^2 n_x)$. This complexity does not depend on the length of the sentence $\bar{y}$ which provides a minimum, and moreover the complexity of the calculation of $\min_{\bar{y} \in L} d(\bar{y}, \bar{x})$ is less than the complexity of the mere calculation $d(\bar{y}, \bar{x})$ for the given sentence $\bar{y} \in L$.

The features quoted in Theorem 9.1 are not intuitively understandable and therefore their proof should satisfy the most demanding requirements possible for formal correctness.

Theorem 9.1 will be proved within the scope of formalism which is similar to the formalism of generalised matrices used in this and the previous lecture. Since the calculus of generalised matrices will not be used now only for a more concise expression of relations already known, but for a proof of a hitherto not proved theorem, we are formulating this calculus more precisely than before.

9.5.7 Generalised convolutions and their properties

Let W be a set. Let $\oplus$ and $\otimes$ be two operations which assign each pair of elements $x \in W$ and $y \in W$ with the respective results $x \oplus y$ and $x \otimes y$ which also belong to the set W. Let the operations $\oplus$ and $\otimes$ along with the set W have the following properties:

1. $x \oplus y = y \oplus x$.
2. $(x \oplus y) \oplus z = x \oplus (y \oplus z)$.
3. $x \otimes y = y \otimes x$.
4. $(x \otimes y) \otimes z = x \otimes (y \otimes z)$.
5. $x \otimes (y \oplus z) = (x \otimes y) \oplus (x \otimes z)$.
6. The set W contains a zero element which will be denoted $0^{\oplus}$. For each $x \in W$ the equalities $x \oplus 0^{\oplus} = x$ and $x \otimes 0^{\oplus} = 0^{\oplus}$ are valid.
7. The set W contains a unitary element which will be denoted $1^{\otimes}$. For each $x \in W$ the equality $x \otimes 1^{\otimes} = x$ is valid.

The set W along with the operations $\oplus$ and $\otimes$ which satisfy the above conditions form a semi-ring with commutative multiplication.

It is decisive for our task that the operation min used as $\oplus$ along with the operation $+$ used as $\otimes$ also form a semi-ring on a set that contains all non-negative real numbers and a particular 'number' (denoted ∞) which is greater than all real numbers. The introduced ∞ has the following features

$$x + \infty = \infty ,$$

$$\min(x, \infty) = x .$$

After extending the set of non-negative numbers by ∞, all the above quoted properties of the semi-ring are satisfied, and therefore the set $R \cup \{\infty\}$ along with the operations min as addition and $+$ as multiplication form a semi-ring. Indeed, there holds that:

1. $\min(x, y) = \min(y, x)$.
2. $\min\big(\min(x, y), z\big) = \min\big(x, \min(y, z)\big)$.
3. $x + y = y + x$.
4. $(x + y) + z = x + (y + z)$.
5. $x + \min(y, z) = \min(x + y, x + z)$.

6. The set $R \cup \{\infty\}$ contains the 'number' ∞, so that for each x the equality $\min(x, \infty) = x$ and $x + \infty = \infty$. The 'number' ∞ is then a zero element with respect to the operation $\oplus$.
7. The set $R \cup \{\infty\}$ contains the number 0, so that for each x the equality $x + 0 = x$ is valid. The number 0 is then a unitary element with respect to the operation $\otimes$.

We will deal with functions that are defined on finite sets and assume their values on a commutative-multiplication semi-ring. The denotation $f[x, y]$ will be interpreted as a function in total, i.e., a mapping from one set to the other. Inside the square brackets identifiers of variables are written on which the function depends. The denotation $f(x, y)$ in round brackets will not mean a function, but it will be the value which the function $f[x, y]$ assumes with certain values of the argument. For example, $f(a, b, c)$ is a value which the function $f[x, y, z]$ assumes when $x = a$, $y = b$, $z = c$ are substituted.

Let X be a set $\{x_1, x_2, \ldots, x_n\}$. The expression $\bigoplus_{x \in X} f(x)$ will be used for a brief expression of the sum $f(x_1) \oplus f(x_2) \oplus \ldots \oplus f(x_n)$.

Let X, Y, Z be three finite sets, $(W, \oplus, \otimes)$ be a commutative semi-ring, and $f_1[x, y] \colon X \times Y \to W$, $f_2[y, z] \colon Y \times Z \to W$ be two functions. The expression

$$f_1[x, y] \bigotimes_y f_2[y, z]$$

will be used for a brief denotation of the function $f[x, z] \colon X \times Z \to W$ values of which are defined by the expression

$$f(x, z) = \bigoplus_{y \in Y} \big(f_1(x, y) \otimes f_2(y, z) \big) . \tag{9.19}$$

The function $f_1[x, y] \bigotimes_y f_2[y, z]$ will be termed the *convolution of functions* $f_1[x, y]$ and $f_2[y, z]$ with respect to the variable y.

Identifiers x, y, z in the definition (9.19) can be understood not only as a denotation of one variable but also as a denotation of groups of variables. A group of variables, say, x, y, z, where $x \in X$, $y \in Y$, $z \in Z$, can be considered as one variable that assumes its values from the set $X \times Y \times Z$. For example, the convolution $f_1[x, y, z] \bigotimes_z f_2[z, y, u]$ is a function $f[x, y, u]$. In the general case, the convolution of two functions f_1 and f_2 is dependent on all variables on which the functions f_1 and f_2 are dependent, except for the variables according to which the convolution is performed. The identifiers x, y, z in the definition (9.19) can be understood as empty groups of variables, and thus they need not be present in (9.19) at all. Therefore we regard as understandable also the convolution $f_1[x, y] \bigotimes_y f_2[y]$ which is a function of one variable x, as well as the convolution $f_1[x] \bigotimes f_2[y]$ which is a function of two variables, x and y, and finally the convolution $f_1[x] \bigotimes_x f_2[x]$ which is no function at all and which is simply an element from the set W.

We can see that convolution expressions are generalised expressions of linear algebra which also include matrix products as a particular case. The definition (9.19) can thus also be understood as multiplication of the matrix f_1 of dimension $|X| \times |Y|$ by the matrix f_2 of dimension $|Y| \times |Z|$. The result of the product is the matrix f of dimension $|X| \times |Z|$. Note however that *convolution expressions* have a certain preference over a matrix product because convolution expressions in a commutative semi-ring *are commutative*, i.e., the following equation is valid

$$f_1[x,y] \bigotimes_y f_2[y,z] = f_2[y,z] \bigotimes_y f_1[x,y] \,.$$

Matrix products are not commutative. In matrix multiplication the variable with respect to which the convolution is performed is implicitly defined. Actually it is always the second variable on which the first function depends, and the first variable on which the second function depends. In a convolution expressions, the respective variable is defined explicitly. Furthermore, convolution expressions can be expressed even for functions which depend on three, four, or more arguments; whereas by means of matrix products only the convolution of functions that depend on two arguments at most can be expressed. And last but not least, convolution expressions are meaningful not only for addition and multiplication in a common sense, but for any pair of operations that form a commutative semi-ring.

Despite all the generality, convolution expressions possess features owing to which they can be equivalently transformed, simplified, and the like. Furthermore the above mentioned commutativity, these are

- associativity

$$\left(f_1[x,y] \bigotimes_y f_2[y,z]\right) \bigotimes_z f_3[z,u] = f_1[x,y] \bigotimes_y \left(f_2[y,z] \bigotimes_z f_3[z,u]\right) , \tag{9.20}$$

- distributivity

$$f_1[x,y] \bigotimes_y \left(f_2[y,z] \bigoplus f_3[y,z]\right)$$
$$= \left(f_1[x,y] \bigotimes_y f_2[y,z]\right) \bigoplus \left(f_1[x,y] \bigotimes_y f_3[y,z]\right) ,$$

- and one more property, which has no name, and results from commutativity and associativity

$$\left(f_1[x,y,z] \bigotimes_y f_2[y]\right) \bigotimes_z f_3[z] = \left(f_1[x,y,z] \bigotimes_z f_3[z]\right) \bigotimes_y f_2[y] \,. \tag{9.21}$$

Let X be some set and $\delta[x,y]$ be a function of the form $X \times X \to W$ for which $\delta(x,y) = 1^{\otimes}$ if $x = y$, and $\delta(x,y) = 0^{\oplus}$ if $x \neq y$. The function will be referred to as a Kronecker function. For a Kronecker function

$$f[x,y] \bigotimes_y \delta[y,z] = f[x,z]$$

also holds, i.e., convolution with a Kronecker function does not transform the function itself and changes only the denotation of its argument.

If the semi-ring is formed by the operations min in the sense of addition, + in the sense of multiplication then the convolution expressions assume additional features resulting from the *idempotent property* of addition which means $f \oplus f = f$. Let $f[x,y]$ be a function of the form $X \times X \to (\mathbb{R} \cup \{\infty\})$. For any function f of this form we will define the function $f^0[x,y]$ as the Kronecker function, and the function $f^n[x,y]$ as the convolution $f[x,z] \bigotimes_z f^{n-1}[z,y]$.

Lemma 9.2 Convergence of a sum. *Let the set X contain k elements. In such a case the sum*

$$f^0[x,y] \oplus f^1[x,y] \oplus f^2[x,y] \oplus \ldots \oplus f^n[x,y]$$

at $n \to \infty$ converges to the function $(\delta \oplus f)^{k-1}$, i.e.,

$$\lim_{n\to\infty} \bigoplus_{i=0}^{n} f^i[x,y] = (\delta \oplus f)^{k-1}[x,y] \,.$$

▲

Proof. We will prove first that for an arbitrary n the following equality holds

$$\bigoplus_{i=0}^{n} f^i = (\delta \oplus f)^n \,. \tag{9.22}$$

Equation (9.22) is evidently correct at $n = 0$ and $n = 1$. If $n = 0$ then the left-hand side is δ and the right-hand side is $(\delta \oplus f)^0$ which is also δ since according to the definition, any function with the zero power is δ. If $n = 1$ then both the left-hand side and right-hand side of Equation (9.22) are equal to $\delta \oplus f$. We will prove that if Equation (9.22) holds for some n then this holds for $n+1$ too.

From idempotent property, i.e., from $f \oplus f = f$ the following derivation follows,

$$\begin{aligned}
\bigoplus_{i=0}^{n+1} f^i[x,y] &= \left(\bigoplus_{i=0}^{n} f^i[x,y]\right) \bigoplus \left(\bigoplus_{i=1}^{n+1} f^i[x,y]\right) \\
&= \left(\bigoplus_{i=0}^{n} f^i[x,y]\right) \bigoplus \left(f[x,z] \bigotimes_z \left(\bigoplus_{i=0}^{n} f^i[z,y]\right)\right) \\
&= \left(\delta[x,z] \bigotimes_z \left(\bigoplus_{i=0}^{n} f^i[z,y]\right)\right) \bigoplus \left(f[x,z] \bigotimes_z \left(\bigoplus_{i=0}^{n} f^i[z,y]\right)\right) \\
&= \Big(\delta[x,z] \bigoplus f[x,z]\Big) \bigotimes_z \left(\bigoplus_{i=0}^{n} f^i[z,y]\right) \\
&= \Big(\delta[x,z] \bigoplus f[x,z]\Big) \bigotimes_z \Big(\delta[z,y] \bigoplus f[z,y]\Big)^n \\
&= \Big(\delta[x,y] \bigoplus f[x,y]\Big)^{n+1} \,.
\end{aligned}$$

So the relation (9.22) is proved for each integer n.

We will now prove the main assertion of Lemma 9.2. The set X will be identified with the graph vertices. The value $f(x,y)$, $x \in X$, $y \in X$ will be identified with the length of the oriented edge (arrow) that starts from the vertex x and points toward the vertex y. The number $f(x,y)$ is the length of the shortest path from the vertex x to the vertex y under the condition that the path consists of a single arrow. The number $f(x,z) \bigotimes_z f(z,y)$ is the length of the shortest path from the vertex x to the vertex y under the condition that the path consists of two arrows. In the general case $f^n(x,y)$ is the length of the shortest path from the vertex x to the vertex y under the condition that the path consists of n arrows. This statement has its sense even when $n = 0$. The shortest path from x to y, which does not consist of any arrow, is evidently $0 = 1^{\otimes}$ if $x = y$, and it is $\infty = 0^{\otimes}$ if $x \neq y$. Thus it is $\delta(x,y)$, i.e., $f^0(x,y)$. The sum $\bigoplus_{i=0}^{n} f^i[x,y]$ is the length of the shortest path from x to y which consists of n arrows at most. This path cannot pass through more than $k-1$ arrows. In the opposite case this path would contain a cycle, and this cannot happen with a non-negatively defined function f. From this it follows that at $n \geq k$ the following equation holds

$$\bigoplus_{i=0}^{n} f^i[x,y] = \bigoplus_{i=0}^{k-1} f^i[x,y]\,.$$

As a result of the already proved Equation (9.22) we have

$$\lim_{n\to\infty} \bigoplus_{i=0}^{n} f^i = (\delta \oplus f)^{k-1}\,.$$

■

The lemma proved shows a constructive way of calculating an infinite polynomial $\bigoplus_{i=0}^{\infty} f^i$. Later on this infinite polynomial will be denoted f^* which will be used for an arbitrary function $f\colon X \times X \to W$.

Based on Lemma 9.2 a constructive way can be shown of calculating convolutions with respect to a variable which assumes values from an infinite set. For example, these may be convolutions with respect to a variable that has the form of a sequence.

Lemma 9.3 Calculating infinite convolution expressions. *Let K and X be two finite sets, $X^* = \bigcup_{i=0}^{\infty} X^i$, $f\colon K \times X \times K \to W$, and $\varphi\colon X \to W$ be two functions. Let # represent an empty sentence, i.e., a sentence of zero length.*

Let the function $F\colon K \times X^* \times K \to W$ satisfy the conditions

$$F(k', \#, k'') = \delta(k', k'')\,, \quad k' \in K\,, \quad k'' \in K\,, \tag{9.23}$$

$$F(k', \bar{x}x, k'') = F(k', \bar{x}, k) \bigotimes_k f(k, x, k'')\,, \quad k' \in K\,,\ k'' \in K\,,\ \bar{x} \in X^*\,,\ x \in X\,.$$

Let the function $\Phi\colon X^* \to W$ satisfy the conditions

$$\Phi(\#) = 1^{\otimes}\,,$$

$$\Phi(\bar{x}x) = \Phi(\bar{x}) \otimes \varphi(x)\,, \quad \bar{x} \in X^*\,, \quad x \in X\,. \tag{9.24}$$

In this case the convolution

$$F[k', \bar{x}, k''] \bigotimes_{\bar{x}} \Phi[\bar{x}] \tag{9.25}$$

is

$$\left(\delta[k', k''] \oplus \left(f[k', x, k''] \bigotimes_{x} \varphi(x) \right) \right)^{|K|-1} .$$

▲

Proof. Immediately from the definition of convolution there follows that the convolution (9.25) is only a brief designation for the function of two variables, k' and k'' the value of which for the given pair $(k', k'') \in K^2$ is the sum

$$\bigoplus_{n=0}^{\infty} \bigoplus_{\bar{x} \in X^n} F(k', \bar{x}, k'') \otimes \Phi(\bar{x})\,, \tag{9.26}$$

where X^n is the set of sequences $x_1, x_2, \ldots, x_n$, $x_i \in X$, of length n. This assertion will be repeated in the proof of this lemma several times. That is why we will express it in the form of equality

$$F[k', \bar{x}, k''] \bigotimes_{\bar{x}} \Phi[\bar{x}] = \bigoplus_{n=0}^{\infty} \bigoplus_{\bar{x} \in X^n} F(k', \bar{x}, k'') \otimes \Phi(\bar{x})\,. \tag{9.27}$$

Strictly speaking the preceding relation is not correct because on the left-hand side of the equality the function $K \times K \to W$ is stated, and on the right-hand side of the equality the value appears which this function assumes for the pair k', k''. In spite of this incorrectness we will use the denotation of (9.27) so that after this explanation no misunderstanding should occur.
The sum

$$\bigoplus_{\bar{x} \in X^0} F(k', \bar{x}, k'') \otimes \Phi(\bar{x})$$

is evidently

$$F(k', \#, k'') \otimes \Phi(\#) = \delta(k', k'') \otimes 1^{\otimes} = \delta(k', k'')\,.$$

According to the definitions (9.23), (9.24) and according to the definition of convolution, the sum

$$\bigoplus_{\bar{x} \in X^1} F(k', \bar{x}, k'') \otimes \Phi(\bar{x})$$

has the form

$$\bigoplus_{\bar{x} \in X^1} F(k', \bar{x}, k'') \otimes \Phi(\bar{x}) = f[k', x, k''] \bigotimes_{x} \varphi(x)\,. \tag{9.28}$$

We will prove that at any $n = 1, 2, \ldots$ the following equality holds

$$\bigoplus_{\bar{x} \in X^n} F(k', \bar{x}, k'') \otimes \Phi(\bar{x}) = \left(f[k', x, k''] \bigotimes_x \varphi[x] \right)^n . \tag{9.29}$$

For $n = 1$ the equality is valid because it is identical with the equality (9.28). We will prove that if the equality (9.29) is valid for some n then it is also valid for $n + 1$. This statement is proved by the following derivation

$$\begin{aligned}
\bigoplus_{\bar{x} \in X^{n+1}} F(k', \bar{x}, k'') \otimes \Phi(\bar{x}) &= \bigoplus_{\bar{x} \in X^n} \bigoplus_{x \in X} F(k', \bar{x}x, k'') \otimes \Phi(\bar{x}x) \\
&= \bigoplus_{\bar{x} \in X^n} \bigoplus_{x \in X} \bigoplus_{k \in K} F(k', \bar{x}, k) \otimes f(k, x, k'') \otimes \Phi(\bar{x}) \otimes \varphi(x) \\
&= \bigoplus_{k \in K} \left(\left(\bigoplus_{\bar{x} \in X^n} F(k', \bar{x}, k) \otimes \Phi(\bar{x}) \right) \otimes \left(\bigoplus_{x \in X} f(k, x, k'') \otimes \varphi(x) \right) \right) \\
&= \left(f[k', x, k] \bigotimes_x \varphi[x] \right)^n \bigotimes_k \left(f[k, x, k''] \bigotimes_x \varphi[x] \right) \\
&= \left(f[k', x, k''] \bigotimes_x \varphi[x] \right)^{n+1} .
\end{aligned}$$

We will insert (9.29) in (9.26) and find that the convolution (9.26) is

$$\bigoplus_{n=0}^{\infty} \left(f[k', x, k''] \bigotimes_x \varphi[x] \right)^n .$$

As a result of Lemma 9.2 there holds

$$\left(\delta[k', k''] \oplus \left(f[k', x, k''] \bigotimes_x \varphi[x] \right) \right)^{|K|-1} .$$

■

We can see that a mere formal analysis of convolution expressions provides rules of their equivalent transformation. It is important that two of the rules quoted in Lemma 9.2 and Lemma 9.3 enable transformation of infinite convolution expressions to their finite equivalents. In the task on Levenstein approximation one of the major difficulties is the requirement for minimisation on the set of all possible sequences of arbitrary length. (Roughly speaking, a function is to be minimised which depends on an infinitely large number of variables.) Such sets cannot be coped with by any computational procedure in a finite number of steps. Therefore in solution of that task we will make use of the last two results which reduce the infinite convolution expressions to finite ones. The task has to be expressed in a convolution form for this purpose. Its previous nonconvolution formulation was given in Subsection 9.5.5.

9.5.8 Formulation of a task and main results in convolution form

Let X and K be two finite sets and let three functions $\varphi\colon K \to \{0,\infty\}$, $P\colon K \times X \times K \to \{0,\infty\}$, $\psi\colon K \to \{0,\infty\}$ determine the language $L \subset X^*$ in such a way that the sequence $\bar{x} = (x_1, x_2, \ldots, x_n)$ belongs to L if and only if

$$F(\bar{x}) = \min_{k_0 \in K} \min_{k_1 \in K} \cdots \min_{k_n \in K} \left(\varphi(k_0) + \sum_{i=1}^{n} P(k_{i-1}, x_i, k_i) + \psi(k_n) \right) = 0 \quad (9.30)$$

which is equivalent to the property that there exists a sequence $k_0, k_1, \ldots, k_n$ for which there holds

$$\left.\begin{aligned} \varphi(k_0) &= 0\,, \\ P(k_{i-1}, x_i, k_i) &= 0\,, \quad i = 1, 2, \ldots, n\,, \\ \psi(k_n) &= 0\,, \end{aligned}\right\} \quad (9.31)$$

because the sum $\varphi(k_0) + \sum_{i=1}^{n} P(k_{i-1}, x_i, k_i) + \psi(k_n)$ for an arbitrary sequence $k_0, k_1, \ldots, k_n$ can be either 0 or ∞. This sum is 0 if a system of conditions (9.31) is satisfied; and is ∞ if at least one condition from (9.31) is not satisfied. It is evident that the language L created in this way belongs among regular languages. And vice versa each regular language can be expressed in the form of (9.31).

Let us have Levenstein dissimilarity $d\colon X^* \times X^* \to \mathbb{R}$. The task (9.14) which has to be solved requires for each given sequence $\bar{x} \in L$ calculation of the number

$$D(\bar{x}) = \min_{\bar{y} \in L} d(\bar{y}, \bar{x}) \quad (9.32)$$

which is Levenstein dissimilarity between the sentence $\bar{x}$ and the language L.

The number (9.30) which depends on the sequence $\bar{x} \in X^*$ will be denoted $F(\bar{x})$. Since $F(\bar{x})$ can be either 0 or ∞, the number $D(\bar{x})$ defined by the relation (9.32) can be expressed as

$$D(\bar{x}) = \min_{\bar{y} \in X^*} \Big(F(\bar{y}) + d(\bar{y}, \bar{x}) \Big)\,. \quad (9.33)$$

The function $D\colon X^* \to \mathbb{R}$ defined by the relation (9.33) can be expressed as a convolution over the set of all possible sequences $\bar{y} \in X^*$

$$D[\bar{x}] = F[\bar{y}] \bigotimes_{\bar{y}} d[\bar{y}, \bar{x}]$$

in a semi-ring with operations min and + on a set of non-negative real numbers, enlarged by the 'number' ∞. According to (9.30) the function $F\colon X^* \to \mathbb{R}$ can be defined as a convolution

$$\begin{aligned} F[\bar{y}] &= F[y_1, y_2, \ldots, y_n] \qquad (9.34) \\ &= \varphi[k_0] \bigotimes_{k_0} P[k_0, y_1, k_1] \bigotimes_{k_1} P[k_1, y_2, k_2] \bigotimes_{k_2} \cdots \bigotimes_{k_{n-1}} P[k_{n-1}, x_n, k_n] \bigotimes_{k_n} \psi[k_n]. \end{aligned}$$

The relation of the preceding expression to an expression used before in the form of a generalised matrix product is obvious. If the function $\varphi[k]$ of the variable $k \in K$ is understood as a $|K|$-dimensional row vector then the function $P[k', y_i, k'']$ of two variables $k' \in K$, $k'' \in K$ will be interpreted as a square matrix P_i of the dimension $|K| \times |K|$, and if the function $\psi[k]$ of one variable $k \in K$ is understood as a $|K|$-dimensional column vector then the convolution (9.34) can be expressed as the matrix product

$$\varphi\, P_1\, P_2\, \cdots\, P_{n-1}\, P_n\, \psi\,.$$

The main expected outcome of this part of the explanation which was stated before as Theorem 9.1 can be now expressed in the convolution form.

Let X and K be two finite sets and let us have three functions $\varphi\colon K \to \{0, \infty\}$, $P\colon K \times X \times K \to \{0, \infty\}$, $\psi\colon K \to \{0, \infty\}$. Let the function $f\colon K \times X^* \times K \to \mathbb{R}$ be defined in the following manner.

If $\bar{x}$ is an empty sentence # then

$$f(k', \bar{x}, k'') = f(k', \#, k'') = \delta(k', k''), \quad k' \in K, \quad k'' \in K\,.$$

If $\bar{x} = \bar{x}_1 x$ then

$$f[k', \bar{x}, k''] = f[k', \bar{x}_1 x, k''] = f[k', \bar{x}_1, k] \bigotimes_k P[k, x, k'']\,.$$

Let the function $F\colon X^* \to \mathbb{R}$ have the form

$$F[\bar{x}] = \varphi[k] \bigotimes_k f[k, \bar{x}, k'] \bigotimes_{k'} \psi[k']\,.$$

Let us have three functions $in\colon X \to \mathbb{R}$, $ch\colon X \times X \to \mathbb{R}$, $de\colon X \to \mathbb{R}$ which determine Levenstein dissimilarity $d[\bar{y}, \bar{x}]\colon X^* \times X^* \to \mathbb{R}$ between the sentences $\bar{x}$ and $\bar{y}$ and Levenstein dissimilarity $D[\bar{x}]\colon X^* \to \mathbb{R}$ between the sentence $\bar{x}$ and the set of sentences $\bar{y} \in X^*$ for which $F(\bar{y}) = 0$ holds, i.e.,

$$D[\bar{x}] = F[\bar{y}] \bigotimes_{\bar{y}} d[\bar{x}, \bar{y}]\,. \tag{9.35}$$

Theorem 9.2 Convolution form of Levenstein similarity. *For each six-tuplet of functions $(\varphi, P, \psi, in, ch, de)$ there exists a pair of functions $P'\colon K \times X \times K \to \mathbb{R}$ and $\psi'\colon K \to \mathbb{R}$ such that the function $D\colon X^* \to \mathbb{R}$ defined by the relation (9.35), will assume the form*

$$D[\bar{x}] = \varphi[k] \bigotimes_k f'[k, \bar{x}, k'] \bigotimes_{k'} \psi'[k']\,, \tag{9.36}$$

where

$$f'(k, \#, k') = \delta(k, k')\,, \quad k \in K\,, \quad k' \in K\,,$$
$$f'[k', \bar{x}x, k''] = f'[k', \bar{x}, k] \bigotimes_k P'[k, x, k'']\,.$$

▲

Theorem 9.2 states that Levenstein dissimilarity D for the sentence $\bar{x} = (x_1, x_2, \ldots, x_n)$ can be calculated as the matrix product

$$\varphi\, P_1'\, P_2' \,\cdots\, P_{n-1}'\, P_n'\, \psi'\,,$$

where the vectors φ and ψ' express the functions φ and ψ' of one variable $k \in K$ and matrices P_i', $i = 1, \ldots, n$, represent functions $P'[k, x_i, k']$ of two variables $k \in K$, $k' \in K$. It results from the Theorem 9.2 that the calculation of Levenstein dissimilarity between the sentence $\bar{x}$ and the language L has a complexity $\mathcal{O}(|K|^2 n)$, where n is the length of the sentence $\bar{x}$. In this way the Levenstein dissimilarity problem has been transferred to simpler problems analysed before.

Validity of the declared main result will be proved in the following subsection.

9.5.9 Proof of the main result of this lecture

Based on the property (9.13) the Levenstein dissimilarity $d(\bar{y}, \bar{x})$ between the sentences $\bar{x}$ and $\bar{y}$ can be expressed in the form of convolution

$$d[\bar{y}, \bar{x}] = d_{in}[\bar{y}, \bar{y}_1] \bigotimes_{\bar{y}_1} d_{ch}[\bar{y}_1, \bar{y}_2] \bigotimes_{\bar{y}_2} d_{de}[\bar{y}_2, \bar{x}]\,.$$

Let us incorporate the function $d[\bar{y}, \bar{x}]$ in this form into the definition (9.35) of Levenstein dissimilarity $D[\bar{x}]$ between the sentence $\bar{x}$ and the language L,

$$D[\bar{x}] = F[\bar{y}] \bigotimes_{\bar{y}} \left(d_{in}[\bar{y}, \bar{y}_1] \bigotimes_{\bar{y}_1} d_{ch}[\bar{y}_1, \bar{y}_2] \bigotimes_{\bar{y}_2} d_{de}[\bar{y}_2, \bar{x}] \right)\,. \tag{9.37}$$

Owing to the associativity of convolution the expression (9.37) can be written as

$$D[\bar{x}] = \left(\left(F[\bar{y}] \bigotimes_{\bar{y}} d_{in}[\bar{y}, \bar{y}_1] \right) \bigotimes_{\bar{y}_1} d_{ch}[\bar{y}_1, \bar{y}_2] \right) \bigotimes_{\bar{y}_2} d_{de}[\bar{y}_2, \bar{x}]\,.$$

The proof of Theorem 9.2 can be reduced to a proof of the next three lemmata by decomposing the problem on optimal sentence transformation into three independent simpler problems of optimal insertion, changing and deleting of symbols.

Lemma 9.4 Optimal transformation of a sentence by inserting symbols. *Let us have two finite sets X, K and three functions $\varphi\colon K \to \mathbb{R}$, $P\colon K \times X \times K \to \mathbb{R}$, $\psi\colon K \to \mathbb{R}$ determining the function $F\colon X^* \to \mathbb{R}$ so that*

$$F[\bar{y}] = \varphi[k'] \bigotimes_{k'} f[k', \bar{y}, k''] \bigotimes_{k''} \psi[k'']\,, \tag{9.38}$$

where

$$\left. \begin{aligned} & f[k, \#, k''] = \delta[k', k''], \\ & f[k', \bar{x}x, k''] = f[k', \bar{x}, k] \bigotimes_{k} P[k, x, k'']\,. \end{aligned} \right\}$$

Let the function $in\colon X \to \mathbb{R}$ *define the function* $d_{in}\colon X^* \times X^* \to \mathbb{R}$ *the value* $d_{in}(\bar{y}, \bar{y}_1)$ *of which is the minimal penalty for transforming the sentence* $\bar{y}$ *to the sentence* $\bar{y}_1$ *by repeatedly inserting symbols from* X *to the sentence* $\bar{y}$.

In this case such a function $P_1\colon K \times X \times K \to \mathbb{R}$ *exists that the function*

$$F_1[\bar{y}_1] = F[\bar{y}] \bigotimes_{\bar{y}} d_{in}[\bar{y}, \bar{y}_1] \tag{9.39}$$

is identical with the function

$$\varphi[k'] \bigotimes_{k'} f_1[k', \bar{y}_1, k''] \bigotimes_{k''} \psi[k''] ,$$

where

$$f_1[k', \#, k''] = \delta[k', k''] , \tag{9.40}$$

$$f_1[k', \bar{y}_1 y_1, k''] = f_1[k', \bar{y}_1, k] \bigotimes_{k} P_1[k, y_1, k''] . \tag{9.41}$$

▲

Lemma 9.5 Optimal sentence transformation by changing symbols. *Let us have two finite sets* X, K *and three functions* $\varphi\colon K \to \mathbb{R}$, $P_1\colon K \times X \times K \to \mathbb{R}$, $\psi\colon K \to \mathbb{R}$ *determining the function* $F_1\colon X^* \to \mathbb{R}$ *so that*

$$F_1[\bar{y}_1] = \varphi[k'] \bigotimes_{k'} f_1[k', \bar{y}_1, k''] \bigotimes_{k''} \psi[k''] , \tag{9.42}$$

where

$$\left.\begin{aligned} f_1[k', \#, k''] &= \delta[k', k''] , \\ f_1[k', \bar{y}_1 y_1, k''] &= f_1[k', \bar{y}_1, k] \otimes_{k} P_1[k, y_1, k''] . \end{aligned}\right\} \tag{9.43}$$

Let the function $ch\colon X \times X \to \mathbb{R}$ *define the function* $d_{ch}\colon X^* \times X^* \to \mathbb{R}$ *the value* $d_{ch}[\bar{y}_1, \bar{y}_2]$ *of which is the minimal penalty for transforming the sentence* $\bar{y}_1$ *to the sentence* $\bar{y}_2$ *by repeatedly changing symbols in the sentence* $\bar{y}_1$.

In this case such a function $P_2\colon K \times X \times K \to \mathbb{R}$ *exists that the function*

$$F_2[\bar{y}_2] = F_1[\bar{y}_1] \bigotimes_{\bar{y}_1} d_{ch}[\bar{y}_1, \bar{y}_2] \tag{9.44}$$

is identical with the function

$$\varphi[k'] \bigotimes_{k'} f_2[k', \bar{y}_2, k''] \bigotimes_{k''} \psi[k''] , \tag{9.45}$$

where

$$f_2[k', \#, k''] = \delta[k', k''] , \tag{9.46}$$

$$f_2[k', \bar{y}_2 y_2, k''] = f_2[k', \bar{y}_2, k] \bigotimes_{k} P_2[k, y_2, k''] . \tag{9.47}$$

▲

Lemma 9.6 Optimal sentence transformation by deleting symbols. *Let us have two finite sets X, K and three functions $\varphi\colon K \to \mathbb{R}$, $P_2\colon K \times X \times K \to \mathbb{R}$, $\psi\colon K \to \mathbb{R}$ determining the function $F_2\colon X^* \to \mathbb{R}$ so that*

$$F_2[\bar{y}_2] = \varphi[k'] \bigotimes_{k'} f_2[k', \bar{y}_2, k''] \bigotimes_{k''} \psi[k''] \,, \tag{9.48}$$

where

$$\left.\begin{aligned} f_2[k', \#, k''] &= \delta[k', k''] \,, \\ f_2[k', \bar{y}_2 y_2, k''] &= f_2[k', \bar{y}_2, k] \bigotimes_{k} P_2[k, y_2, k''] \,. \end{aligned}\right\} \tag{9.49}$$

Let the function $de\colon X \to \mathbb{R}$ define the function $d_{de}\colon X^* \times X^* \to \mathbb{R}$ the value $d_{de}[\bar{y}_2, \bar{x}]$ of which is the minimal penalty for transforming the sentence $\bar{y}_2$ to the sentence $\bar{x}$ by repeatedly deleting symbols from the sequence $\bar{y}_2$.

In this case such functions $P'\colon K \times X \times K \to \mathbb{R}$ and $\psi'\colon K \to \mathbb{R}$ exist that the function

$$D[\bar{x}] = F_2[\bar{y}_2] \bigotimes_{\bar{y}_2} d_{de}[\bar{y}_2, \bar{x}] \tag{9.50}$$

is identical with the function

$$\varphi[k'] \bigotimes_{k'} f'[k', \bar{x}, k''] \bigotimes_{k''} \psi'[k''] \,,$$

where

$$f'[k', \#, k''] = \delta[k', k''], \tag{9.51}$$

$$f'[k', \bar{x}x, k''] = f'[k', \bar{x}, k] \bigotimes_{k} P'[k, x, k''] \,. \tag{9.52}$$

▲

Proof. (to lemma 9.4)

1. For the numbers $d_{in}(\bar{y}, \bar{y}_1)$ following relations are valid. If both $\bar{y}$ and $\bar{y}_1$ are empty then $d_{in}(\#, \#) = 0 = 1^{\otimes}$. If $\bar{y}$ is empty and $\bar{y}_1$ is non-empty then evidently

$$d_{in}(\#, \bar{y}_1' y_1) = d_{in}(\#, \bar{y}_1') + d_{in}(\#, y_1) \,,$$

or in the convolution form

$$d_{in}[\#, \bar{y}_1' y_1] = d_{in}[\#, \bar{y}_1'] \otimes d_{in}[\#, y_1] \,. \tag{9.53}$$

If a sequence $\bar{y}$ consists of a single symbol y and a sequence $\bar{y}_1$ consists of a single symbol y_1 then $d_{in}(\bar{y}, \bar{y}_1) = \infty$ if $(y \neq y_1)$ and $d_{in}(\bar{y}, \bar{y}_1) = 0$ in opposite case. This means in the convolution form

$$d_{in}[y, y_1] = \delta[y, y_1] \,. \tag{9.54}$$

If neither $\bar{y}$ nor $\bar{y}_1$ is empty then value $d_{in}(\bar{y}, \bar{y}_1)$ results from the following considerations. The optimal transformation of the sequence $\bar{y} = \bar{y}'y$ to the sequence $\bar{y}_1 = \bar{y}'_1 y_1$ with insertions can be performed only in two possible ways: either the symbol y_1 was inserted during the transformation or not. In the first case the penalty of transformation will evidently be

$$d_{in}(\bar{y}'y, \bar{y}'_1) + d_{in}(\#, y_1)\,.$$

In the second case the penalty will be

$$d_{in}(\bar{y}', \bar{y}'_1) + d_{in}(y, y_1)\,.$$

The value $d_{in}(\bar{y}'y, \bar{y}'_1 y_1)$ will evidently be the smaller value of these two numbers,

$$d_{in}(\bar{y}'y, \bar{y}'_1 y_1) = \min\big(d_{in}(\bar{y}'y, \bar{y}'_1) + d_{in}(\#, y_1), d_{in}(\bar{y}', \bar{y}'_1) + d_{in}(y, y_1)\big)$$

or in the convolution form

$$d_{in}[\bar{y}'y, \bar{y}'_1 y_1] = \big(d_{in}[\bar{y}'y, \bar{y}'_1] \otimes d_{in}[\#, y_1]\big) \oplus \big(d_{in}[\bar{y}', \bar{y}'_1] \otimes d_{in}[y, y_1]\big)\,. \quad (9.55)$$

2. Let us now derive a more detailed expression for the function

$$F_1[\bar{y}_1] = F[\bar{y}] \bigotimes_{\bar{y}} d_{in}[\bar{y}, \bar{y}_1]$$

which existence states Lemma 9.4. Let us use the expression (9.38) for $F[\bar{y}]$ and obtain

$$F_1[\bar{y}_1] = \varphi[k'] \bigotimes_{k'} f[k', \bar{y}, k''] \bigotimes_{k''} \psi[k''] \bigotimes_{\bar{y}} d_{in}[\bar{y}, \bar{y}_1]\,.$$

Based on the property (9.21) the equivalent expression

$$F_1[\bar{y}_1] = \varphi[k'] \bigotimes_{k'} \left(f[k', \bar{y}, k''] \bigotimes_{\bar{y}} d_{in}[\bar{y}, \bar{y}_1] \right) \bigotimes_{k''} \psi[k'']$$

is valid. In this way we have proved that the function $F_1[\bar{y}_1]$ is of the form

$$\varphi[k'] \bigotimes_{k'} f_1[k', \bar{y}_1, k''] \bigotimes_{k''} \psi[k'']\,,$$

where

$$f_1[k', \bar{y}_1, k''] = f[k', \bar{y}, k''] \bigotimes_{\bar{y}} d_{in}[\bar{y}, \bar{y}_1]\,. \quad (9.56)$$

Now it is to be proved that the function $f_1[k', \bar{y}_1, k'']$ defined by Equation (9.56) satisfies conditions (9.40) and (9.41).

3. The property (9.40) is quite obvious. If $\bar{y}_1 = \#$ then $d_{in}(\bar{y}, \bar{y}_1) = 0 = 1^{\otimes}$ only when $\bar{y} = \#$. It is because a nonempty sequence cannot be changed into an empty one by inserting symbols. Equation (9.56) in this case will obtain the form

$$f_1[k', \#, k''] = f[k', \#, k'']\,.$$

In Lemma 9.4 the condition $f[k', \#, k''] = \delta[k', k'']$ is given and thus also $f_1[k', \#, k''] = \delta[k', k'']$ holds.

4. Let us write a more detailed expression for Equation (9.56) if the sequence $\bar{y}_1$ is nonempty and consequently is of the form $\bar{y}_1 y_1$,

$$f_1(k', \bar{y}_1 y_1, k'') = \bigoplus_{\bar{y} \in X^*} f(k', \bar{y}, k'') \otimes d_{in}(\bar{y}, \bar{y}_1 y_1) .$$

Summation over all possible sequences $\bar{y}$ will be expressed as a summation over two classes of numbers. The first of them consists of a single number $f(k', \bar{y}, k'') \otimes d_{in}(\bar{y}, \bar{y}_1 y_1)$ for the empty sequence $\bar{y} = \#$. The second class of numbers corresponds to all other nonempty sequences of the form $\bar{y}y$. So we can write

$$\begin{aligned} f_1(k', \bar{y}_1 y_1, k'') &= f(k', \#, k'') \otimes d_{in}(\#, \bar{y}_1 y_1) \\ &\oplus \bigoplus_{\bar{y} \in X^*} \bigoplus_{y \in X} f(k', \bar{y}y, k'') \otimes d_{in}(\bar{y}y, \bar{y}_1 y_1) . \end{aligned}$$

Using expressions (9.53) for $d_{in}(\#, \bar{y}_1 y_1)$, (9.39) for $f(k', \bar{y}y, k'')$, and (9.55) for $d_{in}(\bar{y}y, \bar{y}_1 y_1)$, we obtain

$$\begin{aligned} f_1(k', \bar{y}_1 y_1, k'') &= f(k', \#, k'') \otimes d_{in}(\#, \bar{y}_1) \otimes d_{in}(\#, y_1) \\ &\oplus \bigoplus_{\bar{y} \in X^* \setminus \{\#\}} f(k', \bar{y}, k'') \otimes d_{in}(\bar{y}, \bar{y}_1) \otimes d_{in}(\#, y_1) \\ &\oplus \bigoplus_{\bar{y} \in X^*} \bigoplus_{k \in K} \left(f(k', \bar{y}, k) \otimes d(\bar{y}, \bar{y}_1) \otimes \left(\bigoplus_{y \in X} P(k, y, k'') \otimes d_{in}(y, y_1) \right) \right) . \end{aligned} \tag{9.57}$$

The first and the second lines in the preceding expression can be written in one line using the summation over all possible sentences $\bar{y}$ including the empty sentence $\#$. So Equation (9.57) can be rewritten in the form

$$\begin{aligned} f_1(k', \bar{y}_1 y_1, k'') &= \bigoplus_{\bar{y} \in X^*} f(k', \bar{y}, k'') \otimes d_{in}(\bar{y}, \bar{y}_1) \otimes d_{in}(\#, y_1) \\ &\oplus \bigoplus_{\bar{y} \in X^*} \bigoplus_{k \in K} \left(f(k', \bar{y}, k) \otimes d(\bar{y}, \bar{y}_1) \otimes \left(\bigoplus_{y \in X} P(k, y, k'') \otimes d_{in}(y, y_1) \right) \right) . \end{aligned} \tag{9.58}$$

5. The first line in (9.58) can be written owing to the distributive property of multiplication as

$$d_{in}(\#, y_1) \otimes \left(\bigoplus_{\bar{y} \in X^*} f(k', \bar{y}, k'') \otimes d_{in}(\bar{y}, \bar{y}_1) \right)$$

in which the sum in round brackets is $f_1(k', \bar{y}_1, k'')$ according to the definition (9.56). The first line in Equation (9.58) is then

$$f_1(k', \bar{y}_1, k'') \otimes d_{in}(\#, \bar{y}_1) . \tag{9.59}$$

Based on the property of the Kronecker function $\delta[k,k']$ we can write the number (9.59) as

$$\bigoplus_{k\in K} f_1(k',\bar{y}_1,k)\otimes\delta(k,k'')\otimes d_{in}(\#,y_1)\,. \tag{9.60}$$

6. We will examine the second line in (9.58). Owing to Equation (9.54) we can write

$$\bigoplus_{y\in X} P(k,y,k'')\otimes d_{in}(y,y_1) = \bigoplus_{y\in X} P(k,y,k'')\otimes\delta(y,y_1) = P(k',y_1,k'')$$

and the second line in the expression (9.58) becomes

$$\bigoplus_{k\in K} P(k,y_1,k'')\otimes\left(\bigoplus_{\bar{y}\in X^*} f(k',\bar{y},k)\otimes d_{in}(\bar{y},\bar{y}_1)\right).$$

On the base of the definition (9.56) the sum in round parenthesis is equal to

$$\left(\bigoplus_{\bar{y}\in X^*} f(k',\bar{y},k)\otimes d_{in}(\bar{y},\bar{y}_1)\right) = f_1(k',\bar{y}_1,k)$$

and for the second line in (9.58) we can write

$$\bigoplus_{k\in K} f_1(k',\bar{y}_1,k)\otimes P(k,y_1,k'')\,. \tag{9.61}$$

7. If we substitute (9.60) instead of the first line of (9.58) and substitute (9.61) instead of the second line then we obtain

$$f_1(k',\bar{y}_1y_1,k'') = \bigoplus_{k\in K} f_1(k',\bar{y}_1,k)\otimes\Big(\delta(k,k'')\otimes d_{in}(\#,y_1)\oplus P(k,y_1,k'')\Big)\,.$$

This proves that

$$f_1[k',\bar{y}_1y_1,k''] = f_1[k',\bar{y}_1,k]\bigotimes_{k\in K} P_1[k,y_1,k'']\,,$$

where

$$P_1[k,y_1,k''] = P[k,y_1,k'']\oplus\big(\delta[k,k'']\otimes d_{in}[\#,y_1]\big)\,.$$

So it has been proved that the function f_1 defined by the relation (9.56) satisfies the condition (9.41). ■

For completeness we note that $d_{in}[\#,y_1]$ is $in[y_1]$, and thus the function P_1 which is referred to in the Lemma can be expressed in the following simple way,

$$P_1[k',y,k''] = P[k',y,k'']\oplus\big(\delta[k',k'']\otimes in[y]\big)\,.$$

The preceding relation explicitly provides a constructive way for creating the function P_1 at the known functions P and in.

Proof. (of Lemma 9.5)

1. As with the proof of the previous Lemma 9.4 we will substitute the expression (9.42) for the function $F_1[\bar{y}_1]$ into (9.44). We will obtain the expression for the function $F_2[\bar{y}_2]$

$$F_2[\bar{y}_2] = \varphi[k'] \bigotimes_{k'} f_1[k', \bar{y}_1, k''] \bigotimes_{k''} \psi[k''] \bigotimes_{\bar{y}_1} d_{ch}[\bar{y}_1, \bar{y}_2]$$

which owing to the property (9.21), can be written as

$$F_2[\bar{y}_2] = \varphi[k'] \bigotimes_{k'} \left(f_1[k', \bar{y}_1, k''] \bigotimes_{\bar{y}_1} d_{ch}[\bar{y}_1, \bar{y}_2] \right) \bigotimes_{k''} \psi[k''] .$$

The previous expression implies that the function F_2 is of the required form (9.45). The function $f_2[k', \bar{y}_2, k'']$ is the convolution

$$f_2[k', \bar{y}_2, k''] = f_1[k', \bar{y}_1, k''] \bigotimes_{\bar{y}_1} d_{ch}[\bar{y}_1, \bar{y}_2] . \tag{9.62}$$

It is to be proved now that the function defined in this way satisfies the conditions (9.46) and (9.47).

2. The property (9.46) is evidently correct. If $\bar{y}_2 = \#$ then the convolution (9.62) assumes the form

$$f_2[k', \#, k''] = f_1[k', \#, k''] \otimes d_{ch}[\#, \#] ,$$

because no non-empty sentence $\bar{y}_1$ can be transformed to an empty one by changing symbols. The number $f_1(k', \#, k'')$ is $\delta(k', k'')$ according to the assumption. The number $d_{ch}(\#, \#)$ is clearly $0 = 1^{\otimes}$. Therefore $f_2(k', \#, k'')$ is equal to $\delta(k', k'')$ and the function f_2 has been proved to satisfy the condition (9.46).

3. Let the sentence $\bar{y}_2$ be non-empty and of the form $\bar{y}_2 y_2$. Let us write in detail the number $f_2(k', \bar{y}_2 y_2, k'')$ defined by (9.62),

$$f_2(k', \bar{y}_2 y_2, k'') = \bigoplus_{\bar{y}_1 \in X^*} f_1(k', \bar{y}_1, k'') \otimes d_{ch}(\bar{y}_1, \bar{y}_2 y_2) . \tag{9.63}$$

In the preceding sum only sentences $\bar{y}_1$ the lengths of which are equal to the length of the sentence $\bar{y}_2 y_2$ can be considered since no sequence of symbol changes can change the length of a sentence. For the sentences of the form $\bar{y}_1 y_1$ the lengths of which are equal to the length $\bar{y}_2 y_2$ the following holds

$$d_{ch}(\bar{y}_1 y_1, \bar{y}_2 y_2) = d_{ch}(\bar{y}_1, \bar{y}_2) \otimes d_{ch}(y_1, y_2) . \tag{9.64}$$

For the sentences of the form $\bar{y}_1 y_1$ according to assumption (9.43) the following is valid

$$f_1(k', \bar{y}_1 y_1, k'') = \bigoplus_{k \in K} f_1(k', \bar{y}_1, k) \otimes P_1(k, y_1, k'') . \tag{9.65}$$

We will include (9.64) and (9.65) into (9.63) and obtain

$$f_2(k', \bar{y}_2 y_2, k'') = \bigoplus_{\bar{y}_1 \in X^*} \bigoplus_{y_1 \in X} \left(\bigoplus_{k \in K} f_1(k', \bar{y}_1, k) \otimes P_1(k, y_1, k'') \right)$$

$$\otimes\ d_{ch}(\bar{y}_1, \bar{y}_2) \otimes d_{ch}(y_1, y_2)\,.$$

In the preceding sum the order of summation will be altered and we obtain

$$\bigoplus_{k \in K} \left(\bigoplus_{\bar{y}_1 \in X^*} f_1(k', \bar{y}_1, k) \otimes d_{ch}(\bar{y}_1, \bar{y}_2) \right) \otimes \left(\bigoplus_{y_1 \in X} P_1(k, y_1, k'') \otimes d_{ch}(y_1, y_2) \right).$$

The sum $\bigoplus_{\bar{y}_1 \in X^*} f_1(k', \bar{y}_1, k) \otimes d_{ch}(\bar{y}_1, \bar{y}_2)$ is $f_2(k', \bar{y}_2, k)$ according to the definition (9.62). We will denote

$$P_2(k, y_2, k'') = \bigoplus_{y_1 \in X} P_1(k, y_1, k'') \otimes d_{ch}(y_1, y_2) \tag{9.66}$$

and express $f_2(k', \bar{y}_2 y_2, k'')$ in the form

$$f_2(k', \bar{y}_2 y_2, k'') = \bigoplus_{k \in K} f_2(k', \bar{y}_2, k) \otimes P_2(k, y_2, k'')\,.$$

This means that the function f_2 satisfies the condition (9.47). The function P_2 is defined by expression (9.66). ■

For completeness we will express P_2 directly by means of functions that are known. The number $d_{ch}(y_1, y_2)$ is the penalty for the cheapest sequence of symbol changes which transform the symbol y_1 to the symbol y_2. This number can be calculated as the length of the shortest path between two vertices of the graph consisting of $|X|$ vertices that correspond to the symbols $x \in X$, and in which the length of the arrow from the vertex y_1 to the vertex y_2 is $ch(y_1, y_2)$. This length $d_{ch}\colon X \times X \to \mathbb{R}$ is equal to the sum $\bigoplus_{i=0}^{\infty} ch^i$. The sum was proved to equal the function $(\delta \oplus ch)^{|X|-1}$, which was denoted by means of ch^*.

The function P_2 which was stated before by means of (9.66) can be thus expressed in the form

$$P_2[k, y_2, k''] = P_1[k, y_1, k''] \bigotimes_{y_1} ch^*[y_1, y_2]$$

which shows an explicit way of constructively creating the function P_2 provided the functions P_1 and ch are already known.

Proof. (of Lemma 9.6) We will analyse a function $D[\bar{x}]$ which is defined by the expression (9.50). Let $\bar{x}$ be a sequence $x_1, x_2, \ldots, x_n$. The value of the function D for this sequence is

$$D(\bar{x}) = F_2[\bar{y}_2] \bigotimes_{\bar{y}_2} d_{de}[\bar{y}_2, \bar{x}]\,, \tag{9.67}$$

where $d_{de}[\bar{y}_2, \bar{x}]$ is now understood as a function of one variable $\bar{y}_2$ since in all further consideration $\bar{x}$ will be a fixated sequence. In this sum the addition need not be performed over all possible sequences $\bar{y}_2$, but only over those that have the form

$$\bar{y}_2 = \bar{x}_1\ x_1\ \bar{x}_2\ x_2\ \bar{x}_3\ \cdots\ \bar{x}_{n-1}\ x_{n-1}\ \bar{x}_n\ x_n\ \bar{x}_{n+1}\ , \tag{9.68}$$

where $\bar{x}_1, \bar{x}_2, \ldots, \bar{x}_{n+1}$ are sequences. In other words the addition is to be performed only over those sentences $\bar{y}_2$ that can be transformed to the sentence $\bar{x}$ by merely deleting some symbols, i.e., sentences that include the sentence $\bar{x}$ as a subsequence. For the sentence in the form of (9.68) the function $F_2[\bar{y}_2]$ has in agreement with (9.49) the following form

$$\begin{aligned} F_2[\bar{y}_2] \;=\;& \varphi[k_0] \bigotimes_{k_0} f_2[k_0, \bar{x}_1, k'_0] \bigotimes_{k'_0} P_2[k'_0, x_1, k_1] \bigotimes_{k_1} f_2[k_1, \bar{x}_2, k'_1] \\ & \bigotimes_{k'_1} P_2[k'_1, x_2, k_2] \bigotimes_{k_2} \cdots \bigotimes_{k_{n-1}} f_2[k_{n-1}, \bar{x}_n, k'_{n-1}] \\ & \bigotimes_{k'_{n-1}} P_2[k'_{n-1}, x_n, k_n] \bigotimes_{k_n} f_2[k_n, \bar{x}_{n+1}, k'_n] \bigotimes_{k'_n} \psi[k'_n] \,. \end{aligned} \tag{9.69}$$

The number $d_{de}[\bar{y}_2, \bar{x}]$ will be

$$d_{de}[\bar{x}_1, \#] \otimes d_{de}[\bar{x}_2, \#] \otimes \cdots \otimes d_{de}[\bar{x}_n, \#] \otimes d_{de}[\bar{x}_{n+1}, \#] \,. \tag{9.70}$$

If we substitute (9.70) into (9.67) and take into consideration that convolution is calculated only for sentences $\bar{y}_2$ of the form (9.68) then we will obtain

$$\begin{aligned} F_2[\bar{y}_2] \bigotimes_{\bar{y}_2} d_{de}[\bar{y}_2, \bar{x}] \;=\;& F_2[\bar{x}_1 x_1 \bar{x}_2 x_2 \bar{x}_3 \cdots \bar{x}_n x_n \bar{x}_{n+1}] \bigotimes_{\bar{x}_1} d_{de}(\bar{x}_1, \#) \\ & \bigotimes_{\bar{x}_2} d_{de}(\bar{x}_2, \#) \bigotimes_{\bar{x}_3} \cdots \bigotimes_{\bar{x}_{n+1}} d_{de}(\bar{x}_{n+1}, \#) \,. \end{aligned}$$

If we include into the previous expression a detailed statement (9.69) then we will find out that the function $D[\bar{x}] = D[x_1, x_2, \ldots, x_n]$ has the form

$$\begin{aligned} D[\bar{x}] \;=\;& \varphi[k_0] \bigotimes_{k_0} P'[k_0, x_1, k_1] \bigotimes_{k_1} P'[k_1, x_2, k_2] \bigotimes_{k_2} \cdots \\ & \bigotimes_{k_{n-1}} P'[k_{n-1}, x_n, k_n] \bigotimes_{k_n} \psi'[k_n] \,, \end{aligned}$$

where the function $P' \colon K \times X \times K \to \mathbb{R}$ is

$$P'[k', x, k''] = f_2[k', \bar{x}, k] \bigotimes_{\bar{x}} d_{de}[\bar{x}, \#] \bigotimes_{k} P_2[k, x, k''] \,, \tag{9.71}$$

and the function $\psi' \colon K \to \mathbb{R}$ is

$$\psi'[k] = f_2[k', \bar{x}, k''] \bigotimes_{\bar{x}} d_{de}[\bar{x}, \#] \bigotimes_{k''} \psi[k''] \,. \tag{9.72}$$

This concludes the proof of Lemma 9.6. ■

The expressions (9.71) and (9.72), however, do not yield any constructive way for calculating the functions P' and ψ' since they include the convolution $f_2[k', \bar{x}, k''] \bigotimes_{\bar{x}} d_{de}[\bar{x}, \#]$ over the infinite set of all possible sequences $\bar{x}$. Such a constructive calculation is possible owing to Lemma 9.3 which asserts that this infinite convolution is

$$\left(\delta[k', k''] \oplus \left(P_2[k', x, k''] \bigotimes_{x} de[x]\right)\right)^{|K|-1} .$$

Explicit expressions for calculating the functions P' and ψ' thus are

$$\begin{aligned} P'[k', x, k''] &= \left(\delta[k', k] \oplus (P_2[k', x, k] \bigotimes_{x} de[x])\right)^{|K|-1} \bigotimes_{k} P_2[k, x, k''] \,, \\ \psi'[k] &= \left(\delta[k, k'] \oplus (P_2[k, x, k'] \bigotimes_{x} de[x])\right)^{|K|-1} \bigotimes_{k'} \psi[k'] \end{aligned}$$

which show a constructive way how they can be obtained on the basis of the known functions P_2, ψ and de.

The three proved lemmata 9.4, 9.5 and 9.6 prove Theorem 9.2, and thus the equivalent Theorem 9.1 as well. It is because these three lemmata show how to find the functions P' and ψ' the existence of which is mentioned in the above two theorems. The functions P' and ψ' are to be created on the basis of the five functions P, ψ, in, ch, de by means of the following seven steps.

Algorithm 9.1 Constructive calculation of functions P' and ψ'.

1. The function $P_1 \colon K \times X \times K \to \mathbb{R}$ is calculated,

$$P_1[k', y, k''] = P[k', y, k''] \oplus (\delta[k', k''] \otimes in[y]) \,. \tag{9.73}$$

   The complexity of this calculation is $\mathcal{O}(|K|\,|X|\,|K|)$.

2. The function $ch^* \colon X \times X \to \mathbb{R}$ is calculated, i.e., in the following way. First, the function ch^* is expressed as $\delta \oplus ch$ then the following operator is used repeatedly but not more than $\log |X|$-times,

$$ch^*[x, y] ::= ch^*[x, z] \bigotimes_{z} ch^*[z, y] \,. \tag{9.74}$$

This is performed until the function ch^* stops changing. The complexity of a single calculation of (9.74) is obviously $\mathcal{O}(|X|^3)$, and the overall complexity of creating the function ch^* is $\mathcal{O}(|X|^3 \log |X|)$ in the worst case.

3. The function $P_2 \colon K \times X \times K \to \mathbb{R}$ is calculated,

$$P_2[k', x, k''] = P_1[k', y, k''] \bigotimes_y ch^*[y, x] \,. \tag{9.75}$$

The complexity of this calculation is $\mathcal{O}(|K|^2 \, |X|^2)$.

4. An auxiliary function $q \colon K \times K \to \mathbb{R}$ is calculated

$$q[k', k''] = \delta[k', k''] \oplus (P_2[k', x, k''] \bigotimes_x de[x]) \,. \tag{9.76}$$

The complexity of this calculation is $\mathcal{O}(|K|^2 \, |X|)$.

5. The auxiliary function $q^* \colon K \times K \to \mathbb{R}$ is calculated, i.e., in the following way. First, the function q^* is substituted by q. Then repeatedly but not more than $\log |K|$-times the following operator is used

$$q^*[k', k''] ::= q^*[k', k] \bigotimes_k q^*[k, k''] \,, \tag{9.77}$$

until q^* stops changing. The computational complexity is $\mathcal{O}(|K|^3 \log |K|)$.

6. The function $P' \colon K \times X \times K \to \mathbb{R}$ is calculated,

$$P'[k', x, k''] = q^*[k', k] \bigotimes_k P_2[k, x, k''] \,. \tag{9.78}$$

The complexity of calculation is $\mathcal{O}(|K|^3 \, |X|)$.

7. The function $\psi' \colon K \to \mathbb{R}$ is calculated,

$$\psi'[k] = q^*[k, k'] \bigotimes_{k'} \psi[k'] \,. \tag{9.79}$$

The complexity of this calculation is $\mathcal{O}(|K|^2)$.

All calculations quoted here do not depend on the actual sequence $\bar{x}$ for which Levenstein dissimilarity with a known regular language is calculated. If the dissimilarities are sought for various different sentences $\bar{x}$ with the same automaton φ, P, ψ and the same penalty functions *in*, *ch*, *de* then the above relations can be calculated in advance only once for all sentences that will be analysed in the future.

If the functions P' and ψ' are already at our disposal then Levenstein dissimilarity $D(\bar{x})$ is calculated for each sentence $\bar{x} = (x_1, x_2, \ldots, x_n)$ as a convolution expression

$$D(\bar{x}) = \varphi[k_0] \bigotimes_{k_0} P'[k_0, x_1, k_1] \bigotimes_{k_1} \cdots \bigotimes_{k_{n-1}} P'[k_{n-1}, x_n, k_n] \bigotimes_{k_n} \psi'[k_n] \,.$$

One of the possible procedures of this calculation is a gradual building of the functions $f_0, f_1, \ldots, f_n$ according to

$$\left.\begin{aligned} f_0[k_0] &= \varphi[k_0]\,, \\ f_i[k_i] &= f_{i-1}[k_{i-1}] \bigotimes_{k_{i-1}} P'[k_{i-1}, x_i, k_i]\,, \quad i = 1, 2, \ldots, n\,, \\ D(\bar{x}) &= f_n[k_n] \bigotimes_{k_n} \psi'[k_n]\,. \end{aligned}\right\} \tag{9.80}$$

Note that this is by no means the only possibility.

9.5.10 Nonconvolution interpretation of the main result

For a final solution of the Levenstein task the solution in a convolution form is to be transformed to the form in which the task was originally formulated. The original formulation of the task reads as follows.

Let us have two finite sets X and K and three functions $\varphi\colon K \to \{0, \infty\}$, $P\colon K \times X \times K \to \{0, \infty\}$, $\psi\colon K \to \{0, \infty\}$ which determine a language, i.e., a set L of sequences $\bar{x} = (x_1, x_2, \ldots, x_n)$. The sequence $\bar{x} = (x_1, x_2, \ldots, x_n)$ belongs to L if and only if there exists a sequence $k_0, k_1, \ldots, k_n$ for which the following holds

$$\begin{aligned} \varphi(k_0) &= 0\,, \\ P(k_{i-1}, x_i, k_i) &= 0\,, \quad i = 1, \ldots, n\,, \\ \psi(k_n) &= 0\,. \end{aligned}$$

Furthermore let us have three functions $in\colon X \to \mathbb{R}$, $ch\colon X \times X \to \mathbb{R}$, $de\colon X \to \mathbb{R}$ which determine the function $d\colon X^* \times X^* \to \mathbb{R}$ the value $d(\bar{y}, \bar{x})$ of which denotes Levenstein dissimilarity of the sentences $\bar{x}$ and $\bar{y}$. An algorithm is to be found in the task which for each sentence $\bar{x} \in X^*$ will find the number

$$D(\bar{x}) = \min_{\bar{y} \in L} d(\bar{y}, \bar{x}) \tag{9.81}$$

which is called Levenstein dissimilarity between the sentence $\bar{x}$ and the language L.

The algorithm for calculating the number $D(\bar{x})$ consists of two parts. The first part are preliminary calculations the complexity of which is not greater than $\big(\mathcal{O}(|K| \log |K|),\ \mathcal{O}(|X^*| \log |X|),\ \mathcal{O}(|K|^3\, |X|),\ \mathcal{O}(|K|^2\, |X|)\big)$. These calculations do not depend on the input sentence $\bar{x}$ and are calculated only once for the given language L and Levenstein dissimilarity. The second part are calculations which depend on the given sentence and the complexity of which is $\mathcal{O}(|K|^2\, n)$, where n is the length of the sentence $\bar{x}$.

Now we will present the calculations of the first part which were expressed before by means of convolution formulæ (9.73) through (9.79). We will explain the necessary computations. The explanations will rest on the representation of the language L by a finite automaton.

1. A function $P_1\colon K \times X \times K \to \mathbb{R}$ will be created (see formula (9.73)),

$$P_1(k', y, k'') = \begin{cases} \min\big(P(k', y, k''), in(y)\big)\,, & \text{if } k' = k''\,, \\ P(k', y, k'')\,, & \text{if } k' \neq k''\,. \end{cases}$$

The number $P_1(k', y, k'')$ represents the minimal penalty for adding the symbol y to the end of the sentence under the condition that the automaton was in the state k', and reached the state k'' after adding. For $k' \neq k''$ there exists only one way of adding. The symbol y must be generated by the automaton and the penalty in this case is $P(k', y, k'')$. There are two options for how to add a symbol y if $k' = k'' = k$. The first option is that the automaton generates symbol y and is penalised by $P(k, y, k)$. In the second option the operation of the automaton is interrupted, the symbol y is inserted at the end of the already generated sequence and a penalty $in(y)$ is paid. It is quite natural that the cheapest alternative is selected.

2. A function $ch^*\colon X \times X \to \mathbb{R}$ will be created (see formula (9.74)) for instance in this way. First the numbers $ch^*(x, y)$ will be created such that $ch^*(x, y) = 0$ for $x = y$ and $ch^*(x, y) = ch(x, y)$ for $x \neq y$. Then the numbers are transformed many times by the operator

$$ch^*(x, y) ::= \min_{z \in X} \Big(ch^*(x, z) + ch^*(z, y) \Big) .$$

The number $ch^*(x, y)$ corresponds to the penalty for the cheapest chain of changes by which the symbol x is transformed to the symbol y, i.e.,

$$ch^*(x, y) = \min_{x_1, x_2, \ldots, x_n} \left(ch(x, x_1) + \sum_{i=2}^{n} ch(x_{i-1}, x_i) + ch(x_n, y) \right) ,$$

where the length n is not known beforehand and can even be zero.

3. A function $P_2\colon K \times X \times K \to \mathbb{R}$ will be created (see formula (9.75)),

$$P_2(k', x, k'') = \min_{y \in X} \big(P_1(k', y, k'') + ch^*(y, x) \big) .$$

The number $P_2(k', x, k'')$ represents the minimal penalty for adding the symbol x to the end of the sentence under the condition that the automaton was in the state k' before adding, and got to the state k'' after adding. It is a number resembling the number $P_1(k', x, k'')$, but there is an essential difference between them. The added symbol can be either generated by the automaton or inserted at the end of the sequence. Afterwards the added symbol can be changed by an arbitrary long sequence of changes or not changed at all. Thus now the number $P_2(k', x, k'')$ is the result of optimisation on a quite extent set.

4. An auxiliary function $q\colon K \times K \to \mathbb{R}$ is calculated

$$q(k', k'') = \begin{cases} 0 , & \text{if } k' = k'' , \\ \min\limits_{x \in K} \big(P_2(k', x, k'') + de(x) \big) , & \text{if } k' \neq k'' . \end{cases}$$

The number $q(k', k'')$ is the price of the cheapest process by which the automaton passes from the state k' to the state k'', but the generated sentence was not changed as a result of the whole process, though it was changed during the process. Any process of this class consists of:

- Adding a symbol at the end of the sequence of symbols;
- Sequence (even an empty one) of arbitrary changes of newly added symbol;
- Deletion of added and changed symbol from the sentence.

5. An auxiliary function $q^* : K \times K \rightarrow \mathbb{R}$ will be calculated (see formula (9.77)). First is the substitution

$$q^*(k', k'') = \begin{cases} 0\,, & \text{if } k' = k''\,, \\ q(k', k'')\,, & \text{if } k' \neq k''\,. \end{cases}$$

Then the operator will be repeatedly used

$$q^*(k', k'') ::= \min_{k \in K} \left(q^*(k', k) + q^*(k, k'') \right) .$$

The number $q^*(k', k'')$ is similar to the number $q(k', k'')$. There is a difference between these numbers as it was assumed when creating the number $q(k', k'')$ that the automaton generated only one symbol or only one symbol was inserted at the end of the sequence. This symbol was then further manipulated. When the number $q^*(k', k'')$ is being created a situation is taken into account when the automaton can generate any sequence of symbols (it can be an empty sequence as well as a rather long one). The automaton begins to generate symbols in the state k' and finally gets to the state k''. Apart from that any symbols can be inserted to the obtained sequence. The sequence generated is subject to a number of changes until it ends in deleting all generated symbols. The price for the least expensive procedure of this class is $q^*(k', k'')$.

6. A function $P' \colon K \times X \times K \rightarrow \mathbb{R}$ will be calculated (see formula (9.78)),

$$P'(k', x, k'') = \min_{k \in K} \left(q^*(k', k) + P_2(k, x, k'') \right) .$$

The function P' resembles the functions P_1 and P_2. The number $P'(k', x, k'')$ is the price for the least expensive procedure of the following class. The automaton which is in the state k' will generate a sentence. Then each symbol of the sentence is subject to changes and is deleted. Then the automaton will either generate the symbol x' or the symbol will be inserted at the end of the sentence. Being the last symbol, it is subject to repeated changes until it is changed to the symbol x which is no longer deleted. Through the above procedure the automaton gets to the state k''.

7. A number $\psi'(k)$ will be calculated (see formula (9.79)),

$$\psi'(k) = \min_{k' \in K} \left(q^*(k, k') + \psi(k') \right) .$$

The number $\psi'(k)$ represents the price for the least expensive procedure of the following class. The automaton which is in the state k generates a sentence and gets to the state k' where it stops. Then each symbol in the generated sentence is changed until it is finally deleted.

The computational procedure (9.80) has the following form when explained in natural language. For each sequence $\bar{x} = (x_0, x_1, \ldots, x_n)$ the numbers $f_i(k)$, $k \in K$, $i = 0, 1, \ldots, n$, and the number $D(x)$ are to be calculated according to the formulæ

$$\begin{aligned}
f_0(k_0) &= \varphi(k_0)\,, \quad k_0 \in K\,, \\
f_i(k_i) &= \min_{k_{i-1} \in K} \big(f_{i-1}(k_{i-1}) + P'(k_{i-1}, x_i, k_i)\big)\,, \quad k_i \in K\,, \quad i = 1, 2, \ldots, n\,, \\
D(\bar{x}) &= \min_{k_n \in K} \big(f_n(k_n) + \psi'(k_n)\big)\,.
\end{aligned}$$

We can see that mere explanation of the already validated algorithm using natural language is rather difficult. The explanation becomes inevitably lengthy and consequently not transparent and convincing. It could be even worse if the algorithm were not available yet and should be created and validated by so called reasonable consideration. In such a case the risk becomes rather great that some hardly noticeable peculiarities of the problem will be omitted and consequently an erroneous outcome will be obtained.

We choose the formal way in this lecture to construct the algorithm for Levenstein approximation. The formal convolution expressing the problem is equivalently transformed step by step until the convolution expression is obtained that represents the algorithm. Every step in this deducing is guided by formal rules equivalently transforming convolutions and not by a vague reasoning. Such a way is not very amusing but it excludes unfortunate inadvertence.

9.6 Discussion

I have noticed a substantial difference between Lecture 8 and Lecture 9, even when their topics are close, both dealing with recognising sequences. The preceding Lecture 8 actively uses results from the general statistical pattern recognition theory. The present Lecture 9 is quite different. It seems to me as if the problem started being examined from another side and from the very beginning. Well, in the substantial part of the lecture the term 'probability' does not occur even once and the outcomes of the preceding lectures are not made use of. On the whole this lecture could be placed at the beginning of the course and nothing would obstruct understanding it. It seems to me that through this lecture the explanation loses its clearly ordered structure, in which its individual parts clung closely to each other. I see Lecture 9 as if hanging in the air, and so a number of questions arise. Was not the explanation on structural recognition started from some other, nonstatistical standpoint? Or, have not I, perhaps, overlooked an important relationship between current and previously explained matter.

You have hardly overlooked anything very important. It may rather have escaped our notice. But if you feel a little bit confused then it is only because you expect more from the theory of pattern recognition than it can offer you in its present day state of the art. For the time being the theory does not present a well worked out hierarchy of tasks according to which their relationship as a 'task equivalence type' or the 'task A is a particular case of task B' type would be evident, and which would involve the task generalising all the tasks in the hierarchy as well. The present day theory of pattern recognition is still something other than, say, linear programming. The frame of linear programming is defined by a single task, which, moreover, can be expressed in a brief and illustrative way. In pattern recognition, and even in part of it, i.e., structural recognition, several tasks occur.

Two previous lectures were devoted to two groups of related tasks. The basis of the first task group is the assumption of a random character of an observed object, in the given case it is of a random character of the observed sentence in a certain alphabet. A number of pattern recognition tasks can be naturally formulated within the limits of this assumption as a statistical estimate of unobserved parameters.

In the second group the formulation of tasks is not based on the statistical model of an object but on the following two concepts. The first of them is a certain subset L in the set X of all possible observations. In our case it was a regular language. The subset L can be considered as a set of some ideal, non-damaged observations. The algorithm which is to reveal important properties of the observation is assumed to be easily implemented in the case in which the observation belongs to L. For example, it is assumed to be known to which class each observation from the set L is to be included, which is the most frequent assumption. In the case in which the observation is different from the set L rather complicated calculations are needed.

The second important concept is the function $d: X \times X \rightarrow \mathbb{R}$ which for each pair of observations x and y finds the number $d(x, y)$. The function $d(x, y)$ formalises intuitive considerations to which extent the observations x and y are similar to each other.

On the basis of these two concepts the following recognition procedure is defined, i.e., is considered as a postulate. For any observation $x \in X$ an ideal observation $y^* \in L$ is sought that is the most similar to the observation x. Then the observation x is assumed to have the same properties as the ideal observation y^*. It is assumed, for instance, that the observation x belongs to the same class of observations as the observation y^*.

Our Lecture 9 is devoted to the way in which the preceding procedure is to be implemented if L is a regular language and the dissimilarity function belongs to the class of Levenstein functions. Your assumption that this class of tasks seems as if to hang in the air is wrong. In situations when observation is considered as a point in a linear space, these methods are not less known than statistical methods that were discussed in the first part of our course. This class of methods is known as nearest neighbour methods.

Thus we can understand your question as a question about a relation between the statistical methods and the nearest neighbour method. This question is really rather difficult for us. The best answer to the question would then be that we do not know. We know, of course, a number of rather trivial examples in which the statistically formulated task can be simplified to a nearest neighbour method, and also vice versa. The two classes described have a relatively large intersection. On the other hand, for some nontrivial tasks such a reduction has not been found. This also applies to the task of Levenstein matching of a sentence to a regular language which was discussed in Lecture 9.

We would like to add an important comment. For the time being we do not know a universal formulation that would generalise the tasks of Lecture 8 and Lecture 9. But along with you we have become convinced that the algorithms for solving them are more than only being related. The algorithms are, actually, identical and form the link between the subject matter of those two lectures.

Will the following explanation be based on statistical estimation methods or on nearest neighbour methods?

Our explanation will be based on both the approaches as much as possible.

I expect that, beginning with this lecture, the nearest neighbour methods will gradually acquire more weight. But I am surprised that you did not devote at least one lecture to these methods in the first part of the course.

You yourself answered this question at the beginning of the discussion. It is because the structural recognition, based on the nearest neighbour approach, can be dealt with without preliminarily explaining general features of these methods. It naturally does not mean, however, that general features would not be worth their own price. If you did not know them, it would be a serious gap in your education. But there is an extensive literature by other authors at your disposal.

The mutual similarity of algorithms for solving different structural recognition tasks, which you so much draw attention to, is quite clear to me now. I have also noticed that the whole class of mutually similar algorithms also have a common drawback which cannot be overlooked.

I assume that there is a sequence $x_1, x_2, \ldots, x_i, \ldots, x_n$ of observations at my disposal and I am to determine the sequence of directly unobservable states $k_0, k_1, \ldots, k_n$ by means of one of the procedures described in the lectures. All the algorithms described require that first the whole sequence $x_1, \ldots, x_n$ should be examined, and then, at once, the whole sequence $k_0, k_1, \ldots, k_n$ should be decided upon. Even if the correctness of such a procedure is clear to me, it seems to me quite unnatural that the decision about the first element k_0 in the sought sequence can be made only when the last element x_n is arrived at. I have already come across this obstacle even in the simplest applications in which I do not know when I can regard the observation sequence has ended. When I am reading a book, for example, I have to decide, sooner or later, what the first symbol in the book is like. When am I able to do it? After I see over the first word as a whole? Or after I come to the end of a line? Or of a page?

...or at the end of the book? We cannot help turning your question into a joke. When we go further in the direction of your consideration we can find that the

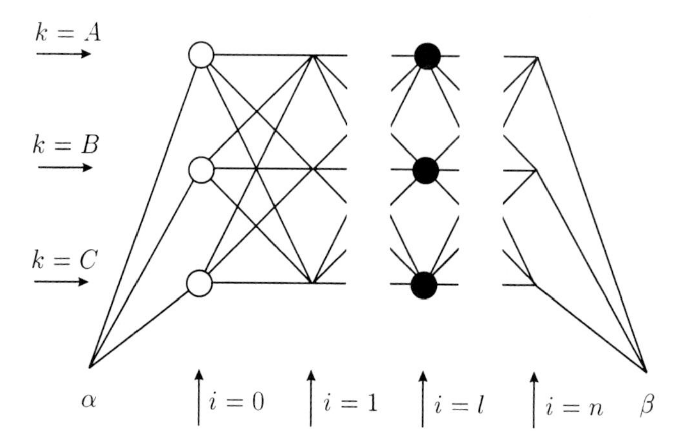

Figure 9.5 Structural analysis of a sequence with not all lengths of graph edges known.

content of one book depends on what has been written in the other book, then consequently you have to read all books first and only then to decide what was the meaning of the first symbol in the first book. One could write poetry on that!

And the content of all such poems can be expressed formally. Let $x_1, x_2, \ldots, x_n$ be a sequence and let $k_0^*, k_1^*, \ldots, k_n^*$ be a sequence defined as follows,

$$(k_i^* \mid i = 0, \ldots, n) = \operatorname*{argmax}_{k_0} \max_{k_1} \cdots \max_{k_n} \left(\sum_{i=1}^{n} f_i(k_{i-1}, x_i, k_i) \right) . \tag{9.82}$$

It is to be discovered under what conditions the first element k_0^* in the sequence $k_0^*, k_1^*, \ldots, k_n^*$ can be uniquely determined if only the initial part $x_1, x_2, \ldots, x_l$, $l < n$, of the sequence $x_1, x_2, \ldots, x_n$ is available.

This question was asked by V.A. Kovalevski of himself immediately after he had designed the first algorithm known for the structural analysis of a sequence [Kovalevski, 1967]. We will illustrate your question by means of a graph in Fig. 9.5 which we have already come across earlier in Fig. 8.2. In the same way as before the graph is formed by the vertices α and β and a group having $|K|\,n$ vertices. Each of vertices is represented as a point (k, i), $k \in K$, $i = 0, \ldots, n$, with the horizontal coordinate i and the vertical coordinate k. The graph presented illustrates a situation where the set K consists of three states A, B, C.

Your question can now be asked in the following way. Assume you have examined the given graph only as far as to a horizontal coordinate l, where $l < n$. This means that you know only the lengths of those edges lying to the left of the coordinate l, and the other lengths are still unknown to you. In such a situation, of course, you cannot find the shortest path from the vertex α to the vertex β, because the path depends on data still unknown to you. But you are not interested in the shortest path as a whole. You are interested only through which leftmost vertex the path passes. In Fig. 9.5 these three vertices are marked by white circles.

You could answer this question after the following considerations. Even if you do not know what the shortest path from α to β will be, one thing is certain. It passes through one of the vertices the coordinate of which in the horizontal direction is l. In Fig. 9.5 these vertices are marked by black circles (A, l), (B, l) and (C, l). You can find the shortest path from the vertex α to every of the vertices marked by black circles. Thus you will find three paths: the shortest path from the vertex α to the vertex (A, l), and similarly those to the vertex (B, l) and the vertex (C, l). Now, you already know that whatever the path from the vertex α to the vertex β may be, its initial section must be one of the three paths, which you already know. For each of the three paths you will find through which vertex with a horizontal coordinate $i = 0$, i.e., through which vertex marked by a white circle in Fig. 9.5, it passes. Finally, you will arrive at the following conclusion. If each of three found paths passes through the same vertex marked by a white circle, e.g., through the vertex $(A, 0)$, then the vertex $(A, 0)$ belongs to the shortest path from α to β.

And what about when the given condition is not satisfied?

This means that a partial knowledge of the graph is not yet sufficient for an unambiguous decision on the state k_0 and you have to continue examining it. Is this informal explanation clear to you?

Yes, it is!

Well, write down, please, an exact answer to your question.

I will introduce a number $F_l(k_l)$

$$F_l(k_l) = \max_{k_0} \max_{k_1} \cdots \max_{k_{l-1}} \sum_{i=1}^{l} f_i(k_{i-1}, x_i, k_i) \tag{9.83}$$

and I will denote as $k_{0l}(k_l)$ the first element in the sequence $k_0^*, k_1^*, \ldots, k_{l-1}^*$, in which the sought maximum (9.83) is reached. The numbers $F_l(k_l)$ and $k_{0l}(k_l)$ are calculated according to the following recurrent relations

$$\begin{aligned}
F_1(k'') &= \max_{k_0} f_1(k_0, x_1, k'') \,, \\
k_{01}(k'') &= \operatorname*{argmax}_{k} f_1(k, x_1, k'') \,, \\
F_l(k'') &= \max_{k \in K} \Big(F_{l-1}(k) + f_l(k, x_{l-1}, k'') \Big) \,, \\
k_{0,l}(k'') &= k_{0,l-1}\left(\operatorname*{argmax}_{k \in K} \Big(F_{l-1}(k) + f_l(k, x_{l-1}, k'') \Big) \right) .
\end{aligned}$$

If for some value l all the values $k_{0l}(k'')$, $k'' \in K$, are the same, and this value is k^*, for example, it means that the initial part of the observation $x_1, x_2, \ldots, x_l$ is sufficient to determine unambiguously the element k_0^* in the sequence $k_0^*, k_1^*, \ldots, k_n^*$ which maximises the sum

$$\sum_{i=1}^{n} f_i(k_{i-1}, x_i, k_i) .$$

And this element is k^*.

Try to imagine now the situation in which you did not start observing the sequence from the very beginning, but from somewhere in the middle. You can assume that the object observed by you started working at the moment $-T$ in the past and will be working until to the moment T in the future. During its functioning it generated a sequence of symbols

$$x_{-T+1}, x_{-T+2}, \ldots, x_{-1}, x_0, x_1, \ldots, x_{T-2}, x_{T-1} . \tag{9.84}$$

If the preceding sequence were known then you could find the most probable sequence of the states

$$k_{-T}^*, k_{-T+1}^*, \ldots, k_{-1}^*, k_0^*, k_1^*, \ldots, k_{T-1}^*, k_T^* \tag{9.85}$$

through which the object had passed. You did not note, however, the moment when the object started working. You only observe a section of the sequence (9.84) from x_{-l+1} till x_{l-1}. You are expected to answer the question whether the information is sufficient for uniquely determining the element k_0^* in the sequence (9.85). In the case of a positive answer, you are also to determine what the element k_0^* is equal to.

I will denote the number $F_{ij}(k_i, k_j)$, $-l \le i < j \le l$, $k_i \in K$, $k_j \in K$,

$$F_{ij}(k_i, k_j) = \max_{k_{i+1}} \max_{k_{i+2}} \cdots \max_{k_{j-2}} \max_{k_{j-1}} \sum_{t=i}^{j} f_t(k_{t-1}, x_t, k_t) . \tag{9.86}$$

I will denote by $k_{ij}(k_i, k_j)$ the element k_0 in the sequence k_{-l+1}, $k_{-l+2}, \ldots$, k_{l-2}, k_{l-1}, which maximises (9.86). The answer to your question is: if the vertex $k_{-l,l}(k_{-l}, k_l)$ is the same for each pair $k_{-l} \in K, k_l \in K$, say k_0^*, then the sequence x_{-l+1}, x_{-l+2}, $\ldots$, x_{l-2}, x_{l-1} is sufficient for uniquely determining the element k_0 in the sequence $k_{1-T}, k_{-T+1}, \ldots, k_{T-1}, k_T$, which maximises the sum $\sum_{i=-T+1}^{T} f_i(k_{i-1}, x_i, k_i)$. This element is just k_0^*. I hope I need not write down the recursive formulæ for calculating the functions F_{ij} and k_{ij}?

No, you need not. It may be evident enough for anybody who has thoroughly studied the two previous lectures. And you have mastered them in an excellent way.

I would like to discuss one more modification of the examined task with you. We have studied a case in which we wanted to determine what was the first element k_0 in the sequence $k_0, k_1, \ldots, k_n$ which was the most probable with respect to the sequence $x_1, x_2, \ldots, x_n$ being observed. We wanted to determine it when we did not have a complete sequence $x_1, x_2, \ldots, x_n$ at our disposal, but only its initial part.

But if we are interested only in the first element and we would like to determine it with the least probability of the wrong decision, we should take quite another approach to the task. You have mentioned this several times and I absolutely agree with you. In this case if we had the complete sequence $x_1, x_2, \ldots, x_n$ at our disposal then we should calculate $|K|$ numbers $p(k_0, x_1, x_2, \ldots, x_n)$, $k_0 \in K$, and select a value k_0 from them corresponding to the largest of these numbers. I wonder what you will advise me to do if I do not have the entire sequence $x_1, x_2, \ldots, x_n$ at my disposal but only its part $x_1, x_2, \ldots, x_l$. On the one hand, I can calculate $|K|$ numbers $p(k_0, x_1, x_2, \ldots, x_l)$, $k_0 \in K$, and evaluate k_0 on the basis of the information I already have. Though the information will be used in an optimal way, the quality of decision will be worse in the general case than that attainable in observing the complete sequence $x_1, x_2, \ldots, x_n$.

I would intend not only to make an optimal use of the available information $x_1, x_2, \ldots, x_l$, but, moreover, to have a criterion which would guarantee that the quality of decision on the basis of the information is not a bit worse than the quality of decisions I would attain through further observations $x_{l+1}, x_{l+2}, \ldots$ In other words, I am looking for an answer to the question whether the infinite part of the sequence of observations $x_{l+1}, x_{l+2}, \ldots, x_i, \ldots$ is negligible from the standpoint of information gain, which it yields as to the state k_0. When the criterion was met I could interrupt my observations and estimate the state k_0. I could be certain that the estimated quality cannot be enhanced by any further (even infinite) observation of the object. If the criterion was not met then the observation should continue.

The question formulated like this evokes an assumption that I should look for an answer within the frame of Wald sequential analysis. I have not thought about it thoroughly enough, but for the time being I have arrived at the conclusion that I will not find the answer there. I suspect that the answer to my question is either negative, or very complicated.

If I denote the probability $p(k_0, x_1, x_2, \ldots, x_l)$ by $p_l(k_0)$ then I can formulate the question as follows. Does a number l exist such that at any $i > l$ the probabilities $p_i(k_0)$ will be identical to the probabilities $p_l(k_0)$? This question can be answered positively in the trivial case only, when the observation x_i, $i > l$, and the state k_0 are statistically independent events. But here I deal with Markovian models, where all parameters are dependent on each other. This case was illustrated in Lecture 8 with a mechanical model. It follows that even with a rather large i each observation x_i yields some, probably small, information on the state k_0. The result of it is quite unfortunate, i.e., I cannot decide on the first symbol in the book before I read through the whole book. Is there any other possibility for me than to call for help?

You have missed significant circumstances from which a constructive answer to your question directly follows. But we are pleased at your asking interesting questions like that and at their excellent formulations.

Based on the observation $x_1, x_2, \ldots, x_l$ you can calculate not only $|K|$ numbers $p(k_0,\ x_1,\ x_2,\ \ldots,\ x_l)$, $k_0 \in K$, but $|K|$ such groups $(p(k_0,\ x_1,\ x_2,\ \ldots,\ x_l, k_l),\ k_0 \in K)$, $k_l \in K$, in which the k_l-th group corresponds to the state k_l in the instant l. These groups do not unambiguously determine the ensemble of numbers $p(k_0, x_1, x_2, \ldots, x_i, \ldots)$ which corresponds to an infinitely long observation. But fortunately they determine quite simply a set that is sure to contain the particular set. For briefness we will introduce the notation q_l for a $|K|$-dimensional vector the coordinates of which are probabilities $p(k_0,\ x_1,\ x_2,\ \ldots,\ x_l)$. Similarly, we will denote by $q_l(k_l)$, $k_l \in K$, $|K|$-dimensional vector the k_0-th coordinate of which is the probability $p(k_0, x_1, x_2, \ldots, x_l, k_l)$. We are interested in the probabilities $p(k_0, x_1, x_2, \ldots, x_n)$, $n > l$, (i.e., a vector q_n) which cannot be calculated because the observations $x_{l+1}, x_{l+2}, \ldots, x_n$ are not known. However, it follows from the Markovian character of the model examined that these probabilities are

$$p(k_0, x_1, x_2, \ldots, x_n) = \sum_{k_l \in K} p(k_0, x_1, x_2, \ldots, x_l, k_l)\, p(x_{l+1}, x_{l+2}, \ldots, x_n \mid k_l)\,.$$

We will write the preceding relation in vector form

$$q_n = \sum_{k_l \in K} q_l(k_l)\, p(x_{l+1}, x_{l+2}, \ldots, x_n \mid k_l)\,. \tag{9.87}$$

The preceding relation states that the vector q_n (i.e., the ensemble $p(k_0, x_1, x_2, \ldots, x_n)$, $k_0 \in K$) belongs to a convex cone the boundary of which is formed by vectors $q_l(k_l)$ (i.e., the ensembles $\big(p(k_0,\ x_1,\ x_2,\ \ldots,\ x_l,\ k_l),\ k_0 \in K\big)$, $k_l \in K$). We will denote the cone by Q_l. The question can then be formulated as follows. When can we unambiguously find what $\operatorname{argmax}_{k_0 \in K} p(k_0, x_1, x_2, \ldots, x_n)$ is equal to if we only know that the ensemble $p(k_0,\ x_1,\ x_2,\ \ldots,\ x_n)$ belongs to the cone Q_l? And the answer to the question is quite evident.

If for any ensemble $p'(k_0)$, $k_0 \in K$, belonging to the cone Q_l the $\operatorname{argmax}_{k_0 \in K} p'(k_0)$ results in the same value k_0^* then $\operatorname{argmax}_{k_0 \in K} p(k_0,\ x_1,\ x_2,\ \ldots, x_n)$ is also k_0^*. It also holds vice versa. If the cone Q_l contains two such ensembles $(p'(k_0) | k_0 \in K)$ and $(p''(k_0) | k_0 \in K)$, that $\operatorname{argmax}_{k_0 \in K} p'(k_0) \neq \operatorname{argmax}_{k_0 \in K} p''(k_0)$, then from the statement that the ensemble $p(k_0,\ x_1,\ x_2,\ \ldots, x_n)$ belongs to the cone no conclusion can be drawn what $\operatorname{argmax}_{k_0 \in K} p(k_0,\ x_1,\ x_2,\ \ldots,\ x_n)$ is equal to.

To find if the conditions of the previous statements are satisfied, not all points of the cone Q_l are to be examined (and they are indefinitely many). It is sufficient to examine only the cone boundaries, i.e., the ensembles $\big(p(k_0, x_1, x_2, \ldots, x_l, k_l),\ k_0 \in K\big)$, $k_l \in K$. You are sure to prove the following two assertions, the latter of which is quite trivial, the former not being very complicated either.

Let a value $k_0^* \in K$ exist such that for each $k_0 \in K$ and $k_l \in K$ the following inequality holds

$$p(k_0^*, x_1, x_2, \ldots, x_l, k_l) \geq p(k_0, x_1, x_2, \ldots, x_l, k_l)\,.$$

Then independently of observations $x_{l+1}, x_{l+2}, \ldots, x_n$ the following inequality holds

$$p(k_0^*, x_1, x_2, \ldots, x_n) \geq p(k_0, x_1, x_2, \ldots, x_n) .$$

Let two values k_l' and k_l'' exist such that

$$\operatorname*{argmax}_{k_0 \in K} p(k_0, x_1, x_2, \ldots, x_l, k_l') \neq \operatorname*{argmax}_{k_0 \in K} p(k_0, x_1, x_2, \ldots, x_l, k_l'') .$$

In this case with some probabilities $p(x_{l+1}, x_{l+2}, \ldots, x_n \mid k_l)$ the value $\operatorname{argmax}_{k_0 \in K} p(k_0, x_1, x_2, \ldots, x_n)$ will vary.

Your explanation is valid for the case in which we intend to determine the most probable value of the state k_0. I would like to know if it is possible to generalise your considerations to a more general case in which a Bayesian risk for an arbitrary, but prior known, penalty function is to be minimised.

Of course, it is. Have a look at the general case now without our help. Recall the theorem on the convex form of classes in the space of probabilities which was introduced as early as in the first lecture. We can see that a rifle loaded in the first act of our explanation has fired at last. We had nearly thought that it would not be needed in our course.

Even when I have understood your explanation I still cannot find where I made a mistake arriving at the conclusion that each observation x_i yielded information on the state k_0.

But you made nearly no mistake! You simply passed from one question to the other, considering them to be equivalent. You are right in that any observation x_i even at large i makes the information on the state k_0 more exact. But we are asking about something else. We are interested if the overall information resulting from the infinitely extended observation is sufficient for flipping the decision on the state k_0 from one class to the other. This, at first glance more difficult, question can be, as you can see, quite easily answered.

There is still an obstacle to my being completely sure about the questions discussed. In our considerations we nowhere referred to Wald's results. Does it mean that the tasks which are objects of Wald sequential analysis can be solved in a far easier way which is just the one you have shown? Why then is Wald sequential analysis so terribly complicated? Did we not once more discover America which had been already found by Wald? Unfortunately, I do not know how to describe my apprehension in a more exact manner since I am not sufficiently familiar with Wald sequential procedures.

You are far better off than those who believe that they know Wald sequential analysis quite well. Your advantage over them is, at least, that you can see not only the ingenious simplicity of Wald's procedures but you fear the complexity of a proof that just these procedures solve some tasks in an optimal way. Wald sequential analysis should not be referred to in a cursory way. That is why we answer your questions, just to allay your quite justifiable doubt. Our brief answer does not make a claim to be exact and comprehensive.

The questions which we discussed with you do not belong to the 'continent' discovered by Wald. Very roughly speaking, Wald's procedures answer the question of how long an object has to be observed so that its state may be evaluated with a previously given quality. Under certain assumptions about the statistical observation model, the observation sequence has been successfully proved to converge in a sense. It is to understand that any quality of estimate is available, even when it can be sometimes attained only through a long observation. And we particularly bring to your attention that this convergence occurs only under certain assumptions concerning the object.

Sometimes, with an inaccurate reference to Wald, it is stated in a vulgar way that by increasing the number of observed features of the object an arbitrarily high recognition quality can be attained. It is, of course, only a negligent manipulation with Wald's results. Just now, we have, together with you, got convinced that in observing Markovian objects and in analysing Markovian sequences it may easily happen that the information so far obtained is not enough for a sufficient quality of recognition. In spite of that, further observation of the object ceases to be decisive. It is because further observation cannot affect the decision, and so not enhance it either.

Together with you we have also examined a task in which the observation of an object can be interrupted. But as a condition for the interruption we did not regard the attainment of a previously given recognition quality (as was in the case of Wald), but a situation in which the quality attained, whatever it may be, cannot improve further, perhaps by an infinitely long observation. There are, therefore, two different conditions for interrupting the observation of an object, and we cannot guarantee that either of them will be satisfied. For some Markovian objects the necessary quality of its state estimate can be unattainable however long the observation may be, and then Wald's condition for interrupting the observation will not be satisfied. For other Markovian objects a situation may occur, even in a quite long observation, that further observation can flip the decision on the state attained on the basis of the already known observations.

And only now can we ask a really interesting question. We are asking how these two conditions for interrupting observation interact. Can one state that the observation of any Markovian object should be sure to interrupt at some time, either because the observations already attained allow us to qualitatively recognise the state, or because further observation cannot improve the already attained quality?

Am I right to understand that an answer to this question is not yet known?

We do not know such an answer, at least.

In our discussion the nearest neighbour method unexpectedly appeared, and this gave rise to many new questions.

Well, go ahead!

When I revived my knowledge of the nearest neighbour method, I found that the algorithms described in the lectures realise only the simplest method. In the set L only one single sequence is sought that is nearest to the observed sequence. In the theoretical examination of the nearest neighbour method and in their practical application emphasis is laid on more complicated methods. In the set L not only the nearest element is sought, but a previously determined number, say d nearest elements. As opposed to the simplest method seeking one single most similar element these methods are called the d-nearest neighbour methods. I think about important advantages of the d-nearest neighbour methods when compared to the method of one nearest method not only thanks to an extent research in this discipline, but also on the basis of my own experience, even when it is not very rich.

I can easily imagine the following practical situation. Assume that an application task can be exactly formalised, but the solution of a task so formalised is very complicated. This means that for the observation x it is difficult to find a k such that it is in agreement with the observation within the framework of the created formal scheme. At the same time it can happen that for each given state k it can be easily found out whether it is in agreement with the observation x. It is a similar situation to that in which a solution of a complicated equation is sought. To find a solution can be a very complicated job, but for each number one can easily verify whether the equation is satisfied.

Assume I can simplify the application task being solved to seeking the shortest path in a graph, but only when I neglect some important properties of the task. This means that the simplification does not represent the application task quite precisely, but it still has something in common with it.

In this situation, which is nothing infrequent in pattern recognition applications, the following procedure is quite natural. First, the shortest path in the graph is sought, i.e., the sequence k^* which is in best agreement with the recognised sequence x. But the satisfaction of some requirements which are important in the original application task is not guaranteed, because for simplification purposes these requirements were not taken into consideration in the first stage. Only in the second stage does one verify whether the sequence k^* found satisfies additional conditions. When it does so the sequence k^* found is then the solution of the original task. In the opposite case the recognition algorithm gives the answer `not known`.

This procedure can be enlarged in such a way that instead of the answer `not known` *the best sequence k_1^* in the set L is sought except for the sequence k^*. With the result that the additional conditions are again verified. If neither this sequence satisfies the condition, still further sequences are sought. The algorithm ends with providing the* `not known` *answer only when, e.g., none out* *of the best one hundred sequences does not satisfy the additional conditions, which in creating the set L were not taken into consideration.*

To be able to use such a technique I need to have an algorithm that is capable of finding d number of the best paths in a graph. I wish to get an effective algorithm the computational complexity of which should not rise too much with increasing d.

All that you have said now is right. We only do not understand what is the core of the question.

I do not know what the algorithm for finding the d-best paths in a graph should look like.

We do not believe that. We can guess that the algorithm for solving that task possesses a complexity which rises with d increasing linearly. This means that seeking the d best paths is not more than d-times more complex than seeking one single best path, and thus, its computational complexity is $\mathcal{O}(|K|^2\, dn)$.

Now it is I who does not believe it. Could not you, please, explain the algorithm in more detail for me?

Let us try it together. But first, tell us about the train of your considerations so that we would not start a wrong way for the second time.

Assume I have a graph G which determines the set of paths L from the vertex α to the vertex β. I assume to have already found the path $k^* \in L$ which is the shortest in the set L. My job now is to find the path k^1 which is the shortest in the set $L \setminus \{k^*\}$. The difficulties are because the subset $L \setminus \{k^*\}$ cannot be represented as a set of paths in the subgraph of G. No edge in the graph G can be excluded, since through each edge some of the paths from the set $L \setminus \{k^*\}$ passes. The set $L \setminus \{k^*\}$ contains not only paths which do not intersect with the paths k^*, but all the paths which diverge from the path k^* in some section of it, at least. In spite of all these difficulties, a new graph G^1 can be created such that the set of paths in it will represent just the set $L \setminus \{k^*\}$. But the new graph will have twice as many vertices as there were in the original graph. This means that seeking two best paths will be three times more difficult than seeking a single best path. That would still do, but I do not think I can continue doing so, since the number of vertices in a graph the paths of which match the set $L \setminus \{k^*, k^1, k^2, \ldots, k^d\}$ (where $k^*, k^1, \ldots, k^d$ are the earlier found, and thus firmly determined, paths), is 2^d-times larger than the number of vertices in the graph G the paths of which match the set L.

We have understood your difficulties and we will show how to get over them. But it will be the last opportunity in our lectures for you to see the fruitful way of stating structural analysis tasks as algebraic expressions in appropriate semi-rings. We believed that we had sufficiently explained the subject matter when we were examining the task of Levenstein matching of a sentence to a regular language. But now we can see we have not convinced you. If you had taken the explained subject matter seriously, you would not have your difficulties now.

I admit I have not yet included the algebraic methods explained amongst other tools I actively use. Levenstein matching did not seem so much convincing since you had deliberately, and as I can see it, rather unnecessarily complicated the task by letting the edit functions in, de and ch be without any restriction. Well, even some weak and quite natural assumptions on edit functions are sufficient to make the task at least reasonably solvable, if not quite simple. I mean, for example, a constraint in the form of triangular inequality. I have heard about Wagner algorithms [Wagner and Fischer, 1974; Wagner and Seiferas, 1978], which solve the task not only for regular languages, but for the context-free ones as well.

And do not the difficulties you see in the task on d-best paths in a graph occur to you as accumulated in an artificial way?

No, they do not. But I already suspect that you know the solution of the task which will be quite unexpected from my part.

Well. Now look how the task on d-best paths is formulated in the form of generalised convolution expressions. You will see that a task formulated like this is not worth mentioning. But be patient, since for briefness' sake the subject matter will be explained in a similar way as can be found in the most indigestible pseudo-mathematical articles, where something is referred to without saying in advance what it is good for.

Let us first write down in a form of enumeration the main notions which are necessary for seeking the d-best paths in a graph.

1. Let R be a set of nonnegative real numbers extended by a particular 'number' ∞, and it is assumed that for an arbitrary $a \in R$ the $\infty + a = \infty$ and $\min(\infty, a) = a$ hold.
2. Let R^d be a set of ordered ensembles of the form $(a_1, a_2, \ldots, a_d)$, where $a_i \in R$, $i = 1, 2, \ldots, d$, $a_1 \le a_2 \le \ldots \le a_d$.
3. On the set $R^d \times R^d$ a function of two variables is defined which assumes values on R^d. The function will be called *addition of ensembles*. For each pair $a \in R^d$ and $b \in R^d$ the function determines the sum $c = a \oplus b$ in the following way.
   If $a = (a_1, a_2, \ldots, a_d)$ and $b = (b_1, b_2, \ldots, b_d)$ then for the calculation of their sum it is needed
   - to create an ensemble $(a_1, a_2, \ldots, a_d,\ b_1, b_2, \ldots, b_d)$ of the length $2d$;
   - to order the ensemble in an ascendant way;
   - to regard the first d numbers in the ensemble as the sum $a \oplus b$.

   Addition defined in this way is an associative and commutative operation with $0^{\oplus}$ which is the ensemble $(\infty, \infty, \ldots \infty)$.

4. On the set $R^d \times R^d$ the function of two variables is defined which assumes its values on R^d. The function will be called *multiplication of ensembles*. For each pair $a = (a_1, a_2, \ldots, a_d)$ and $b = (b_1, b_2, \ldots, b_d)$ the function determines their product $a \otimes b$ such that
   - an ensemble $(a_i + b_j, i = 1, 2, \ldots, d;\ j = 1, 2, \ldots, d)$ of the length d^2 is created;
   - the ensemble is ordered in an ascendant way;
   - the first d numbers in the ensemble are regarded as the product $a \otimes b$.

   Multiplication defined in this way is an associative and commutative operation with $1^{\otimes}$ which is the ensemble $(0, \infty, \ldots, \infty)$.
5. The multiplication introduced is distributive with respect to the addition introduced earlier, i.e., $a \otimes (b \oplus c) = (a \otimes b) \oplus (a \otimes c)$. By the distributivity and also because the product of each ensemble with an introduced zero is also a zero the given operations of addition and multiplication form a semi-ring on the set R^d.
6. Let X be a finite set and f be a non-negatively defined real function $X \to R$. We will denote as f' the following function which assumes values on the set R^d. For $x \in X$ the value $f'(x)$ is an ensemble that consists of d numbers, where the first element is $f(x)$ and further $d - 1$ elements are ∞. There follows from this definition that the product $\bigotimes_{x \in X} f'(x)$ is an ensemble of d numbers in which the first number is $\sum_{x \in X} f(x)$ and further $d - 1$ elements are ∞. The sum $\bigoplus_{x \in X} f'(x)$ is an ordered ensemble which contains d smallest numbers from the ensemble $(f(x), x \in X)$.

Now we can state the *original formulation of the task* seeking the d-best paths in a graph.

Let X and K be finite sets and p_i, $i = 1, 2, \ldots, n$, be n functions of the form $K \times X \times K \to \mathbb{R}$. For any sequence $\bar{x} = (x_1, x_2, \ldots, x_n)$, $x_i \in X$, and any sequence $\bar{k} = (k_0, k_1, \ldots, k_n)$, $k_i \in K$, the number

$$F(\bar{x}, \bar{k}) = \sum_{i=1}^{n} p_i(k_{i-1}, x_i, k_i) \tag{9.88}$$

is defined. The aim of the task is to find an algorithm which for each sequence $\bar{x} \in X^n$ determines d smallest numbers from the ensemble $\left(F(\bar{x}, \bar{k}), \bar{k} \in K^{n+1}\right)$ that consists of $|K|^{n+1}$ elements.

Finally, we can start the *algebraic formulation of the task* seeking the d best paths in a graph and to *its solution.*

By means of the operations introduced that add and multiply ensembles of the length d, the formulated task will be reduced to seeking the ensemble Q given by the formula

$$Q(\bar{x}) = \bigoplus_{\bar{k} \in K^{n+1}} F'(\bar{x}, \bar{k}) = \bigoplus_{\bar{k} \in K^{n+1}} \bigotimes_{i=1}^{n} p_i'(k_{i-1}, x_i, k_i)\,. \tag{9.89}$$

In the preceding formula notations F' and p_i' are used instead of F and p_i in the formula (9.88). The values of the functions F' and p_i' (primed) are ensembles of length d, and not numbers for which the denotations F and p_i (without primes) were used.

In the definition (9.89) nothing will be changed if each summand is multiplied by the ensembles $\varphi'(k_0)$ and $\psi'(k_n)$ which are ones, i.e., ensembles $(0, \infty, \ldots, \infty)$. Thus we have

$$Q(\bar{x}) = \bigoplus_{\bar{k} \in K^{n+1}} \varphi'(k_0) \otimes \left(\bigotimes_{i=1}^{n} p_i'(k_{i-1}, x_i, k_i) \right) \otimes \psi'(k_n) \, . \tag{9.90}$$

The expression (9.90) can be written as

$$Q(\bar{x}) = \bigoplus_{k_0 \in K} \left(\bigoplus_{k_1 \in K} \left(\bigoplus_{k_2 \in K} \cdots \left(\bigoplus_{k_n \in K} \varphi'(k_0) \otimes \left(\bigotimes_{i=1}^{n} p_i'(k_{i-1}, x_i, k_i) \right) \otimes \psi'(k_n) \right) \cdots \right) \right) . \tag{9.91}$$

By the distributivity of multiplication with respect to addition, each coefficient in the formula (9.91) can be factored out before the symbol of addition operation with regard to the variables on which the coefficient does not depend. Thus we obtain

$$Q(\bar{x}) = \bigoplus_{k_0} \varphi'(k_0) \otimes \left(\bigoplus_{k_1} p_1'(k_0, x_1, k_1) \otimes \cdots \left(\bigoplus_{k_n} p_n'(k_{n-1}, x_n, k_n) \otimes \psi'(k_n) \right) \cdots \right) .$$

I can say I already understand it. I am now expected to gradually calculate the sum in the innermost parentheses. The result of each such addition will be $|K|$ ensembles of length d. I will first calculate the ensembles $f_{n-1}(k_{n-1})$ of length d, where $k_{n-1} \in K$. It will be according to the formula

$$f_{n-1}(k_{n-1}) = \bigoplus_{k_n \in K} \left(p_n'(k_{n-1}, x_n, k_n) \otimes \psi'(k_n) \right) , \tag{9.92}$$

and then I will calculate the ensembles $f_{n-2}(k_{n-2})$ for each $k_{n-2} \in K$,

$$f_{n-2}(k_{n-2}) = \bigoplus_{k_{n-1} \in K} \left(p_{n-1}'(k_{n-2}, x_{n-1}, k_{n-1}) \otimes f_{n-1}(k_{n-1}) \right) .$$

I continue gradually for $i = n-3, n-4, \ldots, 2, 1, 0$,

$$f_i(k_i) = \bigoplus_{k_{i+1} \in K} \left(p_{i+1}'(k_i, x_{i+1}, k_{i+1}) \otimes f_{i+1}(k_{i+1}) \right) , \quad k_i \in K \, . \tag{9.93}$$

The ensemble $Q(\bar{x})$ I am seeking will be then found according to the formula

$$Q(\bar{x}) = \bigoplus_{k_0 \in K} \left(\varphi(k_0) \otimes f_0(k_0) \right) = \bigoplus_{k_0 \in K} f_0(k_0) \, . \tag{9.94}$$

You stopped before the last step which would have much pleased us. The calculation according to the formulæ (9.92), (9.93) and (9.94) can be regarded as a calculation of the matrix product

$$\varphi\, P_1\, P_2\, \cdots\, P_{n-1}\, P_n\, \psi \tag{9.95}$$

in a semi-ring which is created on a set of d-dimensional ensembles by the operations of addition and multiplication of ensembles mentioned. In (9.95) the φ represents a row vector and ψ represents a column vector. Both vectors are $|K|$-dimensional. Each component of theirs represents the ensemble $(0, \infty, \infty, \ldots, \infty)$ of the length d. The matrices P_i, $i = 1, \ldots, n$, are to be understood as particular matrices in which each element is a d-dimensional ensemble. The ensemble in the k'-th row and k''-th column of the matrix P_i has the number $p_i(k', x_i, k'')$ in its first element and in all other elements the values ∞.

The solution of the task presented here is the most convincing illustration of the fruitfulness of the algebraic expressions. I do not mean by that the final result (9.95), but I admire how brief and transparent the path from a formal definition of the task (9.89) to the computational procedure in (9.92), (9.93), (9.94) can be. This intelligibility becomes even more impressive when I compare the algebraic formulation of the task with its original graph formulation. The algebraic formulation immediately indicates the direction in which to look for the solution. The object sought (in our case an ensemble of numbers) is defined by the formula (9.95). The task analysis lies in formal transformations of this formula in which the unchanged form of the defined object is guaranteed.

It was quite different with the original task. The task was not defined by a formula but the definition was verbal. It was stated that d-shortest paths in the graph is to be found. This formulation seems to be illustrative only at first glance because it is not supported by any apparatus by which a verbal formulation can be transposed to another verbal formulation. The researcher who tries to solve the task in its verbal formulation can rely only on his/her rational considerations. These considerations can be very extensive and besides leading in the right direction they can put the researcher on many other paths. Just this kind of 'intelligibility' routed me at first toward such task solving algorithms which were not practically implementable. At that time, the 'intelligibility' simply disoriented me.

Now I have understood at last how just the algebraic representation of the task seeking d-best path revealed its simplicity, which was treacherously hidden in graph representation. Graph representation does represent something, but it is not what is to be found in the formulated task. The object sought, i.e., a group of d-best paths cannot be represented in the graph expression at all. When we superimpose the paths from some group on them we obtain a graph which contains not only the paths of the group but also a lot of irrelevant paths.

When I have now seen the actual simplicity of the task examined, I feel I could master even more general tasks. I have in mind, for example, the case of seeking d-nearest neighbours, when instead of being described by a simpler sequence the object observed is described by a more complex arbitrary beforehand known acyclic structure. I would like to generalise the Levenstein matching task so that not only the best approximation, but the d-best approximations should be sought.

We are sure that you will easily generalise the computational procedure (9.95) even for the case of acyclic graph structures. But be careful in generalising Levenstein matching. One of the most significant properties, thanks to which Levenstein approximation was successfully managed, is the idempotent property of addition which was minimisation in our particular case. Only thanks to the idempotent property did we manage to prove that some infinite convolution expressions have finite equivalents. Remember Lemmata 9.2 and 9.3 of the lecture. Adding d-dimensional ensembles, as we have defined it for the task of d-best sequences, is not idempotent. Therefore, you will have to devise something which would be similar to the above mentioned lemmata 9.2 and 9.3 for the given case.

I would like to come back to the computational procedure (9.95) and see what form it will assume when I do not express it by means of macro operations of multiplying and adding of ensembles, but through elementary level operations dealing with individual numbers, i.e., with elements of ensembles.

We will be pleased to go with you through the computational procedure (9.95) because it is interesting from the purely computational standpoint and possesses quite surprising features. Well now, you are to start.

The task is to create an ensemble of the length d which is determined by the matrix product

$$\varphi\, P_1\, P_2\, \cdots\, P_n\, \psi\,. \tag{9.96}$$

A sequence of $|K|$-dimensional row vectors $f_1, f_2, \ldots, f_n$ is to be calculated, where $f_1 = \varphi \cdot P_1$ and $f_i = f_{i-1} \cdot P_i$, $i = 2, \ldots, n$. Then a 'scalar product' $f_n \cdot \psi$ is to be calculated. The computational complexity of (9.96) is n times greater than computational complexity of multiplying the $|K|$-dimensional row vector by a matrix of dimension $|K| \times |K|$. I will examine the complexity of this multiplication. I will take into consideration that the components of the vector f_{i-1} and the matrix P_i are d-dimensional ensembles, not numbers. The component $f_i(k)$ of the vector $f_i = f_{i-1} \cdot P_i$ is an ensemble of the length d that is determined by the sum

$$f_i(k) = \bigoplus_{k' \in K} f_{i-1}(k') \otimes P_i(k', k)\,. \tag{9.97}$$

To create the ensembles $f_i(k)$ for all $k \in K$ means to create $|K|^2$ auxiliary ensembles

$$c(k', k) = f_{i-1}(k') \otimes P_i(k', k)\,, \quad k' \in K\,, \tag{9.98}$$

and then their sum

$$f_i(k) = \bigoplus_{k' \in K} c(k', k) \qquad (9.99)$$

for all $k \in K$. So, multiplication $f_i = f_{i-1} \otimes P_i$ requires $|K|^2$ multiplications and $|K|^2$ additions of ensembles of length d. Let me examine the complexity of these operations with ensembles.

Let $a = (a_1, a_2, \ldots, a_d)$ and $b = (b_1, b_2, \ldots, b_d)$ be two ensembles. To calculate their product in the general case means to create an ensemble $(a_i + b_j,\ 1 \le i \le d,\ 1 \le j \le d)$ and select d smallest numbers from it. In our case one of the ensembles is the ensemble $P_i(k, k')$ which has a specific form. All the numbers in it, except the first one, are ∞, i.e., $b = (b_1, \infty, \infty, \ldots, \infty)$. The product of the ensemble $a = (a_1, a_2, \ldots, a_d)$ with the ensemble of this specific form is simply $(a_1 + b_1,\ a_2 + b_1,\ \ldots,\ a_d + b_1)$. Its calculation obviously has the complexity $\mathcal{O}(d)$.

Let $a = (a_1, a_2, \ldots, a_d)$ and $b = (b_1, b_2, \ldots, b_d)$ be two ensembles. Their addition means a selection of d smallest numbers from the ensemble $(a_1, a_2, \ldots, a_d, b_1, b_2, \ldots, b_d)$. Since the ensembles a and b are in ascending order their addition has complexity $\mathcal{O}(d)$. This addition could be computed by the following program fragment.

```
k = i = j = 1;
while ( k ≤ d ) {
   if ( a_i ≤ b_j )
      { c_k = a_i; i = i + 1; }
   else
      { c_k = b_j; j = j + 1; }
   k = k + 1;
}
```

Thus the multiplication $f_{i-1} \cdot P_i$ is of complexity $\mathcal{O}(|K|^2\, d)$, the multiplication (9.96) being of complexity $\mathcal{O}(|K|^2\, d\, n)$. It is the result you have anticipated from the very beginning of the task analysis. The result states that seeking d-best sequences is only d-times more complicated than seeking the best sequence which can be performed with complexity $\mathcal{O}(|K|^2\, n)$.

You have missed some important properties of ensemble addition. You are right that the addition $a_1 \oplus a_2$ has the complexity $\mathcal{O}(d)$. But it does not follow from that that the complexity of the addition $a_1 \oplus a_2 \oplus \ldots \oplus a_m$ is $\mathcal{O}\big((m-1)d\big)$. Actually, it is substantially less, being just $\mathcal{O}\big((m-1) + (d-1)\log m\big)$. When the calculation procedure is clear to you then you will arrive at the conclusion that the calculation of the product $f_{i-1} \cdot P_i$ is not of complexity $\mathcal{O}(|K|^2\, d)$, but a more favourable value $\mathcal{O}(|K|^2 + d\,|K| \log |K|)$. Think over the program for computing the product of the vector f_{i-1} with the matrix P_i more thoroughly.

Not to deal with unnecessary details, I will introduce simplified notation for calculating the ensemble $f_i(k)$ for one certain $k \in K$ in agreement with the for-

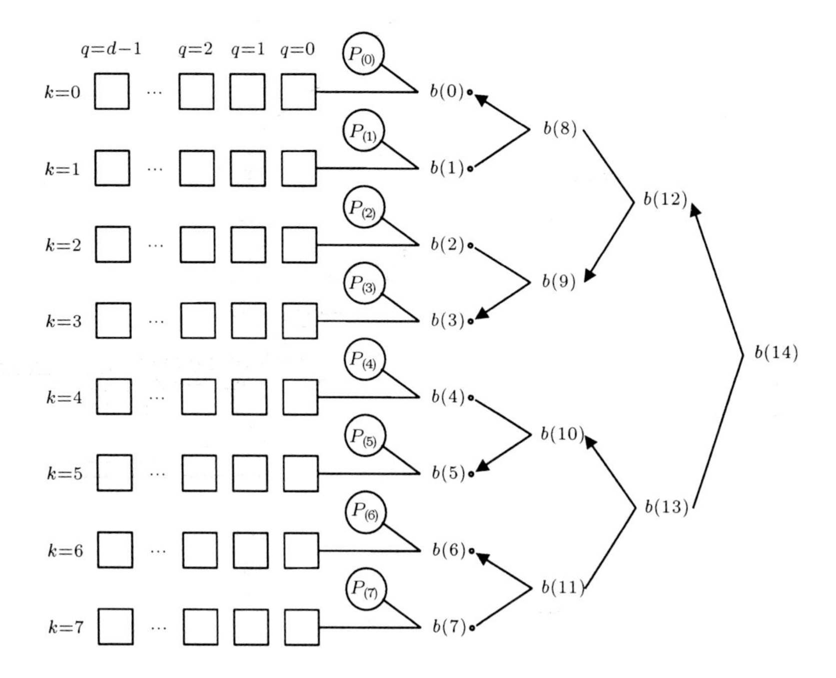

Figure 9.6 Arrangement of data for effective computation of the product of vector f_{i-1} with matrix P_i. The squares contain numbers $f(k,q)$.

mula (9.97). If the value k is fixed then it can be omitted and the formula (9.97) has the form

$$f' = \bigoplus_{k \in K} f(k) \otimes P'(k)\,, \tag{9.100}$$

where f', $f(k)$, $P'(k)$, $k \in K$, are ensembles of d numbers each. Ensembles $P'(k)$, $k \in K$, have a specific form. Only the first number in each ensemble is not ∞. All other numbers are ∞. I denote as $P(k)$ the first number in the ensemble $P'(k)$. The calculation according to (9.100) means, in fact, the following. There are $d|K|$ numbers $f(k,q)$, $k = 0,1,\ldots,|K|-1$, $q = 0,1,\ldots,d-1$. Furthermore, there are also $|K|$ numbers $P(k)$, $k = 0,1,\ldots,|K|-1$. The ensemble of numbers $f(k,q)$ is partially ordered in such a sense that for each triplet k, q_1, q_2, $q_1 < q_2$, the inequality $f(k,q_1) \le f(k,q_2)$ holds. The ensemble of numbers $P(k)$ is arbitrary.

The ensemble $f(k,q)$, $k = 0,1,\ldots,|K|-1$, $q = 0,1,\ldots,d-1$, is represented by a rectangular field of squares in Fig. 9.6. Each of the squares represents a corresponding number $f(k,q)$. The ensemble $P(k)$, $k = 0,1,\ldots,|K|-1$, is depicted in Fig. 9.6 by a column of circles in which the corresponding numbers $P(k)$ are written. The figure shows a case in which $|K| = 8$ and d is arbitrary.

The complexity of selecting the d smallest numbers from the ensemble $\big(f(k,q) + P(k),\ k = 0,1,\ldots,|K|-1,\ q = 0,1,2,\ldots,d-1\big)$ is to be determined.

The first number selected is evidently $\min_k \big(f(k,0) + P(k)\big)$. *To determine this number is not complicated and therefore it is obvious that its computational complexity is* $\mathcal{O}(|K|)$. *However, I do not intend to seek the minimum in a common fashion, but by means of a data structure which is represented by the tree in the right-hand side of Fig. 9.6. The vertices in the tree are labelled by the values* $b(j)$ *in the Fig. 9.6. Their amount is* $2\,|K| - 1$. *The numbers* $b(j)$ *are determined by the following program fragment*

```
for ( k = 0; k < |K|; i++)
  { b(k) = f(k,0) + P(k); q*(k) = 1; }
k = 0; j = |K|;
while ( j ≠ 2|K| - 1) {
  if  ( b(k) ≤ b(k+1))
    { b(j) = b(k); ind(j) = k; }
  else
    { b(j) = b(k+1); ind(j) = k+1; }
  k = k+2; j = j+1;
}                                                    (9.101)
```

When the program finishes then at each tree vertex a number is written which is the smaller of two numbers which are written at the vertices connected by edges with the respective vertex and lie to the left of it. For example, $b(9) = \min\big(b(2), b(3)\big)$ *and* $b(12) = \min\big(b(8), b(9)\big)$. *The numbers* $\text{ind}(j)$, $j = |K|, \ldots, 2\,(|K| - 1)$ *are indices pointing which of the two numbers was overwritten from the left to the right. The indices are represented by arrows in Fig. 9.6.*

For example, the arrow from the ninth vertex points towards the third vertex, which means that $\text{ind}(9) = 3$, *and thus* $b(3) \le b(2)$. *It is evident that with such a data arrangement the number* $b\big(2(|K| - 1)\big)$ *at the root of the tree is the smallest of the numbers* $b(j)$, $j = 0, \ldots, |K| - 1$. *The ensemble of arrows shows the number* j *of the tree leaf where the least number lies. The ensemble of arrows in Fig. 9.6 indicates, for example, that* $b(14) = b(3)$. *The program (9.101) also determines indices* $q^*(k)$ *which indicate how many numbers were taken of the ensemble* $f(k,q)$, $q = 0, \ldots, d-1$, *and were written in the tree leaves.*

The above mentioned program for finding the least number from the ensemble $\big(f(k,0) + P(k),\ k = 0,\ \ldots,\ |K| - 1\big)$, *is exceedingly complicated only at first glance. In fact, its complexity is* $\mathcal{O}(|K| - 1)$, *i.e., of the same order as that with the common procedure seeking the least number by simple examination of the numbers* $f(k,0) + P(k)$, $k = 0, \ldots, |K| - 1$. *An important advantage of the algorithm (9.101) is that in addition to seeking the least number it creates supplementary data which allows us to find the succeeding least number not in* $(|K| - 1)$ *operations, but in substantially fewer* $\log |K|$ *operations. I will show how it occurs.*

Assume that the number written in the output ensemble f' *as the first was a number from the* k^**-th row of the ensemble* $f(k,q)$. *This means that the number* $f(k^*,0) + P(k^*) = f'(0)$ *was the least number in the ensemble*

$\big(f(k,0) + P(k), k = 0, 1, \ldots, |K| - 1\big)$ *which consists of* $|K|$ *numbers. The succeeding number* $f'(1)$ *which is to be written in the output ensemble must be either* $f(k^*, 1) + P(k^*)$ *or the least number of the group* $f(k, 0) + P(k)$, $k \neq k^*$. *So the numbers* $f'(0)$ *and* $f'(1)$ *are the least numbers of two different groups which, however, differ from one other only by a single number. In seeking the number* $f'(1)$ *it is not necessary to examine the whole new group. It is sufficient just to calculate the new values in the ensembles* $b(j)$, $\text{ind}(j)$, $j = |K|, \ldots, 2(|K| - 1)$. *However, the number of values which must be changed is not greater than* $\log |K|$ *numbers. The changed numbers and arrows are just those in the tree vertices which lie on the path from the leaf* k^* *to the root.*

Imagine that after starting the program (9.101) we have obtained all needed data in the tree, i.e., numbers $b(j)$ *and arrows* $\text{ind}(j)$ *for each tree vertex. Assume the arrows have corresponded to those in Fig. 9.6. The ensemble of arrows depicted states that the least of the numbers* $f(k, 0) + P(k)$, $k = 0, \ldots, |K| - 1$, *was the number* $f(3, 0) + P(3)$. *Thus the number* $b(3)$, *which previously was* $f(3, 0) + P(3)$, *is now to be changed into the number* $f(3, 1) + P(3)$. *The data within the tree are also to be changed in a respective way, i.e., numbers* $b(j)$ *and* $\text{ind}(j)$ *will be changed only at* $3 = \log 8$ *vertices 9, 12 and 14. The number* $b(9)$ *is to be changed into the number* $\min(b(2), b(3))$. *Into the vertex number 12 the number* $\min\big(b(9), b(8)\big)$ *will be written. Finally, the number* $\min\big(b(12), b(13)\big)$ *will be written at the root of the tree. Obviously the same way will lead to obtaining new indices* $\text{ind}(j)$ *for* $j = 9, 12, 14$.

I will now deal with the general case. Assume the program has already written q *numbers into the output ensemble* f' *that were the* q *least numbers in the ensemble* $(f(k, q) + P(k), k = 0, 1, \ldots, |K| - 1,\ q = 0, 1, \ldots, d - 1)$. *Assume also that the algorithm has the following data at its disposal. These are numbers* $q^*(k)$, $k = 0, \ldots, |K| - 1$, *which means that the least* $q^*(k)$ *numbers have been already taken of the* k*-th row of the ensemble* $f(k, q)$ *and written in the respective leaves of the tree. So the numbers* $f\big(k, q^*(k) - 1\big) + P(k)$, $k = 0, 1, \ldots, |K| - 1$, *are written just now into the tree's leaves* $b(k)$. *Assume also that all other numbers* $b(j)$ *and indices* $\text{ind}(j)$, $j = |K|, |K| + 1, \ldots, 2(|K| - 1)$, *in the tree vertices are in agreement with the data in the tree leaves.*

The next least number in the remaining part of the ensemble $\big(f(k, l) + P(k), k = 0, 1, \ldots\ldots, |K| - 1, l = q^*(k), q^*(k) + 1, \ldots, d - 1\big)$, *i.e.,* (q)*-th number of the output ensemble, is sought by means of Algorithm 9.2.*

Algorithm 9.2 Seeking a further least number in the remaining part of the ensemble.

1. *It is to find from what tree leaf the last number* $f'(q - 1)$ *was taken,*

$$\left.\begin{array}{l} k^* = \text{ind}(2(|K| - 1)); \\ \texttt{while}\ (k^* \geq |K|) k^* = \text{ind}(k^*); \end{array}\right\} \qquad (9.102)$$

   After executing the previous commands the number k^* *informs us that the last number* $f'(q-1)$ *written to the output ensemble* f' *was the number* $f\big(k^*, q^*(k^*) - 1\big) + P(k^*)$.

2. *The succeeding number of the k^*-th row of the ensemble $f(k,q)$ is taken out and the number in the k^*-th tree leaf is changed.*

$$\left.\begin{aligned} b(k^*) &= f(j^*, q^*(k^*)) + P(k^*); \\ q^*(k^*) &= q^*(k^*) \qquad\qquad + 1\,; \end{aligned}\right\} \tag{9.103}$$

3. *With respect to the change of the number in the k^*-th tree leaf, new data in the tree are computed,*

$$\left.\begin{array}{l} \texttt{while}\ (k^* \neq 2\,(|K|-1))\ \{ \\ \quad j1 = k^* \operatorname{xor} 1\,;\ j2 = |K| + k^*/2\,; \\ \quad \texttt{if}\ (b(k^*) \leq b(j1)) \\ \quad\quad \{\, b(j2) = b(k^*);\ \operatorname{ind}(j2) = k^*; \} \\ \quad \texttt{else} \\ \quad\quad \{\, b(j2) = b(j1);\ \operatorname{ind}(j2) = j1; \} \\ \quad k^* = j2; \\ \} \end{array}\right\} \tag{9.104}$$

4. *The number from the tree root is written at the position q of the output ensemble f' and the number q is incremented by one.*

In Algorithm 9.2 the operation $k^* \operatorname{xor} 1$ represents inversion of the least significant bit in the binary representation of the integer nonnegative number k^*. The operation $k^*/2$ is an integer number division ignoring, at the same time, the information in the least significant bit in the binary representation of the number k^*. The index $j2$ points to the vertex within the paths from the vertex k^*. This is the vertex that is connected with the vertex k^* by an edge. The index $j1$ is the second vertex that is connected with the vertex $j2$ but the distance of which from the tree root is greater than that of $j2$. The meaning of indices k^*, $j1$ and $j2$ can be understood from Fig. 9.7.

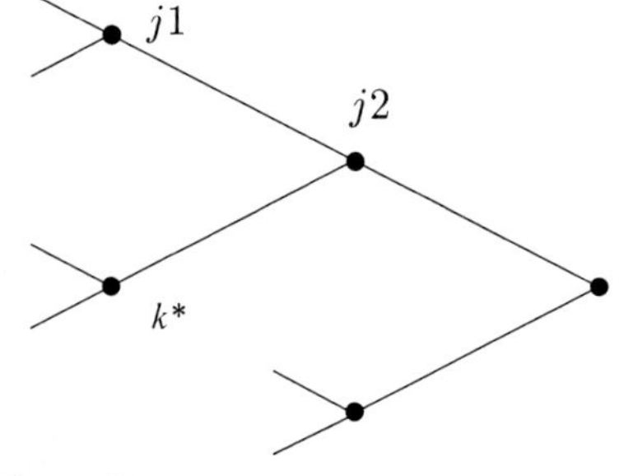

Figure 9.7 Meaning of indices j^*, $j1$ and $j2$ in Algorithm 9.2.

After Algorithm 9.2 has ended, the number of positions already written in the output ensemble has increased by one and data needed for seeking the succeeding number to be written into the output ensemble have been prepared. The complexity of Algorithm 9.2 is given by the number of cycle repetitions in its fragments (9.102) and (9.104). This number is always $\mathcal{O}(\log |K|)$ and it is the complexity of finding one element (but not the first one) in the output ensemble f'. The creating of the whole ensemble of the length d, i.e., the calculation (9.97), consists of the calculation of the first element by means of the program (9.101), having the complexity $|K|$ and the calculation of the $(d-1)$ other elements. The total complexity thus is $\mathcal{O}(|K| + (d-1)\log|K|)$. The ensemble (9.97) is to be calculated for each $k \in K$, and therefore the complexity *of the operation* $f_i = f_{i-1} \cdot P_i$ *is* $\mathcal{O}\big(|K|(|K| + (d-1)\log|K|)\big)$. *It follows from that the calculation complexity of the ensemble (9.96) of the length* d *is*

$$\mathcal{O}\big(|K|^2\, n + |K|(\log|K|)n(d-1)\big)\,. \tag{9.105}$$

This is a damned interesting result. I expected that the search for d-best sequences will be substantially more complicated than the search for one single best sequence. But in reality the algorithm shows quite the opposite property. It is the first best sequence that requires the most of calculations. When we are looking for the first sequence with deliberation then seeking any further sequence is less demanding.

When you understand so well the advantage attained by a carefully thought over computational procedure we can advise you to further continue in analysing our task. In a more detailed examination of task properties you will find that the computational complexity is even less than (9.105). You will see that the complexity is only

$$\mathcal{O}\big(|K|^2 n + (\log|K|)\, n\, (d-1)\big)\,. \tag{9.106}$$

Note that the number $(\log|K|)\, n\, (d-1)$ is simply a number of bits which is needed only for the $(d-1)$ shortest paths found to be stored somewhere. This means that if you create an algorithm of the complexity of (9.106) then the finding of any succeeding shortest path will have a complexity of the same order as simply writing this path to the output memory. It will be an algorithm which seeks the shortest path in such a deliberate way that each succeeding path will look as if the paths were only rewritten from one memory to another.

If I did not have any experience of discussion with you then I would believe that further reducing the computational complexity was not possible. Well now, the matrix product

$$\varphi\, P_1\, P_2\, \cdots\, P_n\, \psi \tag{9.107}$$

cannot be calculated in any other way than by an n*-times repeated product of the vector* f *with the matrix* P*, can it? The complexity of such a calculation cannot be less than* $\mathcal{O}(n\, C)$*, where* C *is the complexity of the calculation* $f\,P$*. But the result of multiplying* f *and* P *is an ensemble consisting of* $|K|$ *ensembles of length* d*. Thus the* $|K|\, d$ *numbers have to be sought and stored. Because of that the calculation of the product* $f\,P$ *must have a component the complexity of which is* $\mathcal{O}(|K|\, d)$*. Furthermore, it must have an* n*-fold product of the vector with the matrix, which is needed for the calculation (9.107), a component with the complexity* $\mathcal{O}(|K|\; d\, n)$*. But a contribution that would refer to the component mentioned here, I cannot find in your estimate of the complexity of (9.106).*

Before showing where you made your mistake let us go back to the computational procedures which we have already dealt with.

You and we clearly understand that adding two ensembles a_1 and a_2 has a complexity $\mathcal{O}(d)$. There is no hope for improvement here, since an ensemble of length d is being created. It would seem to result from it that the complexity of adding n ensembles $a_1, a_2, \ldots, a_n$ cannot be less than $\mathcal{O}\big((n-1)\,d\big)$. As the addition can be done (attention, an error will follow!) only in such a way that a sum of two summands is computed, then of two others, and then to the already computed summands another summand in turn is added, and so forth. It would follow from it that we cannot do without $n-1$ additions.

But just a while ago you proposed an excellent algorithm for calculating the sum $\bigoplus_{k=1}^{n} a_k$ the complexity of which is not $\mathcal{O}\big((n-1)\,d\big)$ at all, but which is substantially less, i.e., $\mathcal{O}\big((n-1)+(d-1)\log n\big)$. Even when you scrutinise your algorithm as best as possible, you will not find anything which would indicate that auxiliary data could be interpreted as partial sums of subsets of summands, in our case the subsets of ensembles. This means that you have managed to create an algorithm for adding a set of ensembles and avoiding the addition of some of their subsets.

Now let us return to calculating an ensemble of length d

$$\varphi\, P_1\, P_2\, \cdots\, P_n\, \psi \tag{9.108}$$

based on the procedure which we already understand well,

$$\left.\begin{aligned} f_0 &= \varphi \\ f_i &= f_{i-1}\, P_i\,, \quad i = 1,2,\ldots,n\,. \end{aligned}\right\} \tag{9.109}$$

The product (9.108) is a product $f_n\, \psi$, i.e., an ensemble of the length d

$$\bigoplus_{k\in K} f_n(k) \otimes \psi(k)\,, \tag{9.110}$$

where $f_n(k)$ and $\psi(k)$ at any $k \in K$ are ensembles of the length d. You examined that calculation and arrived at a correct result that its complexity is $\mathcal{O}\big(|K| + (d-1)\log|K|\big)$. Go thoroughly through your algorithm once more and notice that for the calculation (9.110) you need not have complete information on all $|K|$ ensembles $f_n(k)$, $k \in K$. The complete information about them consists of $|K|\,d$ numbers, but only d of them are necessary for calculating (9.110). Unnecessary numbers need not be calculated.

You can continue in considerations like that. For calculating the vector f_n complete information about the vector f_{n-1} is not needed either. To obtain partial information about the vector f_n, even fewer data of the vector f_{n-1} suffice. The procedure of computing (9.109) is in a sense a bit wasteful. In each step $i = 1,2,\ldots,n$, the procedure creates spare data and a large part of these data will not be ever used. When you design an algorithm such that only what is used in calculation will be included in it then you will see that you will calculate the whole product (9.108) and avoiding the calculation of products $\varphi\, P_1, \varphi\, P_1\, P_2, \varphi\, P_1\, P_2\, P_3$, etc..

I will go back to the graph interpretation of the task. I am going to speak about a graph the vertices of which are α and β (see Fig. 9.5) and further $K\ (n+1)$ vertices of the form (k,i), $k = 0,1,\ldots,K-1$, $i = 0,1,\ldots,n$. I did not denote the set of values of the variable k by the symbol K, but their number, since further on in the algorithm I will use the value k as an integer index.

The graph contains the following oriented edges. The K edges point from the vertex α to vertices of the form $(k,0)$. The lengths of the edge $\big(\alpha,(k,0)\big)$ is determined as $\varphi(k)$. From each edge $(k',i-1)$ the K edges point to the vertices (k'',i). The length of each of these edges is $P_i(k',k'')$. From each vertex (k,n), the single edge points to the vertex β and is of the length $\psi(k)$.

I am interested in the complexity of the algorithm that will find d shortest paths from the vertex α to the vertex β.

For the analysis of the task and the algorithm of its solution I will introduce the following notation. For each vertex γ of the graph which corresponds either to the pair (k,i) or β, I will denote by $C(\gamma)$ the set of paths from the vertex α to the vertex γ. I will order this set in ascendant order by the path length and denote by $c(\gamma,q)$ q-th path in this ordered set. The paths in the ordered set $C(\gamma)$ will be enumerated starting from zero. The shortest path in the ordered set $C(\gamma)$ is therefore $c(\gamma,0)$. In the program presented later the sets $C(\gamma)$ and the paths $c(\gamma,q)$ are not explicitly represented. They do not belong to the objects which the program has to manipulate but they are concepts explaining the sense of the data that the program creates and transforms. These data are subdivided into groups corresponding to the graph vertices. The sense of the data is equal for all vertices as well as for their processing. But formal descriptions of their meaning are different for the vertices of the form (k,i) and β. I will explain the data and rules for their processing which apply only to the vertices of the form (k,i) in a rather detailed way. The processing of the vertex β will be explained in a less detailed way and I hope that the explanation will be clear even without detailed comments.

The most important group of data in the algorithm is the ensemble of numbers $f(k,i,q)$, which stand for the length of the path $c(k,i,q)$, i.e., the q-th path from the vertex α to the vertex (k,i). The number $f(\beta,q)$ stands for the length of the q-th best path from α to β. For a fixed i the ensemble $f(k,i,q)$ is just the the row vector f_i in the algebraic representation of the task, and the ensemble of numbers $f(\beta,0)$, $f(\beta,1)$,...,$f(\beta,d-1)$ is the ensemble expressed by the product in (9.109). If the situation $q \geq |C(k,i)|$ occurs then I consider $f(k,i,q) = \infty$.

For each vertex (k,i) and each number q the indices $k'(k,i,q)$ and $q'(k,i,q)$ are defined. These indices have the following meaning. The path $c(k,i,q)$ consists of the path $c\big(k'(k,i,q),i-1,q'(k,i,q)\big)$ to the end of which the vertex (k,i) is to be appended. By means of indices $k'(k,i,q)$ and $q'(k,i,q)$ any path $c(k,i,q)$ can be created. For example, the q-th path from α to β is

$$\alpha,(k_1,1),(k_2,2),\ldots,(k_i,i),\ldots,(k_n,n),\beta\ ,$$

where

$$k_n = k'(\beta,q), q_n = q'(\beta,q)\ , \quad k_{i-1} = k'(k_i,i,q_i)\ , \quad q_{i-1} = q'(k_i,i,q_i)\ .$$

The algorithm, which will be presented later, calculates the quoted data successively in such a way that in every step the data $f(k,i,q)$, $k'(k,i,q)$, $q'(k,i,q)$ are available only for some triplets (k,i,q), not for all. Which part of the data are already known is determined by numbers $q^*(k,i)$. They indicate for each vertex (k,i) that the data $f(k,i,q)$, $k'(k,i,q)$, $q'(k,i,q)$ are available for $q^*(k,i)$ best paths leading to the vertex (k,i).

Furthermore the quoted data additional data relate to each vertex (k,i) which will be referred to as a tree of the vertex (k,i). These data have the same structure as that which was presented in Fig. 9.6. The vertices of the tree (k,i) correspond to triplets (k,i,j), $j = 0,1,\ldots,2(K-1)$, the numbers $b(k,i,j)$ and indices $\text{ind}(k,i,j)$ being defined for every vertex. The root of the tree (k,i) corresponds to the index $\big(k,i,2(K-1)\big)$. The number $b\big(k,i,2(K-1)\big)$ in the root is the number $f\big(k,i,q^*(k,i)\big)$, i.e., the length of the longest path out of all the already determined paths which is the $q^*(k,i)$-th shortest path of the set $C(k,i)$ out of the paths from α to (k,i). The leaves of the tree (k,i) correspond to the indices (k,i,j), $j=0,\ldots,K-1$. Some numbers $b(k,i,0), b(k,i,1),\ldots,b(k,i,K-1)$ and indices $\text{ind}(k,i,0)$, $\text{ind}(k,i,1)$, ..., $\text{ind}(k,i,K-1)$ are stored in the leaves. The number $b(k,i,k')$ is equal to number $f(k',i-1,q) + P_i(k',k)$ at some q the value of which is stored as $\text{ind}(k,i,k')$.

The numbers $b(k,i,j)$ and indices $\text{ind}(k,i,j)$ in all the other vertices of the tree (k,i) have the same sense which I mentioned before when explaining them in Fig. 9.6. So the indices $\text{ind}(k,i,j)$ in the leaves of the tree have a slightly different meaning than in other vertices of the tree.

At first glance the data manipulated by the algorithm seem to be quite numerous. But fortunately only at first glance. The data are subdivided into two groups. The former contains quantities $f(k,i,q)$, $k'(k,i,q)$ and $q'(k,i,q)$. The required size of memory for storing these data depends on the numbers $q^*(k,i)$, because it is proportional to $\sum_{k,i} q^*(k,i)$. Further on we will see that $\sum_{k=0}^{K-1} q^*(k,i)$ is not greater than the number $q^*(\beta)$. This number $q^*(\beta)$ indicates how many shortest paths from the vertex α to the vertex β have been created. The total size of the memory for $f(k,i,q)$, $k^*(k,i,q)$ and $q'(k,i,q)$ is then not greater than $\sum_{i=0}^{n} q^*(\beta) = (n+1)\,q^*(\beta)$. It is the memory of the same order of size as the memory needed for storing final results.

The other data group is formed by trees for each graph vertex. The total size of the memory for storing trees is $\mathcal{O}(K^2 n)$, which corresponds to the size of memory needed for storing the input data of the task, i.e., for storing information of edge lengths $P_i(k',k)$. It can be seen that neither the demands for memory are exaggerated. The required memory is of the same order as the memory size necessitated for storing the input data of the task and the results of its solution. If these data are exceedingly many nothing can be done, and it means that the task under solution is really too extensive.

Quite a different question is that the data are quite diverse and cannot be overlooked at one sight to see all mutual relationships among the data. These are the programmer's trouble, whose job is to write a program so that the required relation between data should not be violated in their transforming or supplementing the data.

In observing the accumulated data I am once more aware of the appropriateness of the algebraic construction that was used before. It enabled me to get an overall view of the task at the level of macro-concepts without being forced to annoyingly fiddle with individual odds and ends. Though at the level of macro-concepts a number of small items are ignored, a program has been successfully created for solving the task, which, on the whole, was not the worst one. Now, I intend to take advantage of everything out of the task that can add to speeding up the algorithm, and thus I cannot help fiddling with odds and ends. I believe I have mastered it, but I would not have managed it had I not had a firm basis which was the solution of the task at the algebraic macro-level.

Finally, our patience has been rewarded and you well understand the advantages of the algebraic expression of tasks we have been dealing with.

I will present an auxiliary algorithm which will be the most substantial part of the algorithm solving the problem. I will call the algorithm NEXT(k,i) *and define its functions as follows. The algorithm changes the given data in such a way that the number $q^*(k,i)$ is increased by one and all other data will adapt to this new value. Before starting the algorithm* NEXT(k,i) *the data yielded information on q-best paths from the vertex α to the vertex (k,i). When the algorithm* NEXT(k,i) *stops then the data have been transformed so that they yield information on the $(q+1)$-th paths.*

As a prototype for the algorithm NEXT(k,i) *the algorithms (9.102), (9.103) and (9.104) will serve for calculating the ensemble $\bigoplus_{k\in K} f(k) \otimes P(k)$, where $f(k)$ and $P(k)$ are ensembles. The prototype has to be modified with respect to the fact that the algorithm* NEXT(k,i) *has to transform different data according to input arguments (k,i). The algorithm will also include a command which will state what has to be done if the data the algorithm is to use are not yet available.*

Algorithm 9.3 **NEXT(k,i)**

1. *First it has to be found which number was taken out of the ensemble $f(k',i-1,q)$ when the number $f(k,i,q^*(k,i)-1)$ was being computed, i.e., the length of the $q^*(k,i)$-th path from the vertex α to the vertex (k,i). In the prototype algorithm, the command (9.102) was used for this purpose. Now it is no longer needed as the necessary information is stored in the data $k'(k,i,q^*(k,i)-1)$ and $q'(k,i,q^*(k,i)-1)$. Instead of the command (9.102) the following command is performed*

$$k^* = k'\big(k,i,q^*(k,i)-1\big); \quad q^* = q'\big(k,i,q^*(k,i)-1\big)\,.$$

   The indices k^* and q^* indicate that the number $f\big(k,i,q^*(k,i)\big)$ was obtained as the sum of numbers $f(k^*,i-1,q^*)$ and $P_i(k^*,k)$.

2. *Now the length $f(k^*,i-1,q^*)$ of the succeeding (q^*+1)-th path from the set $C(k^*,i-1)$ is to be taken out, the number $P_i(k^*,k)$ added to it and place the*

obtained sum in the k^*-th leaf of the (k,i)-th tree.

$$\left.\begin{array}{l} \texttt{if} \ \left(q^*(k^*,i-1) = q^* + 1\right) \ \mathrm{NEXT}(k^*,i-1)\,; \\ b(k,i,k^*) = f(k^*,i-1,q^*+1) + P_i(k^*,k)\,; \end{array}\right\}. \tag{9.111}$$

The preceding operation resembles the command (9.103) from the prototype program with that important difference that in the command (9.111) the first line is also present which the command (9.103) did not contain. The number $f(k^*,i-1,q^*+1)$, which is necessary for satisfying the second line of the command (9.111), may not be yet available. It will be so when $q^*+2 > q^*(k^*,i-1)$. But before the command (9.111) was carried out the number $f(k^*,i-1,q^*)$ had already been available. This means that $q^*(k^*,i-1) \geq q^*+1$. It follows from this that if the number $f(k^*,i-1,q^*+1)$ is not available then $q^*(k^*,i-1)$ is exactly q^*+1. To create the number $f(k^*,i-1,q^*+1)$ it is sufficient to let the program $\mathrm{NEXT}(k^*,i-1)$ *run just once which will occur in the first line of the command (9.111).*

3. *The data in the tree (k,i) are transformed with respect to the change of the number $b(k,i,k^*)$ in the tree leaf, which is done by means of a program which only slightly differs from the prototype program (9.104).*

$$\left.\begin{array}{l} \texttt{while}\ (k^* \neq 2(K-1))\ \{ \\ \quad j1 = k^* \operatorname{xor} 1\,;\ j2 = K + k^*/2\,; \\ \quad \texttt{if}\ (b(k,i,k^*) \leq b(k,i,j1)) \\ \quad\quad \{\ b(k,i,j2) = b(k,i,k^*);\ \mathrm{ind}(k,i,j2) = k^*;\ \} \\ \quad \texttt{else} \\ \quad\quad \{\ b(k,i,j2) = b(k,i,j1);\ \mathrm{ind}(k,i,j2) = j1;\ \} \\ \quad k^* = j2; \\ \} \end{array}\right\} \tag{9.112}$$

4. *The numbers $f\big(k,i,q^*(k,i)+1\big)$ and indices $k'\big(k,\ i,\ q^*(k,i)+1\big)$ and $q'\big(k,\ i,\ q^*(k,i)+1\big)$ corresponding to the path $c\big(k,i,q^*(k,i)+1\big)$ just found are to be stored in the respective memory cells. The value $f(k,i,q^*(k,i)+1)$ is obtained from the the cell $b(k,i,2(K-1))$ and the indices $k'(k,i,q^*(k,i)+1)$, $q'(k,i,q^*(k,i)+1)$ will be determined by the program, which resembles the prototype fragment (9.102).*

$$\left.\begin{array}{l} q^*(k,i) = q^*(k,i) + 1\,; \\ j^* = 2\,(K-1)\,; \\ f(k,i,q^*(k,i)) = b(k,i,j^*)\,; \\ \texttt{while}\ (j^* \geq K)\ j^* = \mathrm{ind}(k,i,j^*)\,; \\ k'(k,i,q^*(k,i)) = j^*\ ;\ q'(k,i,q^*(k,i)) = \mathrm{ind}(k,i,j^*)\,; \end{array}\right\} \tag{9.113}$$

An important property of the auxiliary algorithm $\mathrm{NEXT}(k,i)$ *defined is that in performing it the* $\mathrm{NEXT}(k',i-1)$ *is called not more than once (it may be called not even once). And if the program* $\mathrm{NEXT}(k',i-1)$ *is called then it is called only for one of the values k'. This means that the computation of the function* $\mathrm{NEXT}(k,i)$ *consists of computations complexity of which is $\mathcal{O}(\log K)$ (which is owed to the* `while` *cycles in the commands (9.112), (9.113)) and of computations of the function* $\mathrm{NEXT}(k',i-1)$ *for one certain k'. The computation of the function* $\mathrm{NEXT}(k',i-1)$ *consists of computations the complexity* *of which is $\mathcal{O}(\log K)$, and perhaps also of the computations of the function* NEXT$(k'', i-2)$) *for one certain k''. Since the function* NEXT$(k, 0)$ *will be called not even once, the overall complexity of the algorithm* NEXT(k, i) *is $\mathcal{O}(i \log K)$.*

It can be easily understood that the algorithm NEXT(k, i) *presented can be changed through a slight modification so that to its domain of definition not only vertices of the form (k, i) is included but also the target vertex β. For this modification in the programs (9.111), (9.112) and (9.113), it is sufficient to write β instead of every pair (k, i) as well as n instead of $i - 1$. Execution the algorithm* NEXT(β) *will have a complexity of at most $\mathcal{O}(n \log K)$.*

It is self-evident that before the first starting the program NEXT(β) *the data must be initialised to become consistent.*

Algorithm 9.4 Data initiation before starting **NEXT(β).**

1. *For the vertices of the form $(k, 0)$ the number $q^*(k, 0)$ is to be equal to one, which means that the information about two shortest paths from the vertex α to the vertex $(k, 0)$ is available. The length of the shortest path is $\varphi(k)$, i.e., $f(k, 0, 0) = \varphi(k)$ is substituted. The length of the succeeding path is ∞, i.e., $f(k, 0, 1) = \infty$, which means that the second path from α to $(k, 0)$ does not exist. The other data for the vertices $(k, 0)$ are not defined.*

2. *For all other vertices $q^*(\gamma) = 0$ is substituted. This means that only information on the best path from the vertex α to the vertex γ is available. The number $f(\gamma, 0)$ is the length of that best path and the numbers $f(\gamma, q)$ for $q > 0$ are not defined, since in this case $q > q^*(\gamma)$ holds.*

3. *The index $k'(k, i, 0)$ is* $\text{argmax}_{k'} \big(f(k', i-1, 0) + P_i(k', k)\big)$, *and the index $q'(k, i, 0)$ is 0.*

4. *The index $k'(\beta, 0)$ is* $\text{argmax}_{k'} f(k', n, 0)$ *and the index $q'(\beta, 0)$ is 0.*

5. *In the leaves of (k, i)-th tree the numbers $f(k', i-1, 0)$ are entered, i.e., $b(k, i, k') = f(k', i-1, 0)$ and* $\text{ind}(k, i, k') = 0$, *$k' = 0, 1, \ldots, K-1$. The information on all other tree vertices, including the tree root, should correspond to the information in the tree leaves.*

6. *Similarly as the initiation of the tree (k, i), the initiation of the tree (β) is to be carried out.*

The initiation presented contains, in fact, all computations which are needed to seek the shortest path from α to β. The results of seeking and further auxiliary data are stored in such a form that the program NEXT *may be used. The complexity of the initiation is $\mathcal{O}(K^2 n)$.*

In this way I have arrived at a result in which seeking the d-shortest path from α to the vertex β consists of the initiation having the complexity $\mathcal{O}(K^2 n)$ and $(d-1)$-fold running of the program NEXT(β) *having the complexity $\mathcal{O}(n \log K)$. The overall complexity of the task solution is not greater than*

$$\mathcal{O}\big(K^2 n + (d-1)\, n \log K\big) .$$

I cannot but close the question that I mentioned only cursorily before. The matter is that the memory needed for storing the data $f(k, i, q)$, $k'(k, i, q)$

and $q'(k,i,q)$, is actually much smaller than the size of $\sum_k \sum_i \left(q^*(k,i)+1\right)$, which it would seem to be at first glance. Based on the knowledge of a concrete algorithm we can claim that immediately after the initiation the sum $\sum_k \sum_i \left(q^*(k,i)+1\right)$ slightly differs from the number Kn. Further on, in each application of the program NEXT(β) the sum does not increase by more than n, because for each i no more than one of the numbers $q^*(k,i)$ is changed. When the sum changes then it is increased only by one. It follows from this that after a d-fold application of the program NEXT(β) the sum of the memory sizes will not be greater than $\mathcal{O}(Kn+nd)$.

Well, this may be all with respect to the d-shortest paths in the graph and, consequently, with respect to search for the d-best approximations of given sentence with sentences of given regular language. Certainly, the procedure described can be easily generalized to the situation in which the object under approximation is of more complicated acyclic structure than a sequence.

The algorithm for solving the task on d shortest paths in a graph begins to gradually assume an appropriate form. You might publish your results, since they have a significance on their own, and not only in the pattern recognition domain. On the whole we can see that you have excellently mastered the subject matter of the last two lectures.

The analysis was quite instructive for me because I became convinced once more how substantially the efficiency of a procedure carefully thought over may differ from the one which occurs to me at first and seems to be self-evident.

Allow me to say frankly that in one item of your lecture I saw quite some negligence in the estimate of computational complexity. It concerns the complexity of the matrix polynomial $\bigoplus_{i=0}^{\infty} A^i$ in a semi-ring with idempotent addition. Without any hesitation you wrote that the complexity of that operation was $\mathcal{O}(k^3 \log k)$ for the matrix A of the dimension $k \times k$. But that is the complexity of the most primitive algorithm which occurs to anybody in the first place. For this task there exists a long known algorithm owed to Floyd [Floyd, 1962] complexity of which is $\mathcal{O}(k^3)$. How shall I come to terms with it?

Simply by forgiving the negligence. We did not intend to interrupt the firm and purposeful storming of the main aim of the lecture.

I am thankful that you have assessed my understanding of the subject matter presented as being quite satisfactory. Without pretending mock modesty, I feel that I have got oriented in the subject matter has been dealt with. In spite of that, I still miss the last item to be supported in writing an actual program for my task on text-line pattern recognition. I miss something significant for being able to represent my actual problem by abstract concepts which were used in the lectures. I do not ask questions only to enquire about something more. I tried to settle the problems by myself, but now I am arriving at the idea that the theoretical results of the last two lectures have nothing in common with even the simplest practical task, which I would like to solve at all costs.

Well, will you get on with it and formulate the question in a more concrete fashion?

I cannot get on quickly, I am afraid, because the question cannot be expressed briefly.

After Lecture 7 we thoroughly discussed a number of questions related to practical aspects of recognising text documents. At the end of the discussion you anticipated what difficulties were still in store for me after I started recognising a whole line of symbols, even when having an appropriate algorithm for recognising individual isolated symbols at my disposal. You promised that I would find the key to meeting the difficulties in Lectures 8 and 9. Your prediction came true only partially. The difficulties arose, that is true, but the subject matter presented is still of no help to me, though I understand it. At least, it seems to me that I understand it, and even you have appreciated it. I have an impression that practical difficulties were about one thing and the lectures were about something quite different.

The lectures were about how, according to the sequence of images, $x_1, x_2, \ldots, x_n$ *to find the sequence of character labels* $k_1, k_2, \ldots, k_n$*, which correspond to the images. Even the way how to make use of the knowledge of mutual dependence of two neighbouring character labels was referred to. This model situation is substantially different from what I actually have.*

Input data for recognition do not have, at all, the form of a sequence $x_1, x_2, \ldots, x_n$*, in which* x_i *should depict the i-th character with an unknown label* k_i*. The input information has the form of one non-segmented image* x *that corresponds to even an entire line of characters. Furthermore the images* x *there is nothing to tell me where one character ends and the other starts. To use the subject matter of the lectures I must first segment my image into individual characters in some way. This problem again lets me remain without help. If I knew how to segment the input image* x *into individual characters then it would mean that I had got over the greatest obstacle in solving the task.*

Assume that I had managed to segment the image with text into individual characters in some way. What can the theoretical exposition offer to me? Only that I can use the information on mutual dependence of neighbouring characters and so improve the quality of their recognition. But the structural relationships of individual characters in a natural language and finally even in a formalised language, are far more complicated than those which can be expressed by virtue of such primitive means as are regular languages and their stochastic generalisation. By these simplest tools the results of recognising isolated symbols can be substantially improved only in the case in which results are quite bad.

Based on ideas of the discussion after Lecture 7 I wrote a program which recognises isolated characters quite well. But wrong answers in recognising a text line do not occur as a result of not respecting the dependency between neighbouring symbols in the text, but mainly because an error had already occurred in segmenting the line into individual characters. For segmentation I used an algorithm which I had devised on my own without any theory. I admit that it yields bad results. But the theory does not tell me anything about what a segmentation algorithm should look like. But image segmentation does belong to pattern recognition, does it not?

The situation as a whole seems to me as if a person promised to serve me any kind of meal, and only when I really wanted to get a meal, did I find that I had first to prepare it myself. Then independently of what kind of meal it was, that person could bring and serve it to me. He had kept his promise, but I saw that it had been a pure hoax.

Here I can see a clear gap between the theory explained and my actual practical task. The theory deals with making use of the dependency between symbols which in my practical task cannot substantially improve the results of recognition. The major difficulty of my task is how the originally compact observation x *is to be transformed into a sequence of observations* $x_1, x_2, \ldots, x_n$, *in which each observation corresponds to one character. The theory is silent about that. In the theory the sequence* $x_1, x_2, \ldots, x_n$ *is already assumed as given. To worry about how to get it is to be my job.*

Do I see the gap in the right place? Could not you, perhaps, help me in analysing my task of recognising an image with a line of text in which segmentation problems appear? These problems seem to me the most significant. The relationship between neighbouring characters could be ignored, as it seems less substantial to me.

First you created the gap yourself and now you can clearly see it. You have assumed since the very beginning that the alphabet K of states is identical with the alphabet of characters. This conception of the set of states immediately results in the idea that the sequence of observations must have the form $x_1, x_2, \ldots, x_n$, in which x_i is a part of the image containing the i-th character and no part of any other character. But for using the theory it is not at all inevitable for the sequences $k_1, k_2, \ldots, k_n$ and $x_1, x_2, \ldots, x_n$ to have exactly this meaning.

We will examine your task quoting a form of data which can be assumed to undoubtedly correspond to input data. In any case we can assume that the input data have the form of a two-dimensional array $\bigl(x(i,j),\ 1 \le i \le n,\ 1 \le j \le m\bigr)$ which consists of n columns and m rows. As usual, when referring to images, the value $x(i,j)$ is considered to be the brightness of the image at a point with integer coordinates (i,j). These data can be interpreted even as a sequence $x_1, x_2, \ldots, x_n$, where x_i is a one-dimensional ensemble of the length m; simply speaking, it is the i-th column in an original two-dimensional array.

But in such a case the set X, *from which the quantities* x_i *assume their values, is extremely extensive.*

Yes, it is. But for the time being do not worry about it. Now it is more important to reveal what this multi-dimensional, and therefore so complicated, operation x_i depends on. Recall that i indicates the number of the column being processed in the whole line of the text.

The observation x_i depends only on two quantities. One is the number q, i.e., the counter of columns from the beginning of a character, since the first column in the character being processed is the column $i - q$. The other quantity is the name k of the character being processed which is a name from the alphabet K. For each $k \in K$ the number $Q(k)$ will be introduced which means the width of the symbol k, i.e., it indicates of how many columns the image, representing the character, consists. The pair (k, q), on which the column in the observed two-dimensional array $\big(x(i,j),\ 1 \le i \le n,\ j \le m\big)$ depends, belongs to the set $\{(k,q) |\ |k \in K,\ q = 0, 1, \ldots, Q(k) - 1\}$. This set will be considered as a set of states of an automaton which generates images with lines of text.

I have caught it! The main thing is that the automaton states need not correspond just to what is to be recognised, but they may be something more detailed. I remember that we were discussing something similar to it after Lecture 7. Then you directed my attention to a situation in which in creating a model of a recognised object some artificially created parameters are sometimes to be added to the hidden parameters of a natural kind. By extending the set of hidden parameters the task does not become more complicated. Just the opposite, it becomes simpler.

Here just such a situation has occurred. We are surprised that you did not arrived at it sooner. Well, continue by yourself.

The automaton generates a sequence of states (k_1, q_1), (k_2, q_2),...,(k_n, q_n) and a sequence $e_1, e_2, \ldots, e_n$ of columns, each of which consists of m elements, appears at the output. The sequence $e_1, e_2, \ldots, e_n$ then forms a two-dimensional ensemble $\big(e(i,j),\ i = 1, 2, \ldots, n,\ j = 1, 2, \ldots, m\big)$ which can be considered as an image. For the image to correspond to an ideal, undamaged text line, certain constraints are to be satisfied as to what the state (k_i, q_i) can be in the i-th moment in dependence on what the state (k_{i-1}, q_{i-1}) was in the preceding moment e_i on the state (k_i, q_i) is to be determined. For simplicity I will not take into consideration that the labels of characters in the text are mutually dependent. The automaton generating an image of a text line in which mutually independent symbols occur can be, for example, defined as follows.

1. *The set of initial automaton states is the set $\{(k, 0) \mid k \in K\}$ by which a clearly understandable property is stated that the generation of an image with text begins from generating the initial (zero) column of some of the symbols.*
2. *The set of target states is $\{(k, Q(k) - 1) \mid k \in K\}$. This means that the generation of an image with text can end only in those states when generating some of the symbols is finished.*
3. *If the automaton is in some state (k, q), $k \in K$, $q \neq Q(k) - 1$, (i.e., if the state labelled k has not been processed as a whole) then the succeeding state must be $(k, q+1)$ (i.e., it must continue generating the same character k).*

4. *If the automaton is in the state* $\big(k, Q(k) - 1\big)$, $k \in K$, *(i.e., when the generation of the state labelled* k *has already ended) then the automaton can either stop generating the text line or pass on to the state* $(k', 0)$, $k' \in K$ *(i.e., to start generating the succeeding state).*
5. *If the automaton is in the state* (k_i, q_i) *at the* i*-th moment then to an earlier constructed ensemble* $e(i', j)$, $i' = 1, 2, \ldots, i-1$, $j = 1, 2, \ldots, m$, *a column* e_i *is added. The column* e_i *is defined by the label of the character* k_i *which is just being constructed, and by the number* q_i *of the column in the character.*

We will denote by $E(k, q)$ *the set of all columns which can be regarded as ideal representatives of the* q*-th column of the character named* k*. The set* $E(k, q)$ *need not be very extensive. In the simplest case it can consist only of one column. The diversity of all possible columns* x_i *which can actually occur in the position of the* q*-th column of the character* k *can be expressed not only by the extension of the set* $E(k, q)$*, but also by the introduction of a similarity function* $d(k, q, x)$*. This function indicates to what extent the column* x *can be considered as the representation of the* q*-th column of the character named* k*.*

The concepts presented here serve as a basis for the following formulation of a task the solution of which is the segmentation of the ensemble $\big(x(i,j), 1 \le i \le n,\ 1 \le j \le m\big)$ *into individual characters as well as the recognition of individual characters.*

The ensemble $\big(x(i,j), 1 \le i \le n,\ 1 \le j \le m\big)$ *will be considered as a sequence of columns* $x_1, x_2, \ldots, x_n$*, and the task will be formulated as seeking the sequence* (k_1^*, q_1^*), (k_2^*, q_2^*), $\ldots$, (k_n^*, q_n^*) *which is*

$$\big((k_i^*)\,,\ i = 1, 2, \ldots, n\big) = \operatorname*{argmax}_{\big((k_i, q_i),\ i=1,2,\ldots,n\big)} \sum_{i=1}^{n} d(k_i, q_i, x_i)$$

under the conditions

$$(k_1, q_1) \in \{(k, 0) \mid k \in K\}, \quad (k_n, q_n) \in \{(k, Q(k) - 1) \mid k \in K\}, \quad k_i = k_{i-1},$$

$$q_i = \begin{cases} q_{i-1} + 1\,, & \text{if} \quad q_{i-1} \neq Q(k_{i-1}) - 1\,, \\ 0\,, & \text{if} \quad q_{i-1} = Q(k_{i-1}) - 1\,. \end{cases}$$

The sequence obtained (k_i^*, q_i^*), $i = 1, \ldots, n$, *determines the sequence* $i_1, i_2, \ldots, i_M$ *of indices, where* $q_{i_m}^* = 0$*. The index* i_m *provides the horizontal coordinate of a point in which the* m*-th character begins,* $k_{i_m}^*$ *denotes the name of the character, and* M *is the number of of characters in the text line which is being recognised.*

What you have designed is one of the simplest approaches. As we know you, some more accomplished algorithms will occur to you after a period of thinking it over.

July 1998

9.7 Link to a toolbox

The public domain demonstration software related to recognition in regular languages was written in C language by P. Soukup as a diploma thesis in Summer 2001. It can be downloaded from the website `http://cmp.felk.cvut.cz/cmp/cmp_software.html`. The library implementing generalised convolution and several tasks solving the best matching problem are available including the source code.

9.8 Bibliographical notes

The chapter is based on the formulation of languages and grammars in the sense of Chomsky [Chomsky, 1957; Chomsky et al., 1971] which in its significance far exceeds pattern recognition. The concept of the fuzzy grammar alternative comes from [Zadeh, 1965; Zimmermann et al., 1984].

A large group of tasks in structural pattern recognition deals with comparing an object with the ideal object (exemplar). The objects are usually of a more complex structure than that of a sequence. They are mostly graphs. The comparison criterion usually is a modified Levenstein dissimilarity [Levenstein, 1965; Bunke, 1996]. A further step towards the statistical interpretation of Levenstein tasks was made by Kashyap and Oommen [Kashyap and Oommen, 1984; Oommen, 1987] when they formulated nontrivial pattern recognition tasks and solved them.

In the lecture another approach was applied in which the object examined is compared with a set of objects (with a regular language). In the case in which Levenstein dissimilarity is the metric the exact solution of the best matching problem belongs to Wagner [Wagner and Fischer, 1974; Wagner and Seiferas, 1978]. Efficient algorithms for solving tasks of this kind were designed by [Amengual and Vidal, 1996]. The solution for a more general case in which Levenstein similarity is not the metric has been presented in this lecture.

Jiří Pecha brought to our attention in the discussion Floyd's excellent algorithm [Floyd, 1962].

Jiří Pecha is not the only one whose interest was attracted by the beautiful task of searching for d-best derivations of a formal language sentence, in particular, its computational aspects. After the Czech version of this monograph was published we learned about the paper [Jimenez and Marzal, 2000] which solves the problem seeking d-best derivation of a sentence in a context-free language. The paper solves the more general version of the problem compared to the one we have analysed with Jiří Pecha. Even so, the motivation and results obtained are close to the conclusions of Jiří Pecha. It seems that there are more of us who are interested in these nice and not very simple problems.

The algebraic constructions for solving optimisation tasks of structural pattern recognition, quoted in the lecture, were presented by Schlesinger [Schlesinger, 1989; Schlesinger, 1994; Schlesinger, 1997].

Lecture 10

Context-free languages, their two-dimensional generalisation, related tasks

10.1 Introductory notes

From time to time scientific terminology seems to make fun of a trusting reader, deliberately wanting to confuse him or her. It happens that scientific concepts are used which are common in everyday life but denote something quite different. For example, the theory of catastrophes does not deal with what we normally consider a catastrophe, a disaster. Similarly, games theory has nothing to do with what is happening on a football ground or on a chess board.

The concept 'context-free language' is one of the examples of such a perfidious concept. According to the name it could be supposed to mean manipulating a language in such a way that one wants only to switch from one topic to another until the sentence resembles a chaotic chain of mutually independent fragments. In fact, a context-free language is determined by precise definitions. If the definitions were not known then one could not guess what the particular term might mean. This is the case in which the application of a familiar and expressive concept for a concrete idea brings about only disorientation.

This lecture is devoted to a formalism with the aid of which sets of images and a probability distribution on them are constructively defined. Based on the definitions, different pattern recognition tasks are solved resembling those analysed in the previous two lectures 8 and 9. This lecture differs from the previous two lectures in that the objects recognised will not only have the form of one-dimensional sequences but primarily the form of two-dimensional and multi-dimensional arrays. We will see that the formalism proposed is a natural generalisation of context-free grammars and languages according to N. Chomsky's hierarchy. In its turn, the Chomsky's context-free grammars and languages are generalizations of regular grammars and languages.

10.2 Informal explanation of two-dimensional grammars and languages

Imagine a dialogue between a human and an artificially produced device, say, a computer. The computer is expected to make use of the dialogue and to learn to recognise images of a class. The dialogue starts by presenting an image to the computer and asking the computer if the image belongs to the class of images which denote, e.g., the Russian letter *SH* pronounced as 'sh", which will be for typographical reasons denoted here as *SH*. Its shape can be seen in Fig. 10.1.

The computer scans its library of programs and checks if it contains the pertinent program. If it does then it replies to the question and finishes the dialogue. If not then the computer tells the user that it cannot answer the question. A human can react to the information by inserting a program into the computer enabling it to answer. Such a method of communication between a human and the computer seems natural from the present day point of view even when it greatly differs from a dialogue in which the partner would be another human. The human partner could be explained in a way the meaning that the image is called *SH*. This explanation would not have the form of a program, i.e., a sequence of instructions. Such non-procedural definitions are possible even in the human/computer dialogue. In this case the definition of the objects concerned must be provided in a form given beforehand which is understandable even to the computer. Assume that three admissible definition forms were to be given beforehand.

1. The image may be labelled s, i.e., may have a name s, if it can be divided into two parts by a horizontal line so that the image in the upper part is labelled s_{u}, and the image in the lower part is labelled s_{d}.
2. The image may be labelled s if it can be divided into two parts by a vertical line so that the image in the left part is labelled s_{l} and the image in the right part is labelled s_{r}.
3. The image may be labelled s if it has another label s'.

The quoted form of definitions can be understood as metarules, i.e., the rules for formulating other rules serving for recognising whether the image has a certain label. At the same time, the metarules can be understood as rules for formulating other rules for the generation of images with given labels.

The first metarule can be interpreted like this: To find whether the image may have a label s the image is to be divided by a horizontal line into two parts in all possible ways, and after each division a program is to be run which finds whether the upper part is labelled s_{u} and the label of the lower part is s_{d}. When a positive answer occurs with at least one decomposition of the image then it is decided that the image presented is labelled s.

The first metarule can be interpreted in another way, i.e., as a rule for generating images labelled s. In order to make the drawing of an image labelled s possible, two images are to be drawn consisting of the same number of columns. The first of them can be any of the images labelled s_{u}, and the second can be any of the images labelled s_d. The images drawn are to be arranged to form one image so that the image labelled s_u forms its upper part and the image labelled s_d forms its lower part. The second and third metarules can be similarly interpreted.

Let us show now how the class of the images with label *SH* can be defined using the quoted metarules.

By means of the third and second metarules the following definition of the set of images labelled *SH* can be stated: The image may have a label *SH* if its label is *SH1* or if it is composed of two parts divided by a vertical line, the left part being labelled *SH1* and the right one *WR* (white rectangle, i.e., an image all pixels of which are white). The first part of the definition is represented by the left picture in Fig. 10.1 and the second part by the right picture in Fig. 10.1. If we deleted all useless words from the definition, and kept only what makes the definitions differ from each other then the definitions could be written in the following brief form:

$$\left.\begin{aligned} &SH ::= SH1 \mid WR\,; \\ &SH ::= SH1\,. \end{aligned}\right\} \tag{10.1}$$

In this definition *SH1* is the label of the image in which *SH* is closely adjacent to the right side of the rectangle (of the field of view) on which it is drawn. If the computer understood this situation, i.e., if it had a program for recognising white rectangles and letters *SH* closely adjacent to the right side of the field of view then the given definition would suffice for automatically creating a program which would recognise even the letters that are not adjacent to the right side of the field of view. If the computer does not have programs of this type then it asks additional questions as to what the *SH1* and *WR* mean. A human answers and uses additional labels in his answer. The dialogue continues until the concept 'the image is labelled *SH*' is explained by means of concepts which the computer already knows. We will show one of the possible definitions of the concept 'the image is labelled *SH*' for the case in which the computer knows two concepts only which need not be defined. These are *WP* (white pixel) and *BP* (black pixel). We will need some auxiliary concepts too, namely *WR* (white rectangle), *BR* (black rectangle), *U* (a shape resembling letter U), *I* (a shape resembling letter I), *L* (a shape resembling letter L). The definition can be visualised by the corresponding figures.

$$\left.\begin{aligned} &SH1 ::= \frac{WR}{SH2}\,; \\ &SH1 ::= SH2\,. \end{aligned}\right\} \tag{10.2}$$

$$\left.\begin{aligned} &SH2 ::= SH3\,; \\ &SH2 ::= WR \mid SH3\,. \end{aligned}\right\} \tag{10.3}$$

$$\left.\begin{aligned} &SH3 ::= SH4\,; \\ &SH3 ::= \frac{SH4}{WR}\,. \end{aligned}\right\} \tag{10.4}$$

$$SH4 ::= U \mid BR\,. \tag{10.5}$$

$$U ::= L \mid L\,; \tag{10.6}$$

$$L ::= \frac{I}{BR}\,; \tag{10.7}$$

$$I ::= BR \mid WR\,. \tag{10.8}$$

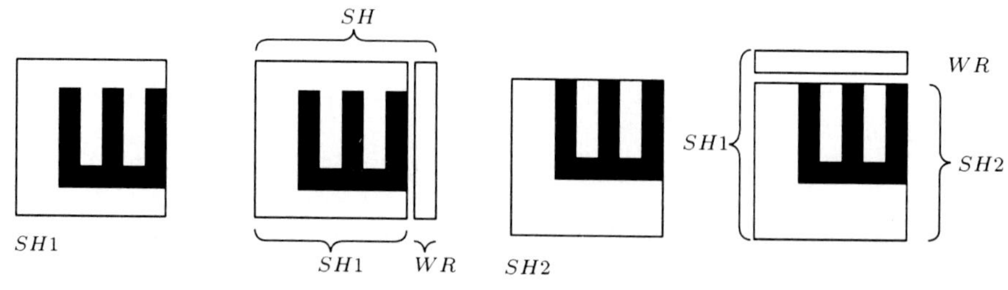

Figure 10.1 Generating letter *SH* according to rule (10.1), i.e., separating the image margin at the right side of the letter.

Figure 10.2 Illustration of the (10.2), i.e., separation of the image margin above the letter.

$$\left.\begin{aligned} BR &::= BR \mid BR\,; \\ BR &::= \frac{BR}{BR}\,; \\ BR &::= BP\,. \end{aligned}\right\} \qquad (10.9) \qquad \left.\begin{aligned} WR &::= WR \mid WR\,; \\ WR &::= \frac{WR}{WR}\,; \\ WR &::= WP\,. \end{aligned}\right\} \qquad (10.10)$$

The rules (10.1) through (10.10) form the definition of the concept 'the image is labelled *SH*', i.e., a definition of a set of images. The information obtained by the computer from the human during the imaginary dialogue can be arranged in a shape of a six-tuplet

$$G = \langle X, K, k_0, P_v, P_s, P_r \rangle \tag{10.11}$$

which is an example of the two-dimensional context-free grammar. Later we will introduce it in a more precise form. In the six-tuplet X represents a terminal alphabet. In our interpretation it is a finite set of labels assumed to be understood by the computer without additional explanation. In the above example concerning the letter *SH* the terminal alphabet X consists of two terminal symbols, i.e., *WP* (white pixel) and *BP* (black pixel).

The set K is called a non-terminal alphabet. This alphabet contains a finite number of labels. In our example these are *BR*, *WR*, *I*, *L*, *U*, *SH4*, *SH3*, *SH2*, *SH1* and *SH*. One of the non-terminal names which is in the grammar G denoted as k_0 is called an axiom. It is a concept for which the grammar G was actually created. In our example the axiom is the symbol *SH*.

Three additional concepts of the grammar are three relations $P_{\rm h}$, $P_{\rm v}$, $P_{\rm r}$. The first two relations are subsets of the triplets of non-terminal symbols, i.e., $P_{\rm h} \subset K \times K \times K$, $P_{\rm v} \subset K \times K \times K$. The relation $P_{\rm r}$ is a subset of the pairs $P_{\rm r} \subset K \times (K \cup X)$. Indices h, v, r in the denotations $P_{\rm h}$, $P_{\rm v}$ and $P_{\rm r}$ are the first letters of the words 'horizontal', 'vertical', and 'renaming'. The relations $P_{\rm h}$, $P_{\rm v}$ and $P_{\rm r}$ determine the rules by which images are generated and which belong to the set generated by the grammar. In our case these are the rules

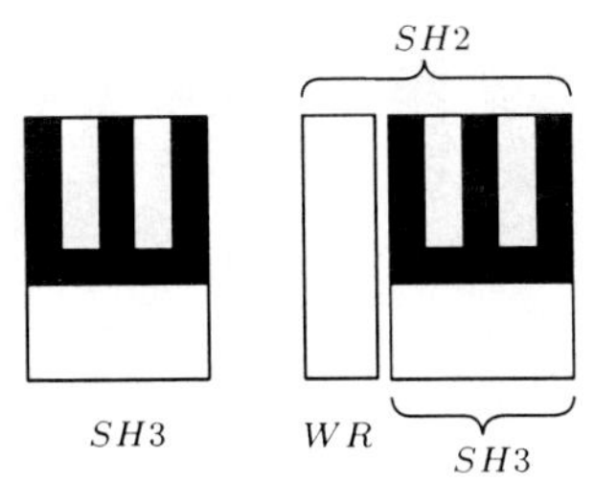

Figure 10.3 Illustration of the rule (10.3), i.e., separation of the margin at the right side of the letter.

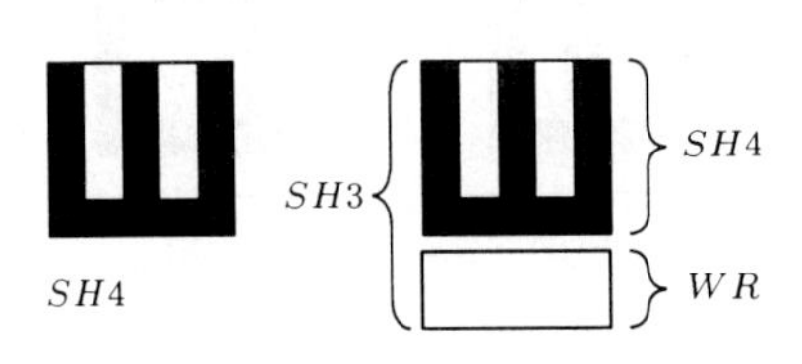

Figure 10.4 Illustration of the rule (10.4), i.e., separation of the margin below the letter.

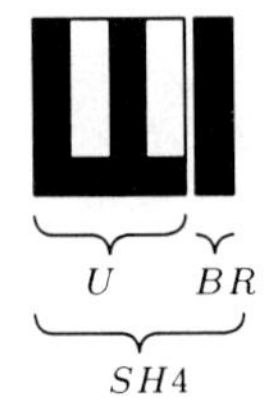

Figure 10.5 Illustration of the rule (10.5), i.e., separation of the black rectangle (BR) at the right side.

Figure 10.6 Illustration of the rule (10.6), i.e., decomposition of the remaining part of the letter into two shapes resembling letter L.

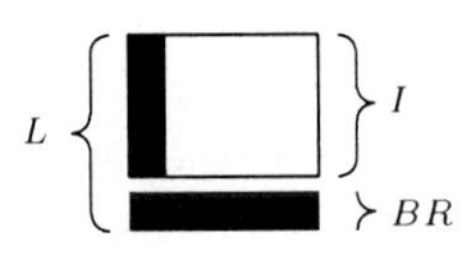

Figure 10.7 Illustration of the rule (10.7), i.e., decomposition of the shape resembling letter L into two parts: the black rectangle at the bottom and a shape resembling letter I, where the black rectangle is at the left and the white rectangle is at the right.

Figure 10.8 Illustration of the rule (10.8), i.e., decomposition of the shape resembling letter I into a black rectangle and a white rectangle.

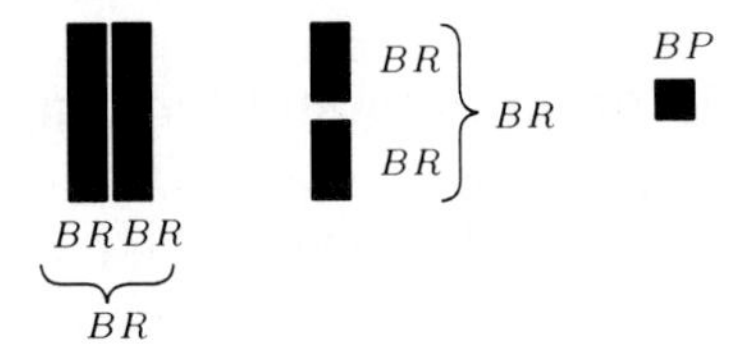

Figure 10.9 Illustration of the rule (10.9), i.e., the black rectangle can be created by concatenating of these black rectangles only.

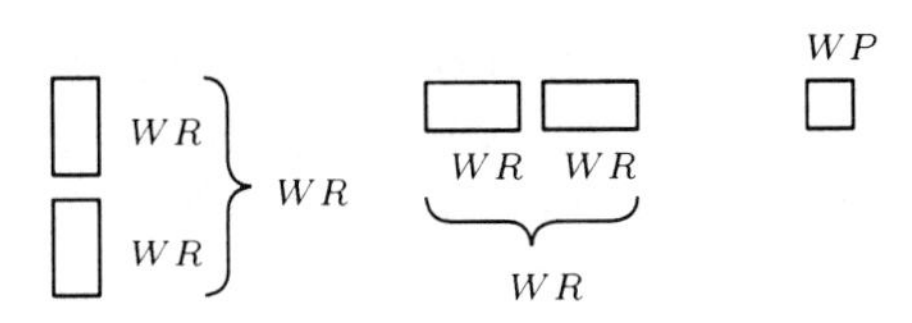

Figure 10.10 Illustration of the rule (10.10), i.e., the white rectangle can be composed of these white rectangles.

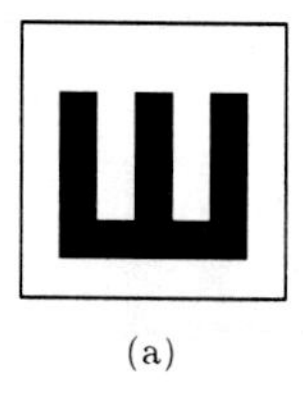

(a)

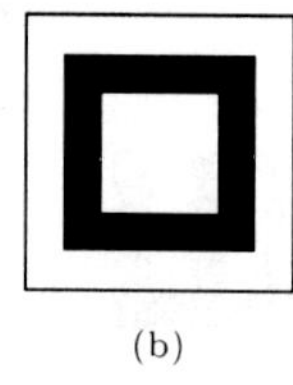

(b)

Figure 10.11 (a) An example of the image with a letter resembling *SH* which can be created by the introduced rules. (b) An example of the image not resembling *SH* which cannot be created by the introduced rules.

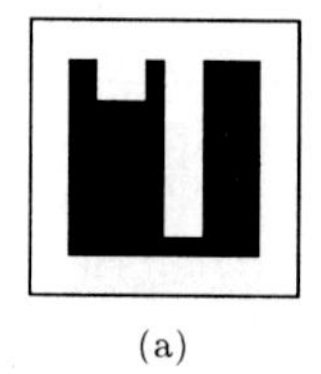

(a)

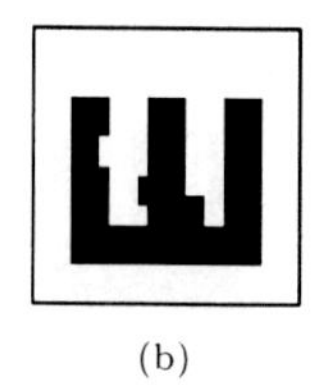

(b)

Figure 10.12 (a) An example of the image with a letter not resembling *SH* which can be created by introduced rules. (b) An example of the letter resembling *SH* which cannot be created by introduced rules.

(10.1) through (10.10), and thus they are the relations

$$\left.\begin{aligned} P_{\mathrm{h}} &= \{(SH, SH1, WR), (SH2, WR, SH3), (SH4, U, BR), (U, L, L),\\ &\qquad (I, BR, WR), (BR, BR, BR), (WR, WR, WR)\}\,,\\ P_{\mathrm{v}} &= \{(SH1, WR, SH2), (SH3, SH4, WR),\\ &\qquad (L, I, WR), (WR, WR, WR), (WR, WR, WR)\}\,,\\ P_{\mathrm{r}} &= \{(SH, SH1), (SH1, SH2), (SH2, SH3), (SH3, SH4),\\ &\qquad (BR, BP), (WR, WP)\}\,. \end{aligned}\right\} \qquad (10.12)$$

By the relations (10.12), or equivalently by the rules (10.1) to (10.10), images can be generated which resemble the letter *SH*, such as the letter in Fig. 10.11(a). By applying this procedure other images which do not resemble the letter *SH* cannot be generated, see Fig. 10.11(b). However, one can make sure that by the quoted rules even images which are not the letters *SH* can be created, e.g., see Fig. 10.12(a). On the other hand, some images which, except for small deformations, correspond to nearly satisfactorily shaped letters *SH*, e.g., see Fig. 10.12(b), cannot be generated by the rules quoted. Thus the grammar G introduced defines the set of images which only remotely resembles the set of all images that could be understood as letters *SH*. The reason for quoting the example was not our intention to define a set of images for actual application. We intended only to elucidate the main tool by which sets of images are syntactically defined. In the succeeding section we will present a formal definition of two-dimensional context-free grammars and languages.

10.3 Two-dimensional context-free grammars and languages

Let X be a finite terminal alphabet the elements of which are called terminal symbols. For every two positive integers m (the number of rows) and n (the number of columns) the notation $T(m,n)$ introduces a rectangle in a two-dimensional integer grid, i.e., $T(m,n) = \{(i,j) \mid 1 \le i \le m,\ 1 \le j \le n\}$. The image x is understood to be a pair of numbers m, n and a function of the form $T(m,n) \to X$, i.e., $x = \langle m, n, T(m,n) \to X\rangle$. The value (i.e., intensity, brightness) of the image in the point $(i,j) \in T(m,n)$ will be denoted $x(i,j)$.

The set of all possible images will be denoted X^* and each subset $L \subset X^*$ will be called the two-dimensional language in the alphabet X.

We will introduce an operation of horizontal and vertical concatenation of images. Let $x_1 = \langle m, n_1, T(m, n_1) \to X\rangle$ and $x_2 = \langle m, n_2, T(m, n_2) \to X\rangle$ be two images which have the same number of rows m. The horizontal concatenation of the images x_1 and x_2 means the image $x = \langle m, n_1 + n_2, T(m, n_1 + n_2) \to X\rangle$ which will be denoted $x = x_1 \mid x_2$ and for which the following holds,

$$x(i,j) = \begin{cases} x_1(i,j)\,, & \text{if } 1 \le j \le n_1\,, \quad i = 1,2,\ldots,m\,, \\ x_2(i, j - n_1)\,, & \text{if } n_1 < j \le n_1 + n_2\,, \quad i = 1,2,\ldots,m\,. \end{cases}$$

Let $x_1 = \langle m_1, n, T(m_1, n) \to X\rangle$ and $x_2 = \langle m_2, n, T(m_2, n) \to X\rangle$ be two images with the same number of columns n. The vertical concatenation of the images x_1 and x_2 means the image $x = \langle m_1 + m_2, n, T(m_1 + m_2, n) \to X\rangle$, which will be denoted $x = \frac{x_1}{x_2}$, and for which the following holds,

$$x(i,j) = \begin{cases} x_1(i,j)\,, & \text{if } 1 \le i \le m_1\,, \quad j = 1,2,\ldots,n\,, \\ x_2(i - m_1, j)\,, & \text{if } m_1 < i \le m_1 + m_2\,, \quad j = 1,2,\ldots,n\,. \end{cases}$$

Let us note that the concatenations introduced in this way are associative, but they are not commutative operations for the subsets of image pairs.

Let K be a finite set of labels, i.e., a non-terminal alphabet, and P_h, P_v, P_r be three relations of the form $P_\mathrm{h} \subset K \times K \times K$, $P_\mathrm{v} \subset K \times K \times K$, $P_\mathrm{r} \subset K \times (K \cup X)$. These three relations determine $|K|$ two-dimensional languages L_k, $k \in K$, which are defined in the following recursive way.

Definition 10.1 Two-dimensional languages.

1. *For the selected $k \in K$ the image $x = \langle 1, 1, T(1,1) \to X\rangle$ belongs to the language L_k if $\big(k, x(1,1)\big) \in P_\mathrm{r}$. Explained descriptively, an elementary image consisting of one single pixel can be called k if a rule from the set P_r says that a symbol written in this pixel can be renamed to k. In addition to the simplest images of such a form even other images can be labelled k in agreement with additional metarules.*
2. *For $k \in K$ the image x belongs to the language L_k if there exists a non-terminal symbol k' for which $x \in L_{k'}$ and $(k, k') \in P_\mathrm{r}$.*
3. *For $k \in K$ the image x belongs to the language L_k if there exist images x_t (top), x_b (bottom) and symbols $k_\mathrm{t} \in K$, $k_\mathrm{b} \in K$, so that $x = x_\mathrm{t}/x_b$, $x_\mathrm{t} \in L_{k_\mathrm{t}}$, $x_\mathrm{b} \in L_{k_\mathrm{b}}$, and $(k, k_\mathrm{t}, k_\mathrm{b}) \in P_\mathrm{v}$ are valid.*
4. *For $k \in K$ the image x belongs to the language L_k if there exist images x_l (left), x_r (right) and symbols $k_\mathrm{l} \in K$, $k_\mathrm{r} \in K$, so that $x = x_\mathrm{l} \mid x_\mathrm{r}$, $x_\mathrm{l} \in L_{k_\mathrm{l}}$, $x_\mathrm{r} \in L_{k_\mathrm{r}}$, and $(k, k_\mathrm{l}, k_\mathrm{r}) \in P_\mathrm{h}$ are valid.*

▲

The presented concepts defining the group of languages L_k, $k \in K$, can be expressed in the form of a six-tuplet

$$G = \langle X, K, k_0, P_\mathrm{h}, P_\mathrm{v}, P_\mathrm{r}\rangle$$

which will be called the two-dimensional context-free grammar where:

X – finite terminal alphabet,

K – finite non-terminal alphabet,

k_0 – axiom as one of the non-terminal symbols, $k_0 \in K$,

$P_h \subset K \times K \times K$ – collection of rules for horizontal concatenation,

$P_v \subset K \times K \times K$ – collection of rules for vertical concatenation,

$P_r \subset K \times (K \cup X)$ – collection of rules for renaming.

The language L_{k_0}, in which k_0 is an axiom of the grammar G, will be called the language of the grammar G, and the denotation for it will be $L(G)$. Images belonging to the language $L(G)$ will be said to be admissible images in the grammar G.

A grammar in which one of the sets P_h or P_v is empty is an ordinary, i.e., one-dimensional context-free grammar expressed in a canonical form according to N. Chomsky. The language of such a grammar contains 'images' which consist of a single column or a single row. Actually, they are no longer two-dimensional images, but sequences. It is generally known and can be easily demonstrated that every regular language can be defined by means of a context-free grammar. This means that the class of context-free languages contains all regular languages the properties of which were analysed in the two preceding lectures. In this lecture we will formulate pattern recognition tasks for context-free languages resembling those we examined before for regular languages. First we will deal with the exact matching problem of an image with the two-dimensional context-free language.

10.4 Exact matching problem. Generalised algorithm of Cocke–Younger–Kasami

Let $G = \langle X, K, k_0, P_h, P_v, P_r \rangle$ be a two-dimensional context-free grammar and $x = \langle m, n, T(m,n) \to X \rangle$ be an image. In the exact matching problem an algorithm is to be created which for each grammar G and each image x finds whether $x \in L(G)$ is valid.

The algorithm which solves the exact matching problem is based on the following simple considerations. Let $R(i_t, i_b, j_l, j_r)$, $1 \le i_t \le i_b \le m$, $1 \le j_l \le j_r \le n$, be a rectangle which is a subset in the rectangle $T(m,n)$, i.e., $R(i_t, i_b, j_l, j_r) = \{(i,j) \mid i_t \le i \le i_b,\ j_l \le j \le j_r\}$. The set of all such rectangles will be denoted $\mathcal{R}$. This set is partially ordered by the relation $\subset$. The set $\mathcal{R}$ can be, therefore, ordered into a one-dimensional sequence in such a way that if $R' \subset R''$ and $R' \neq R''$ then R' appears in the sequence of rectangles before R''.

Let $x(R)$, $R \in \mathcal{R}$, be a contraction of the image x to the subset R. Let us denote $f\big(i_t, i_b,\ j_l, j_r, k\big)$, $k \in K$, the number which assumes value 1 if $x\big(R(i_t, i_b, j_l, j_r)\big) \in L_k$, and assumes value 0 otherwise. The equality $f(1, m, 1, n, k_0) = 1$ is thus equivalent to the statement $x \in L(G)$. In the task being solved one single bit $f(1, m, 1, n, k_0)$ is to be computed.

Here and later on we will use the denotation $P_{\rm h}, P_{\rm v}, P_{\rm r}$ in two senses. Sometimes these letters will be considered as before to denote relations, i.e., subsets $P_{\rm h} \subset K \times K \times K$, $P_{\rm v} \subset K \times K \times K$ and $P_{\rm r} \subset K \times (K \cup X)$. Other times the denotations will be understood as functions $P_{\rm h}\colon K \times K \times K \to \{0,1\}$, $P_{\rm v}\colon K \times K \times K \to \{0,1\}$, $P_{\rm r}\colon K \times (K \cup X) \to \{0,1\}$. The relations $(k_1,k_2,k_3) \in P_{\rm h}$ and $P_{\rm h}(k_1,k_2,k_3) = 1$, $(k_1,k_2,k_3) \in P_{\rm v}$, $P_{\rm v}(k_1,k_2,k_3) = 1$, and others will be considered equivalent. With the newly introduced denotations we can write the following recursive relations for the values $f(i_{\rm t}, i_{\rm b}, j_{\rm l}, j_{\rm r}, k)$, $1 \le i_{\rm t} \le \le i_{\rm b} \le m$, $1 \le j_{\rm l} \le j_{\rm r} \le n$, $k \in K$,

$$
\begin{aligned}
&f(i_{\rm t}, i_{\rm b}, j_{\rm l}, j_{\rm r}, k) \\
&= \left[\bigvee_{j=j_{\rm l}}^{j_p-1} \bigvee_{k_l} \bigvee_{k_r} \Big(f(i_{\rm t}, i_{\rm b}, j_{\rm l}, j, k_l) \wedge P_{\rm h}(k, k_l, k_r) \wedge f(i_{\rm t}, i_{\rm b}, j+1, j_{\rm r}, k_r) \Big) \right] \\
&\vee \left[\bigvee_{i=i_{\rm t}}^{i_d-1} \bigvee_{k_t} \bigvee_{k_b} \Big(f(i_{\rm t}, i, j_{\rm l}, j_{\rm r}, k_t) \wedge P_{\rm v}(k, k_t, k_b) \wedge f(i+1, i_{\rm b}, j_{\rm l}, j_{\rm r}, k_b) \Big) \right] \\
&\vee \left[\bigvee_{k' \in K \cup X} \Big(f(i_{\rm t}, i_{\rm b}, j_{\rm l}, j_{\rm r}, k') \wedge P_{\rm r}(k, k') \Big) \right] .
\end{aligned}
\qquad (10.13)
$$

The previous relation precisely expresses the assertions 1–4 from the earlier introduced Definition 10.1 which define images belonging to the language L_k. The last term (in square brackets) in the relation (10.13) corresponds to the first and second assertion of the definition. The second term corresponds to the third assertion, and the first term corresponds to the fourth assertion of the definition. The relation (10.13), and thus also the Definition 10.1 of the language L_k, $k \in K$, form a basis for the following algorithm recognising whether the image $x = \langle m, n, T(m,n) \to X \rangle$ belongs to the language $L(G)$.

Algorithm 10.1 Two-dimensional generalisation of the Cocke–Younger–Kasami algorithm

1. For all pairs (i,j), $1 \le i \le m$, $1 \le j \le n$, the values of the function f is initialised,

$$f\big(i, i, j, j, x(i,j)\big) := 1 .$$

   The other values of f are filled with zeroes.

2. All rectangles $R \in \mathcal{R}$, i.e., quadruplets $\langle i_1, i_2, j_1, j_2 \rangle$ are examined in proper order in such a way that if $R' \subset R''$ then R' is processed before processing R''. This was mentioned already that such a one-dimensional ordering exists. The following three steps are performed for each rectangle $\langle i_1, i_2, j_1, j_2 \rangle$ and each $k \in K$.

   (a) The following condition is verified

$$\bigvee_{j \mid j_1 \le j < j_2} \; \bigvee_{k_l \in K} \bigvee_{k_r \in K} \big(f(i_1, i_2, j_1, j, k_l) \wedge P_{\rm h}(k, k_l, k_r) \wedge f(i_1, i_2, j+1, j_2, k_r) \big) . \qquad (10.14)$$

If the condition is satisfied then $f(i_1, i_2, j_1, j_2, k) := 1$ is substituted.

(b) If the preceding condition (10.14) is not satisfied then the following condition is verified,

$$\bigvee_{i|i_t \le i < i_1} \bigvee_{k_t \in K} \bigvee_{k_b \in K} \left(f(i_t, i, j_1, j_2, k_t) \wedge P_v(k, k_t, k_b) \wedge f(i+1, i_2, j_1, j_2, k_b) \right). \tag{10.15}$$

If the condition (10.15) is satisfied then $f(i_1, i_2, j_1, j_2, k) := 1$ is substituted.

(c) For values of k for which $f(i_1, i_2, j_1, j_2, k)$ continues to be zero the following condition is verified

$$\bigvee_{k' \in K} f(i_1, i_2, j_1, j_2, k) \wedge P_r(k, k'). \tag{10.16}$$

If the condition (10.16) is satisfied then $f(i_1, i_2, j_1, j_2, k) := 1$ is substituted.

3. The solution of the task is the bit $f(1, m, 1, n, k_0)$.

The presented algorithm is a direct two-dimensional generalisation of the known Cocke–Younger–Kasami algorithm which decides whether the sequence of symbols from X belongs to the given context-free (one-dimensional) language. The original Cocke–Younger–Kasami algorithm is a special case of Algorithm 10.1 when one of the two relations P_h, P_v is empty.

Even when Algorithm 10.1 is the two-dimensional generalisation of the known Cocke–Younger–Kasami algorithm, it does not follow from it that the computational complexity of recognising two-dimensional images is greater than that of one-dimensional sequences.

The computational complexity of the calculation using the formulæ (10.14), (10.15) and (10.16) depends on the size of the image and is $\mathcal{O}\big(m^2 n^2 (m+n)\big)$. From this the known estimate of the complexity of the original one-dimensional alternative of the Cocke–Younger–Kasami algorithm follows. If the length of the analysed sequence is l then the computational complexity is $\mathcal{O}(l^3)$.

We have arrived at a rather surprising, but now quite understandable property that recognition of a (two-dimensional) image consisting of $(m \times n)$ pixels is realised order of magnitude faster than that of the (one-dimensional) sequence containing the same number of mn symbols. The complexity of the two-dimensional case is $\mathcal{O}\big(m^2 n^2 (m+n)\big)$. However, the complexity in the one-dimensional case is $\mathcal{O}\big((mn)^3\big)$. The result can be explained so that in generalising the concept of the context-free language to the two-dimensional case, other languages were added to the original class, but within the extended class the most difficult cases are still those which the class contained before its extension. All pattern recognition tasks which had been added to the previous tasks appeared easier. It is quite an unexpected feature because one could suppose that the transition from recognising sequences to recognising two-dimensional structures might require calculations of markedly greater complexity.

We can only admire the thoroughness and ingeniousness with which Nature or evolution (or something else) has discovered that for the transfer of a large amount of information an image is far more suitable than a one-dimensional sequence of signals. Nature seems to have deliberately seen to it that the analysis of a vast amount of information may be at least a little facilitated by an appropriate choice of its structure.

10.5 General structural construction

We have dealt so far with two mechanisms for formulating a model of a recognised object. The former mechanism was based on the concept of a regular language including its various stochastic and fuzzy modifications. This mechanism has been subject to our analysis in the two preceding lectures. In the subject matter exposed so far in this lecture we have outlined a more general mechanism based on the context-free language and its two-dimensional generalization. Now we are going to introduce the general construction including the two already explained mechanisms as special cases.

This general approach will be called a *structural construction.* We have introduced it not only because of the possibility to briefly express the already explained subject matter but owing to other reasons. After we unify the two explained mechanisms into one consistent formal construction, we will see that within this frame even models of a recognised object can be formulated which cannot be expressed by means of regular or context-free languages. In this way holes and gaps which have been looming in the so far explained subject matter will be filled up. We will bring to your attention the most interesting points.

1. Even if it is obvious that the class of context-free languages contains all regular languages, we can easily notice a great logical jump when passing from regular languages to context-free ones. It can be seen when we realise the fast and vehemently increasing complexity of recognising sentences in context-free languages compared with regular languages. Various tasks have been discussed for regular languages. Some of them seemed complicated at first glance, but we always finally managed to find a solution. The complexity of the solution depended linearly on the length of the analysed sequence.
   We have analysed only one single task for context-free languages so far which concerns exact match. The complexity itself for solving this simplest task is cubic with respect to the length of the sequence analysed. This is a substantial jump in complexity. An assumption stealthily occurs to us about the existence of an interlinking class of languages that are more complicated than the regular ones, but not so complicated as are the context-free languages in their complete generality.
2. Between the formalism of context-free languages, both one-dimensional and two-dimensional, and real pattern recognition tasks a gap starts to appear. It was quite evident in Section 10.2 in which we informally elucidated the fundamental concepts of context-free languages on the example of a Russian letter *SH*. The only characteristic of the fragment of an image which determines its dependence on other parts of the image is the label of the particular fragment. This label must belong to a finite alphabet of labels (which must be a small one in practically applicable cases). With such weak

means it is rather difficult to state actual constraints with respect to admissible images of a certain class. It is hard to state, for example, that a letter can be only an image of a certain size, or an image with a certain ratio of dimensions pertaining to its fragments, and the like. Therefore an effort is justified to add additional means to context-free grammars by which pure geometrical relations and constrains could be expressed.

3. We have been dealing only with the simplest task for context-free grammars, i.e., with an exact matching problem. It is natural to expect that even within the frame of context-free languages the best matching problem as well as its stochastic and fuzzy modifications should be formulated and analysed.

Now we can start introducing a system of concepts which will be called a structural construction. Formal definitions of fundamental concepts in the structural construction will be interlaced with examples which will illustrate the introduced notions. The introduced concepts have their importance beyond the cases in which the formalised objects are images and sets of images. Nevertheless, for these concepts we will use the terminology which have their origin in recognising images.

10.5.1 Structural construction defining observed sets

Let T be a set which will be called the *observed field* or field of observation (in the same sense as the viewing field of a photographic camera or a telescope). Let V be a finite alphabet of symbols. The elements of the set T will be called *pixels* and denoted by t. Let 2^T be a set of all subsets of T, and $\mathcal{T} \subset 2^T$ be a set of some subsets of T. The set $\mathcal{T}$ will be called a *structure of the observed field* and its elements will be called the *fragments of the observed field*, or simply fragments. This means if $T' \in \mathcal{T}$ then $T' \subset T$, but not vice versa. The structure $\mathcal{T}$ does not contain all the subsets of the observed field. Later on we will deal only with such structures in which for each $t \in T$ the relation $\{t\} \in \mathcal{T}$ holds. Roughly speaking we will assume that each individual pixel forms one of the elements of the structure.

The *observation* x, i.e., the image x will be said to be given in the field T if the fragment $T_0 \in \mathcal{T}$ and function $v\colon T_0 \to V$ are defined. The observation is thus expressed as a pair of the form (T_0, v), $T_0 \in \mathcal{T}$.

Example 10.1 Finite sequence of letters. *Let T be a set of positive integers and the structure $\mathcal{T}$ contain all intervals of the form $\{t \mid 1 \le t \le n\}$, $n \in T$, and all sets of the form $\{t\}$, $t \in T$. Let the set V consist of letters A, B, C. An observation is thus a sequence of finite length composed of letters A, B, C.* ▲

Example 10.2 Binary image. *Let T be a set of pairs of positive integers, and the structure $\mathcal{T}$ contain sets of the form $\{(i,j) \mid m_1 \le i \le m_2,\ n_1 \le j \le n_2\}$, $1 \le m_1 \le m_2$, $1 \le n_1 \le n_2$. If $V = \{0,1\}$ then the observation can be considered as a binary image of size $(m_2 - m_1 + 1) \times (n_2 - n_1 + 1)$.* ▲

Example 10.3 Binary image varying in time from the instant t_1 to the instant t_2. *Let T be a set of triplets of positive integers and the structure $\mathcal{T}$ contain sets of the form $\{(i,j,t) \mid m_1 \le i \le m_2,\ n_1 \le j \le n_2,\ t_1 \le t \le t_2\}$,*

$1 \leq m_1 \leq m_2$, $1 \leq n_1 \leq n_2$, $1 \leq t_1 \leq t_2$. *At* $V = \{0,1\}$ *the observation can be considered as a time varying binary image which is observed in the interval from* t_1 *to* t_2. ▲

Example 10.4 Observed field with a more complicated structure. *The observed field* T *can have a more complicated structure. It can be, for example, a set of vertices of an acyclic graph or that of a Cartesian product of such graphs.* ▲

Remark 10.1 *The alphabet of symbols* V *corresponds to terminal and non-terminal alphabets in formal grammars. For our later purposes these two types of symbols need not be differentiated.* ▲

The set $\mathcal{T} \times V$ will be called a set of *structural elements* and will be denoted S. Individual structural elements will generally be denoted by a lower-case letter s distinguished by indices. A structural element is a certain fragment from the structure $\mathcal{T}$ marked by a symbol from V.

Example 10.5 Structural element. *If a pixel is said to be black then it determines the structural element. If a set of pixels is said to form an abscissa then it determines the structural element too. If a rectangle in the observed field set is said to contain an image representing the letter* A *then it is also referred to as a structural element.* ▲

A four additional important concepts are segmentation of a fragment, hierarchical segmentation, map, and hierarchical map of the structural element.

The *segmentation of the fragment* $T_0 \in \mathcal{T}$ is a subset $R \subset \mathcal{T}$ of fragments which contains the fragment T_0 and some other fragments $R \setminus \{T_0\}$ which form the decomposition of the fragment T_0. In other words if $T_0 = \bigcup_{i=1}^{m} T_i$, $T_i \in \mathcal{T}$, and $T_i \cap T_j = \emptyset$ for any $i > 0$, $j > 0$, $i \neq j$ then $R = \{T_i \mid i = 0, 1, \ldots, m\}$ is the segmentation of the fragment T_0.

Any function $m \colon R \to V$ in which R is the segmentation of the fragment T_0 defines the *map* of the structural element $\big(T_0, m(T_0)\big)$. Each map is a subset of labelled fragments, i.e., a subset of structural elements.

We will introduce two important particular cases of maps. The map defined on segmentation which consists of three fragments is called a *rule.* For this particular case the denotation π will be used and different sets of rules will be denoted by Π with different indices and arguments.

The other particular case of a map is the *labelled image.* The information contained in this map consists of the definition of the fragment T_0, label v which characterises the image as a whole, and an ensemble of labels which characterise individual pixels in the fragment T_0. Formally speaking, it is the segmentation $R = \{T_0\} \cup \big(\bigcup_{t \in T_0} \{\{t\}\}\big)$ of the fragment T_0 to individual pixels and the function $x \colon R \to V$. Thus, this pair is a map as well. The labelled image is also a subset of labelled fragments of a certain form, i.e., the set $\{(T_0, v_0)\} \cup \big\{\big(\{t\}, x(t)\big) \mid t \in T_0\big\}$.

Hierarchical segmentation is defined in the following recursive way.

1. The segmentation R which consists of three fragments is the hierarchical segmentation.

2. Let R be a hierarchical segmentation and $T' \in R$ be a fragment which does not contain any other fragment from R. Let $T'' \in \mathcal{T}$ and $T''' \in \mathcal{T}$ be a decomposition of the fragment T'. In this case $R \cup \{T'', T'''\}$ is also a hierarchical segmentation.

The *hierarchical map* is a hierarchical segmentation of R supplied by the function $R \to V$ which assigns a label to every fragment of R. The hierarchical map will be denoted H and different sets of hierarchical maps will be denoted $\mathcal{H}$ with different arguments. Both the map and the hierarchical map are understood as a set of structural elements of a certain form.

From the definition presented there immediately follows that a subset of rules $\Pi(H)$ corresponds to each hierarchical map H. Moreover, this set is the only possible set for every hierarchical map.

For the labelled image m we will denote $\mathcal{H}(m)$ the set of all hierarchical maps containing m.

Let W be a commutative semi-ring where $\oplus$ and $\otimes$ denote addition and multiplication performed on that semi-ring. We will denote by $0^{\oplus}$ an element from W for which $w \oplus 0^{\oplus} = w$ and $w \otimes 0^{\oplus} = 0^{\oplus}$ holds for each $w \in W$. We will denote by $1^{\otimes}$ an element from W for which $w \otimes 1^{\otimes} = w$ holds at any arbitrary $w \in W$. Let P be a function which for any rule π defines the quantity $P(\pi) \in W$.

Example 10.6 *The function P is a tool by means of which the generation of images on the fragment T_0 labelled v_0 is controlled. The image is generated by means of $|T_0| - 1$ steps. The result of each i-th step is a segmentation of the fragment T_0 to the $i+1$ labelled fragments. In each step a fragment is selected from the already created map which consists of more than one pixel and does not contain any other fragment of the current map. The fragment selected is decomposed into two smaller fragments which are labelled. Thus the decomposition implements some rule and the function P controls this procedure in different ways presented in later examples.* ▲

For each hierarchical map H the quantity $g(H)$ will be introduced,

$$g(H) = \bigotimes_{\pi \in \Pi(H)} P(\pi) . \tag{10.17}$$

For each labelled image m the quantity $G(m)$ will be introduced,

$$G(m) = \bigoplus_{H \in \mathcal{H}(m)} g(H) = \bigoplus_{H \in \mathcal{H}(m)} \bigotimes_{\pi \in \Pi(H)} P(\pi) . \tag{10.18}$$

The basic problem of structural pattern recognition is to create an algorithm which will calculate the quantity $G(m)$ for each labelled image m.

Example 10.7 *Let $W = \{0, 1\}$ and let the operation $\oplus$ represent disjunction and $\otimes$ represent conjunction. In this case the function P defines a subset of rules which are admissible in that particular task being solved. Starting from the fragment T_0 labelled by the symbol v_0 only some admissible images can be* *generated not arbitrary ones. The quantity (10.18) states whether the particular image is admissible.* ▲

Example 10.8 *Let W be a completely ordered set. Let $\oplus$ mean* max *and $\otimes$ mean* min. *In this case the function P defines an element from W for each rule π which indicates the degree of confidence in the rule. At the same time the degree of confidence in generating the image is assumed not to be less than θ if the degree of confidence in each rule applied in generation is not less than θ either. The expression (10.18) in this case determines the safest way of generating the particular image, i.e., roughly speaking finding the most convincing reason that the image presented is labelled v_0.* ▲

Example 10.9 *Let W be a set of nonnegative numbers. Let $\oplus$ mean* min *and $\otimes$ mean addition in the common sense. In this case the function P can be understood as stating a penalty for the application of the rule. The number (10.17) represents the total penalty for the actual procedure in generating the image and the number (10.18) means the least possible total penalty with which it is still possible to generate the image presented.* ▲

Example 10.10 *Let W be a set of nonnegative numbers. Let the operations $\oplus$ and $\otimes$ correspond to addition and multiplication in their common sense. If for an arbitrary s the function P satisfies the equality $\sum_{s_1,s_2} P(s_1, s_2 \mid s) = 1$ then the function P can be understood as the determination of a random mechanism in generating the image. In this case the quantity (10.18) states the probability of occurrence of the presented image within a group of images defined on the given fragment T_0 which can be assigned the label v_0.* ▲

10.5.2 Formulation of the basic problem in structural recognition of images

Let the following be given:

set T,

set V,

structure $\mathcal{T} \subset 2^T$,

function $P\colon (\mathcal{T} \times V) \times (\mathcal{T} \times V) \times (\mathcal{T} \times V) \to W$,

operations $\oplus\colon W \times W \to W$ and $\otimes\colon W \times W \to W$, zero element $0^\oplus \in W$ and unit element $1^\otimes \in W$,

fragment $T_0 \in \mathcal{T}$, image $x\colon T_0 \to V$ and symbol $v_0 \in V$.

Based on this input data the following quantity is to be calculated

$$G(T_0, v_0, x) = \bigoplus_{H \in \mathcal{H}(T_0, v_0, x)} \bigotimes_{(s_1,s_2,s_3) \in \Pi(H)} P(s_1, s_2, s_3)\,. \tag{10.19}$$

In the preceding formula $\mathcal{H}(T_0, v_0, x)$ is a set of hierarchical maps which contain structural elements (T_0, v_0) and $\big(\{t\}, x(t)\big)$, $t \in T_0$, whereas $\Pi(H)$ is a set of rules (s_1, s_2, s_3) contained in the hierarchical map H.

10.5.3 Computational procedure for solving the basic problem

For the given observation T_0 and $x\colon T_0 \to V$ we will take into consideration an ensemble of quantities

$$G\Big(T', v', x(T')\Big) = \bigoplus_{H\in\mathcal{H}(T',v',x(T'))} \; \bigotimes_{(s_1,s_2,s_3)\in\Pi(H)} P(s_1,s_2,s_3) \qquad (10.20)$$

which are to be calculated for each structural element (T', v'), $T' \subset T_0$, $T' \in \mathcal{T}$, $v' \in V$. In the formula (10.20) $x(T')$ means contraction of the analysed image x to the fragment T'.

If we compute all the quantities (10.20) then we will solve the task (10.19) as well, because the quantity (10.19) is one of the quantities given by (10.20).

In each actual task, the image $x\colon T_0 \to V$, as well as its contraction to different fragments, is constant. Therefore to make later formulæ brief, we will neither refer to the denotation of the image x, nor to the denotation for its contraction $x(T')$. Recall the previously introduced denotations s for structural elements, i.e., pairs of the form (T', v'), $T' \subset T^0$, $T' \in \mathcal{T}$, $v' \in V$, and the denotation π for rules, i.e., triplets (s_1, s_2, s_3) of a certain form. In using these denotations the formula (10.20) will have a shorter form

$$G(s) = \bigoplus_{H\in\mathcal{H}(s)} \; \bigotimes_{\pi\in\Pi(H)} P(\pi)\,. \qquad (10.21)$$

In the preceding formula the quantity $G(s)$, $s = (T', v')$, characterises a map which consists of the contraction of the presented image $x\colon T_0 \to V$ to the fragment $T' \subset T_0$ and the label v' of the fragment T'. The set $\mathcal{H}(s)$, $s = (T', v')$, is the set of all hierarchical maps which contain the structural element (T', v') and the contraction of the presented image x to the fragment T'. The set $\Pi(H)$ is the set of rules in the hierarchical map H.

The quantities $G(s)$, $s = (t', v')$, $T' \subset T_0$, $T' \in \mathcal{T}$, $v' \in V$, cannot be arbitrary. They satisfy certain relations which make their constructive calculation possible. Let us examine these relations.

For each structural element s the set $\mathcal{H}(s)$ contains s. For each pair s_1, s_2, which makes up a rule with the element s, we will denote by $\mathcal{H}(s, s_1, s_2)$ the subset of the set of hierarchical maps $\mathcal{H}(s)$ which contain elements s_1 and s_2. The relation $\mathcal{H}(s) = \bigcup_{s_1} \bigcup_{s_2} \mathcal{H}(s, s_1, s_2)$ is evident and thanks to this we can write the quantity (10.21) in the form

$$G(s) = \bigoplus_{s_1} \bigoplus_{s_2} \bigoplus_{H\in\mathcal{H}(s,s_1,s_2)} \; \bigotimes_{\pi\in\Pi(H)} P(\pi)\,. \qquad (10.22)$$

We will examine the sum

$$\bigoplus_{H\in\mathcal{H}(s,s_1,s_2)} \; \bigotimes_{\pi\in\Pi(H)} P(\pi) \qquad (10.23)$$

which will be denoted $G(s, s_1, s_2)$. The rule s, s_1, s_2 is present in each hierarchical map H from the set $\mathcal{H}(s, s_1, s_2)$, and thus the factor $P(s, s_1, s_2)$ is present in each product $\bigotimes_{\pi \in \Pi(H)} P(\pi)$. It can be factored out behind the addition

$$G(s, s_1, s_2) = P(s, s_1, s_2) \otimes \left(\bigoplus_{H \in \mathcal{H}(s,s_1,s_2)} \bigotimes_{\pi \in \Pi(H) \setminus \{(s,s_1,s_2)\}} P(\pi) \right) . \tag{10.24}$$

For each hierarchical map $H \in \mathcal{H}(s, s_1, s_2)$ the set $\Pi(H) \setminus \{(s, s_1, s_2)\}$ is decomposed into two subsets. One is the hierarchical map H_1 from the set $\mathcal{H}(s_1)$, and the other is the hierarchical map H_2 from the set $\mathcal{H}(s_2)$. Therefore the product $\bigotimes_{\pi \in \Pi(H) \setminus \{(s,s_1,s_2)\}} P(\pi)$ means $\left(\bigotimes_{\pi \in \Pi(H_1)} P(\pi)\right) \otimes \left(\bigotimes_{\pi \in \Pi(H_2)} P(\pi)\right)$, and addition along all the maps from the set $\mathcal{H}(s, s_1, s_2)$ means addition along all the maps from the Cartesian product $\mathcal{H}(s_1) \times \mathcal{H}(s_2)$. Thus the formula (10.24) assumes the form

$$\begin{aligned} & G(s, s_1, s_2) \\ & = P(s, s_1, s_2) \otimes \left(\bigoplus_{H_1 \in \mathcal{H}(s_1)} \bigoplus_{H_2 \in \mathcal{H}(s_2)} \left(\bigotimes_{\pi \in \Pi(H_1)} P(\pi) \right) \otimes \left(\bigotimes_{\pi \in \Pi(H_2)} P(\pi) \right) \right) . \end{aligned} \tag{10.25}$$

On the basis of the evident equality $\bigoplus_{i \in I} \bigoplus_{j \in J} (\beta_i \otimes \gamma_j) = \left(\bigoplus_{i \in I} \beta_i\right) \otimes \left(\bigoplus_{j \in J} \gamma_j\right)$ we further have

$$\begin{aligned} & G(s, s_1, s_2) \\ & = P(s, s_1, s_2) \otimes \left(\bigoplus_{H \in \mathcal{H}(s_1)} \bigotimes_{\pi \in \Pi(H)} P(\pi) \right) \otimes \left(\bigoplus_{H \in \mathcal{H}(s_2)} \bigotimes_{\pi \in \Pi(H)} P(\pi) \right) . \end{aligned} \tag{10.26}$$

This expression is a product of three terms. According to the definition (10.21) the second term is nothing else than $G(s_1)$. The third term is $G(s_2)$. Thus the formula (10.26) assumes a brief form

$$G(s, s_1, s_2) = G(s_1) \otimes P(s, s_1, s_2) \otimes G(s_2) ,$$

and it is the quantity (10.23). If we substitute it into (10.22) then we will obtain a recursive formula

$$G(s) = \bigoplus_{s_1, s_2} \left(G(s_1) \otimes P(s, s_1, s_2) \otimes G(s_2) \right) \tag{10.27}$$

which is the tool for solving the basic problem. We are now going to demonstrate it.

The quantity $G(s)$, as we defined it before and as can be seen from the formula (10.27), is relevant only for such structural elements $s = (T', v')$ which are decomposed into other elements, i.e., when $T' \geq 2$. We will extend this definition even to the cases in which the structural element is an individually labelled pixel. For the presented image $x\colon T_0 \to V$ we define $G(s)$, $s = (\{t\}, v)$, $t \in T_0$, $v \in V$, in such a way that

$$G(\{t\}, v) = \begin{cases} 0^{\oplus}, \text{ if } x(t) \neq v\,, \\ 1^{\otimes}, \text{ if } x(t) = v\,. \end{cases} \tag{10.28}$$

We will order all the structural elements into a one-dimensional sequence (it does not depend on the dimensions of the observed field T which are by no means taken into consideration) so that the element $s' = (T', v')$ precedes the element $s'' = (T'', v'')$, if $T' \neq T''$ and $T' \subset T''$. If $T' = T''$ then the elements are arranged in the sequence in an arbitrary order which will later be considered as fixed. For the given observation $x\colon T_0 \to V$ the quantities $G(\{t\}, v)$, $t \in T_0$, $v \in V$, will be defined in accordance with (10.28). Then we stepwise examine all the elements in the before settled order and for each of them, assume for the element s, we will calculate the value $G(s)$ according to the formula (10.27). All the data is already at hand for the calculation because only the values of $G(s')$ for those elements s' are needed which occurred in the ordered sequence of elements before. In this procedure the values $G(s)$ as well as those of the fragment (T_0, v), $v \in V$, are calculated. In this way it can be stated to what extent the assertion is valid that the presented image can be labelled v. The total number of operations in the calculation is proportional to the number of rules (s, s_1, s_2), $s = (T', v')$, $s_1 = (T'_1, v'_1)$, $s_2 = (T'_2, v'_2)$, $T' \subset T_0$, for which $P(s, s_1, s_2) \neq 0^{\oplus}$.

Example 10.11 Regular language and structural construction. *In the case of regular languages and their stochastic and fuzzy generalisations, the set T is a set of positive integer numbers. The structure $\mathcal{T}$ contains fragments of the form $\{t \mid 1 \leq t \leq n\}$, $n \in T$, and fragments of the form $\{t\}$, $t \in T$. Let a sequence of the length n be given which is to be processed, i.e., let a set $T_0 = \{1, 2, \ldots, n\}$ and the function $x\colon T_0 \to V$ be given. For the set T_0 only $n-1$ triplets of fragments (T_1, T_2, T_3), $T_i \subset T_0$, $i = 1, 2, 3$, exist in which T_2 and T_3 constitute the decomposition of the fragment T_1. The reason is that the triplet (T_1, T_2, T_3) can have only the form $(T_1, T_1 \backslash \{t\}, \{t\})$ in which t is the last pixel in the fragment T_1. Because the structural element s_3 can only have the form $(x(t), \{t\})$ (where $x(t)$ is the t-th symbol in the presented sequence), to each triplet of fragments at most $|V|^2$ rules $\pi = (s_1, s_2, s_3)$ correspond for which $P(\pi) \neq 0$. Thus the complexity of the analysis of a sequence with the created construction is $\mathcal{O}\big(|V|^2 (n-1)\big)$ which is of the same order as that for algorithms analysing sequences which refer solely to regular languages.* ▲

Example 10.12 Context-free language and structural construction. *In defining a context-free language by means of the created structural construction, the set T is a set of integer numbers as it was in the preceding example. The structure $\mathcal{T}$ contains all intervals of the form $\{t \mid i \leq t \leq j\}$, $i \in T$, $j \in T$, $i \leq j$. If the recognised sequence $T_0 = \{1, 2, \ldots, n\}$ is of the length n then the number of segmentations of the form $\{T_1, T_2, T_3\}$, $T_i \subset T_0$, is of the order n^3 which is just the complexity of sequence recognition by means of the known algorithms which have been created solely for the case of context-free languages.* ▲

Now we can more clearly understand what causes such a steep increase in the complexity of the pattern recognition task in passing from regular to context-free languages if intermediate levels were not considered. It is owed to the difference between the used structures which in the theory of formal grammars are not quoted under one separate notion at all. Nevertheless, the definition of a class of languages is based on the application of a certain structure, even if it is not explicitly described. A structure of regular languages contains only such fragments which can be decomposed into other fragments in a single way. The result is that one single hierarchical segmentation corresponds to each sequence presented for recognition which is easy to find. It is natural that the complexity of the syntactic analysis falls rapidly in this case.

The situation is different in the case of context-free languages. Here each fragment of the length n can be decomposed in all the $(n-1)$ ways into other two fragments. Thanks to the greater freedom in decomposition of the fragment into two parts, the greater amount of hierarchical segmentations corresponds to the sequence.

Now when the fundamental factor that determines the complexity of syntactic analysis of the sentence has been revealed, languages can be constructed which, strictly speaking, are not regular but for which the complexity of the pattern recognition task is not greater than that for regular languages.

Example 10.13 A language between the regular and the context-free language. *Let the structure $\mathcal{T}$ be defined in the same way as it was in Example 10.12 for context-free grammars. The function P, however, has been chosen so that for each fragment T_1 the quantity $P\big((T_1, v_1), (T_2, v_2), (T_3, v_3)\big)$ is non-zero only in the case of one single decomposition of fragment T_1 into fragments T_2 and T_3. For example, it can be done in such a way that only those pairs of fragments are to be taken into consideration the lengths of which do not differ from each other by more than 1. The language defined by such a structure and by the function P will be no longer a regular one. Nevertheless, the complexity of recognising sequence in this language remains linearly dependent on the sentence length as was the case in regular languages.*

Thus the class of languages can be constructed which ranks in a sense between regular and context-free languages. Eventually languages can be constructed which, strictly speaking, do not belong to the class of context-free languages, and at the same time the complexity of sequence analysis with these languages rises slower than the third degree polynomial (cubic) of the sentence length. Examples of such languages were given at the beginning of the lecture where they were called two-dimensional context-free languages. ▲

The presented structural construction for observed sets is, therefore, a generalisation of the known formalisms, such as the formal regular or context-free grammars including their stochastic and fuzzy modifications. The generalisation consists in that no 'one-dimensional' property of observed data is assumed. The construction is based on other, more general means, with the aid of which sets of other forms than those of sequences are defined.

10.6 Discussion

In the lecture you introduced a structural construction the particular cases of which are regular and context-free grammars. Compared with the grammars, the construction has additional tools at hand, by which not only regular and context-free languages, but also other sets of various forms can be defined. In the first place I would like to ask you what tools have made such generalisation possible. Then I will ask other questions.

The first step to generalisation is that the function $P\colon K \times K \times K \to \{0, 1\}$ expressing rules in common grammars is replaced by a function of a more general form $P\colon K \times K \times K \to W$ for an appropriate semi-ring W. The result is that not only languages can be defined, but even functions which are defined on a set of sequences.

You have already presented this generalisation and used it in previous lectures. I am interested in further steps in generalisation which appeared in this lecture.

It is important that we have created the construction without applying the concept of a sequence. In this way we have achieved that the sets of more diverse mathematical objects can be defined. These objects are not obligatory sequences, but they can be sequences too. Together with the generalisation, which was our first step, we obtained a construction not only for the definition of admissible observed sets, but also for the definition of certain functions on observed sets. Thus the construction has assumed a quite strong and general power.

I would like to be more at home with the problem. I do not understand properly why it is necessary to formalise observation in any other way than as a sequence. Of course, I understand that an image is something different than a sentence. However, I do not know why one should formalise these two representations of information in different manner. Well, even when in terms of the general structural construction I say that an observation is an ensemble of structural elements $s_1, s_2, \ldots, s_n$, I still write it down as a sequence. Why could not the expression I have just written be called a sequence?

You have come across the same difficulties that pattern recognition encountered in the 1960s, when formalisation of an observation by a point in a multi-dimensional space seemed to be universally applicable. It was found even at that time that it was necessary to make a break with that charming idea. Actually, it is not important how the observation $s_1, s_2, \ldots, s_n$ is called. An observation can be called a vector, a sequence, a set, etc.. It is essential what operations on this object are considered understandable in a particular application. If years ago it was found that the formalisation of observation by means of the multi-dimensional vector is not convenient for the purpose then it resulted in something more serious than in replacing the word vector by another word.

It meant that operations and concepts resulting from formalisation by means of vectors (such as vector addition, vector scalar product, hyperplane, convex subset, etc.) did not correspond to some applied problems. In some applications nothing understandable corresponds to concepts related to vectors. Therefore in the formulation of such tasks and their solutions vectors should not be used. Finally, when nothing but the word vector itself remained from vector formulation then it was evident that another formalisation of observation should be introduced as well.

The case with sequences is similar. The question is not whether the observation $s_1, s_2, \ldots \ldots, s_n$ is called a sequence. It is of importance if the operations on sequences (such as concatenation, iteration, deleting a connected subsequence, etc.) state something that is understandable even in the concepts of the applied problem you are to master. And vice versa, if all properties of observation in your application can be easily expressed by means of operations that are natural for sequences. From the point of view of structural construction, a sequence is nothing else but an auxiliary concept representing structures in an illustrative way in that precise meaning in which structures were used at the lecture. For some structures the concept of sequence is beneficial, for others it is not. When saying that the observation in a particular application is not a sequence then we mean that the structure of the observation is expressed clumsily, not illustratively, through operations pertaining for sequences. You yourself can admit that some sets T and some structures $\mathcal{T}$ lose their entire illustrative character when they are represented as a set of integers.

Everything depends on representation, i.e., how the elements of a set T are numbered.

Not everything. There are sets T having a completely clear structure $\mathcal{T}$, which loses its lucidity with any mapping of the set T to a set of integers.

For example?

An example is a two-dimensional integer-number lattice, i.e., a structure which is defined as a set of rectangles of finite dimensions.

I understand that the correct definition of the set T and its corresponding structure $\mathcal{T}$ is the most important step in representing an applied problem by means of a structural construction.

We would still add to it the selection of an alphabet V for the labelling of structural elements.

Let me put the alphabet aside, for a while. Let me even put aside the structure $\mathcal{T}$, because I agree with you that the words by which I will determine what set T is referred to will immediately delimit the structure $\mathcal{T}$ natural for a set.

Now I am coming to the main question. The entire structural construction is a tool for defining some sets. But at least one element in the construction is *again a set. It is the set T, with respect to it nothing at all is stated, neither how it should be defined nor what its form could be. Briefly speaking, nothing is said about it and therefore I can consider is as an arbitrary set. Thus, the entire structural construction stops being constructive. To define a set (here a set of admissible observations) I must have defined another set. And this set is just the set T.*

Indeed, you have revealed the weakest point in the proposed structural construction. The definition of a set of admissible observations cannot start with the words 'let T be a set' because it is a too general sentence. At least, we should say 'let T be a finite set' or otherwise strongly limit the set T, so that further doing with respect to construction might become correct. We did not do it for different reasons, and so understand the sentence in the following informal sense 'let T be a set which is easy to define and quite obviously results from the applied problem being solved'.

But still I do not understand why the form of the set T could not be reasonably limited for the whole construction to become correct and in spite of it to contain all possible sets that can practically occur. Well, the variety of the sets T which are of practical interest is not very large. It can be a completely ordered set, e.g., a set of integers, or it can be a Cartesian product of a finite number of such sets. What else do I dare to ask?

And situations should be added in which the observation is a function defined on the vertices of an acyclic graph and also on the Cartesian product of a finite number of such graphs.

And would that really be all for the present?

Hardly all. Objects in images are commonly represented by contours. It is a function the domain of definition of which is a closed curve (a cycle), i.e., something which is not covered by the two previous cases. We do not intend to exclude beforehand the analysis of such objects.

Let us sum up what has been discussed by admitting that something is still missing in the proposed structural construction to make it constructive. But in actual application, the construction can be made precise to such an extent to become constructive. It is possible with the sets T at least which have been just mentioned. We do not intend just now to deal with describing the form of all observed fields which can occur in practice. The first reason is that we simply do not know the form. Every time we tried to limit the form of sets T, after some time we encountered a new practical problem in which the limitation had not been satisfied. But the main idea remained valid in the proposed construction. Secondly (and this is essential), in a general view of the task, not confused by useless details, the real simplicity of fundamental concepts of structural pattern recognition is being revealed.

These are really simple. Further procedure in structural construction seems to me so unsophisticated that I have been in fear so far that I might again not understand something important. I would like to make sure that I understand the simplicity in a correct way. I will quote one of the possible implementations of the computational procedure for solving the basic problem.

Do it, but only after a while. We have not answered your first question yet. You asked us by means of what additional tools it was possible to use the extended potentiality of structural construction when compared with formal grammars.

You have certainly noticed that rules in the structural construction have another form than those in formal grammars. In the structural construction the rule is a triplet of structural elements (s_1, s_2, s_3), where each element is a labelled fragment, i.e., a pair of the form (T, v), $T' \in \mathcal{T}$, $v' \in V$. Thus a rule is the six-tuplet $\langle T_1, v_1,\ T_2, v_2,\ T_3, v_3 \rangle$. A rule in formal grammars has a simpler form. It is a triplet of labels and each grammar is characterised by a subset of triplets that is determined by the function $V \times V \times V \rightarrow \{0, 1\}$ which will be denoted PV. If the language of a classical formal grammar is expressed by means of a general structural construction then the function $P \colon \mathcal{T} \times V \times \mathcal{T} \times V \times \mathcal{T} \times V \rightarrow \{0, 1\}$ will have the form

$$P(T_1, v_1, T_2, v_2, T_3, v_3) = PT(T_1, T_2, T_3) \times PV(v_1, v_2, v_3)\,. \tag{10.29}$$

Formal grammars are, therefore, a particular case of the structural construction in which the function P has the form of (10.29). Moreover, one single function $PT \colon \mathcal{T} \times \mathcal{T} \times \mathcal{T} \rightarrow \{0, 1\}$ must be used for any regular grammars and another single function for any context-free grammar.

With grammars the applicability of a rule does not depend on fragments, whereas with the structural construction it does. The triplet of labels v_1, v_2, v_3 can be admissible for one selected triplet of fragments T_1, T_2, T_3 and can be inadmissible for some other triple. Do I understand it correctly?

Yes, you do!

I will show how I understand the computational procedure solving the basic problem. I will present it for the case in which the function P assumes only two values, 0 and 1, and these values can be subject to logical addition and multiplication. During your lectures I got used to all other cases, seemingly more complicated, being in fact exactly as simple as this lucid one.

The structural recognition task is understood by me in such a way that a set of objects $s_1, s_2, \ldots, s_n$ (I have nearly said a sequence!) is given. It is to be found whether this set is admissible. In other words, it is to be checked whether the objects presented can be understood as parts of a composed object. I imagine the following procedure for solving the task. The set S^i, $i = 0, 1, 2, \ldots$ of objects is being created step by step which are regarded as examined. At the beginning the set S^0 is represented by the set $\{s_1, s_2, \ldots, s_n\}$ which was presented for analysis. Let a set S^{i-1} be created after the step $(i-1)$. Then a *triplet of objects* (s', s'', s''') *is sought for which* $P(s', s'', s''') = 1$, $s' \notin S^{i-1}$, $s'' \in S^{i-1}$, $s''' \in S^{i-1}$, *holds, and the set* $S^i = S^{i-1} \cup \{s'\}$ *is being created. Simply speaking, in each step the set of already found partial objects is increased by one more object the existence of which was proved in that step. This procedure of creating the set* S *continues until it is possible with respect to the function* P.

I regard the gradual growing of the set presented as the most essential part of structural recognition. I would even say that it is its property. After the set S *is created in the manner described some details are to be completed for its interpretation which I do not consider important. If I did not make any mistake somewhere then the simplicity is all too much remarkable. Moreover, I would say that all pattern recognition algorithms which we have discussed so far since Lecture 8 on Markovian sequences have been successfully packed into a simple procedure.*

Do not wonder at it. A general view of the class of tasks (if possible at all) allows to see the properties of the tasks which are difficult to observe when the tasks are analysed apart. Well, usually if you intend to know, e.g., a large building, you had better move away from it a little than come nearer to it. It is similar to the situation we have already spoken about. As long as people have counted one type of objects by pairs, another by dozens, the third by tens, and the fourth by three scores, the manipulation with quantities seemed to be very complicated. Only since the time when unified representation of quantities was proposed, counting has been accessible to every child.

You have grasped well that part in structural recognition which does not change in nearly every application. You have used correct expressions, except for using the concept of object instead of the concept of structural element.

It seems to me that in this way the main idea was pointed out more illustratively.

It may be so, but not to get confused let us go back to the terminology introduced in the lecture.

The procedure you presented can be formulated more concretely to add to the subject matter already grasped an elucidation concerning the computational complexity of the algorithm, and to take into consideration factors that influence the complexity. Moreover, we will more precisely state which part of the algorithm is changed when one goes on from one application to the another.

We will create the universal algorithm. This means that its input are observations $\{s_1, s_2, \ldots, s_n\}$ and the function P determining which structural analysis is to be performed on the particular observation. The function takes its values from the set W. Two operations $\oplus$ and $\otimes$ of the form $W \times W \to W$ are given which form a semi-ring on the set W. These operations are also defined in input data.

We will order structural elements in such a way that the structural element $s' = (T', v')$ precedes the structural element $s'' = (T'', v'')$ if $T' \subset T''$ and

$T' \neq T''$. The ordering will be denoted $s' \prec s''$. If $T' = T''$ then the elements s' and s'' will be arbitrarily ordered, either as $s' \prec s''$ or $s'' \prec s'$. The order defined will be regarded as fixed. We will order the rules π, i.e., triplets of structural elements (s_1, s_2, s_3) so that the rule $\pi' = (s'_1, s'_2, s'_3)$ precedes the rule $\pi'' = (s''_1, s''_2, s''_3)$, $\pi' \prec \pi''$, if $s'_1 \prec s''_1$. The algorithm consists of the following operations.

Algorithm 10.2 Structural construction.

1. For each element s, the $G(s) = 0^{\oplus}$ is defined, where the quantity $0^{\oplus}$ is taken from the input data.
2. For each element s_i from the input data $\{s_1, s_2, \ldots, s_m\}$ the $G(s_i) = 1^{\otimes}$ is defined, where the quantity $1^{\otimes}$ is also taken from the input data.
3. The rules π are examined in a beforehand given order, and for each $\pi = (s_1, s_2, s_3)$ the quantity $G(s_1)$ is modified by the operator

$$G(s_1) := G(s_1) \oplus \big(G(s_2) \otimes P(s_1, s_2, s_3) \otimes G(s_3)\big) ,$$

where the operations $\oplus$ and $\otimes$ are determined from the input data.

From the expression for the algorithm there immediately follows that its computational time linearly depends on the number of the rules π for which $P(\pi) \neq 0^{\oplus}$, since each such rule is applied only once.

I can now see from the description that in every actual application of the structural construction I have to do a lot of work. Its result can be considered as a program which creates a sequence of triplets of structural elements (s_1, s_2, s_3). *These triplets together with the quantities* $P(s_1, s_2, s_3)$ *are then provided to the program which already is an invariant, i.e., it does not depend on the selected application. It is painful and often annoying work. It seems to me that structural analysis of data is simple only under the condition that somebody has already done all the unpleasant work beforehand. And this 'somebody' may be I, myself!*

This is usual in applied informatics. We will remind you once more that no formalisation, including formal methods of pattern recognition, is a magical means for lazybones like The Magic Table fairy tales. No matter how well elaborated and lucid the formalism may be, it does not relieve the researcher of the pains of representing an informally conceived task in the particular formalism.

I seem to be closer to the lazybones dreaming of that magic means. Is it not possible to formalise this painful work in a narrower domain at least, for example, for one-dimensional context-free grammars? I am speaking about a system such that for the given set of sequences would be either capable of creating a context-free grammar for generating that set of sequences, or would assert with certainty that such a grammar does not exist. I have noticed several articles that aim at solving tasks of such a form.

Yes, of course. But realise please that we all (by which not only the three of us but the whole pattern recognition community are meant) are only at the beginning of a long path.

And what might the first steps along this path be like?

Keeping in line with our course, it is quite natural that first of all the learning task should be formulated correctly. This means that in formulating the task the insolvability of one task and the uselessness of others should be taken into account. The analysis of the learning task for regular languages we were dealing with in Lecture 8 can be regarded as the zero step in the due direction.

And now, have another considered look at the structural construction which we have proposed in this lecture. Different generalisations of regular and context-free languages which can be defined by means of construction also contain a stochastic generalisation of context-free languages. This means that by means of structural construction not only a certain context-free language can be defined but also the corresponding probability distribution on such a language. Spare some good thought for this as it is by no means trivial. Stochastic modification of context-free languages is not as simple as that in the case of regular languages. The probability distribution on a set of rules applied in grammars hardly ever determines the probability distribution on a set of sequences. It is a known problem, which within structural construction can be overcome thanks to the rule not being considered as a triplet of labels, as it is in grammars, but as a triplet of labelled fragments. Think it out yourself because it is worth considering. Now, however, it is essential that by means of structural construction varied probability distributions on a set of observations can be defined. A particular case is the probability distribution of a certain form on the context-free language. We regard it as a basis when formulating a task of a statistical estimation of a stochastic context-free grammar with respect to the observation of a finite set of random sequences. The first step in solving a learning task for structural recognition could be that for context-free grammars all the results should be repeated which were demonstrated for regular grammars in Lecture 8.

I am nearly sure that I have understood you in the right way, but still I would like to be certain about it. Structural construction is based on one type of concept and the formulation of statistical learning tasks is based on other concepts. In my opinion their mutual correspondence is as follows. The observed parameter of an object is an image, and the hidden parameter is a hierarchical map. An unknown parameter that determines the joint probability of the image and the hierarchical map is the probability distribution on a set of rules, where each rule is understood as a triplet of labelled structural elements and not as a triplet of labels.

You have understood it in the correct way.

It seems to me that now I could be able to create an understandable formulation of a learning task in structural pattern recognition and to find a practically applicable algorithm for its solution.

We have had no doubts about it, but in spite of that we are glad to hear it from you. We would like to thank you for your patience and the ideas you have contributed.

January 12, 1999.

10.7 Bibliographical notes

The subject matter explained in this lecture on context-free grammars is to a great extent original. The Cocke–Younger–Kasami algorithm is described in [Aho and Ullman, 1971] as well as in the original publications [Kasami, 1965; Younger, 1967].

The design of two-dimensional context-free grammars is owed to Schlesinger [Schlesinger, 1989].

Bibliography

Aho, A., Hopcroft, J., and Ullman, J. (1975). *The design and analysis of computer algorithms.* Addison-Wesley, Reading, Mass.

Aho, A. and Ullman, J. (1971). *The theory of parsing, translation, and compiling,* volume 1 – *Parsing.* Prentice-Hall, Englewood Cliff, New Jersey.

Ajzerman, M., Braverman, E., and Rozoner, L. (1970). *Metod potencialnych funkcij v teorii obucenia mashin*; in Russian (*The method of potential functions in machine learninng theory*). Nauka, Moskva.

Amengual, J. and Vidal, E. (1996). Two different approaches for cost-efficient Viterbi parsing with error correction. In *Proceedings of the 5th International Workshop Advances in Structural and Syntactical Pattern Recognition, Leipzig,* pages 30–39, Heidelberg, Germany. Springer-Verlag, Lecture Notes in Computer Science 1121.

Anderson, T. (1958). *An introduction to multivariate statistical analysis.* John Wiley, New York, USA.

Anderson, T. and Bahadur, R. (1962). Classification into two multivariate normal distributions with different covariance matrices. *Annals of Mathematical Statistics,* 33:420–431.

Ball, G. and Hall, D. (1967). A clustering technique for summarizing multivariate data. *Behavioral Science,* 12:153–155.

Baum, L., Petrie, T., Soules, G., and Weiss, M. (1970). A maximization technique occuring in the statistical analysis of probabilistic functions of Markov chains. *Annals of Mathematical Statistics,* 41:164–167.

Bayes, T. (1763). An essay towards solving a problem in the doctrine of chance. *Philosophical Transactions of the Royal Society,* London. Reprinted in *Biometrika,* 45:298–315, 1958.

Bellman, R. and Dreyfus, S. E. (1962). *Applied dynamic programming.* Princeton University Press, Princeton, New Jersey.

Beymer, D. and Poggio, T. (1996). Image representations for visual learning. *Science,* 272:1905–1909.

Bishop, C. (1996). *Neural networks for pattern recognition.* Oxford University Press.

Borůvka, O. (1926). O jistém problému minimálním; in Czech, (On a minimal problem). *Práce moravské přírodovědecké společnosti*, III(3):37–58.

Boser, B., Guyon, I., and Vapnik, V. (1992). A training algorithm for optimal margin classifiers. In D.Haussler, editor, *5th Annual ACM Workshop on COLT*, pages 144–152, Pittsburgh, PA. ACM Press.

Bunke, H. (1996). Structural and syntactic pattern recognition. In Chen, C., Pau, L., and Wang, P., editors, *Handbook of Pattern Recognition and Computer Vision*, chapter 1.5, pages 163–209. World Scientific, Singapore.

Cayley, A. (1889). A theory on trees. *Quarterly Journal of Pure and Applied Mathematics*, 23:376–378.

Chazen, E. (1968). *Metody optimalnych statisticheskich reshenij i zadachi optimalnogo upravlenia*; in Russian (*Methods of optimal statistical sulutions in optimal control tasks*). Sovetskoe radio, Moskva.

Chen, C., Pau, L., and Wang, P., editors (1993). *Handbook of Pattern Recognition and Computer Vision*, Singapore. World Scientific.

Chomsky, N. (1957). *Syntactic structures.* Mouton, The Hague.

Chomsky, N., Allen, J., and Van Buren, P. (1971). *Chomsky: Selected Readings.* Oxford University Press, London-New York.

Chow, C. (1965). Statistical independence and threshold functions. *IEEE Transactions on Computers*, 14:247–252.

Chow, C. and Liu, C. (1968). Approximating discrete probability distributions with dependence trees. *IEEE Transactions on Information Theory*, 14:462–467.

Cooper, D. and Cooper, P. (1964). Nonsupervised adaptive signal detection and pattern recognition. *Information and Control*, 7:416–444.

Demster, A., Laird, N., and Rubin, D. (1977). Maximum likelihood from incomplete data via the EM algorithm. *Journal of the Royal Statistic Society*, B39:1–38.

Devijver, P. and Kittler, J. (1982). *Pattern recognition: A statistical approach.* Prentice-Hall, Englewood Cliffs, NJ.

Devroye, L., Györfi, L., and Lugosi, G. (1996). *A probabilistic theory of pattern recognition.* Springer-Verlag, New York.

Duda, R. and Hart, P. (1973). *Pattern classification and scene analysis.* John Willey and Sons, New York.

Duda, R. O., Hart, P. E., and Stork, D. G. (2001). *Pattern Classification.* John Wiley & Sons, New York, USA.

Fisher, R. (1936). The use of multiple measurements in taxonomic problems. *Annals of Eugenics*, 7, Part II:179–188.

Floyd, R. (1962). Algorithm 97, Shortest path. *Communications of the ACM*, 5(6):345.

Franc, V. and Hlaváč, V. (2001). A simple learning algorithm for maximal margin classifier. In Leonardis, A. and Bischof, H., editors, *Proceedings of the Workshop on Kernel & Subspace Methods for Computer Vision, Wien, Austria, August 25, 2001, adjoint to the International Conference on Artificial*

Neural Networks 2001, pages 1–11. Technische Universität Wien, Austria.

Fu, K. (1974). *Syntactic methods in pattern recognition*. Academic Press, New York.

Fukunaga, K. (1990). *Introduction to statistical pattern recognition*. Academic Press, Boston, 2nd edition.

Glushkov, V. (1962a). K voprosu o samoobuchenii v perceptrone; in russian (On the question of self-learning in the perceptron). *Zhurnal matemematiki i matematicheskoj fiziki*, (6):1102–1110.

Glushkov, V. (1962b). Teoria obuchenia odnogo klassa diskretnych perceptronov; in Russian (Theory of learning of one class of discrete perceptrons). *Zhurnal matemematiki i matematiceskoj fiziki*, (2):317–335.

Grim, J. (1986). On numerical evaluation of maximum likelihood estimates for finite mixture of distributions. *Kybernetika*, 18:173–190.

Halmos, P. R. (1971). *Kak pisaty matematiceskie texty*; in Russian (*How to write mathematical texts*). Number 5. Uspechi matematiceskich nauk, Moskva.

Jakubovich, V. A. (1966). Rekurrentnyje konechno schodjashchijesja algoritmy reshenija sistem neravenstv; in Russian (Recurrent finite converging algorithms solving a system of inequalities). *Doklady Akademii nauk*, 166(6): 1308–1311.

Jakubovich, V. A. (1969). Ob odnoj zadache samoobuchenija celesoobraznomu povedeniju; in Russian (On self-learning task related to goal driven behaviour). *Doklady Akademii nauk*, 189(3):495–498.

Jimenez, V. and Marzal, A. (2000). Computation of the n best parse trees for weighted and stochastic context-free grammars. In Ferri, F., editor, *Proceedings of the Joint IAPR Workshops Structural and Syntactic Pattern Recognition and Statistical Pattern Recogntion, Alicante, Spain, August 30–September 1, 2000*, volume Lecture Notes in Computer Science 1876, pages 183–192, Berlin. Springer-Verlag.

Kasami, T. (1965). An efficient recognition and syntax analysis algorithm for context-free languages. Scientific report AFCLR-65-758, Air Force Cambridge Research Laboratory, Bedford, Mass., USA.

Kashyap, L. and Oommen, B. (1984). String correction using probabilistic methods. *Pattern Recognition Letters*, 2(3):147–154.

Kovalevski, V. (1965). Zadacha rospoznavania obrazov s tochki zrenia matematicheskoj statistiki; in Russian (Pattern recognition tasks from the standpoint of mathematical statistics). In *Chitajushchije automaty* (*Reading automata*), pages 3–41. Naukova Dumka, Kiev.

Kovalevski, V. (1967). Optimalnyj algoritm rozpoznavania nekotorych posledovatelnostij izobrazhenij; in russian (Optimal algorithm recognizing some sequences of images). *Kibernetika*, (4):75–80.

Kovalevski, V. (1969). Sequential optimization in pattern recognition and pattern description. In *Proceedings of the IFIP Congress, Amsterdam, 1968*, New York. Academic Press. Earlier version in Russian: Posledovatelynaja optimizacija v zadachach raspoznavanija i opisanija izobrazhenij, in proceedings: Raspoznavanije obrazov i konstruirovanije chitajushchich automatov, Institut Kibernetiki Academy of Sciences USSR, Kiev 1967, pages 3–26.

Kozinec, B. (1973). Rekurentnyj algoritm razdelenia vypuklych obolochek dvuch mnozhestv; in Russian (Recurrent algorithm separating convex hulls of two sets). In Vapnik, V., editor, *Algoritmy obuchenia raspoznavania (Learning algorithms in pattern recognition)*, pages 43–50. Sovetskoje radio, Moskva.

Kruskal, J. (1956). On the shortest spanning subtree of a graph and the traveling salesman problem. *Proceedings of the American Mathematical Society*, 7:48–50.

Kuhn, H. and Tucker, A. (1950). Nonlinear programming. In *Proceedings of the Second Berkeley Symposium on Mathematical Statistics and Probability*, pages 481–492, Berkeley, Calif.

Lehmann, E. (1959). *Testing statistical hypotheses.* John Willey, New York.

Levenstein, V. (1965). Dvojichnyje kody s ispravlenijem vypadenij, vstavok i zameshchenij simvolov; in Russian (Binary coded correcting deletions, insertions and replaces of symbols). *Doklady Akademii nauk SSSR*, 163(4):840–850.

Linnik, J. (1966). *Statisticheskie zadachi s meshajushchimi parametrami*; in Russian (*Statistical tasks with intervening parameters*). Nauka, Moskva.

Markov, A. (1916). Ob odnom primenenii statisticeskogo metoda; in Russian (An application of statistical method). *Izvestia imperialisticeskoj akademii nauk*, 6(4):239–242.

Minsky, M. and Papert, S. (1969). *Perceptrons: An introduction to computational geometry.* MIT Press, Cambridge, Mass., USA. 2nd edition in 1988.

Nadler, M. and Smith, E. (1993). *Pattern recognition engineering.* John Wiley and sons, New York, USA.

Neyman, J. (1962). Two breakthroughs in the theory of statistical decision making. *Review de l'Inst. Intern. de Stat.*, 30(1):11–27.

Neyman, J. and Pearson, E. (1928). On the use and interpretation of certain test criteria for purposes of statistical inference. *Biometrica*, 20A:175–240.

Neyman, J. and Pearson, E. (1933). On the problem of the most efficient tests of statistical hypotheses. *Phil. Trans. Royal Soc. London*, 231:289–337.

Nilsson, N. (1965). *Learning machine: Foundation of trainable pattern recognition classifying systems.* McGraw-Hill, New York.

Novikoff, A. (1962). On convergence proofs for perceptrons. In *Proceedings of the Symposium on Mathematical Theory of Automata*, volume 12, pages 615–622, Brooklyn, New York. Polytechnic Institute of Brooklyn.

Oommen, B. (1987). Recognition of noisy subsequences using constrained edit distances. *IEEE Transactions on Pattern Analysis and Machine Intelligence*, 9:676–685.

Pavel, M. (1993). *Fundamentals of pattern recognition.* Marcel Dekker, Inc., New York, USA.

Raudys, S. and Pikelis, V. (1980). On dimensionality, sample size, classification error, and complexity of classification algorithm in pattern recognition. *IEEE Transactions on Pattern Analysis and Machine Intelligence*, 1:7–13.

Renyi, A. (1972). *Briefeüber die Wahrscheinlichkeit*; in German (*Letters on probability*). VEB Deutscher Verlag der Wissenschaften, Berlin, 2nd edition.

Robbins, H. (1951). Asymptotically subminimax solutions of compound statistical decision problems. In *Proceedings of the second Berkeley symposium on mathematical statistics and probability*, pages 131–148, Los Angeles. University of California Press.

Robbins, H. (1956). An empirical Bayes approach to statistics. In *Proceedings of the Third Berkeley Symposium on Mathematical Statistics and Probability*, pages 157–163, Los Angeles. University of California Press.

Rosenblatt, F. (1957). The perceptron: The perceiving and recognizing automaton. Technical Report 85-460-1, Cornell University, Aeronautical Lab., USA. Project PARA.

Rosenblatt, F. (1959). *Two theorems of statistical separability in the perceptron.* H.M. Stat. Office, London.

Rosenblatt, F. (1962). *Principles of neurodynamiscs: Perceptron and theory of brain mechanisms.* Spartan Books, Washington, D.C.

Schlesinger, M. (1965). O samoproizvolnom razlichenii obrazov; in Russian (On automatic separation of patterns). In *Chitajushchie avtomaty (Reading Automata)*, pages 38–45. Naukova Dumka, Kiev.

Schlesinger, M. (1968). Vzaimosvjaz obuchenija i samoobuchenija v raspoznavaniji obrazov; in Russian (Relation between learning and self-learning in pattern recognition). *Kibernetika*, (2):81–88.

Schlesinger, M. (1972a). Issledovanie odnogo klassa razpoznajushich ustrojstv reshajushchich zadach proverki slozhnych hipotez; in Russian (Study of one class of recognition devices solving tasks of analysis of complex hypotheses). *Avtomatika*, (2):38–42.

Schlesinger, M. (1972b). Sintez linejnogo reshajushego pravila dla odnogo klassa zadach; in Russian (Synthesis of linear discrimination rule for one class of tasks). *Izdatelstvo Akademii nauk*, (5):157–160.

Schlesinger, M. (1979a). Dopolnitelnyje sledstvia teorii dvojstvennosti v nebayesovskich zadachach raspoznavania; in Russian (Additional consequences of the duality theory in non-Bayesian recognition tasks). In *Raspoznavanie graficheskich i zvukovych signalov (Recognition of graphic and sound signals)*, pages 36–47, Kiev. Institut Kibernetiki AV USSR.

Schlesinger, M. (1979b). Teoria dvojstvennosti v nebayesovskich zadachach raspoznavania; in Russian (Theory of duality in non-Bayesian recognition tasks). In *Raspoznavanie graficheskich i zvukovych signalov (Recognition of graphic and sound signals)*, pages 21–35, Kiev. Institut Kibernetiki AV USSR.

Schlesinger, M. (1989). *Matematiceskie sredstva obrabotki izobrazenij*; in Russian (*Mathematical tools for image processing*). Naukova Dumka, Kiev.

Schlesinger, M. (1994). Systeme von Funktionsoperationen angewendet auf eine Aufgabe der besten bereinstimmung; in German (Systems of function operations applied to one task of the best tuning). *Wissenschaftliche Beitraege zur Informatik,* Fakultaet fuer Informatik Technische Universitaet Dresden, Germany, 7(3):62–79.

Schlesinger, M. (1997). Algebraic method for solution of some best matching problems. In *Advances in Computer Vision, Proceedings of the Dagstuhl*

Seminar, Saarland, Germany, March 1997, pages 201–210, Wien. Springer Verlag.

Schlesinger, M. and Gimmel'farb, G. (1987). Vozmozhnosti sovremennogo raspoznavanija obrazov v prikladnom raspoznavaniji izobrazhenij; in Russian (Possibilities of current pattern recognition in applied image analysis). *Upravljajuchshije sistemy i mashiny*, (6):21–28.

Schlesinger, M., Kalmykov, V., and Suchorukov, A. (1981). Sravnitelnyj analiz algoritmov sinteza linejnogo reshajushchego pravila dlja proverki slozhnych gipotez; in Russian (Comparative analysis of algorithms synthesising linear decision rule for analysis of complex hypotheses). *Automatika*, (1):3–9.

Schlesinger, M. and Svjatogor, L. (1967). O postrojenii etalonov dlja korelacionnych chitajushchich automatov; in Russian (On creating etalons for correlation-based reading automata). In *III. vsjesojuznaja konferencija po informacionno poiskovym sistemam i automatirizovannoj obrabotke nauchnotechnicheskoj informacii* (*3rd all-Soviet Union conference on information-searching systems and automatic processing of scientific and technological information*), volume 3, pages 129–139, Moskva. Vsesojuznyj institut nauchnoj i technicheskoj informacii.

Shor, N. (1979). *Metody minimizacii nedifferenciruemych funkcij i ich prilozenia*; in Russian (*Methods of non-differentiable functions minimizations and their applications*). Naukova Dumka, Kiev, Ukrajine.

Shor, N. (1998). *Nondifferentiable optimization and polynomial problems*. Kluwer Academic Publisher, Dordrecht, The Netherlands.

Theodoridis, S. and Koutroumbas, K. (1999). *Pattern Recognition*. Academic Press, San Diego, USA.

Vapnik, V. (1995). *The nature of statistical learning theory*. Springer-Verlag, New York.

Vapnik, V. (1998). *Statistical learning theory*. Adaptive and Learning Systems. Wiley, New York, New York, USA.

Vapnik, V. and Chervonenkis, A. (1974). *Teoria raspoznavania obrazov, statisticheskie problemy obuchenia*; in Russian (*Pattern recognition theory, statistical learning problems*). Nauka, Moskva.

Vidyasagar, M. (1996). *Theory of Learning and Generalization. With Application to Neural Networks and Control Systems*. Springer.

Vincjuk, T. Rospoznavanie ustnoj rechi metodami dinamicheskogo programmirovania; in Russian (Speech recognition by means of dynamic programming).

Viterbi, A. (1967). Convolutional codes and their performance in communications systems. *IEEE Transactions on Communications Technology*, 13(2):260–269.

Waerden, B. v. d. (1957). *Mathematische Statistik*; in German(*Mathematical statistics*). Springer-Verlag, Berlin-Goettingen-Heidelberg.

Wagner, R. and Fischer, M. (1974). The string-to-string correction problem. *Journal of ACM*, 21(1):168–173.

Wagner, R. and Seiferas, J. (1978). Correcting counter-automaton-recognizable languages. *SIAM Journal of Computing*, 7(3):357–375.

Wald, A. (1947). *Sequential analysis*. John Wiley, New York.

Wald, A. (1950). Basic ideas of a general theory of statistical decision rules,. In *Proceedings of the International Congress of Mathematicians*, volume I. Russian translation, A. Wald: Posledovatelnyj analiz, Gosudarstvenoe izdatelstvo fiziko-matematiceskoj literatury, Moskva 1960, paper in Appendix., pages 308-325.

Wald, A. and Wolfowitz, J. (1948). Optimum character of the sequential ratio test. *Ann. Math. Stat.*, 19(3):326–339.

Wu, C. (1983). On the convergence properties of the EM algorithm. *Annals of Statistics*, 11:95–103.

Young, T., editor (1994). *Handbook of Pattern Recognition and Image Processing: Computer Vision*, volume 2, San Diego, USA. Academic Press.

Younger, D. (1967). Recognition of context-free languages in time n^3. *Information and Control*, 10:189–208.

Zadeh, L. (1965). Fuzzy sets. *Information and Control*, 8:338–353.

Zagorujko, N. (1999). *Prikladnyje metoda nalaliza danych i znanij*; in Russian (*Applied methods of data and knowledge analysis*). Izdatelstvo Instituta Matematiki, Novosibirsk, Russia.

Zimmermann, H., Zadeh, L., and Gaines, B. (1984). *Fuzzy sets and decision analysis*. North Holland, Amsterdam-New York.

Zuchovickij, S. and Avdejeva, L. (1967). *Linejnoje i vypukloje programmirovanije*; in Russian (*Linenar and convex programming*). Nauka, Moskva.

Index

Computational Imaging and Vision

1. B.M. ter Haar Romeny (ed.): *Geometry-Driven Diffusion in Computer Vision.* 1994 ISBN 0-7923-3087-0
2. J. Serra and P. Soille (eds.): *Mathematical Morphology and Its Applications to Image Processing.* 1994 ISBN 0-7923-3093-5
3. Y. Bizais, C. Barillot, and R. Di Paola (eds.): *Information Processing in Medical Imaging.* 1995 ISBN 0-7923-3593-7
4. P. Grangeat and J.-L. Amans (eds.): *Three-Dimensional Image Reconstruction in Radiology and Nuclear Medicine.* 1996 ISBN 0-7923-4129-5
5. P. Maragos, R.W. Schafer and M.A. Butt (eds.): *Mathematical Morphology and Its Applications to Image and Signal Processing.* 1996 ISBN 0-7923-9733-9
6. G. Xu and Z. Zhang: *Epipolar Geometry in Stereo, Motion and Object Recognition.* A Unified Approach. 1996 ISBN 0-7923-4199-6
7. D. Eberly: *Ridges in Image and Data Analysis.* 1996 ISBN 0-7923-4268-2
8. J. Sporring, M. Nielsen, L. Florack and P. Johansen (eds.): *Gaussian Scale-Space Theory.* 1997 ISBN 0-7923-4561-4
9. M. Shah and R. Jain (eds.): *Motion-Based Recognition.* 1997 ISBN 0-7923-4618-1
10. L. Florack: *Image Structure.* 1997 ISBN 0-7923-4808-7
11. L.J. Latecki: *Discrete Representation of Spatial Objects in Computer Vision.* 1998 ISBN 0-7923-4912-1
12. H.J.A.M. Heijmans and J.B.T.M. Roerdink (eds.): *Mathematical Morphology and its Applications to Image and Signal Processing.* 1998 ISBN 0-7923-5133-9
13. N. Karssemeijer, M. Thijssen, J. Hendriks and L. van Erning (eds.): *Digital Mammography.* 1998 ISBN 0-7923-5274-2
14. R. Highnam and M. Brady: *Mammographic Image Analysis.* 1999 ISBN 0-7923-5620-9
15. I. Amidror: *The Theory of the Moiré Phenomenon.* 2000 ISBN 0-7923-5949-6; Pb: ISBN 0-7923-5950-x
16. G.L. Gimel'farb: *Image Textures and Gibbs Random Fields.* 1999 ISBN 0-7923-5961
17. R. Klette, H.S. Stiehl, M.A. Viergever and K.L. Vincken (eds.): *Performance Characterization in Computer Vision.* 2000 ISBN 0-7923-6374-4
18. J. Goutsias, L. Vincent and D.S. Bloomberg (eds.): *Mathematical Morphology and Its Applications to Image and Signal Processing.* 2000 ISBN 0-7923-7862-8
19. A.A. Petrosian and F.G. Meyer (eds.): *Wavelets in Signal and Image Analysis.* From Theory to Practice. 2001 ISBN 1-4020-0053-7
20. A. Jaklič, A. Leonardis and F. Solina: *Segmentation and Recovery of Superquadrics.* 2000 ISBN 0-7923-6601-8
21. K. Rohr: *Landmark-Based Image Analysis.* Using Geometric and Intensity Models. 2001 ISBN 0-7923-6751-0
22. R.C. Veltkamp, H. Burkhardt and H.-P. Kriegel (eds.): *State-of-the-Art in Content-Based Image and Video Retrieval.* 2001 ISBN 1-4020-0109-6
23. A.A. Amini and J.L. Prince (eds.): *Measurement of Cardiac Deformations from MRI: Physical and Mathematical Models.* 2001 ISBN 1-4020-0222-X

Computational Imaging and Vision

24. M.I. Schlesinger and V. Hlaváč: *Ten Lectures on Statistical and Structural Pattern Recognition.* 2002 ISBN 1-4020-0642-X

Kluwer Academic Publishers – Dordrecht / Boston / London